HERPETOLOGICAL
HISTORY
OF THE
ZOO AND AQUARIUM WORLD

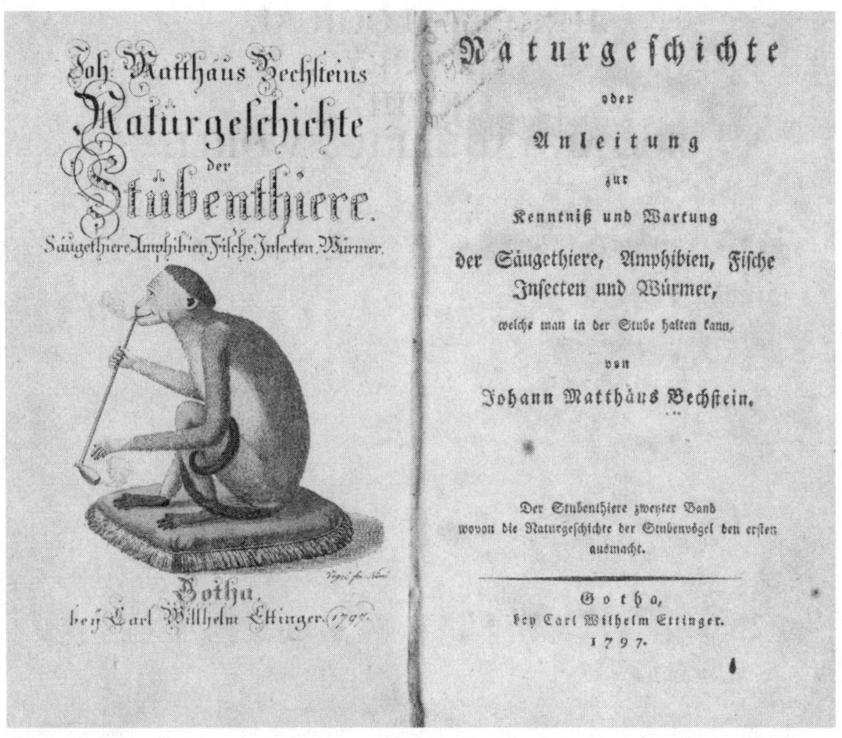

Fig. 1. The first book on captive care of reptiles and amphibians was titled *Naturgeschichte; oder, Anleitung zur Kenntniss und Wartung der Säugethiere, Amphibien, Fische, Insecten und Würmer, welche man in der Stube halten kann* by Johann Matthaeus Bechstein in 1797. *Courtesy of Smithsonian Institution Libraries, Washington, DC.*

HERPETOLOGICAL HISTORY OF THE ZOO AND AQUARIUM WORLD

James B. Murphy

KRIEGER PUBLISHING COMPANY
Malabar, Florida
2007

Original Edition 2007

Printed and Published by
KRIEGER PUBLISHING COMPANY
KRIEGER DRIVE
MALABAR, FLORIDA 32950

FROM A DECLARATION OF PRINCIPLES JOINTLY ADOPTED BY A COMMITTEE OF THE AMERICAN BAR ASSOCIATION AND A COMMITTEE OF PUBLISHERS:
This publication is designed to provide accurate and authoritative information in regard to the subject matter covered. It is sold with the understanding that the publisher is not engaged in rendering legal, accounting, or other professional service. If legal advice or other expert assistance is required, the services of a competent professional person should be sought.

Library of Congress Cataloging-in-Publication Data

Murphy, James B. (James Bernard), 1939-
  Herpetological history of the zoo and aquarium world / by James B. Murphy.
       p. cm.
   Includes bibliographical references.
   ISBN-13: 978-1-57524-285-9 (alk. paper)
   ISBN-10: 1-57524-285-0 (alk. paper)
  1. Captive reptiles. 2. Captive amphibians. 3. Herpetology—History. I. Title.
   SF515.M87 2007
   597.9073—dc22

                                                                                    2007012526

10 9 8 7 6 5 4 3 2

Dedication

As a signal of my respect and gratitude,
I am pleased to dedicate this book to
Kraig Adler, Charles Carpenter, David Chiszar, and Warren Iliff,
knowing that each of them
has made a profound difference in my life.

Reptile Panorama

Contents

Foreword

When I entered the zoo profession as Curator of Reptiles at the Toledo Zoo in 1929 and James B. Murphy began his professional career thirty-six years later, chiefly at the Dallas Zoo, the world was a very different place. Our challenges centered on keeping amphibians and reptiles alive in our zoos, exhibiting them in attractive settings, and doing research on their biology and natural history.

Several decades ago, as human pressures on biological communities increased exponentially and the list of threatened and endangered herpetofauna expanded dramatically, it was clear that the mission and direction of our profession were changing. It became necessary to try to preserve these interesting creatures by developing viable captive colonies, and we needed to become major players with *in situ* initiatives.

In this book, Murphy follows the changes in our zoo and aquarium communities by looking at the development and expansion of our discipline, the evolution of ideas which led to greater conservation awareness and activity, vignettes of interesting historical moments, and pioneers in zoo herpetology. He presents portraits of a selected number of zoos and aquariums throughout the world to show the chronology of herpetological discovery, the people who worked at those places, and the breadth of the programs that were put in place.

Murphy and I have witnessed astounding changes, both good and bad. Future zoo and aquarium workers are likely to be confronted with a suite of enormous challenges, responsibilities, and pressures that are unimaginable to us now.

ROGER CONANT

South American Iguana (*Iguana iguana*)

Acknowledgments

For various courtesies, I thank Fred Antonio, Barry Armstrong, John Arnett, Ray Ashton, Kurt Auffenberg, Dave Ball, Jonathan Ballou, Chris Banks, David Barker, David Barnaby, Susan Barnard, Breck Bartholomew, Richard Bartlett, Benjamin Beck, Charles Beck, the late John Behler, the late A. d'A. Bellairs, Jean-Luc Berthier, David Blody, Quentin Bloxam, Wolfgang Böhme, Martina Borchert, Donal Boyer, Thomas Boyer, Kelly Bradley, Bill Branch, Peter Brazaitis, Paul Breese, Ed Bronikowski, Eugène Bruins, Richard Buickerood, Kevin Buley, Vincent Burke, Robert Butsch, Jonathan Campbell, Winston Card, Charles Carpenter, Jodie Chapman, Don Church, Claudio Ciofi, Scott Citino, Joseph Collins, Steve Connors, William Cooper Jr, Jack Cover, M. Damm, Michael Davenport, Béla Demeter, Kevin de Queiroz, Scott Derrickson, Michael Dloogatch, J. S. Dobbs, Herndon Dowling, William Duellman, John Edwards, Serena Eldridge, Zhang Enquan, Jeff Ettling, Gary Ferguson, Lee Fikes, Bill Flanagan, the late Pierre Fontaine, Jack Frazier, Paul Freed, Darrel Frost, Fredric Frye, Jin Zhong Fu, Carl Gans, Gerardo Garcia, Clay Garrett, Claude Gascon, William Gehrmann, Margie Gibson, Richard Gibson, James Gillingham, the late Ron Goellner, Melissa Goodman, David Green, Harry Greene, Richard Griffiths, David Grow, Ray Guese, Rick Haeffner, Heather Hall, Martha Hall, Tim Halliday, James Hanken, Robert Hansen, Ruston Hartdegen, Robert Hoage, Mats Höggren, Monika Holland, William Holmstrom, Jaren Horsley, Rick Hudson, Jeremy Huff, R. Howard Hunt, Victor Hutchison, Michael Hutchins, Ghislaine Iliff, the late Warren Iliff, Frank Indiviglio, the late Steve Irwin, Elliott Jacobson, David Jacques, Robert Jaeger, David Jenkins, Harald Jes, Greg Johnstone, Robert Johnson, Ron Kagan, Eugeny Kamelin, Ken Kawata, C. H. Keeling, Wataru Kimura, the late Dennis King, Devra Kleiman, Jiro Kobara, Steve Kohl, Rana Kozouz, Karl Kranz, William Lamar, William Lamoreaux, Jeff Lang, Jürgen Lange, James Langhammer, Bert Langerwerf, Michael Lannoo, Annabelle de La Panouse, Colomba de La Panouse, Paul de La Panouse, Fred LaRue, Dwight Lawson, Annabelle Lea, Rob Lewis, Ernie Liner, Yuri Lukin, Sergei Mamet, Dale Marcellini, Edward Maruska, Ken McCloud, Rose McCloud, Roy McDiarmid, Wendy McKeown, Marie McLaughlin, Sergei Mamet, Michael Meadows, Joseph Mendelson III, the late Madge Minton, the late Sherman Minton, Robin Moore, the late David Morafka, John Moriarty, Terry Morley, Jack Muntzer, John Murphy, Robert Murphy, Frank Mutschmann, Hosein Nahidian, Akiyoshi Nawa, Donald Nichols, R. Andrew Odum, Ted Papenfuss, Ray Pawley, Clyde Peeling, Janice Perry (Johnson), Tim Perry, Jim Pether, Gisela Petzold, Scott Pfaff, Hans-Dieter Philippen, John (Andy) Phillips, Eric Pianka, Charles Pickett, George Pisani, Louis Porras, Gracia Gonzalez Porter, Peter C. H. Pritchard, Wang qiang, Hugh Quinn, George Rabb, Charles Radcliffe, Steve Reichling, David Roberts, John Romano, Richard Ross, Roger Rosscoe, Manny Rubio, Matt Russell, Richard Sajdak, Robin Saunders, Jay Savage, Neil Schlager, David Schleser, Christian Schmidt, Gowri Shankar, Diane Shapiro, Wade Sherbrooke, Chuck Siegel, Brandie Smith, Brian Smith, Edwin Smith, Hobart Smith, LouAnne Smith, Andrew Snider, Brint Spencer, Vladimir Spitsin, Lisa Stevens, the late Margaret Stewart, Simon Stuart, Raquel Tabat, John Tashjian, Oksana Tishenko, Takeyoshi Tochimoto, Peter Tolson, Barney Tomberlin, Simon Tonge, Linda Trueb, Bern Tryon, Christina Unger, Kristin Vehrs, Kristin Visger, David Wake, Sally Walker, Trooper Walsh, John Weigel, Paul Weldon, Christen Wemmer, Romulus Whitaker, Judy White, Jon Whitehead, John Wilkerson, David Xanten, William Xanten, Robert Zappalorti, Er-mi Zhao, William Zeigler, Kevin Zippel, and Conrad Ziyad.

The Smithsonian Institution's Behind-the-Scenes Volunteer Program, through the efforts of Janeen Haase, Amy Lemon, Janet Miyazaki, Celeste Moneme, and Eriko Osugi, assisted with the translations. Translators Hela Finberg, Maiko Hsu, Hope Patterson, Ann Schauerte, Angela Stavropoulos, and Ellen Unumb graciously lent their linguistic skills to provide equivalents for the non-English titles. Special thanks are extended to Lucian Heichler, a thorough and careful translator, who has helped throughout the completion of this book. I am particularly grateful to NZP Librarians Kay Kenyon, Alvin Hutchinson, Courtney Shaw, and Polly Lasker, who assisted in locating obscure books and papers and arranging loans for these materials. In addition, they contacted their colleagues at the United States Library of Congress and Smithsonian

National Museum of Natural History to allow access to their collections. Ronald Crombie, W. Ronald Heyer and George Zug from the Division of Amphibians and Reptiles at the National Museum of Natural History suggested titles and searched for these books and reprints in the library. Smithsonian Institution (SI) Archivist Bill Cox and Permissions Officers Martin Kalfatovic and Dale Miller allowed access to materials under their care. Leslie Overstreet and Daria Wingreen from the SI Joseph F. Cullman 3rd Library of Natural History were helpful in locating classic herpetological publications, ranging from Charles Owen's essay on a natural history of serpents, published in 1742, to Johann Matthaeus Bechstein's book on captive care of domestic animals and pets, written in 1797. Since I never imagined that these original tomes would ever be available to me, the opportunity to see them was a great pleasure.

There are beautiful plates in Alfred E. Brehm's *Les Merveilles de la Nature. Les Reptiles et les Batraciens* (Librairie J.-B. Baillière et Fils, Paris, 1883), some of which appear at the beginning of each chapter of my book. Brehm was Director of the Hamburg Zoo in Germany from 1863 onward and founder of the Aquarium Unter den Linden (Berlin's main street) in 1869. Technical names are current.

Several of my foreign colleagues deserve special mention for checking references and spelling of non-English titles: Falk Dathe, Wolf-Eberhard Engelmann, Hans-Werner Herrmann, René Honegger, Sergei Kudryavstev, Gisela Petzold, Ivan Rehák, Sergei Ryabov, and Gerard Visser. Allison Alberts, Gordon Burghardt, David Chiszar, Jon Coote, and Kevin Wright have tirelessly reviewed the entire manuscript without resorting to tranquilizers or amphetamines. I am particularly indebted to the late Marvin Jones for his help in checking facts, providing interesting anecdotes, and assuring that my recollections of people, places, dates and times were accurate.

Smithsonian Institution's Office of Imaging and Photographic Services prepared the Bechstein frontispiece for publication. Ann Batdorf, Jessie Cohen, and J'nie Woosley from the Graphics Department at NZP spent many hours sorting through the Zoo's archival assemblage, preparing images and instructing me about photographic software. Former SSAR webmaster Dennis Desmond was most generous in assisting with preparation of photographs into a digital format and instructing me about the intricacies of cyberspace.

I am indebted to the late Roger Conant for penning his thoughtful foreword as he was the premier zoo herpetologist. His collective body of work has advanced our profession immeasurably and he was my valued mentor and friend since I entered the zoo community.

With minimal pressure from me, Kraig Adler agreed to edit the entire manuscript and help locate photographs and historical literature. I am grateful for his editorial skills, encouragement and friendship.

My wife Judith Block, retired Registrar at NZP, spent hours discussing history and evolution of herpetological collections, based on her years of experience in the zoo community. Further, she listened patiently to my incessant whining about computer malfunctions, lack of typing skills, need to purchase even more herpetological literature, and desire to travel to faraway places to look at zoos and aquariums. Judith reviewed the entire manuscript, inserted a multitude of editorial changes to clarify my fractured narrative, and bolstered my waning enthusiasm when the daunting reality of completing this project overwhelmed me.

Robert Krieger, Bill Page, Elaine Rudd, and Shannon Ryder assisted me in many ways to prepare this book for publication by making valuable suggestions as to content, format and presentation.

Finally, I would like to thank my late mother and father, sisters Susan and Patricia and son James who endured crocodilians in the bathtub, snakes in the basement, mealworms in the refrigerator, minnows in the laundry room, box turtles in the fireplace, aquariums in the living room, lizards on the curtains, crickets in the bedroom, aquatic turtles in the sink, salamanders in the kitchen, and frog enclosures on the fireplace mantle. It certainly is not easy to explain to friends and acquaintances why a family member's interests are so bizarre and hardly acceptable in polite society, especially when one of his favorite creatures chooses to escape from its home. Their patience was admirable during my formative years although it must be said that when my maiden aunt took my mealworms which were nestled in bran and ate the lot, thinking that these were cereal, their mettle was tested!

Prologue

This publication has been produced in cooperation with the Society for the Study of Amphibians and Reptiles.

Wild reptiles have been held in captivity for at least five millennia. Crocodilians were kept in temples and sacred lakes for religious purposes during the third to second millennia B.C.E.—gharials in India and crocodiles in Egypt. Later, the maintenance of wild animals shifted from serving religious functions to being demonstrations of the majesty and power of rulers. The zoos of the pharaohs in Egypt, the Chinese emperors, and more recently, the Aztec kings in Mexico are notable examples. The menageries of the Middle Ages served a similar social purpose, and only much later were collections of wild animals including reptiles opened to the general public.

Today, virtually every major city in the world has a public zoo and many have public aquariums. Both typically include amphibians and reptiles. The "Reptile House" is often the single most popular attraction at zoos because of the public's fascination with these animals, especially constricting and venomous snakes, but also large lizards, turtles, and crocodilians. The first public zoo to have more than a simple exhibit function was the menagerie associated with the Muséum d'Histoire Naturelle in Paris. For centuries prior to the French Revolution, the kings of France kept live animals at their palaces in the Louvre and Versailles to show to their guests. Soon after the execution of Louis XVI in January 1793, many of the animals held at Versailles were moved to the gardens adjacent to the newly created museum and placed under its jurisdiction. Significantly, the first director of this menagerie was a 22-year-old professor of anatomy at the museum, Étienne Geoffroy Saint-Hilaire, who would become a leading anatomist and evolutionary theorist as well as the colleague and rival of Georges Cuvier. This became the first public zoo, and from the beginning its functions were closely tied to research and education at the museum. In the 1860s, Auguste Duméril, another professor at the museum, conducted his famous experiments on metamorphosis of a salamander--the axolotl—at the menagerie, a feat that won for him a gold medal at the International Exhibition of 1867, held in Paris. This linkage between zoo and museum continued for over a century and led to other important discoveries in amphibian and reptile biology.

The first independent Reptile House was established in London's Regent's Park in 1849, 21 years after that zoo was founded; however, there was no formal relationship with a museum, university, or any other academic institution. Nevertheless, staff at the zoo did conduct research on the anatomy, systematics, and behavior of amphibians and reptiles. Their findings were published in the zoo's journal, the *Proceedings of the Zoological Society of London*, which soon became one of the leading zoological serials in the world. Beginning in 1864, the society also sponsored the publication of *Zoological Record*, an annual listing and cross index to all articles and books on animals. The research-oriented activities of the Paris and London zoos soon were imitated by many others. These, together with the conservation programs at many zoos and aquariums, have established these institutions as key participants in research on and conservation of amphibians and reptiles. Yet only now has the intriguing history of these important contributions to the discipline of herpetology been written.

No one is better qualified to write the history of herpetology at zoos and aquariums than James B. Murphy. For more than 40 years he has been one of the leaders in the effort to link reptile programs at zoos to broader research and conservation efforts. As Curator of Herpetology at the Dallas Zoo for 30 years, he developed a model program for research and conservation, and trained numerous assistants who became curators at other major parks. At the same time, Murphy has been active in the academic herpetological community. He chaired several committees and served as a member of the governing board of the Society for the Study of Amphibians and Reptiles and was president of the society in 1981. Leading by example, Murphy authored dozens of papers on the behavior and reproductive biology of amphibians and reptiles in the major herpetological journals. Many of these studies were co-authored with collaborators at universities across the United States and have become an exemplar of what can be achieved by combining the different talents and perspectives of academic and zoo personnel. In addition, Murphy has become a major participant in the international effort to conserve amphibians, largely by working behind the scenes to raise funds and coordinate programs. He has co-written or co-edited several books on the reproductive biology, diseases, captive maintenance, and conservation of amphibians and reptiles. He founded and continues to edit the department "Zoo View" in the quarterly news-

journal, *Herpetological Review*, as a vehicle to better integrate the zoo and academic communities. His broad and significant contributions to the partnership between researchers at zoos and universities were recognized in 1989 by the University of Colorado, which conferred upon him the honorary degree of Doctor of Science.

This book chronicles the rich and often unique contributions that the staffs of the world's zoos and aquariums have made to our understanding of the biology of amphibians and reptiles, and to the development of herpetology as a discipline. Those of us in the academic profession can only hope that wise leadership at zoos and aquariums will continue to encourage and invest in these activities and, indeed, even bolster these efforts. At a time when so many amphibians and reptiles are threatened with extinction, zoos and aquariums play a pivotal role in educating the general public about the plight of these animals. Their active participation in captive reproduction and reintroduction programs may be the last best hope to rescue many of these magnificent creatures.

Kraig Adler
11 February 2007

Introduction

One of the most interesting and one of the most fascinating additions of recent years to zoological exhibits is the modern reptile house. Both the animal and the vegetable world are represented in it. It is a whole world of forests, dusky swamps, barren deserts, and tangled jungles brought from North and South and East and West and crowded under one roof. And under the roof, caged among giant palms, hanging bushrope, and drooping resurrection ferns, there squirm and coil the most mysterious and uncanny and merciless of all of God's creatures, housed according to the most advanced ideas of animal science, with all the surroundings peculiar to the wild and natural state of each. This novelty of showing the animals, and the modern facilities for feeding them as once they fed themselves, for studying reptile diseases, and even for operating with the surgeon's knife, have practically revolutionized the snake house of old into the marvelous reptile house of to-day.

A. W. Rolker (1956)

The *International Zoo Yearbook* periodically publishes a list of zoos and aquariums of the world, with information on herpetological collections. To begin this project, I prepared a list of those institutions with sizeable collections from the most recent volume. To assess their histories and scientific contributions, I sent questionnaires to staff members in all herpetological departments worldwide requesting information. As is the danger whenever questionnaires are sent, the response was mixed. Some of my colleagues were marvelous, sending reprints, résumés, animal inventories, historical portraits and reviewing drafts of my accounts. Others were nonresponsive and indifferent, even after several reminder letters were sent over many months. Accordingly, the following chapters are slanted toward those institutions that provided relevant materials.

While waiting for information to arrive, I began perusing the literature to see if my perception about the importance of herpetology in zoos and aquariums was correct. My presupposition was that my colleagues over the years have contributed significantly to herpetology. If one tracks the development of thought in our profession, major trends are apparent, reinforced by reading the first four chapters in this book. Since there is a great deal of information published by these workers on captive management in periodicals and magazines, I decided to not include these articles unless there was a historical point which needed to be stressed. It should be clear that this is not an exhaustive compilation of all publications by zoo and aquarium people but rather a selection of titles which, in my view, has advanced our profession. Hence, my personal bias intrudes mightily throughout the book.

Each non-English citation is either followed by the English equivalent in parentheses or has been translated directly into English. Additional information which expands on the title may be presented in brackets. The name of each of the institutions chosen by me to be included is followed at the top of the account by its opening date (in a few instances a closing date as well) in parentheses.

Phylogenetic arrangements follow Zug et al. (2001). In most cases, common names compiled by Frank and Ramus (1995) are used in the text. In older literature where there might be confusion, the technical name used in the original publication is listed, followed by the current name. For these recent nomenclatural conventions, I used two sources that record all known taxa: Amphibian Species of the World Database (http://research.amnh.org/herpetology/amphibia/index.html) and EMBL Reptile Database (http://www.reptileweb.org). Generally, journal abbreviations follow BIOSIS Zoological Record Serial Sources (1992). Books, monographs and papers published after January 2006 are not included.

After reviewing materials, I decided that those institutions to be highlighted needed to have at least three of the following characteristics. First, there must be a structure or building devoted to amphibians and reptiles. Second, there must be a sustained published scientific output. Third, there must be a strong programmatic or conservation component. Fourth, the history of the institution and its importance to herpetology could be tracked over time, either by consulting the written record or through my personal experiences over a span of 50 years. Most importantly, my colleagues needed to interact with me, especially since I had not visited all of the zoos on my list.

It became clear that some operations, even though not considered in the strictest sense to be zoos or aquariums, either were or are major players in our

xv

understanding of captive amphibians and reptiles. Some of these include Ross Allen's Reptile Institute, Madras Crocodile Bank, Institute for Herpetological Research, Reptile Breeding Foundation and several other private operations throughout the world. These have been included since it is only right to do so; their collective efforts deserve recognition.

There were four reasons why I was motivated to write this book. The first was my fear that much of our history, beginning with the first reptile building at the London Zoo in 1849, was in danger of being lost. The second was that many of my zoo and aquarium colleagues, especially those new to the profession, did not have a sense of the unique contributions of their predecessors; their accomplishments should be celebrated. The third was that some of my academic and museum associates had the perception that the work done in zoos and aquariums was not very important. Finally and most importantly, I am concerned that many zoo administrators view zoo and aquarium herpetological collections and buildings as a relict from the past; as a result, there has been a significant decline in new facilities, emphasis and financial support.

Chapter 1
Evolution of a Discipline

Sailfin Lizard *(Hydrosaurus amboinensis)*

REPTILE KEEPING HAS DEVELOPED INTO A SCIENCE DURING THE PAST TWO DECADES, AND THE ADVANCEMENTS, ESPECIALLY IN THE FIELDS OF MEDICATION AND CAPTIVE BREEDING, ARE LITTLE SHORT OF PHENOMENAL. COMPARED WITH THE SOPHISTICATED TECHNIQUES THAT ARE NOW AVAILABLE, OUR EFFORTS OF HALF A CENTURY AGO SEEM CRUDE IN THE EXTREME. IN THOSE DAYS WE OFTEN CONSIDERED OURSELVES LUCKY IF A MAJORITY OF OUR ANIMALS LIVED A YEAR OR TWO, AND WHAT LITTLE CAPTIVE BREEDING OCCURRED WAS FORTUITOUS, NOT PLANNED.

ROGER CONANT (1980)

This chapter is a portrait of the history of captive amphibians and reptiles compiled in part by consulting the descriptions of Loisel (1912), both papers by Flower (1925), Petzold (1984), Bell (2001), Coote (2001), and Kisling (2001). Publications on specific zoo histories, systematics, taxonomy and morphology are contained in the individual zoo accounts.

To understand the history and evolution of the art and science of keeping amphibians and reptiles in captivity it is important to recognize the contributions made by private fanciers and herpetoculturists. In fact, the first book on herpetoculture in English was the Reverend Gregory Climenson Bateman's *The Vivarium*, published in 1897. Prior to the nineteenth century, detailed observations and recommendations for captive care were limited. Many of the techniques currently used were developed by enthusiasts in the private sector and the astounding array of captive-bred taxa now available reflects their collective commitment and skill. In the 1970s, these practitioners started propagating a variety of boas and pythons, various colubrid snakes, geckos, monitor lizards, water and bearded dragons, poison dart frogs, Malagasy poison frogs and many hylid frogs; many of these species are still available in the trade today. Since thousands have been produced annually, Thorogood and Whimster (1979) considered the leopard gecko (*Eublepharis macularius*) to be a laboratory animal. Extraordinary numbers of books and booklets on the proper care of these herps were offered for sale in pet shops; there were so many that a significant number of these publications were repetitious. Since many of these species were also in zoo collections, zoo herpetologists began to shift focus to species less well known and threatened by mostly human causes. As an example, kingsnakes were produced in large numbers at the Dallas Zoo in order to follow reproductive biology and behavior as well as distribute them to other zoos. It soon became clear that this no longer made sense, given that kingsnakes were already a staple item in the private arena so virtually all kingsnakes were de-accessioned from the collection. Zoo collections were now being shaped in part by the private herpetoculturist. This transition was not always an easy one for some zoo workers felt that there were competing agendas between the private enthusiasts and the zoo community. Tensions increased and an unhealthy competition followed which continues to this day (see Murphy et al., 1997).

In early days, there was closer cooperation between zoo workers and private herpetoculturists. Until 1903, keepers at the London Zoo were able to trade surplus animals to the general public and supplied herps to amateur British herpetoculturists (Coote, 2001). Amazingly, Bateman purchased a pair of tuatara from the Zoo for £2.00 ($3.60) and other species were offered for sale in London (Table 1.1). Today, zoos tend to exchange surplus animals with other institutions which are permanent rather than with individuals who may be transitory. It is easier to keep track of the transactions.

Coote (2001) suggested that the earliest record reflecting the captive maintenance of reptiles such as cobras and crocodiles, is from the pictographs and hieroglyphs of 2500 BC, found at the Saqqara cemetery near Memphis in Egypt. Egyptians kept a wide range of exotic species, some of which were used in religious ceremonies. To collect these creatures, expeditions were mounted; for example, Queen Hatsheput, daughter of Thutmose I of the Eighteenth Dynasty, sent an expedition to the "Land of Punt" (probably Somalia) to obtain many species. Egyptian kings received and dispersed a variety of exotic animals in the form of tribute with other nations. Ptolemy I of Egypt (323–285 BC) created a zoo in Alexandria and his successor Ptolemy II Philadelphus enlarged it. On the Feast of Dionysus in the year 285 BC, snakes such as cobras and large crocodiles were displayed in processions. The private collection of Ptolemy II included an enormous python, collected during a trip to the Upper Nile sometime between 282–246 BC. The kings of Ur kept wild animals from about 2000 BC in large walled gardens in the Babylonian region of Persia. Around 1100 BC, the Egyptian pharaoh sent a large crocodile to the Assyrian king Tiglathpileser I.

In ancient Greece and Rome, there were scattered animal collections which included reptiles. The Romans used crocodiles and large snakes from Africa, as well as other species from Northern Europe and Asia, for the gladiatorial games in 186 BC. Holding facilities

called 'vivaria' were constructed, usually next to the arenas. For example, in the 3rd century AD, one of the vivaria was located just outside the Praenestine Gate. This vivarium, 440 yards long and 70 yards wide, and with one long wall adjacent to the city wall, was likely viewable by the public. At the beginning of the Christian era between 29 BC and 14 AD, the Roman emperor Octavius Augustus had more than 3,500 wild and tamed exotic animals from his vivaria killed in 26 exhibitions, including 420 tigers, 260 lions, 36 crocodiles, and a large snake (Coote, 2001).

In 1519, Hernando Cortés and his band of soldiers arrived at the Aztec capital Tenochtitlan (now Mexico City) where they found the Aztec emperor Montezuma's royal menagerie. In this animal assemblage were ". . . vipers and poisonous snakes which had on their tails things that sound like bells. These are the worst vipers of all, and they keep them in jars and great pottery vessels . . . and there they lay their eggs and rear their young," described by Diaz del Castillo (in Coote, 2001). Although snakes were often used by the Aztecs for sacrificial purposes to their gods, could Montezuma have been our first herpetoculturist?

Crocodilians

CROCODILIANS MAY PERHAPS LIVE TO A GREAT AGE: PROBABLY LONGER IN THE SHELTERED CONDITIONS OF CAPTIVITY THAN WHEN EXPOSED TO THE ACTIVE, COMBATANT, COMPETITIVE CAREER THAT IS THEIRS IN NATURE.

MAJOR STANLEY SMYTH FLOWER (1925)

SUCH IS THE AVERAGE CROCODILE—AN ACTIVE, VICIOUS AND, ABOVE ALL, TREACHEROUS BRUTE. WHEN THE KEEPERS OF THE REPTILE HOUSE IN THE NEW YORK ZOOLOGICAL PARK CLEAN OUT THE BIG POOL FOR CROCODILIANS, THEY ACTUALLY WALK OVER THE BACKS OF SOME OF THE BIG 'GATORS, SO TAME ARE THESE. THEY NEVER BECOME UNDULY FAMILIAR WITH THE CROCODILES, FINDING IT NECESSARY TO PEN THE LATTER BEHIND HEAVY BARRED GATES—AND IN THE PROCESS THE MEN ARE OFTEN CHASED FROM THE ENCLOSURE.

RAYMOND L. DITMARS (1933)

History

Crocodiles may have been the first zoo animals. The Roman, Aemilius Scaurus, built a tank filled with water for his five crocodiles in 58 BC and the 3rd century Roman emperor Heliogabalus (=Elagabalus, 204–222, Rom. Emp. 218–222) kept one of these reptiles, said to be tame, in his palace. Nile crocodiles were a popular display in Europe (Scherpner, 1975). In ancient Oriental societies, they were kept in temple precincts. Egyptian priests in the 3rd millennium held "sacred crocodile feedings" and there are Nile crocodile mummies (Guggisberg, 1972). Alligators were held in China as religious symbols. In Asia, the Arabs and Turks maintained crocodiles. Gavials could be found in the menageries of Buddhist convents in Myanmar (Burma) and Thailand (Siam). At the Buddhist Theyboo Monastery near Mandalay, these crocodilians were seen in the early 1900s. There was a pit for mugger crocodiles near Karachi, Pakistan. In 1860, Andrew Leith Adams described this "mugger-peer,"a pond roughly 300 yards in circumference, as a place filled with crocodiles of all sizes, basking on many small grassy islands; visitors were expected to feed them (Guggisberg, 1972; Chapter 4, Stanley Smyth Flower account). In 1882, the Trivandrum Zoo in India had four mugger crocodiles on exhibit.

In the Menagerie at the Botanic Gardens in Singapore, a saltwater crocodile between 6–7 feet escaped into the zoo lake and began eating waterfowl and stalking coolies. Many methods were tried to kill it. It was twice shot, arsenic and strychnine were put into chickens, dynamite was tried; finally the lake was virtually drained with shots and spears aimed at the crocodile but it buried itself into the mud and escaped that night (Ridley, 1906). Talk about survival of the fittest; Darwin would be proud!

Stanley S. Flower (1925) kept gharials, Nile and mugger crocodiles, and American alligators at the Giza Zoological Gardens in Egypt beginning in 1909. His charges were kept outdoors in large paddocks with pools at least three feet deep which allowed the reptiles to retreat into the water when occasional frosts occurred in the region.

In Europe during the Renaissance, crocodiles were kept in menageries by Italian princes. In other menageries through the Renaissance, a few crocodiles and chelonians were exhibited but no real information is available as to how these animals were maintained.

In France, crocodiles were held at the Royal Menagerie of Louis IV at Versailles as early as 1664 and the arrival of a crocodile at the Vineuil Menagerie in 1783 caused much public excitement. The Menagerie du Jardin des Plantes of the Museum of Natural History in Paris was created in 1793 and was the first of the national menageries. An American crocodile was received in 1851 and Émile Wapler from New Orleans donated six American alligators to the Menagerie the next year where they lived for many years (Duméril, 1854–1855, 1858–1861). In 1870–1874, Emile Blanchard designed a reptile menagerie with a pavilion. The facility was 30 meters long, with two halls which contained smaller squamates and chelonians. Two large center exhibition halls were called "Crocodile Hall" and "Aquarium Hall;" the latter contained freshwater fishes and amphibians. Bronze sculptures were placed throughout the menagerie: "The Snake Charmer" and "The Crocodile Hunter" by Arthur Bourgeois in front of the reptile pavilion and "Eve" by Guitton near the outdoor crocodile pool. A Nile crocodile was added to the collection in 1892. The Zoological Garden of Lyon (Jardin Zoologique des la Ville de Lyon) contained a few large American alligators in 1907.

At the Menagerie in the Imperial Hofburg [Castle] Garden in Vienna, an American crocodile was in the collection in 1821. Gens (1861) briefly discussed the crocodilians in the Promenade at the zoological gardens (Jardin Zoologique d'Anvers) in Antwerp, Belgium. An African slender-snouted crocodile and alligators were on exhibit at the Reptile House in the Royal Zoological Gardens in Amsterdam in 1907.

Fig. 2. Mixed exhibit of turtles and crocodilians at Rotterdam Zoo in 1925. *Provided by Gerard Visser, Rotterdam Zoo.*

The Berlin Zoo, always known for its extensive crocodilian collection, had no less than eight species between 1860–1875. These reptiles were acquired through several sources including Duméril from the Jardin des Plantes in Paris who supplied Morelet's crocodiles. A Chinese alligator was received in 1888 but was later transferred to the Frankfurt Zoological Garden which also maintained a collection of small crocodilians. The Berlin Aquarium, built at the Zoo in 1913, housed an outstanding collection of crocodilians in the first walk-through display in any zoo, but this edifice and most of the collection were destroyed during War II. Two false gavials and several alligators survived. The aquarium was later rebuilt and renovated several times and now crocodilians are on display in an enormous greenhouse located on the top of the building. Glass cages contained some large crocodiles at the Zoological Garden at Hamburg in the early days.

Huish (1830:51-52) described the alligator, housed in the Royal Menagerie in the Tower of London with a nicely rendered plate of a specimen with a seemingly smallish head. There were repeated attempts to keep gharials in the 1870s at the London Zoo but these efforts were in vain until two specimens that arrived from the Calcutta Zoo in 1912 were still alive in 1925. In 1872, the collection included the Nile, Cuban, slender-snouted, mugger and American crocodiles, as well as the Yacare caiman and American alligator (Sclater, 1872). C. J. Cornish (1894) described crocodilians living in the reptile building at the London Zoo. During the period between 1872–1914, the collection also included the African slender-snouted, American, Nile, estuarine, Cuban and West African dwarf crocodiles, as well as the dwarf, spectacled and broad-nosed caiman. American alligators were on display as well as a Chinese alligator, placed by the Hon. Walter Rothschild. Zoos in Brighton, Clifton, Dublin, and Manchester exhibited American alligators in the early 20th century.

At the Adelaide Zoological Gardens in Australia, the first crocodilian was a mugger acquired in 1886 and others were subsequently exhibited until 1908. Two American alligators were received in 1914 and one was still alive on 30 June 1977.

Crocodilians have been exhibited in a number of zoos and aquariums in the United States for many years (Shepstone, 1932). At the New York Zoological Park (Wildlife Conservation Society/Bronx Zoo), several large naturalistic exhibits housing many different species over the years may be found at one end of the reptile building. The Philadelphia Zoo has several impressive indoor exhibits. At the Smithsonian National Zoological Park, three large enclosures at one end of the building currently hold Cuban crocodiles and gharials. During earlier times, an enormous and very aggressive estuarine crocodile was living in the exhibit where it regularly exploded from the water and snatched food from the hands of very agile and alert keepers. Zoo Atlanta has several large solariums with living plants available for large crocodilians, including Morelet's crocodiles which have bred regularly for many years. Ross Allen's Reptile Institute, established in 1931 in Silver Springs, Florida, was a mixture of exhibits and shows designed to appeal to tourists. There were many large outdoor enclosures with a nice collection of native and exotic crocodilians, many of which were obtained from George Campbell, who had a private collection second only to the Berlin Aquarium

in the 1960s. The Miami Metrozoo has several large outdoor exhibits where slender-snouted and Siamese crocodiles have reproduced.

At the Lincoln Park Zoo in Chicago, a modified aquarium opened in the early 1930s as the new reptile building. Because the converted exhibit cages were originally spacious aquariums with concrete dioramas and natural lighting, many crocodilians were on exhibit, including a large specimen said to be an Orinoco crocodile. In the basement were several large aquatic exhibits housing a number of crocodilians. Across town at the Chicago Zoological Park, American alligators have been kept for many years in a spacious greenhouse on one end of the building. A smaller exhibit has been the home for many years of a dwarf crocodile. In 1927, a modern reptile building was constructed at the St. Louis Zoo which features several spacious crocodilian exhibits with living tropical plants. At the Detroit Zoological Institute, the Holden Museum of Living Reptiles which opened in 1960, has elaborate planted crocodilian exhibits.

It is surprising that crocodilians are not particularly well represented in zoos of the southwest, particularly since the climate is less harsh than in zoos in the northeastern United States. A few may be found in Dallas, Oklahoma City, Houston, Ft. Worth, Brownsville, and San Antonio. At Ft. Worth, a large outdoor exhibit features alligators and there are several indoor enclosures with caimans and dwarf crocodiles. Near the reptile building in Houston, an impressive outdoor exhibit contains alligators and alligator snapping turtles. The Gladys Porter Zoo in Brownsville has featured several crocodilian species.

The San Diego Zoo, with marvelous climate, has maintained crocodilians in well-planted outdoor facilities for many years and recently has added a gharial exhibit.

Reproduction

A number of studies and surveys have been published on reproductive behavior and successful breedings (Honegger, 1972, 1975, 1982; Honegger and Hunt, 1990; King and Dobbs, 1975). Two problems over the years had been the inability to accurately identify and sex crocodilians, until the papers published by Brazaitis appeared (1969, 1973). Lists of crocodilians bred in zoos and aquariums may be found in annual volumes of *International Zoo Yearbook* (*IZY*).

Schneider (1941) published his observations about the reproduction of the American alligator (*Alligator mississippiensis*) at the Leipzig Zoo. In 1965, Burrage analyzed courtship behavior of American alligators at the San Diego Zoo. Chaffee (1969) artificially incubated alligator eggs at the Fresno Zoo; five young hatched. Legge (1969) outlined mating behavior at the now defunct Belle Vue Zoo in Manchester, England. Whitworth (1971) described growth and mating at the Cannon Aquarium, Manchester Museum. Wright (1981) bred the American alligator at the Tulsa Zoological Park and Green (1981) described the second hatching of the American alligator at the Australian Reptile Park in Gosford. Alligators have bred many times in captivity.

In an early account, Lederer (1941) described care of the Chinese alligator (*Alligator sinensis*) at the Frankfurt Zoo. Müller (1970) mentioned reproduction at the Leipzig Zoo. Behler (1977) described the excellent propagation program for Chinese alligators which involved a number of zoos and the Rockefeller Wildlife Refuge in Grand Chenier, Louisiana where the alligators were maintained in a semi-captive environment. Behler et al. (1982) and Brazaitis and Joanen (1984) documented its status and propagation in captivity. Davenport (1982) discussed reproduction and aging and Watanabe (1986) was optimistic about its future.

Hunt (1969) and Alvarez del Toro (1969) at the Tuxtla Gutiérrez Zoo in Mexico observed reproduction in the spectacled caiman (*Caiman crocodilus*) and other zoos have recorded successful breedings (IZY). Other caimans have reproduced in zoo settings (IZY). The brown caiman reproduced at the San Antonio Zoo (Holmback, 1981). The first captive breeding worldwide of the dwarf caiman (*Paleosuchus palpebrosus*) occurred at the Cologne Zoo (Jes, 1997). This taxon also was bred at the Rio Grande Zoo (Belcher et al., 1987), Wildlife Conservation Society in 2005, and possible sperm storage was documented by Davenport (1995). Reproduction in the broad-nosed caiman (*Caiman latirostris*) has been reported (Widholzer et al., 1986). This species has reproduced at the Atagawa Tropical Garden and Alligator Farm in Japan for many years. Brazaitis (1986) outlined management, reproduction and growth of the Yacare caiman (*Caiman crocodilus yacare*) in the reptile building at the New York Zoological Park. The first successful captive propagation of Schneider's smooth-fronted caiman (*Paleosuchus trigonatus*) occurred at the Cincinnati Zoo (Jardine, 1981). Herron et al. (1990) observed reproduction in the black caiman (*Melanosuchus niger*).

Tryon (1980) documented the complex behavioral repertoire in the West African dwarf crocodile (*Osteolaemus t. tetraspis*) at the Fort Worth Zoo, a taxon which has bred in many zoos (Beck, 1978; Engelmann, 2001, Hara and Kikuchi, 1978; Payne, 1989; Sims and

Singh, 1978; Teichner, 1978). Falk Dathe (2000) outlined the care of dwarf crocodiles in the Berlin Tierpark from 1956 to 2000; there is a building for crocodiles, hummingbirds and turtles (Dathe, 1990). In 2004, two females hatched at the Toronto Zoo.

R. Howard Hunt from Zoo Atlanta has documented maternal (1975), aggressive (1987) and reproductive behavior (1973) in Morelet's crocodiles (*Crocodylus moreletii*). Pérez-Higareda et al. (1995) followed growth during the first three years of life in a semi-captive setting and compared their findings with zoo-raised specimens in Tuxtla Gutierrez (Alvarez del Toro, 1974).

Yadav (1969, 1979) described breeding of mugger crocodiles (*Crocodylus palustris*) at the Jaipur Zoo in India. Romulus Whitaker (1992) described the extraordinary achievements at the Madras Crocodile Bank in India with the mugger crocodile and gharial (see Chapter 10 for additional information and references). David (1970) bred the mugger crocodile at India's Ahmedabad Zoo. In India, they have reproduced at the Delhi Zoo since 1976, Vizag Zoo since 1978, Hyderabad Zoo since 1980, and recently reproduced at the Baroda Zoo (Sharma, 1998).

Fig. 3. Present crocodilian enclosure at Madras Crocodile Park Trust in India. *Photograph provided by Romulus Whitaker at Madras Crocodile Park Trust.*

Schürer and Bürger (1998) propagated the New Guinea crocodile (*Crocodylus novaeguineae*) in the Wuppertal Zoological Garden where two neonates hatched for the first time in captivity in a zoo. In 2003, the Fort Worth Zoo hatched one Philippine crocodile (*Crocodylus mindorensis*).

Dunn (1977, 1981) detailed reproduction of the Johnstone's crocodile (*Crocodylus johnsoni*) at the Melbourne Zoo in Australia. The taxon has also bred at the Atagawa Tropical Garden and Alligator Farm in Japan and National Aquarium in Baltimore.

Jes (1974) and Balcar (1996) commented on Nile crocodile (*Crocodylus niloticus*) reproduction at the Cologne Zoo and Brno Zoo in the Czech Republic, respectively. Celler (1977) described breeding these crocodiles in zoo conditions. IZY reported many breedings in zoos.

Cuban crocodiles (*Crocodylus rhombifer*) have bred at several zoos: Atagawa Tropical Garden and Alligator Farm, Havana Zoological Garden in Cuba in 1965, and the Smithsonian National Zoological Park. Larsson and Wihman (1989) were successful at reproducing the Cuban crocodile at the Skansen Aquarium in Stockholm, Sweden.

Davenport (1982) documented reproduction and aging in the slender-snouted crocodile (*Crocodylus cataphractus*). Magill (1984) documented breeding at the Miami Metrozoo.

Magill (1982) bred the Siamese crocodile (*Crocodylus siamensis*) at Metrozoo which has large outdoor enclosures for crocodilians. Frolov et al. (1989) bred them in the Îiscow Zoo.

Behler (1978) discussed the likelihood establishing captive-breeding populations for the American crocodile (*Crocodylus acutus*). Schubert and Santana (1996) provided a description of the excellent headstarting program at ZOODOM (Santo Domingo Zoo) for conservation of the crocodile in the Dominican Republic.

Dunn (1981) reproduced the estuarine crocodile (*Crocodylus porosus*) at the Melbourne Zoo. It has bred at Atagawa Tropical Garden and Alligator Farm in Japan in 1971 and Singapore Zoological Garden in 1975 (IZY).

Waturu Kimura (1976, 1978) described artificial egg incubation techniques, morphology and development of neonates, and abnormalities at Atagawa. Brazaitis and Watanabe (1983) used ultrasound scanning for Siamese crocodile eggs.

Behavior

Oskar Heinroth (1941) observed sound production in the American alligator at the Berlin Zoo and interpreted this behavior as communication under water. Garrick (1975) investigated sound production in the Chinese alligator at the New York Zoological Park. Campbell (1973) studied acoustic behavior in the American alligator and crocodile, and black and spectacled caimans. Brazaitis (1969) observed ingestion of gastroliths in two crocodilians. Petzold (1959) noticed pellets which were regurgitated in crocodiles. Asa et al. (1998) studied thermoregulatory behavior of captive American alligators at the Saint Louis Zoo. Hunt (1973, 1975, 1987) from Zoo Atlanta described reproductive, maternal and aggressive behavior in Morelet's crocodiles.

Medical and Captive Management

Several studies on medical and captive management deserve mention. Harwell (1986) at the Houston Zoo listed concerns associated with captive management. Davenport (1979) from the Smithsonian National Zoological Park reviewed husbandry and propagation procedures for captive crocodilians. Brazaitis and Watanabe (1982) proposed using "The Doppler," for reptile and amphibian hematological studies. Almandarz (1975) used chilled water to transfer adult crocodilians at the Lincoln Park Zoo and Wise (1994) reviewed appropriate procedures for capturing and restraining crocodilians, based on his extensive experience at the St. Augustine Alligator Farm in Florida.

Lloyd and Morris (1999) explored blood collection and hematology techniques in crocodilians. Birkett and McCraken (1992) treated a Johnstone's crocodile with bilateral mandibular fractures at the Melbourne Zoo. Brazaitis (1981) reported maxillary regeneration in a mugger crocodile. Heinrich Dathe (1960) documented tail regeneration in caimans. Frelier et al. (1985) described mycotic pneumonia caused by *Fusarium moniliforme* in an alligator. Ippen and Konstantinov (1981) underscored kidney changes in an Indian gavial caused by lack of vitamins. Valentine Lance of the San Diego Zoo has published a number of papers on stress with hormonal implications (Lance, 1990; Lance and Elsey, 1986, 1999; Lance and Lauren, 1984). Elsey et al. (1990a, 1990b) recorded growth rates and plasma corticosterone levels in juvenile alligators maintained at different stocking densities. Loveridge (1979), Bonath (1979) and Lloyd (1999) covered anesthesia and Lloyd et al. (1994) described Gallamine reversal in Cuban crocodiles using data accumulated in a joint study at Roger Williams Park and Toledo Zoos. Lloyd and Morris (1999) used phlebotomy techniques in crocodilians. Misra et al. (1993) listed bacterial isolates recovered from gharials. Ocholi and Enurah (1989) described salmonellosis in a captive Nile crocodile due to *Salmonella choleraesuis*. At the St. Louis Zoo, Wallach et al. (1967) diagnosed hypoglycemic shock in captive alligators and Wallach and Hoessle (1968a) found steatitis in captive crocodilians. Vetesi et al. (1981) described pockmarks resembling a rash in caimans.

There are recognized American Zoo and Aquarium

(AZA) studbooks for the Chinese alligator, African slender-snouted crocodile, Cuban crocodile, Morelet's crocodile, Siamese crocodile, and false gavial (see Read, 1996, for additional information). Conway (1985) presented the AZA Species Survival Plans which included the Chinese alligator and Cuban crocodile. In 2002, the AZA Crocodilian Advisory Group prepared a manual "Crocodilian Biology & Captive Management" for use in its training program.

Longevity

Crocodilians have been long-lived in captivity (Guggisberg, 1972). An American alligator lived for at least 56 years and other species have survived for over 30 years. A Chinese alligator lived for 50 years and a Nile crocodile for 25 years in Berlin. One mugger crocodile reached the age of 31 years in Trivandrum and a gavial lived for nearly 29 years in London. Flower (1925) listed longevities as follows: American alligator–40 years and perhaps longer than 73 but record can not be verified with certainty; mugger crocodile–31 years and still alive; Nile crocodile–17 years and still alive; African dwarf and slender-snouted crocodiles–15 years; Cuban crocodile–13 years; and American crocodile–10 years. Snider and Bowler (1992) listed the following records in North American collections: American alligator over 73 years; Chinese alligator over 46 years; six caiman taxa between 15–30 years; 14 crocodile taxa between 3–50 years and gharial over 27 years.

The Most Significant Zoo Program with Crocodilians

There have been a prodigious number of crocodilian papers on a variety of topics emanating from the Bronx Zoo/Wildlife Conservation Society over many years, mostly by F. Wayne King, Herndon Dowling, John Behler, Peter Brazaitis, George Amato, and John Thorbjarnarson (see Chapter 8, Wildlife Conservation Park/Bronx Zoo account). One example is the publication by Dowling and Brazaitis (1966), who recorded size and growth of the American and Chinese alligator, and black caiman, with extensive data on the Nile crocodile. They provided a table of weight-length measurements for 14 species; the crocodilians were housed in the reptile building at the Bronx Zoo which was built in 1898 and renovated in 1954.

Snakes

THESE VIPERS ARE USUALLY PUT AND KEPT IN A BOX WITH BRAN OR MOSS; NOT THAT THESE INGREDIENTS SERVE TO FEED UPON,

Fig. 4. Naturalistic crocodilian enclosure in Asian Exhibit at New York Zoological Society during 1990s. This is one of the most spectacular zoo exhibits for reptiles. *Photograph courtesy of Wildlife Conservation Society, headquartered at Bronx Zoo.*

AS SOME MAY FANCY; BECAUSE 'TIS SAID, *THEY NEVER EAT AFTER THEY ARE TAKEN* AND CONFIN'D, BUT LIVE ON THE AIR, AND WILL LIVE SO, MANY MONTHS: BUT MORE HEREAFTER . . .

CHARLES OWEN
ON EUROPEAN VIPERS IN 1742

SCIENCE AND ART SHOULD BE MUTUAL PATHWAYS TO UNDERSTANDING: WE NEED SOME THING TO CONFRONT SENSUALLY, SOMETHING ABOUT WHICH TO FEEL, IN ORDER TO PLACE VALUE ON NATURE; WE NEED THE OPPORTUNITY TO INTERPRET, TO REFLECT ON OUR OWN EXPERIENCES AND THOSE OF OTHERS IN THE CONTEXT OF NEW DETAILS AND NEW PERSPECTIVES. WHEN DESCRIBING THE BIOLOGY OF SNAKES, I ALSO WANTED TO EXPLORE THEIR BEAUTY AND MYSTIQUE, THEIR ROLES IN HUMAN EXPERIENCE.

HARRY GREENE
Snakes, The Evolution of Mystery in Nature (1997)

SNAKES APPEAR TO REACH THEIR POTENTIAL PERIOD OF LONGEVITY SOON AFTER TWENTY YEARS.

MAJOR STANLEY SMYTH FLOWER (1925)

History

The Aztecs had a large collection of vipers and other venomous snakes housed in a wild animal house in Tenochtitlan. Also in the New World, the Incas maintained lizards and snakes at the royal palace in Cuzco.

Fig. 5. Plate of European viper from *An Essay Towards the Natural History of Serpents* . . . by Charles Owen in 1742. *Courtesy of Smithsonian Institution Libraries, Washington, DC.*

Wajid Ali Shah kept a large menagerie in Calcutta which included tortoises and a large enclosure containing thousands of snakes (Walker, 2001).

The first published observations on captive snakes were recorded by Georg Seger (Segerus in Latin) in Germany, who was born in 1629 and died in 1678 (see Daszkiewicz for his biography, 2001). He kept snakes for scientific purposes and recorded shedding and egg-laying in "Aesculapian snakes" for the first time, but it is not clear whether the species described was the ratsnake (*Elaphe longissima*) or the grass snake (*Natrix natrix*). In any event, his husbandry techniques needed improvement since he removed their tongues initially as this organ was thought to inflict mortal wounds. Noticing that the snakes were less vigorous after this primitive surgery, he discontinued this practice. He assisted one of his snakes during oviposition and attributed her success in laying 13 eggs to his helpful intervention. In another case, he helped his snake during ecdysis by holding its skin.

In 1872, Charles Darwin published *The Expressions of the Emotions in Man and Animals.* In Chapter IV, he included a section entitled "The inflation of the body, and other means of exciting fear in an enemy." Based on his observations in the London Zoo, literature accounts and personal field experiences, he summarized his findings on defensive behaviors of amphibians and reptiles. Included in his analysis were descriptions of these behaviors in the puff adder, cobra, egg-eating snake, South American pitvipers, several colubrid snakes, saw-scaled viper, and rattlesnake. He concluded that "In the Zoological Gardens, when the rattlesnakes and puff-adders were greatly excited at the same time, I was much struck at the similarity of the sound produced by them; and although that made by the rattle-snake is louder and shriller than the hissing of the puff-adder, yet when standing at some yards distance I could scarcely distinguish the two. For whatever purpose the sound is produced by the one species, I can scarcely doubt that it serves for the same purpose in the other species; and I conclude from the threatening gestures made at the same time by many snakes—, that their hissing,—the rattling of the rattlesnake and of the tail of the Trigonocephalus,—the grating of the scales of the Echis,—and the dilation of the hood of the Cobra,—all subserve the same end, namely, to make them appear terrible to their enemies."

In 1882, Catherine G. Hopley published an interesting book called *Snakes: Curiosities and Wonders of Serpent Life.* Her book covers herpetoculture at the London Zoo and there are many interesting anecdotes, two of which are included here. In Chapter XII called "Ophidian Acrobats: Construction and Constriction," she wrote ". . . So did some young boa constrictors, born alive at the Gardens, June 30th 1877. They were from fifteen to twenty inches in length, and had teeth sufficiently developed to draw blood from Holland's hand, showing fight and ingratitude at the same time. They were exceedingly active, and fed on young mice, which they constricted instinctively. One of them, known as 'Totsey,' subsequently *hung* for her portrait on page 201." Holland was the reptile keeper at London Zoo at that time. The portrait, a line drawing, depicts Totsey coming down from an overhead branch to constrict one of three live sparrows on the floor.

In Chapter XXIV, "Do Snakes Incubate Their Eggs?" Hopley provides data on the snakes at the London Zoo: getting over 100 eggs ('more than a bushel') from a python in 1862, breeding five 'seven-banded snakes' (*Tropidonotus leberis* now *Regina septemvittata*) in August 1872, breeding seven *Coluber natrix* (=*Natrix natrix*) in June 1873, 14 (of which 10 survived) yellow Jamaican boas (*Chilobothrus* (*Epicrates*) *inornatus*) which crawled to the top of their cage in August 1873, and 20 young from a Panamanian boa constrictor in July 1877, ". . . which displayed ability to take care of themselves forthwith by leaving the marks of their teeth on Holland's fingers."

There is validity in tracking a single species over an extended period of time rather than reporting on a single interesting happening. Frederick William FitzSimons was the Director of the Port Elizabeth Museum and Snake Park in South Africa when he published his book *Pythons and Their Ways* in 1930. In this book he described many facets of the reproduction, brooding behavior, and maintenance of African rock pythons. He provided clutch sizes, egg measurements and weights, incubation durations, and neonatal care, based on data from 11 clutches.

In a book chapter entitled "Behavioural Consequences of Husbandry Manipulations: Indicators of Arousal, Quiescence and Environmental Awareness," Chiszar et al. (1995) evaluated the effect that captivity has on the behavior of snakes. Topics covered were cage cleaning and exploratory behavior, use of familiar artificial chemical cues, chemical recognition of self, and sensitivity to spatial configurations. Their conclusions are as follows: "(1) snakes are sensitive not only to the chemical content of their living spaces, but also to the spatial arrangement of objects in it; (2) changes in either dimension result in elevated investigatory behaviour; (3) these changes might have stressful concomitants." There is no question that future research along these lines will prove to be fruitful.

Feeding Captive Snakes

Reverend Bateman (1897) summarized feeding difficulties when snakes are kept in captivity: "No animals, in their manner of taking food, are more capricious than Snakes. Sometimes their desire to feed is so great, that they will eat a meal out of all proportion to their size; and, sometimes, without any apparent cause, they will refuse food, even the most suitable, until they die of starvation. As a rule, these creatures cannot be persuaded to seize any prey but that which is natural to them. For example, a Snake which lives upon frogs will not on any account, as a general thing, take lizards, though the latter might be just as nourishing to the Ophidian as the former, and possibly even more so, nor will a lizard-eating Snake be tempted to swallow a newt.

Even if Snakes did never refuse their natural food, the supplying of it is not always convenient, nor is it, to most people, pleasant. Young mice and rats, frogs

Fig. 6. Force-feeding technique used for large constrictors at New York Zoological Society, possibly around 1913. *Undated photograph courtesy of Wildlife Conservation Society, headquartered at Bronx Zoo.*

and lizards are not continually at hand, and it is not, by any means, an attractive sight to see a Snake, at evident discomfort to itself, swallowing a live animal or one recently constricted or poisoned." In support of the practice of assisting snakes to feed artificially, Bateman continues . . . "I believe many keepers of reptiles in Zoological Gardens adopt, to a greater or lesser extent, the artificial system of feeding their charges. The largest Snake in the Reptile House, Regent's Park, London, and which is also perhaps the largest Snake in captivity anywhere, has been fed by hand for the last three years at least, and is in splendid condition, weighing probably something like 18 st. This Snake, the Reticulated Python (*Python reticulatus*), receives its food regularly once a week . . . . There are several methods of administering food to Snakes. For example (1) some Snakes will swallow a dead animal, or even a piece of meat, when it has been simply placed between their jaws. (2) The Snake's mouth is forcibly opened, and a dead small animal, dipped in milk, is pushed down the throat, and then worked down the gullet by the manipulation of the fingers outside the Snake's body. (3) Pieces of meat, or portions of animals, dipped in milk, are pushed sufficiently far down the opened mouth of a Snake by means of a smooth stick. (4) A tube is filled with suitable food, and passed down the gullet of the reptile, and then contents of the tube are discharged by means of a piece of cane used as a ramrod."

Force-feeding large snakes can be a difficult business as is evident from this excerpt about R. L. Ditmars and his keepers by A. W. Rolker in his colorful article

"The Story of the Snake" in 1956: "A most famous instance of snake feeding happened a year ago in the New York Zoölogical Park when Czarina, a twenty-foot regal python, one of the largest specimens in captivity and known all over among naturalists, was fed against her will . . . For more than six weeks Czarina had not touched food, when Curator R. L. Ditmars, in charge of the reptile house, decided to capture the monster, to take her out of her cage, to stretch her to full length, and to force food down her throat . . . Every keeper in every department of the Park was notified to report to the reptile house that evening, and the snake meal was prepared. Thirty pounds of rabbits were killed, bound together with twine, and fastened to a bamboo pole ten feet long . . . That evening when the visitors had gone the fight began—twelve men against the snake . . .Through a slot in the big door at the back of the cage the Curator peered to locate the snake. Then he threw open the door . . . Then the Curator, only a burlap bag in hand and trusting the eleven to follow, threw himself toward the monster and in an instant the bag was over her head. Blinded and with the hands of a man clutching her throat, like a giant spring hurled herself loose among the struggling men who had piled into the cage and were trying to clutch the wonderful coils. Through the cloud of dust within appeared glimpses of the fighting keepers and flashes of the lashing body, while the scuffle of feet and cries of the men sounded over all. The Curator, still clutching the throat of his great adversary, staggered about, whipped back and forth like a jackstraw. Grabbing, groping, hugging, and tugging, the others were tossed like chaff here and there only to fall back to their work which now had grown a matter of life and death, for to release the snake and permit all hands to escape in safety would now have been impossible. Through the dust were seen five or six men lying side by side on their bellies trying to pin the snake to earth, but the next instant they were up in the air, tossed by the dreadful spring of those giant coils . . . But even Czarina's tremendous strength was unable to withstand indefinitely the assault of the small army of hardened and trained animal men . . . At the end of another five minutes they had pinned down the snake . . . Mr. Ditmars, still clutching the wicked head, was the first to stagger out of the cage. One by one the others followed, each lugging a share of the snake. The python was stretched full length on the floor of the house. The custody of the head was turned over to an assistant, and the Curator prepared to administer the food. The great jaws were forced apart, and the bamboo pole with its bologna of rabbits was shoved

into the mouth and down the endless throat. . . . Still, as if lifeless, she remained, while down, down went the food, until, when nearly the entire pole was buried within the reptile, the rabbits were wriggled loose, and the great snake had been forced to feed." Clearly, this was not a task for the faint-hearted!

Fig. 7. "The Ophiophagus in the Zoological Gardens of London knows its keeper and feeding-time. When a Snake is put into the cage it is immediately on the alert, and the victim tries to escape. But the attack is commenced at once, and the prey is seized behind the head and dragged on to the floor, and gradually swallowed head first." Illustration from *Cassell's Natural History* published in 1877-1882. *Courtesy of Smithsonian Institution Libraries, Washington, DC.*

In the mid-1970s, one of the most sustained and productive research initiatives on the feeding behavior of snakes was started at the University of Colorado by David Chiszar and Charles Radcliffe of the Denver Zoo (see Chapter 8, Denver Zoological Gardens account for list of titles). Many of their studies centered on the feeding behavior of pitvipers, especially rattlesnakes. These contributions were and are crucial references for zoo workers. In another important study, Alving and Kardong (1994) discovered that rattlesnake predatory behavior was relatively unaffected by captivity or aging. Troncone and Silveira (2001) chronicled predatory behavior in the captive environment by testing over 70 recently collected pitvipers *Bothrops jararaca* at Instituto Butantan in Brazil.

In zoo collections, snakes which have specialized diets, such as snail-eaters (*Dipsas*), slug-eaters (*Duberria*), egg-eaters (*Dasypeltis*), Indian egg-eating snakes (*Elachistodon*), centipede eaters (*Aparallactus,*

Tantilla), frog-eaters (*Diaphorolepis*, *Stegonotus*), crayfish snakes (*Regina*), snail suckers (*Sibon*), worm-eating snakes (*Trachischium*), and blind worm snakes of the family Typhlopidae are rarely kept so information derived from observations in zoos and aquariums is lacking. Many zoos tend to maintain and exhibit the large, dangerous and spectacular forms, because the public demands it and the caretakers prefer it.

Reproduction

Pythons and Boas: This group of snakes, hardy as captives, has been featured in zoo collections for centuries so it is not surprising that many papers have been published on reproductive biology. Pierre Bernard mentioned that a female python laid 15 eggs, 8 of which hatched in Le Jardin des Plantes in Paris (Bernard, 1842-1843). According to Edward Turner Bennett (1829), in 1828, a python laid and attended 15 eggs in the Tower Menagerie in London which did not successfully hatch. Sclater (1862) recorded notes on egg incubation in the African rock python (*Python sebae*) in the London Zoo. Benedict (1932) studied the reproductive biology and maternal brooding behavior of the African rock python by comparing environmental and skin temperatures of a captive female at the Smithsonian National Zoological Park. See Chapter 3 for historical portrait of experiments to determine whether female pythons produce heat when brooding egg clutches.

Indian and Burmese pythons (*Python m. molurus* and *P. m. bivittatus*) have been studied in many zoos. Forbes (1881) observed incubation of the Indian python at the London Zoo and addressed the issue of increasing temperatures during brooding. Lederer (1956) investigated the reproductive biology and development of the Indian and Burmese python at the Frankfurt Zoo. Hutchison et al. (1966) documented thermoregulation in a brooding female Indian python at the New York Zoological Park. Takahashi (1967) propagated the Indian python at the Ueno Zoo.

Müller (1970) and Engelmann (1986) documented breeding behavior of the Burmese python at the Leipzig Zoo. Acharjyo and R. Misra (1976) studied reproduction and growth of the Indian python at the Nandankanan Biological Park, Orissa, India. A subsequent study (Acharjyo and Misra, 1980) described growth rate and age at first egg-laying. Vyas (1996) recorded breeding of the Indian python at the Sayaji Baug Zoo, Vadodara, India. Townson (1980) observed reproduction of the Indian and Burmese python, with interbreeding of the two subspecies. *Python molurus* has been bred to the 3rd generation in Tokyo's Ueno

Zoo (Sugiura, 1977). Rehák (1993, 1994) outlined captive care and breeding of the Ceylon python (*Python molurus "pimbura"*) and Indian pythons at the Prague Zoo. Baskar et al. (1999) documented incubation, feeding and growth of Indian pythons at the Arignar Anna Zoological Park, Madras, India. Later, Walsh and Murphy (2003) recorded observations on husbandry, breeding and behavior at the Smithsonian National Zoological Park which included neonates with aberrant color patterns, possibly due to improper egg incubation temperatures.

Other pythons have been studied in zoos as well. In two papers, Murphy et al. (1978, 1981) recorded the reproductive biology of 19 species of boas and pythons, genera *Acrantophis, Aspidites, Candoia, Corallus, Epicrates, Liasis* and *Python*. Lederer (1944) wrote an extensive paper on food supply, development, mating and brood care of the reticulated python (*Python reticulatus*) at the Frankfurt Zoo. Honegger (1970) published his study on the reproductive biology. Panouse and Pellier (1973) recorded oviposition and incubation of the eggs in a reticulated python housed in the vivarium of the Château de Thoiry-en-Yvelines near Paris. Araki et al. (1982) reproduced the reticulated python at the Takarazuka Zoological and Botanical Gardens in Japan and included photographs of the facility and reproductive events.

Boos (1979) from the Trinidad Zoo listed breeding records of Australian pythons. Barker (1984) reproduced black-headed pythons (*Aspidites melanocephalus*) at the Dallas Zoo. In 2005, twin black-headed pythons hatched at the Houston Zoo. Ross (1973) successfully encouraged mating and hatching of Children's python (*Liasis* now *Antaresia childreni*) and Ross and Larman (1977) bred two species, white-lipped or D'Albert's (*Liasis* now *Leiopython albertisii*) and Macklot's pythons (*Liasis mackloti*), at the Institute for Herpetological Research. Dunn (1979) bred the Children's python and Banks (1985) the white-lipped python at the Melbourne Zoo. Ryabov (1999) bred the Stimson's python (*Antaresia stimsoni orientalis*) at the Tula Exotarium. Several studies by Walsh and associates at the Smithsonian National Zoological Park (Van Mierop, et al. 1982; Walsh, 1977, 1980) documented behavior, husbandry and breeding of the green tree python (*Chondropython viridis* now *Morelia viridis*) and Brazilian rainbow boa (*Epicrates c. cenchria*) (Walsh, 1994; Walsh and Davis, 1983). Bels and Van den Sande (1986) bred the Australian carpet python (*Morelia spilota*) at the Antwerp Zoo. Grow et al. (1988) reproduced the amethystine python (*Python =Morelia amethystinus kinghorni*) and Tarbet (1983) the

D'Albert's python (white-lipped python) at the Oklahoma City Zoo. Korzhov et al. (1995) bred the Indonesian scrub python (*Liasis a. amethistinus*) at the Moscow Zoo. Mannion (1996) propagated the green tree python at the Queensland Reptile and Fauna Park.

Ettling (1991) elaborated on captive husbandry and reproduction of the blood python (*Python curtus*) at the St. Louis Zoo. Rehák (1994) bred blood pythons (*Python curtus breitensteini*) at the Prague Zoo. Branch and Griffin (1996) outlined zoo breeding programs for Angolan pythons (*Python anchietae*) in the United States and Transvaal Snake Park, South Africa. Patterson (1974) hatched the African python and Angolan python (1978). At the St. Louis Zoo, Martin de Camilo et al. (1999) described comparative follicular dynamics between the Brazilian rainbow boa and the ball python (*Python regius*).

Müller (1970) bred the boa constrictor (*Boa constrictor*) at the Leipzig Zoo. Honegger (1970) published a paper on its reproductive biology. Gensch (1969) reported on boa hybrids *Constrictor* (=*Boa*) *c. constrictor* x *C. c. imperator* at the Dresden Zoo. Blody and Mehaffey (1989) documented the reproductive biology of the annulated boa (*Corallus annulatus*) at the Fort Worth Zoo. Groves and Mellendick (1973) bred the Madagascan tree boa (*Sanzinia madagascariensis*) at the Baltimore Zoo and McLain (1983) at the Houston Zoo. Branch and Erasmus (1976) recorded reproduction in Madagascar ground (*Acrantophis*) and tree boas (*Sanzinia*). Huff (1984) outlined husbandry and propagation of the Madagascar ground boa with additional

Fig. 8. Boa constrictor (*Boa constrictor*) gave birth to 20 young in 1868 which stayed together in "nest" at Natura Artis Magistra in Amsterdam. *Photograph provided by Eugène Bruins, Natura Artis Magistra Archives.*

notes on other Malagasy boids at the Reptile Breeding Foundation. Reichling (1983) reproduced the Dumeril's boa (*Acrantophis dumerili*) at the Memphis Zoo.

Lederer (1942) described reproduction and development in the yellow anaconda (*Eunectes notaeus*) at the Frankfurt Zoo. Belluomini and Veinert (1967) had breeding anacondas (*Eunectes murinus*) at the Sao Paulo Zoo and Deschanel (1978) at the Lyons Zoo. Holmstrom (1981) propagated the yellow anaconda and Holmstrom and Behler (1981) recorded post-parturient behavior of the common anaconda at the New York Zoological Park. Frolov et al. (1986) reproduced the Paraguayan anaconda in the Moscow Zoo, including second generation births (Kudryavtsev and Odinchenko, 1986).

Bloxam (1983) reported on the captive management and reproduction of the endangered Round Island boa (*Casarea dussumieri*) at the Jersey Wildlife Preservation Trust which included a plan to eradicate the introduced mammals to restore native vegetation. Rehák (1990, 1991) bred the Cuban dwarf boa (*Tropidophis feicki*) and the javelin sand boa (*Eryx j. jaculus*) at the Prague Zoo.

Thomas Huff from the Reptile Breeding Foundation published a number of papers on insular boid reproduction of the genus *Epicrates*: Cuban boa (1976), Puerto Rican boa (1978), Jamaican boa (1979). Murphy and Guese (1977) observed reproduction in the Hispaniolan boa (*Epicrates f. fordii*) at the Dallas Zoo. In my view, Peter Tolson of the Toledo Zoo has the most dynamic and creative program ever developed for Neotropical boids of the genus *Epicrates* (Tolson, 1980, 1982, 1989, 1992, 1994; Tolson and Teubner, 1987; Tolson, et al., 1983). His commitment to the conservation of this snake assemblage, which includes captive-breeding, reintroduction, and field studies, has spanned over two decades and is a superb model for other zoo workers.

Colubrids: North American queen snakes (*Regina septemvittata*) produced 48 young between 1872 and 1880 at the London Zoo (Coote, 2001). Roger Conant recorded reproduction in the black king snake (*Lampropeltis getulus nigra*) (1934), Madagascar hognosed snake (*Leioheterodon madagascariensis*) (1938), southern black racer, Texas patchnose snake and Brazos watersnake (1942), and Arid Land ribbon snake and Aguanaval watersnake (1965). Lederer (1949) published a lengthy study on the viperine snake (*Natrix maura*) and hybrid ratsnakes (1950). Frank Groves from the Baltimore Zoo described eggs and young of the corn snake (*Elaphe g. guttata*) in Maryland (1957) and indigo snake (*Drymarchon corais couperi*) (1960). He also discussed reproduction and venom in the Blanding's tree snake (*Boiga blandingi*) (1973). At the Tula Exotarium, Orlov and Ryabov (2002) elucidated on reproduction of the black mangrove snake (*Boiga dendrophila gemmicincta*) from Sulawesi (Indonesia). Foley (1998) prepared a description on the captive maintenance and reproduction of Oates' twig snake (*Thelotornis capensis oatesi*) at the Riverbanks Zoo. Petzold bred the variable swamp snake (*Tretanorhinus variabilis*) (1967), cat-eyed snake (*Leptodeira annulata*) (1969), Cuban racer (*Alsophis cantherigerus*) (1978), and Indian striped snake (*Natrix (Amphiesma) stolata*) (Petzold and Stettler, 1972) at the Tierpark Berlin. Walsh and Davis (1978) documented husbandry and breeding of the rufous-beaked snake (*Rhamphiophis oxyrhynchus rostratus*) at the Smithsonian National Zoological Park.

Simmons (1977) reproduced the Chinese red snake (*Dinodon rufozonatum*) and Campbell and Murphy (1984) outlined reproduction in five species of Paraguayan colubrids housed at the Fort Worth Zoo. At the Fort Worth and Dallas Zoos, Cover and Boyer (1988) bred the endangered San Francisco garter snake (*Thamnophis sirtalis tetrataenia*). In 2005, the eastern plains garter snake (*Thamnophis radix radix*) reproduced at the Cleveland Metroparks Zoo. The same year, the exiled garter snake (*Thamnophis exsul*) produced a litter at the San Antonio Zoo. At the Houston Zoo, Freed (1989) documented delayed fertilization in the African egg-eating snake (*Dasypeltis scabra*), a species rarely kept in zoos. At the Memphis Zoo, Reichling has embarked on a long-term study of the Louisiana pine snake (*Pituophis melanoleucus ruthveni*) (1986, 1988, 1989, 1990) and the black pine snake (*Pituophis melanoleucus lodingi*). At the Cheyenne Mountain Zoological Park, Connors (1986) described a captive breeding of the Great Basin gopher snake (*Pituophis melanoleucus deserticola*). Bels (1987) observed courtship and mating behavior in the snake *Hydrodynastes gigas*. Dathe and Dedekind explained care and breeding of Baird's ratsnakes (1985), horse shoe snakes (*Coluber hippocrepis*) (1988), and Madagascar hog-nosed snakes (1996) in the Berlin Tierpark. McGeorge (1997) bred the sunbeam snake (*Xenopeltis unicolor*) at the Chester Zoo. Dunn et al. (1987) were successful in exhibiting and breeding the Arafura file snake (*Acrochordus arafurae*) at the Melbourne Zoo.

Kingsnakes (*Lampropeltis*) and ratsnakes (*Elaphe*) have been popular zoo reptiles, due to their beauty and hardiness in captivity. Tryon and Hulsey (1976) noted reproduction in captive Nelson's kingsnakes (*Lampropeltis triangulum nelsoni*) at the Ft. Worth Zoo. Murphy et al. (1978) provided an inventory of reproduction and social behavior in gray-banded kingsnakes (*L. alterna*). Herman (1979) bred the Jaliscan milk snake (*L. t. arcifera*) and scarlet kingsnake (*L. t. elapsoides*) at Zoo Atlanta. Kardon (1979) noted breeding in three Mexican milk snake subspecies (*L. t. polyzona, L. t. nelsoni* and *L. t. sinaloae*) at the San Antonio Zoo. Hosono (1982) bred the Florida kingsnake (*Lampropeltis getula floridana*) at the Kyoto Municipal Zoo in Japan. Hammack (1989) reproduced the Colombian milk snake (*L. t. andesiana*) at the Dallas Zoo. Korinek (1997) bred Campbell's milksnake (*Lampropeltis triangulum campbelli*) in the terrarium of Olomouc Zoo. Tryon and Murphy (1982) investigated the reproductive biology of 13 varieties of kingsnakes (species *mexicana, triangulum* and *zonata*). In 2004, five eastern milksnakes (*Lampropeltis t. triangulum*) hatched at the Detroit Zoological Institute.

At the Ft. Worth Zoo, Campbell (1972) bred the Trans-Pecos ratsnake (*Elaphe* now *Bogertophis subocularis*) and Tryon (1976) reported on second generation reproduction and courtship behavior. Brecke et al. (1976) documented reproduction and social behavior in the Baird's ratsnake (*Elaphe bairdi*) at the Dallas Zoo. Tremper (1981) propagated the Neotropical rat snake (*E. f. flavirufa*) at the Chaffee Zoo. Engelmann (1984) explained egg-laying behavior of the yellow ratsnake (*E. obsoleta quadrivittata*) at the Leipzig Zoo. Frolov and Kudryavtsev (1985) and Kudryavtsev and Frolov (1984) kept and bred Russian ratsnakes (*E. s. schrencki*) at the Moscow Zoo. Ìàmàt (1989) outlined the reproductive biology of ratsnakes of the *longissima*-complex from Southern Azerbaijan. Mamet and Kudryavtsev (1997) propagated the Mandarin ratsnake (*E. mandarina*) and Mamet and Kudryavtsev (1997) provided notes on the reproductive biology of the Asiatic ratsnake (*Elaphe persica*) at the Moscow Zoo. Herrmann (1998) reported on husbandry and reproduction in the black and yellow ratsnake (*Spilotes pullatus*) at the Cologne Zoo. At the Tula Exotarium, Ryabov (1997) explored the extraordinary high productivity in the radiated ratsnake (*E. radiata*), natural history, keeping and breeding the Persian ratsnake (*E. persica*) (2001), and Ryabov and Popovskaya (2000) provided some comparative data on the breeding of four subspecies of the stripe-tailed rat- snake (*Elaphe taeniura*).

Elapids: Oliver (1956) propagated the king cobra (*Ophiophagus hannah*) at the New York Zoological Park and Burchfield (1977) bred the taxon at the Gladys

Porter Zoo. Petzold (1968) described the reproductive biology of Asiatic cobras (*Naja*) at the Berlin Tierpark. Campbell and Quinn (1975) observed reproduction in a pair of Asiatic cobras, (*Naja naja*) and Quinn and Hulsey (1978) followed growth at the Ft. Worth Zoo. Olexa (1975) reproduced the Indian cobra at the Prague Zoo. Another species, the African forest cobra (*Naja melanoleuca*) was bred at the Ft. Worth Zoo (Tryon, 1979). Behler and Brazaitis (1974) bred the Egyptian cobra (*Naja haje*) at the New York Zoological Park. Kopczynski (1993) described reproduction in the monocellate cobra (*Naja kaouthia*) at the Plock Zoo. Vassiliev (1996) reproduced the Ringhals (Rinkals) cobra (*Hemachatus haemachatus*) in the Moscow Zoo. Falk Dathe (1997) outlined care and reproduction of the red spitting cobra (*Naja pallida*) in the Berlin Tierpark. The Central Asia cobra (*Naja oxiana*) was bred at the Leningrad Zoo (Kamelin and Lukin, 2000).

Banks (1983, 1984) described reproduction in the taipan (*Oxyuranus scutellatus*) and two species of brown snakes (*Pseudonaja*) at the Melbourne Zoo. Haagner and Morgan propagated the eastern green mamba (*Dendroaspis angusticeps*) (1989) and black mamba (*Dendroaspis polylepis*) (1993) at the Manyeleti Reptile Centre, Eastern Transvaal, Africa. Leloup (1964) recorded reproduction in Jameson's mamba (*Dendroaspis jamesoni kaimosae*).

Campbell (1973) and Quinn (1979) reproduced the Texas coral snake (*Micrurus fulvius tenere*) at the Ft. Worth Zoo.

True Vipers: Snedigar and Rokosky (1949) reported on new-born Gaboon vipers (*Bitis gabonica*) at the Brookfield Zoo. Biella et al. (1989) documented the reproductive biology of the puff adder (*Bitis arietans*) and gaboon viper along with observations regarding requirements for nourishment, growth, and care. Jacobs and Belcher (1983) noted captive reproduction in the West African gaboon viper (*B. g. rhinoceros*) at the Rio Grande Zoological Park. Amazingly, one puff adder gave birth to 156 young, two of which were stillborn (Janacek, 1976).

Goode (1979) described reproduction in the saw-scaled viper (*Echis colorata*) at the Columbus Zoo. Cherlin (1985) explored reproduction of the Iranian saw-scaled viper (*Echis multisquamatus*) in the Tashkent Zoo.

There have been a number of papers on Mid-Eastern and European vipers (*Vipera lebetina*). Y. A. Lukin from the Leningrad Zoo (1977) described reproduction in the Levantine viper. Petzold (1980) presented statistical data concerning birth weights of the European

Fig. 9. Illustration of rhinoceros viper (*Bitis nasicornis*) and young from London Zoo in Catherine G. Hopley's *Snakes: Curiosities and Wonders of Serpent Life* in 1882. *Courtesy of Smithsonian Institution Libraries, Washington, DC.*

viper (*Vipera b. berus*) at the Berlin Tierpark. In two studies in 1986, Bozhansky and Kudrjavtsev at the Moscow Zoo discussed breeding and ecology of the Caucasian viper. Perry and Blody (1986) reported courtship and reproduction in Cretan vipers (*V. l. schweizeri*) at the Dallas Zoo. Using data accumulated from the Leningrad and Kaliningrad Zoos, we may infer that 121 Levantine vipers (*Vipera l. turanica* and *Vipera l. obtusa*) were produced over 20 yrs, 55 of which were captive born (Igolkina, 1989). Mamet and Kudryavtsev (1996) observed reproduction in Latifi's viper (*Vipera latifiii*) at the Moscow Zoo. Elements of natural history, husbandry, and captive reproduction of two mountain vipers (*Vipera bornmuelleri* and *Vipera wagneri*) at the St. Louis Zoo were described by Ettling (1996). Kudryavtsev and Mamet (1991) outlined the husbandry and propagation of the Radde's viper (*Vipera r. raddei*) and Mamet and Kudryavtsev (1997) bred the Caucasian viper in the Moscow Zoo. Kamelin et al. (1997) described hybridization of *Vipera schweizeri* and *Vipera lebetina obtusa*.

Pitvipers: Rattlesnakes have always been popular zoo ophidians; for example, approximately one hundred rattlesnakes lived in the London Tower Menagerie (Bennett, 1829; Chapter 3). Many papers have been published on their biology by zoo workers (see Murphy and Armstrong, 1978; Armstrong and Murphy, 1979 for examples). Grace Olive Wiley (1929) described reproduction in the western diamondback rattlesnake

(*Crotalus atrox*) in captivity; she was known for taming venomous snakes and described her techniques in this account (see Chapter 4). Hoessle (1963) observed a breeding pair of western diamondback rattlesnakes. Petzold (1963) noted reproduction and development in the western diamondback rattlesnake and cottonmouth moccasin (*Agkistrodon p. piscivorus*) at the Berlin Tierpark. Murphy and Shadduck (1976) reproduced the eastern diamondback rattlesnake (*Crotalus adamanteus*) and described a two-headed neonate at the Dallas Zoo. Tryon and Radcliffe (1977) contributed reproductive data from captive Lower California rattlesnakes (*Crotalus enyo*) at the Fort Worth Zoo and Denver Zoological Gardens. Tryon (1978) bred a pair of captive Arizona ridge-nosed rattlesnakes (*Crotalus willardi*). Carl et al. (1982) documented reproduction in captive Aruba Island rattlesnakes (*Crotalus unicolor*) at the Ft. Worth and Houston Zoos. Peterson (1982) bred speckled and Neotropical rattlesnakes (*Crotalus m. mitchelli* and *Crotalus durissus*) at the Houston Zoo. Loomis and Smith (1987) found that an Aruba Island rattlesnake (*Crotalus unicolor*) was fertile following caesarean section.

Water moccasins or cottonmouths (*Agkistrodon piscivorus*) were recorded as breeding at the London Zoo between 1860 and 1861 (Coote, 2001). Conant (1933) reported on three generations of cottonmouths and the Moscow Zoo had two (Kudryavtsev et al., 1986). Peters (1979) had two generations of the cantil (*Agkistrodon bilineatus*) at the Taronga Zoo. In 2004, five Taylor's cantils (*Agkistrodon taylori*) were born at the San Antonio Zoo. Rokosky (1941) had newborn jumping vipers (*Bothrops* [now *Porthidium*] *nummifer*) at the Brookfield Zoo. Antonio (1980), Cover at the National Aquarium in Baltimore (1982), Blody at the Ft. Worth Zoo (1983) and Hitchiner (1987) at the Rio Grande Zoo in Albuquerque, New Mexico, chronicled mating behavior and reproduction of the eyelash viper (*Bothrops* [now *Bothriechis*] *schlegelii*). Murphy and Mitchell (1984) recorded the reproductive biology of 13 varieties of the genus *Bothrops* at the Dallas and Houston Zoos. Peterson (1992) reproduced the yellow-lined palm viper (*Bothriechis lateralis*) and Peterson and Odum (1986) provided an overview of their techniques for breeding and maintaining arboreal and terrestrial *Bothrops* at the Houston Zoo. Leloup (1975) observed reproduction in the Brazilian lancehead pitviper (*Bothrops moojeni*). In 2005, Dunn's pitviper (*Porthidium dunni*) reproduced at the San Antonio Zoo.

Boyer et al. (1989) propagated bushmasters (*Lachesis m. muta*) at the Dallas Zoo. Kudryavtsev (1985) described the reproductive biology of the Malayan pitviper (*Calloselasma rhodostoma*) and the Brazilian lancehead pitviper (*B. moojeni*) (Kudryavtsev, 1986) in the Moscow Zoo. Reporting on an impressive accomplishment, Kudryavtsev and Frolov (1984) bred three generations and attempted sex determination of the bamboo pitviper (*Trimeresurus gramineus*) at the Moscow Zoo. In 2005, 22 Sumatran pitvipers (*Trimeresurus sumatranus*) hatched at the Denver Zoo. Roberts and Hammack (1995) documented reproduction of the speckled forest-pitviper (*Bothriopsis taeniata*) at the Dallas Zoo. The giant pitviper (*Ermia mangshanensis*) reproduced at the San Diego Zoo.

Behavior

Boas and Pythons: Barker et al. (1979) documented formation of a linear social hierarchy in a captive group of Indian pythons. Marcellini and Peters (1982) reported endogenous heat production after feeding in the Indian python at the Smithsonian National Zoological Park. Madagascan boas (*Sanzinia madagascariensis*) used pelvic spurs to stab other males during combat at the Dallas Zoo (Carpenter et al., 1978). At the Dallas Zoo, Murphy et al. (1978) observed caudal luring in young green tree pythons and Garrett and Smith (1994) found that when juveniles were given a choice between dark and light perches, they more often chose the former as a resting site. Timmis (1969) observed Pacific boas (*Candoia* spp.) at the Taronga Zoo.

Colubrids: Gillingham et al. (1977) created new terminology for a generalized colubrid pattern by using the courtship and copulatory behavior in the Mexican milk snake (*Lampropeltis triangulum sinaloae*) at the Dallas Zoo. Murphy (1977) recorded an unusual method of immobilizing avian prey by the dog-tooth cat snake (*Boiga cynodon*) at the Dallas Zoo and Quinn and Neitman (1978) reproduced the snake at the Houston Zoo.

Elapids: Death adders (*Acanthophis antarcticus*) lured prey with their tails at the Dallas Zoo (Carpenter et al., 1978; Chiszar et al., 1990). At the Central Florida Zoo, Chiszar et al. (1994) discovered strike-induced chemosensory searching in eastern green mambas (*Dendroaspis angusticeps*).

True Vipers: At the Dallas Zoo, Murphy and Barker (1980) saw courtship and copulation of the Ottoman viper (*Vipera xanthina*) and referred to use of hemipenes. Ettling and Marfisi (2002) noticed ritualized male combat in two species of mountain vipers

(*Montivipera raddei* and *M. wagneri*) at the St. Louis Zoo.

Pitvipers: Allen (1949) followed caudal luring in the juvenile cantil (*Agkistrodon bilineatus*). Carpenter et al. (1976) catalogued the combat ritual of the rock rattlesnake (*Crotalus lepidus*) at the Dallas Zoo and introduced new nomenclature for various postures of combating male snakes. Johnson (1988) observed combat and courtship of the eastern massasauga rattlesnake (*Sistrurus c. catenatus*) by comparing field and captive behavior at the Metro Toronto Zoo. Lindsey (1979) documented combat behavior in the dusky pygmy rattlesnake (*Sistrurus miliarius barbouri*) at the Monroe Zoo. Gillingham et al. (1983) noticed courtship, male combat and dominance in the western diamondback rattlesnake.

Chiszar et al. (1985) discovered that duration of strike-induced chemosensory searching was not compromised by feeding dead prey to long-time captive snakes at the Smithsonian National Zoological Park, Audubon Zoo in New Orleans, and San Diego Zoo. Jumping vipers (*Porthidium nummifera*) and bushmasters covered eyes and pits by moving facial skin while holding rodent prey (Chiszar and Radcliffe, 1989; Chiszar et al., 1989). Troncone and Silveira (2001) catalogued predatory behavior of captive *Bothrops jararaca*. Boyer et al. (1995) studied facultative strike-induced chemosensory searching and trail following behavior of bushmasters at the Dallas Zoo and discovered that snakes treated rodents differently based on prey biomass. Boyer and Roberts (1998) reported on the predatory strike of the temple viper (*Tropidolaemus wagleri*) at the Dallas Zoo.

Captive and Medical Management

Berg (1901) discussed Indian Dryophiden in the terrarium. Campden-Main from the Louisiana Purchase Zoo in Monroe wrote a paper on the care of newborn snakes (Campden-Main and Campden-Main, 1982). Banks (1989) described skin lesions in *Acrochordus* and *Erpeton,* perhaps caused by excessively bright lighting. Wright (1992) addressed dysecdysis in boid snakes with neurologic diseases. Brannian et al. (1980) from the Kansas City Zoological Gardens focused on mortality in captive green tree pythons.

Lloyd at the Roger Williams Park Zoo in Rhode Island published three papers in 1991 and 1992 on ophidian paramyxovirus and reptilian dystocias. Ungureanu et al. (1972) identified hemorrhagic septicemia in snakes in captivity at the Zoological Garden of Bucharest. Aronson (1929) described spontaneous

tuberculosis (*Mycobacterium thamnopheos*) in garter snakes at the Philadelphia Zoo. Adamy (1966) investigated the therapy of salmonellosis. Bush and associates from the Smithsonian National Zoological Park published papers on biological half-life of Gentamicin in gopher snakes (1978), blood collection and injection techniques in snakes (1978) and the use of antibiotics in snakes (1976). Hilf et al. (1989) investigated pharmacokinetics of Gentamicin and Piperacillin in blood pythons and Hilf et al. (1990) did a comparative study of upper airway flora in healthy boid snakes and snakes with pneumonia at the Pittsburgh Zoo. Ramsay et al. (1996) from the Knoxville Zoo diagnosed *Salmonella arizona* osteomyelitis in a colony of Arizona ridge-nosed rattlesnakes. Velayan and Ambu (1991) studied bacterial infections in reptiles at the National Zoo, Malaysia. Flanagan and Harwell (1983) from the Houston Zoo discussed pathobiology and management of chronic regurgitation in snakes. At the Baltimore Zoo, Cranfield and Graczyk (1995) updated information on ophidian cryptosporidiosis. Ippen (1980) described mycosis in snakes.

Barnard from Zoo Atlanta and her associates have published a number of papers on parasitism in snakes (see Chapter 8, Zoo Atlanta account for references). Ovezmukhammedov et al. (1975) diagnosed parasites of the reticulated python from the Ashkhabad Zoo. Camin and associates (1948, 1953, 1964) documented the ravages caused by the snake mite (*Ophionyssus natricis*) in captive reptile collections.

Anderson (1986) provided information on snake anatomy for the clinician. Nichols and Lamirande (1994) used methohexital sodium as an anesthetic for colubrid snakes. Glenn et al. (1973) described the influence of Ketamine HCL on rattlesnakes during shipment. Vandeventer and Schmidt (1977) performed a caesarean section on a western gaboon viper. Anderson et al. (1999) tested propofol as an anesthetic agent in brown tree snakes (*Boiga irregularis*) at the Columbus Zoo. Wallach (1975) performed a partial gastrectomy in a Timor python (*Python timoriensis*). Mulder et al. (1979) from the Audubon Zoo surgically removed retained eggs from a kingsnake (*Lampropeltis getula*).

Baumgartner et al. (1987) diagnosed tubular adeno-cancer in the colon of a reticulated python. Garner et al. (1995) diagnosed vertebral chondrosarcoma in a corn snake. Robinson et al. (1978) used radiation therapy for treatment of an intraoral malignant lymphoma in an Indian rock python. Bryant et al. (1997) used cryosurgery in a diamond python (*Morelia s. spilota*) with fibrosarcoma and radiotherapy in a common death adder (*Acanthophis antarcticus*) with melanoma at the Taronga Zoo. Zachiesche et al. (1988) diagnosed lymphoid leukemia with presence of type C virus particles in a yellow ratsnake (*Elaphe obsoleta quadrivittata*).

Longevity

A male ball python was acquired at the Philadelphia Zoo on 26 April 1945 and died on 7 October 1992, thus reaching the approximate age of forty-seven and one-half years (Conant, 1993).

General

Boas and Pythons: Barton and Allen (1961) recorded feeding, shedding and growth rates of captive boid snakes at the Highland Park Zoological Gardens in Pittsburgh, Pennsylvania. Golding (1965) followed growth rate of a captive Indian python at the University of Ibadan Zoo, Nigeria. Bloxam (1982) reported on the care of the Jamaican boa in the Jersey Wildlife Preservation Trust. Arnett et al. (1992) described husbandry techniques for the eyelash boa (*Trachyboa*) at the Cincinnati Zoo. Sekar and Jagnnadha-Rao (1995) outlined management of the Indian rock python at the Arignar Anna Zoological Park, Vandalur. Whitworth (1974) followed growth of an African python and an Indian python at the Cannon Aquarium and Vivarium, Manchester Museum. Edwards (1969) covered care of some tropidophid snakes at the Dallas Zoo. Moore and McLaughlin (1981) surveyed boid care in American zoos.

Colubrids: Barnard et al. (1979) from Zoo Atlanta followed growth and food consumption in the corn snake. Honegger (1984) from the Zürich Zoo and Korinek (1997) at Olomouc Zoo investigated the biology of the Brazilian false water cobra (*Hydrodynastes gigas*) in the terrarium. Cahill (1971) provided observations on a captive beak-nosed snake (*Scaphiophis albopunctatus*) at the University of Ife Zoo in Nigeria. Clarke (1973) recorded longevity in a Great Plains rat snake (*Elaphe guttata emoryi*).

Elapids: Pawley (1982) elaborated on his techniques for keeping the eastern coral snake (*Micrurus f. fulvius*) at the Brookfield Zoo. Dobbs (1967) from Zoo Atlanta discovered that the king cobra would only accept dead food snakes. One black-banded sea krait (*Laticauda laticaudata*) was alive after 5 years, 2 months; the shedding intervals ranged between 1 month, 25 days to 5 months, mean was 3 months, 7 days (Klemmer, 1967). At the San Diego Zoo, Shaw (1962) set the longevity record for yellow-bellied sea snake (*Pelamis platurus*)

Fig. 10. Sumatran black-hooded cobra (called *Naja trupudians sumatrana*) on exhibit at Smithsonian National Zoological Park in Spring 1934. *Photograph by E. Hardy, National Zoological Park Photo Archives.*

which was alive after two years, 4 months. The shedding interval ranged from 7 days to a maximum of 27 days, average 13.7.

True Vipers: Mehrtens (1950) outlined his techniques for keeping the saw-scaled viper at the Ft. Worth Zoo. Naulleau (1973) maintained the asp viper (*Vipera aspis*) by investigating feeding, food, shedding, reproduction, and neonatal husbandry. Naulleau and van den Brule (1981) observed feeding, growth, molt and venom production in the Russell's viper (*Vipera russelli* now *Daboia russelli*).

Pitvipers: Antonio and Barker (1983) followed phenotypic aberrancies in the eastern diamondback rattlesnake (*Crotalus adamanteus*). Armstrong and Murphy (1979) explored the natural history of Mexican rattlesnakes by using a captive and field component. Murphy and Armstrong (1978) described techniques for maintaining rattlesnakes at the Dallas Zoo. Patrick Burchfield (1982) from the Gladys Porter Zoo published a major paper on the natural history of the cantil (*Agkistrodon bilineatus taylori*) as well as several earlier contributions on husbandry of the fer-de-lance, eastern diamondback rattlesnake and bushmaster in 1975. At the Dallas Zoo, Card (1994) recorded the natural history and husbandry of the temple viper (*Tropidolaemus wagleri*). Greenhall (1936) outlined the care of the bushmaster in the New York Zoological Park.

Fig. 11. Eastern and western diamondback rattlesnakes (*Crotalus adamanteus* and *C. atrox*) on exhibit at Smithsonian National Zoological Park in Spring 1934. *Photograph by E. Hardy, National Zoological Park Photo Archives.*

General: Ashley and Burchfield (1966) described procedures for maintaining a snake colony for venom collection. Backhaus (1972) from the Frankfurt a. M. Zoo discussed the geographical variation of venom toxicity and antivenin efficacy in saw-scaled and Russell's vipers. Banks (1985) observed feeding and sloughing in a collection of captive snakes. Greene (1983) collected data from a number of zoos to strengthen his analysis of the dietary correlates of the origin and radiation of snakes. Bloxam and Tonge (1986) appraised the breeding program for reptiles at the Jersey Zoo.

Pestinsky (1939) offered suggestions for venomous snake capture and keeping. Gans and Taub (1964) suggested precautions for keeping venomous snakes in captivity. Fowler (1979) described a cobra snakebite inflicted upon a reptile keeper. Minton (1975) discussed snakebite in zoos. Leloup (1980) in a chapter entitled "Prophylaxie Elementaire des Accidents dans un Elevage de *Bothrops atrox*" covered venomous snakebite maintenance and first aid procedures. Wright (1996) outlined procedures to train keepers to safely handle venomous snakes. Altimari (1998) from the Arizona-Sonora Desert Museum published a safety guide for venomous snakes. Boyer (1995) at the San Diego Zoo elaborated on techniques for venomous snake management. Card and Roberts (1996) documented the incidence of bites from venomous reptiles in North American zoos. David Hardy, who serves as a snakebite consultant for a number of zoos in the United States, offered some current views on management in his 1985 paper on venomous snakebite in North America.

Lizards

Looking down upon the vast Order now engaging the student's attention, even the most passive of observers cannot refrain from expressing amazement at the array of varied forms. In a subdivision, the *Sauria*, we shall consider creatures twelve feet long, with claws as long as those of a leopard—animals strong and active enough to leap at the throat of a young gazelle, tear, dismember and devour the prey; and passing such we stop to realize that tiny, limbness and worm-like, slow-moving things, burrowing their life away deep in the ground where they need no eyes—in fact, have none—are also true lizards.

Raymond L. Ditmars (1933)

As far as the evidence goes, Chameleons appear to be the only short-lived Reptiles, not attaining to even five years. Lizards present varying lengths of life not correlated with their bodily size.

Major Stanley Smyth Flower (1925)

Fig. 12. Beaded lizard (*Heloderma horridum*) and Gila monster (*H. suspectum*) on exhibit at Smithsonian National Zoological Park in Spring 1934. *Photograph by E. Hardy, National Zoological Park Photo Archives.*

History

In early days, Coote (2001) speculated that many reptiles, especially lizards, arrived at the London Zoo in good condition because those responsible for transporting them were given specific instructions for their care: "Correspondents should engage some individual of the ship's company to take charge of the animals on board and guarantee him a handsome recompense on bringing them safely to their destination . . ." Food was also important for reptiles on the long sea journeys and correspondents were advised: "ants eggs, which are abundant in tropical climates, may be preserved in a jar, well tied down and with the addition of the Blattae or cock-roaches so generally obtainable on board in all their stages of growth, and of meal worms, which are equally abundant in the bread room. . . ."

In 1818, the Menagerie of the Imperial Cabinet (Vienna) had chameleons from North Africa and eggs were laid in 1845 but did not hatch. Between 1818–1851, the menagerie received many chameleons as gifts from various sources. Duméril (1854–1855) knew that captive chameleons fared poorly between 1851 and 1855 at the Jardin des Plantes, Paris for although more than 160 were kept, none lived more than 13 months. Females often died from reproductive complications and no eggs were ever hatched.

Johann von Fischer was the first to document copulatory behavior and reproduction in common chameleons (*Chamaeleo vulgaris* now *C. chamaeleon*) in 1882 and 1884. He attempted to incubate nearly 800 eggs but most never hatched. The first veiled chameleon (*Chamaeleo calyptratus*) obtained by the London Zoo was caught in Aden, South Yemen, on 15 March 1885 and donated on 3 June 1885 by Lt.-Col. J. W. Yerbury (Coote, 2001). At the now-defunct Belle Vue Zoo in England, average longevity for 24 chameleons between the years 1898–1901 was slightly over three months and none survived for one year (Flower, 1925).

Raymond L. Ditmars, curator of herpetology at the New York Zoological Park, had problems with chameleons which rarely lasted longer than five or six months (Ditmars, 1910).

Flower (1925) found that chameleons rarely lived one year in captivity. In 55 years, only 15 chameleons out of nearly 770 survived over one year, except the South African dwarf chameleon (*Bradypodion pumilum*) which have live young. By 1928, sixteen different species of chameleons had been exhibited alive in London but mortality rates were high. At the Giza Zoological Gardens in Egypt, 200 chameleons were maintained at the zoo between 1898–1924 but none lived for as long as four years. See Ferguson et al. (2004) for a history of captive chameleons.

A Jamaican land iguana (*Cyclura collei*) was donated to the London Zoo on 31 July 1849 and died on 29 December 1852, thus living in the Zoo for 3 years, 4 months and 29 days (Coote, 2001). Concern about the iguana's future accelerated following the introduction of the Indian mongoose on Jamaica in 1872. In 1940, a total of 22 specimens were brought into captivity to try to save the species from extinction but none ever reproduced. There is now a headstarting program at the Hope Zoo in Kingston which involves other zoos as well (Vogel et al., 1996).

Knowsley Park (1834–1851), the home of the Earls of Derby, housed the giant or Lord Derby's girdled lizard (described as *Zonurus derbianus*, now known as *Cordylus giganteus*) (Keeling, 1984). At the London Zoo, the first New Guinea blue-tongued skinks (*Tiliqua gigas*) were purchased on 17 June 1852 and gave birth to 27 neonates between 1866 and 1893 (Coote, 2001).

A number of papers and books on lizard biology and captive management have been published by staff at the Center for the Reproduction of Endangered Species (CRES) at the San Diego Zoo and the Smithsonian National Zoological Park, (see Chapter 8 for references).

Lizards may prove to be interesting subjects for future behavioral studies. At the Smithsonian National Zoological Park, a female Komodo dragon interacted with inanimate objects and human caretakers (Burghardt et al., 2002). Dennis Desmond sent a photograph to me showing a female green iguana (*Iguana iguana*) sleeping with a furry toy virtually every night and the lizard often wrapped its forelimbs around this unusual object (Fig. 13). Often weird things happen with living animals and sometimes these weird things end up teaching us profound new truths about animals we thought we understood. A careful systematic study to elucidate the causes for this unexpected behavioral response would likely be fruitful.

Fig. 13. Five-year old female green iguana (*Iguana iguana*) often slept with furry toy. See text. *Photograph by Dennis Desmond.*

Some lizard families have been only superficially investigated in zoos: Anguids, Agamids, Dibamids, Amphisbaenids, Bipeds, Rhineurids, Trogonophids, Lacertids, Gymnophthalmids, Teiids, Cordylids, and Xantusids.

Behavior

Gekkonids: Dale Marcellini from the Smithsonian National Zoological Park published several papers on vocal and visual displays of geckos (1977, 1978). Demeter and Marcellini (1981) recorded courtship and aggressive behavior of the streak lizard (*Gonatodes vittatus*) at the Smithsonian National Zoological Park.

Varanids: Honegger and Heusser (1969) contributed an analysis of the behavior inventory of the water monitor (*Varanus salvator*). Davis et al. (1986) recorded ritualized combat in captive Dumeril's monitors (*V. dumerili*). Male pygmy mulga monitors (*V. gilleni*) embraced conspecifics in an arching posture during combat rituals at the Dallas Zoo (Murphy and Mitchell, 1974; Carpenter et al., 1976). Irwin (1994) reported on the behavior and diet of the Queensland tree monitor (*V. teriae*) at the Queensland Reptile and Fauna Park (now Australia Zoo). Hartdegen et al. (1999) observed feeding behavior of the black tree monitor (*V. beccari*) at the Dallas Zoo. A series of studies at the Smithsonian National Zoological Park confirmed that Komodo dragons (*V. komodoensis*) indeed **play** with inanimate objects and interact with caretakers (see Chapter 3 and Burghardt et al., 2002).

Iguanids: Carpenter and Murphy (1978) noticed aggressive behavior and color change between males in the Fiji Island iguana (*Brachylophus fasciatus*). Marcellini and Jenssen (1991) documented avoidance learning by the curly-tailed lizard (*Leiocephalus schreibersi*).

Helodermatids: Demeter (1986) described combat behavior in the Gila monster (*Heloderma suspectum cinctum*) at the Smithsonian National Zoological Park. At the Dallas Zoo, Hartdegen and Chiszar (2001) reported on discrimination of prey-derived chemical cues by the Gila monster and lack of effect of a putative masking odor.

Agamids: Murphy et al. (1978) observed defensive behavior in the angle-headed dragon (*Goniocephalus dilophus* now *Hysilurus dilophus*) at the Dallas Zoo.

Chamaeleonids: Ferguson et al. (2004) described color variation, natural history, conservation and captive management of the Malagasy panther chameleon (*Furcifer pardalis*) which included reproductive and nutritional data on five captive generations at Texas Christian University.

Scincids: Carpenter and Murphy (1978) saw tongue displays directed toward conspecifics by the common bluetongue skink (*Tiliqua scincoides*) at the Dallas Zoo.

Reproduction

Gekkonids: Heinrich Dathe (1942) recorded the birth of a panther or tokay gecko (*Gecko gecko*) at the Leipzig Zoo. Petzold (1963) bred the tokay gecko (*Gekko gecko*) at Berlin Tierpark. Frolow (1981, 1987) outlined reproduction of the skink gecko (*Teratoscincus scincus*) in the Moscow Zoo. Benefield et al. (1981) propagated western banded geckos (*Coleonyx variegatus*) at the Tulsa Zoo. Risley (1989) observed breeding of the Namib gecko (*Chondrodactylus angulifer*) at the London Zoo. Kaverkin et al. (1994) described husbandry and reproduction of the Eublepharid gecko (*Goniurosaurus kuroiwae splendens*) at the Moscow Zoo. Astreiko and Popovskaya (1999) from the Tula Exotarium investigated the potential of breeding and raising the African fat-tailed gecko (*Hemitheconyx caudicinctus*) under laboratory conditions.

Several papers on the long-term Madagascar giant day gecko (*Phelsuma madagascariensis*) colony at the Smithsonian National Zoological Park have been published by Béla Demeter and colleagues (1976, 1994). Bloxam and Townson (1980), Bloxam and Vokins (1978), Langebaek (1979), Wheler and Fa (1995) and Cooper et al. (1998) described breeding, research and medical management at the Jersey Wildlife Preservation Trust with the endangered Round Island gecko (*P. guentheri*), the largest in the genus. Howard (1980) bred the flat-tailed day gecko (*P. laticauda*) at the Twycross Zoo. Rundquist (1980) reproduced Koch's day gecko (*P. madagascariensis kochi*) at the Oklahoma City Zoo. Malagasy flat-tailed geckos (*P. serraticauda*) reproduced at the Central Park Zoo in 2003. In 2005, the endangered yellow-throated day gecko (*Phelsuma flavigularis*) hatched at the Wildlife Conservation Society.

Banks et al. (1999) managed and bred the striped legless lizard (*Delma impar*) at the Melbourne Zoo.

Varanids: Brotzler (1965) outlined Mertens' water monitor (*Varanus mertensi*) breeding in the Wilhelma Zoo (Stuttgart). In 2005, this species reproduced at the Wildlife Conservation Society. Müller (1970) bred the water monitor (*V. salvator*) at the Leipzig Zoo, David

(1970) at the Ahmedabad Zoo, and Hairston and Burchfield (1992) at the Gladys Porter Zoo. Herrmann (1999) documented husbandry and captive breeding of the water monitor in the Cologne Aquarium at the Zoo.

Judd et al. (1977), Morris and Alberts (1996) and Morris et al. (1996) explored sex determination in the Komodo dragon (*V. komodoensis*) and white-throated monitor (*V. albigularis*) at the San Diego Zoo. The Gembira Loka Zoo in Jogjakarta, Indonesia, has successfully bred the Komodo dragon for many years, producing well over 100 neonates (Walsh et al., 1998). The offspring are maintained in large groups in movable enclosures and are placed in sunlight daily (Busono, 1974). Komodo dragon eggs were deposited in the Zoological and Botanical Garden of Jakarta (Gaulstaun, 1973). Osman (1967) prepared a note on the breeding behavior of the Komodo dragons at the Jogjakarta Zoo. Walsh and Rosscoe (1993) and Walsh et al. (1993) reported on the history, husbandry, and breeding of Komodo dragons at the Smithsonian National Zoological Park, the first time in the New World. Birchard et al. (1995) followed oxygen uptake in these Komodo dragon eggs and the energetics of prolonged development.

Perry et al. (1993) described the first captive reproduction of the desert monitor (*V. g. griseus*) at the Research Zoo of Tel Aviv University. Visser (1981, 1985) provided notes on the breeding of the white-throated monitor (*V. albigularis*) and yellow monitor (*V. flavescens*) in the Rotterdam Zoo. Barker (1984) reproduced green tree monitors (*V. prasinus*), Boyer and Lamoreaux (1983) bred the pygmy mulga monitor (*V. gilleni*), Mitchell (1990) reported on reproduction of Gould's monitors (*V. gouldii*), and Card (1994) reported double clutching in Gould's monitors and Gray's monitors (*V. olivaceus*) at the Dallas Zoo. One black tree monitor (*Varanus beccari*) hatched at the Buffalo Zoo in 2005. Radford and Paine (1989) hatched five Dumeril's monitors (*V. dumerili*) at the Buffalo Zoo. Irwin (1996) described courtship, mating and egg-deposition by the perentie (*V. giganteus*) at the Queensland Reptile and Fauna Park (now Australia Zoo). Rehák (1996) followed the reproductive biology of the mangrove monitor (*V. indicus*) in the Prague Zoo. Stirnberg (1997) described care and breeding of the Australian lace monitor (*V. varius*) in the Bochum Zoo which led to 2nd generation offspring.

Eidenmüller and Wicker (1991) covered captive breeding, artificial egg incubation, development and surgical removal of impacted eggs in *Varanus timorensis similis* at the Frankfurt Zoo.

Iguanids: Petzold (1962) successfully bred the northern curlytail lizard (*Leiocephalus carinatus*) at the Berlin Tierpark. Gibson and Buley (1996) described captive management and breeding of Madagascar spiny iguanas (*Oplurus c. cuvieri*) at the Durrell Wildlife Conservation Trust. Vogel (1992) outlined care and breeding of the serrated casqueheaded iguana (*Laemanctus serratus*) over several generations in the Exotarium of the Frankfurt Zoo.

Reproduction and behavior of the green crested basilisk (*Basiliscus plumifrons*) were described by Pawley (1972) at the Brookfield Zoo, Blake and Stewart (1980) at the Edinburgh Zoo, and Banks (1983) at the Melbourne Zoo.

Dedekind (1977) recorded his observations made during hatching and raising of a common or green iguana (*Iguana iguana*) in the Berlin Tierpark. Howard (1980) bred them at the Twycross Zoo and Banks (1984) at the Melbourne Zoo. Arnett (1979) reproduced the Fiji banded iguana at the Knoxville Zoo and over 100 young have been produced at the San Diego Zoo. Boylan (1989) bred the Fijian crested iguana (*Brachylophus vitiensis*) at the Taronga Zoo. Pawley (1965, 1966, 1969) successfully maintained a captive colony of marine iguanas (*Amblyrhynchus cristatus*) at the Brookfield Zoo.

Noegel (1989) reproduced five species of West Indian rock iguanas (*Cyclura*) at the Life Fellowship Bird Sanctuary, Seffner, Florida. The reproductive biology of Cuban rock iguanas (*C. nubila*) has been covered in detail at the San Diego Zoo (Shaw, 1954; Alberts, 1995, 2002; Alberts et al., 1997) and the Prague Zoo (Rehák and Král, 1993; Rehák, 1994, 1995 [1997]; Rehák and Velenský, 1997, 2001). Haast (1969) hatched rhinoceros iguanas (*C. cornuta*) at the Miami Serpentarium and Rehák (1996) documented the first captive breeding of the rhinoceros iguana in the Czech Republic. In two papers, Boylan (1984, 1985) described captive management of a population of rhinoceros iguanas at the Taronga Zoo in Sydney, Australia. Two critically endangered Grand Cayman Island blue iguanas (*Cyclura lewisi*) hatched at the Indianapolis Zoo in 2005.

The reproductive biology of green (*Iguana*) and desert (*Dipsosaurus*) iguanas has received attention (Judd et al., 1976; see Chapter 8, San Diego Zoo account). Montanucci (1984, 1989) outlined breeding, captive care and longevity of the shorthorned lizard (*Phrynosoma douglassi*) and Ditmars' horned lizard (*P. ditmarsi*). In 2005, the Texas horned lizard (*Phrynosoma cornutum*) reproduced at the Fort Worth Zoo.

Helodermatids: Anstandig (1983) reported on the breeding and rearing of the Mexican beaded lizard (*Heloderma horridum*) and Connors (1993) presented the status of the breeding program at the Detroit Zoo a decade later. Honegger (1998) outlined upkeep and breeding in the Zürich Zoo. Wright et al. (1995) introduced ultrasonographic sexing of helodermatid lizards and Morris and Alberts (1998) and Morris and Henderson (1998) used two dimensional ultrasound imaging at the San Diego Zoo. Card and Mehaffey (1994) used radiographs to sex Gila monsters. Peterson (1982) from the Houston Zoo and Grow and Branham (1996) at the Oklahoma City Zoo reproduced the Gila monster. Lawler and Wintin (1987) described captive management and propagation of the reticulated Gila monster.

Anguids: Lawler and Norris (1980) bred the Haitian giant galliwasp (*Diploglossus warreni*) at the Knoxville Zoo and over 300 offspring have been produced at the Nashville Zoo since 2000. One Central American giant galliwasp (*Diploglossus monotropis*) hatched at the Nashville Zoo in 2005. Claffey and Johnson (1983) reproduced the sheltopusik (*Ophisaurus apodus* now *Pseudopus apodus*) at the Metro Toronto Zoo.

Agamids: Almandarz (1969) placed bearded dragon (*Amphibolurus barbatus*) eggs in a portable baby incubator and seven hatched at the Lincoln Park Zoo. Van Aperen (1969) bred them at the Melbourne Zoo. Sherriff (1989) outlined the maintenance and breeding of the inland bearded dragon (*Pogona vitticeps*) at the Edinburgh Zoo. Morley (1992) bred the Nobbi (*Amphibolurus nobbi*) at the Adelaide Zoo. In 2004, nine Javan hump headed lizards (= chameleon forest dragon, *Gonocephalus chamaeleontinus*) hatched at the Denver Zoo.

Demeter (1981) reported on husbandry and breeding of the Chinese water dragon (*Physignathus cocincinus*) at the Smithsonian National Zoological Park, Dedekind and Petzold (1982) bred the species in the Berlin Tierpark, and Frolov and Sanoyan (1986) in the Moscow Zoo. Visser (1984) described husbandry and reproduction of the sailfin lizard (*Hydrosaurus amboinensis*) at the Rotterdam Zoo. Mitchell (1986) reproduced the Philippine water dragon (*Hydrosaurus pustulatus*) at the Dallas Zoo. In 2004, 20 Mali spiny-tailed lizards (*Uromastyx maliensis*) hatched at the Detroit Zoological Institute.

Chamaeleonids: Charles Shaw recorded his observations on the reproductive biology of *Chamaeleo basiliscus* (now *C. africanus*) (Shaw, 1959). Castle (1990, 1991) described husbandry and breeding of several chameleons (*Chamaeleo* spp.) at the Oklahoma City Zoo.

Xantusids: Holmback (1984) discovered parthenogenesis in the Central American night lizard (*Lepidophyma flavimaculatum*) at the San Antonio Zoo.

Teiids: At the Smithsonian National Zoological Park, Hall (1978) documented husbandry, behavior, and breeding of tegu lizards (*Tupinambis tequixin*). McCrystal and Behler (1982) described husbandry and reproduction of captive giant ameiva lizards (*Ameiva ameiva*) at the New York Zoological Park. At the Prague Zoo, Rehák (1999) bred the Guyana caiman lizard (*Dracaena guianensis*). Three neonates hatched at the Shedd Aquarium in 2005.

Scincids: Tschambers (1949) from the Brookfield Zoo documented birth of the Australian blue-tongued skink (*Tiliqua scincoides*). Schildger and Wicker (1987) used endoscopic gender determination in the shingleback skink (*Trachydosaurus rugosus*) at the Frankfurt Zoo.

Groves (1994) described husbandry and reproduction of the colony of prehensile-tailed skinks (*Corucia zebrata*) at the Philadelphia Zoo. Honegger (1975, 1985) from the Zürich Zoo detailed the management and biology of the prehensile-tailed skink. Bloxam (1976) discussed maintenance and breeding of the Round Island skink (*Leiolopisma telfairii*) at Jersey Wildlife Preservation Trust. Russell (1996) explored the natural history, captive husbandry, and reproduction of crocodile skinks (*Tribolonotus gracilis*) at the Dallas Zoo. Morley 1992) reproduced the narrow-banded sand swimmer (*Eremiascincus richardsoni*) and Post (2000) outlined captive husbandry and reproduction of the Hosmer's skink (*Egernia hosmeri*) at the Adelaide Zoo. Dathe and Dedekind (2002) reproduced fire skinks (*Riopa fernandi*) in the Berlin Tierpark.

Lacertids: Herrmann (1998) from the Cologne Zoo described reproduction in the East African fringe-tailed forest lizard (*Holaspis guentheri laevis*).

Xenosaurids: Visser (1989) and Kudryavtsev and Vassiliev (1998) reproduced the Chinese crocodile lizard (*Shinisaurus crocodilurus*) at the Rotterdam and Moscow Zoos, respectively. At Tiergarten Nürnberg, Germany, Mágdefrau (1997) described biology, sex determination using head morphology, care and breeding of the Chinese crocodile lizard. In 2005, the San Antonio Zoo reproduced the flathead knob-scaled lizard (*Xenosaurus platyceps*).

Husbandry
Gekkonids: From the Houston Zoo, Neitman (1983) described captive husbandry of the tuberculate geckos of the genus *Coleonyx*.

Varanids: Murphy (1971, 1972) provided data on Indo-Australian varanids at the Dallas Zoo. Christie (1982) successfully introduced three incompatible male lace monitors at the Indianapolis Zoo.

Iguanids: Bosch and Werning (1996) discussed the captive management of green iguanas and other iguanids. Wallach (1966) described hypervitaminosis D in green iguanas.

Duval (1982) offered suggestions for the captive management of West Indian rock iguanas (*Cyclura*) and Duval and Christie (1990) for the Cuban ground iguana (*Cyclura n. nubila*), based on their experiences at the Indianapolis Zoo. Bosch (1991) discussed ground iguanas (*Cyclura cychlura figginsi*) in the Löbbecke Museum (Dusseldorf).

Lawler and Jarchow (1986) presented a captive management plan, Lawler et al. (1994) outlined the ecological and nutritional management, and Densmore et al. (1994) described a molecular approach for determining genetic variation in captive and natural populations of the piebald chuckwalla (*Sauromalus varius*) at the Arizona-Sonora Desert Museum.

Agamids: Watson (1969) outlined the care of mastigure lizards (*Uromastyx*) at the Jersey Wildlife Preservation Trust. Wheeler (1990, 1991) described husbandry of mastigures at the Oklahoma City Zoo.

Teiids: At the American Museum of Natural History in New York, Townsend (1979) and Townsend and Cole (1985) established colonies of parthenogenetic whiptail lizards (*Cnemidophorus* spp.); a colony of parthenogenetic *C. exsanguis* reached seven generations.

Scincids: Wright (1993) outlined his husbandry techniques for the colony of the Solomon Island prehensile-tailed skinks (*Corucia zebrata*) at the Philadelphia Zoo.

Medical Management
Morales and Dunker (2001) found tuberculosis, *Mycobacterium marinum*, in a group of Egyptian mastigures (*Uromastyx aegyptius*) at the Steinhart Aquarium. At the Smithsonian National Zoological Park, Gray et al. (1966) discovered amoebiasis and Spelman et al. (1996) described anesthesia in the Komodo dragon, Montali et al. (1975) diagnosed dermatophilosis in Australian bearded lizards, and Linnetz et al. (1996) found gout in giant day geckoes. There is an extensive section on diagnosis and treatment of diseases in Komodo dragons, based on information from the Smithsonian National Zoological Park (Murphy et al., 2002).

Wallach and Hoessle (1968) documented fibrous osteodystrophy in green iguanas. Zwart and Van de Watering (1969) described pathology and etiology of abnormal bone formation in the green iguana. Raphael et al. (1999) evaluated vitamin D concentrations in *Uromastyx* spp. with and without radiographic evidence of dystrophic mineralization at the New York Zoological Park.

At the Metro Toronto Zoo, Barker and Cranfield (1988/1989) identified *Schellackia* (*Lainsonia*) *sp.* in chuckwallas. Stunkard and Gandal (1961) identified nematodes and cestodes from the Gould's monitor and five years later, a digenetic trematode, *Parahaplometroides basiliscae* from the mouth of the basilisk at the New York Zoological Park. Susan Barnard from Zoo Atlanta and her associates have published a number of papers on parasitism in lizards (see Zoo Atlanta account for references).

Backues and Ramsey (1994) performed an ovariectomy for treatment of follicular stasis in lizards. Coburn (1975) described postmortem removal and artificial hatching of rainbow lizard eggs (*Agama agama*) at the Cotswold Wildlife Park, Oxfordshire UK.

Tocidlowski et al. (1997) from the North Carolina Zoological Park diagnosed teratoma in the desert grassland whiptail (*Cnemidophorus uniparens*). Whiteside and Garner (2001) found a thyroid adenocarcinoma in a crocodile lizard.

Gamble et al. (1996) investigated plasma Itraconazole pharmacokinetics in spiny lizards (*Sceloporus* spp.) at the Dallas Zoo.

Wright (1992, 1993) and Wright and Skeba (1992) explored medical management of the Solomon Island prehensile-tailed skink at the Philadelphia Zoo. Barrie et al. (1993) documented diseases of chameleons at the Oklahoma City Zoo.

General
William Cooper Jr. has used zoo collections as a resource for his studies on lizard feeding behavior for many years. Examples include varanoid lizards *Heloderma suspectum* and *Varanus exanthematicus*, gec-

kos (*Rhacodactylus*), and skinks (*Tiliqua scincoides* and *Trachydosaurus rugosus*).

Gekkonids: Tytle (1994) and McKeown (1993) outlined husbandry protocols and the captive history of day geckos (*Phelsuma*).

Varanids: Horn and Visser (1989) reviewed reproduction of 20 spp. of monitor lizards in captivity and data on the biology of monitors (1991). The same authors (Horn and Visser, 1997) provided an updated account which covered behavior, diet, egg incubation, light, reproduction, sexing, and taxonomy.

Murphy et al. (2002) investigated factors necessary for the captive management and conservation of Komodo dragons.

Helodermatids: Espinosa et al. (1996) described a cytogenetic technique for karyotyping beaded lizards in the Guadalajara Zoo. Morris and Henderson (1998) determined gender in Gila monsters and beaded lizards by ultrasound imaging of the ventral tail at the San Diego Zoo.

Chamaeleonids: Gary Ferguson from Texas Christian University has worked in a zoo setting for many years to investigate the captive biology of chameleons (Ferguson, 1994; Ferguson et al., 1996, 2003, 2004).

Iguanids: Charles Knapp from the Department of Conservation at the John G. Shedd Aquarium in Chicago has published field studies on several West Indian iguanas (*Cyclura*).

Turtles and Tortoises

TORTOISES, AS IS A MATTER OF COMMON KNOWLEDGE, LIVE TO AGES EXCEEDING THOSE, AS FAR AS OUR PRESENT INFORMATION GOES, OF ALL OTHER VERTEBRATE ANIMALS.

MAJOR STANLEY SMYTH FLOWER (1925)

PITIABLE SEEMS THE CONDITION OF THIS POOR EMBARRASSED REPTILE: TO BE CASED IN A SUIT OF PONDEROUS ARMOUR, WHICH HE CANNOT LAY ASIDE; TO BE IMPRISONED, AS IT WERE, WITHIN HIS OWN SHELL, MUST PRECLUDE, WE SHOULD SUPPOSE, ALL ACTIVITY AND DISPOSITION FOR ENTERPRIZE. YET THERE IS A SEASON OF THE YEAR (USUALLY THE BEGINNING OF JUNE) WHEN HIS EXERTIONS ARE REMARKABLE. HE THEN WALKS ON TIPTOE, AND IS STIRRING BY FIVE IN THE MORNING; AND, TRAVERSING THE GARDEN, EXAMINES EVERY WICKET AND INTERSTICE IN THE FENCES, THROUGH WHICH HE WILL ESCAPE IF POSSIBLE; AND OFTEN HAS ELUDED THE CARE OF THE GARDENER, AND WANDERED TO SOME DISTANT FIELD. THE MOTIVES THAT IMPEL HIM TO UNDERTAKE THESE RAMBLES SEEM TO BE OF THE AMOROUS KIND: HIS FANCY THEN BECOMES INTENT ON SEXUAL ATTACHMENTS, WHICH TRANSPORT HIM BEYOND HIS USUAL GRAVITY, AND INDUCE HIM TO FORGET FOR A TIME HIS ORDINARY SOLEMN DEPORTMENT.

GILBERT WHITE DISCUSSING HIS PET GREEK TORTOISE TIMOTHY IN HIS BOOK *THE NATURAL HISTORY OF SELBORNE* IN 1780

History and Longevity

Coote (2001) followed the history of captive turtles and tortoises in England, beginning with what was perhaps a very large marginated tortoise (*Testudo marginata*) acquired in 1601. Tortoises were the passion of a succession of seven Bishops of Peterborough, recorded from the first one in 1601 until the death of the last one in 1821, which lived in a garden throughout the year and hibernated in the winter. During warmer periods, this last animal, recorded as weighing 13.5 pounds, ate dandelion flowers, strawberries, gooseberries, lettuce, currants, raspberries, pears, plums, apples, peaches and nectarines, endive, green peas and leeks, oranges, but disliked cherries, asparagus, parsley, carrots, turnips, and spinach. This tortoise was offered more interesting and expensive fare than I ever had during my entire life. Whatever species this chelonian was, it certainly was strong, as it was said to crawl with a weight of 18 stone (252 lb) upon its shell.

William Laud bought a spur-thighed tortoise (*Testudo graeca*) in 1625, where it lived at the Palace of Fulham. Eight years later, Laud became the Archbishop of Canterbury and he carried his tortoise to his new and improved abode, Lambeth Palace, where it survived for another 120 years. In 1753, an under-gardener accidentally beheaded the reptile but the shell remains in the library at Lambeth Palace (Coote, 2001).

Another tortoise called "Timothy" was originally found on board a captured Portuguese pirate ship and rescued by Captain John Guy Courtenay Everard of Her Majesty's Ship HMS *Queen* in 1854 (Coote, 2001). After a series of voyages on various British naval vessels, the tortoise was permanently deposited in 1914 at Powderham Castle, home to the Earls of Devon since 1325, where it lived outdoors and hibernated each winter. A noble life style must have suited Timothy — a life span of at least 148 years or as many as 164 years. This tortoise outlived seven or eight Earls. The tortoise dragged a sign (affixed to the rear margin of its shell with a string) which said "My name is Timothy. I am very old. Please do not pick me up." Timothy died during April 2004.

On 4 December 1849, the London Zoo received its first pair of alligator snapping turtles (*Macroclemmys temminckii*). Painted turtles (*Chrysemys picta*), acquired in 1838, were bred for the first time in Great Britain between 1860 and 1861 (Coote, 2001).

Reverend Gregory Bateman (1897) said that various European tortoises could be reproduced in Great Britain; eggs hatch within 8 to 10 weeks if incubated at a temperature of 75°F, but not greater than 90°F. Aquatic turtle eggs take longer to incubate. Even in those early days, chicken egg incubators were used for reptile eggs. Incubation temperatures were rather high for most taxa (up to 96°F). Eggs were placed in damp moss to avoid desiccation.

Fig. 14. Official postcard of Zoological Society of London depicting Charles Island [Santa María or Floreana] tortoise (*Geochelone elephantopus*, now *Geochelone n. nigra*) at London Zoo from series of postcards available to the public in October 1904. This species is now extinct. *Photographed by W. P. Dando, provided by John Edwards.*

A female Galápagos tortoise (*Geochelone nigra porteri*) named Harriet lives at the Queensland Reptile Park. This tortoise was thought to be collected by Charles Darwin on 17 August 1835 on the Galápagos island then called Charles Island (now Islas Santa Maria or Floreana), and taken back to England in 1836 on his ship, the HMS. *Beagle* (Coote, 2001). A detailed history of this tortoise, now over one and a half centuries in age, is beyond the scope of this book but see Coote (2001) and Thomson et al. (1998). Living at the Life Fellowship Bird Sanctuary in Florida since 1960, a male tortoise (*G. n. porteri*), appropriately named Goliath, is the heaviest captive tortoise, weighing nearly 400 kg (pictured in Pritchard, 1996, fig. 23).

Taxonomy of the giant tortoises from the

Fig. 15. Female Galápagos tortoise named "Harriet" at Australia Zoo in Queensland, was thought to be collected by Charles Darwin on 17 August 1835 but recent evidence suggests that this is not true. In 1841, Harriet was sent to Brisbane Botanical Gardens until leaving in 1950. Harriet lived at Fleay's Fauna Sanctuary from 1950 to 1987 and was moved to Australia Zoo. *Photograph by Dennis Desmond.*

Seychelles, Aldabra and Madagascar has long been confused due to the origin of many specimens bred in captivity, lack of reliable collecting data, and loss of specimen labels and tags affixed to museum specimens (Gerlach and Canning, 1998). Indian Ocean giant tortoises (*Geochelone gigantea*) are often kept in zoos but the provenance of most of these chelonians is unknown, hampering the creation of meaningful captive breeding populations. See Swingland (1994) and Swingland and Klemens (1989) for discussion of issues related to international conservation and captive management of tortoises.

A private "Tortoise House" was built between 1820 and 1830 in the south-west corner of the grounds of Wotton House in Surrey, owned by the Evelyn family (Coote, 2001). This edifice was built by either George Evelyn, or his son William John. Tortoise lovers were so civilized in those days. An open upper floor supported a wooden pedimented "summer-house" where visitors were served tea and watched aquatic turtles swimming in a pool. What a lovely way to wile away an afternoon! In 1897, a tortoise house was built at the London Zoo.

Behavior

Burghardt et al. (1996) presented evidence of play behavior in a large captive Nile soft-shelled turtle

(*Trionyx triunguis*) at the Smithsonian National Zoological Park. Frazier and Peters (1981) recorded the call of the Aldabran tortoise. Evans and Quaranta (1951) studied social behavior in a captive herd of giant tortoises, largely from southern Albemarle Island, at the New York Zoological Park. Murphy and Lamoreaux (1978) described mating behavior in three Australian chelid turtles at the Dallas Zoo.

Reproduction

At the Gladys Porter Zoo in Texas, a number of papers have treated the reproductive biology of Galápagos and radiated tortoises (*Geochelone radiata*) (Burchfield, 1975; Burchfield et al., 1980, 1987; Hairston and Burchfield, 1989; Robeck et al., 1990; Hairston, 1991; Rostal et al., 1998). Hatt and Honegger (1997) and Honegger and Rübel (1991) described experiences and observations in raising Galápagos tortoises (*Geochelone (elephantopus) nigra*) in the Zürich Zoo; their accounts covered the husbandry protocols, diseases and the reproductive program which resulted in 29 hatchlings. The Galápagos tortoise has been successfully bred and its reproductive biology investigated over many years at the San Diego Zoo (Shaw, 1961, 1962; Ryder et al., 1989; Schafer and Krekorian, 1983; Schramm et al., 1999; 2000). Over 400 Galápagos tortoises have been hatched at the Life Fellowship Bird Sanctuary in Florida (Pritchard, 1996). At this facility, Noegel and Moss (1989) recorded the first hatching of *G. e. guntheri* from adults acquired on the Townsend Expedition in 1928. Throp (1969, 1975) bred the Galápagos tortoise at the Honolulu Zoo.

Snedigar and Rokosky (1950) described courtship and egg-laying of the yellow-legged tortoise (*Geochelone denticulata*) at the Brookfield Zoo. Vokins (1977) bred the red-footed tortoise (*G. carbonaria*) at the Jersey Wildlife Preservation Trust. Davis (1979) documented husbandry and breeding of the red-footed tortoise at the Smithsonian National Zoological Park. Desert tortoises (*Gopherus agassizii*) have been studied extensively by scientists at the San Diego Zoo (Alberts et al., 1994; Lance et al., 1995; Rostal et al., 1994). Van Devender from the Arizona-Sonoran Desert Museum published a comprehensive book (2002) on the desert tortoise which covers natural history, biology, and conservation. Stearns at the Institute for Herpetological Research published papers on captive husbandry and propagation of tortoises, such as the Aldabra tortoise and African spurred tortoise (*G. sulcata*) (1985, 1987, 1989). Collins (1984) elaborated on breeding and management of the Aldabra tortoise at the Jacksonville Zoo. Peters (1969) bred the radi-

ated tortoise (*Testudo radiata* now *Geochelone radiata*) and Aldabran tortoise (Peters and Finnie, 1979) at the Taronga Zoo. Czernay and Praedicow (1988) reported on care and breeding of the African spurred tortoise in the Thuringia Zoo at Erfurt. Visser and Zwartepoorte (1989) bred the leopard tortoise (*G. pardalis babcocki*) at the Rotterdam Zoo. Yanaga et al. (1984) artificially incubated eggs and reared the leopard tortoise at the Miyazaki Safari Park in Japan and Poglayen-Neuwall (1983) successfully bred this tortoise. Dunn (1976) reproduced the elongate tortoise (*G. elongata*) at the Melbourne Zoo. McDougal and Castellano (1996) elaborated on husbandry protocols which led to captive breeding of the Travancore tortoise (*Indotestudo travancorica*) at the Wildlife Conservation Society (WCS). In 2003, two critically endangered Burmese star tortoises (*G. platynota*) hatched at the WCS's Wildlife Survival Center. Wilke (1984) bred the pancake tortoise (*Malacochersus tornieri*) at the Frankfurt am Main Zoo, Dickson (1992) at the Bristol Zoo, and Darlington and Davis (1990) described reproduction. Later, temperature-dependent sex determination (TSD) and ontogeny were investigated at the Columbus Zoo (Ewert et al., 2004).

In 2004, five tortoises hatched at the Detroit Zoological Institute. Louwman (1982) propagated the Asian brown tortoise (*Manouria emys*) at the Wassenaar Zoo and the Fort Worth Zoo hatched sixteen in 2003. Matsuoka (1994) bred and followed growth of Egyptian tortoises (*Testudo kleinmanni*), an endangered species, at the Ueno Zoo and Dathe (2003) bred this taxon at the Berlin Tierpark (see Buskirk, 1986 for status in Israel and Egypt). Loehr (1999) explained husbandry procedures and behavior which led to captive breeding of the Namaqualand speckled padloper (*Homopus s. signatus*), also described by Morgan (1993) at the Tygerberg Zoopark, Kraaifontein, South Africa. Casares (1995) investigated the reproductive processes in tortoises based on ultrasound diagnostic methods and steroid analyses in the Kot Zoological Garden. Casares et al. (1994) also investigated non-invasive assessments of reproductive patterns in tortoises. As a result of the program supported by Durrell Wildlife Conservation Trust, Gibson and Buley (2004) covered reproductive biology, captive husbandry, and conservation of the critically endangered Malagasy flat-tailed tortoise, *Pyxis planicauda*. In 2004, the Memphis Zoo hatched the Chaco tortoise (*Geochelone chilensis*).

Bonefield (1979) hatched the Argentine snake-necked turtle (*Hydromedusa tectifera*), Kardon (1981) bred Geoffrey's side-necked turtle (*Phrynops g.*

geoffroanus), and Holmback (1987) described reproduction of the New Guinea side-necked turtle (*Emydura australis albertisii*) at the San Antonio Zoo. Corwin (1986) studied the reproductive behavior of two Australian chelid turtles, *Emydura macquarii* and *Elseya latisternum*, at the Dallas Zoo. Fritz and Jauch (1989) explained mating behavior and reproduction of Parker's snake neck turtle (*Chelodina parkeri*) at the Leipzig Zoo which included courtship, breeding, development, and ontogenetic color pattern change. Fritz (1993) provided notes on the courtship behavior of the Australian snake necked turtle (*Chelodina expansa*) at the Wilhelma Zoo. Wicker (1984) followed captive breeding, nesting, incubation and hatching over several generations in Geoffrey's side-necked turtle at the Frankfurt Zoo. Fritz et al. (1991) provided long-term observations concerning the care and breeding of the red-bellied sharp-snouted turtle (*Emydura albertisii*) at the Wilhelma and Cologne Zoos. Richter (1989) bred the matamata turtle (*Chelus fimbriatus*) at the Hamburg Troparium. At the Smithsonian National Zoological Park and New York Zoological Park, matamata eggs were successfully incubated (Rosscoe and Holmstrom, 1996).

Irwin and Thomson (1995) described captive breeding in the alligator snapping turtle (*Macroclemys temminckii*) at the Queensland Reptile and Fauna Park (now Australia Zoo). Netten and Zuurmond (1985) mentioned offspring of the common snapping turtle (*Chelydra serpentina*) in the reptile zoo Iguana.

Herman (1993) described reproduction and management of the southeastern Asian spiny turtle (*Heosemys spinosa*) and Herman and George (1986) outlined research, husbandry, and propagation of the bog turtle (*Clemmys muhlenbergii*) at Zoo Atlanta. Tryon and Hulsey (1977) described breeding and rearing of the bog turtle at the Fort Worth Zoo. Collins (1989) from the Burnet Park Zoo, Syracuse, New York offered a perspective on the captive propagation of bog turtles from western New York State. In 2004, spotted turtles (*Clemmys guttata*) hatched for the third time at Detroit Zoological Institute.

Müller (1970) bred the red-eared slider (*Trachemys scripta elegans*) at the Leipzig Zoo. Kramer and Fritz (1989) outlined courtship in the Florida redbelly turtle (*Pseudemys nelsoni*). Connaughton and Paine (1989) described captive management and reproduction in the Venezuelan slider turtle (*Pseudemys scripta chichiriviche*) at the Buffalo Zoo. Fritz (1990) provided an extensive overview of the care and breeding of the Jamaican pseudemid turtle (*Trachemys terrapen*) with additional notes on the reproductive strategy of Neotropical pseudemid turtles of the genus *Trachemys*. His account described courtship, breeding, nest construction, hatching, reproduction, and teratology. In 2004, Barbour's map turtle (*Graptemys barbouri*) reproduced at the John G. Shedd Aquarium in Chicago.

Murphy and Mitchell (1984) reproduced the aquatic Coahuilan box turtle (*Terrapene coahuila*) in a large outdoor exhibit at the Dallas Zoo. Cerda and Waugh (1992) reviewed status and management of this endangered species at the Jersey Wildlife Preservation Trust.

Honegger (1986) published on the care and long-term reproduction of the black marsh turtle (*Siebenrockiella crassicollis*). Falk Dathe (2001) documented care and reproduction of Amboina-hinge turtles (*Cuora amboinensis*) in the Berlin Tierpark. Lucia Da Silveira (1986) followed the birth and growth of the Neotropical wood turtle (*Rhinoclemmys punctularia*) at the Fundaqão Rio Zoo, Rio de Janeiro in Brazil.

Blanco et al. (1991) propagated the batagurine turtles *Batagur baska* and *Callagur borneoensis* at the Bronx Zoo. Since 2001, three endangered Asian turtles have reproduced at the Metro Toronto Zoo: Vietnamese box turtle (*Cuora g. galbinifrons*), black-breasted leaf turtle (*Geoemyda s. spengleri*), and Malayan river turtle (*Callagur borneoensis*). One McCord's box turtle (*Cuora mccordi*) hatched at the Detroit Zoological Institute in 2004.

Ten Fly River turtles (*Carettochelys insculpta*) hatched at the Miami Metrozoo in early September 2004 in a large outside pool.

Sea turtles rarely breed in captivity. Yoshioka and Samejima (1989) propagated the loggerhead sea turtle (*Caretta caretta*) at the Nagasakibana Parking Garden, Kagoshima, Japan. Remarkably, Uchida (1996) described indoor breeding of loggerhead turtles at the Port of Nagoya Public Aquarium in Japan. Nuitja and Uchida (1982) followed growth and food consumption in juveniles. Oka et al. (1983) from the Shimonoseki Municipal Aquarium in Japan kept the leather-back sea turtle (*Dermochelys coriacea*), a difficult species to maintain due to a specialized diet and pelagic habits. In 2003, SeaWorld San Diego hatched 21 green sea turtles (*Chelonia mydas*).

Goode started an impressive aquatic turtle program at the Columbus Zoo (1988, 1991, 1994) which resulted a significant number of successful reproductive events.

Medical Management

Otis and Behler (1973) surveyed the incidence of Salmonellae and *Edwardsiella* in the turtles of the New

York Zoological Park. Lucia Da Silveira and Andre (1986) published preliminary notes concerning lesions to the plastron of the Gibba turtle (*Phrynops gibbus*) caused by fungi and bacteria. West (2001) performed an endoscopic hepatic biopsy in a Coahuilan box turtle at the San Antonio Zoo. Raphael et al. (1994) described pharmacokinetics of enrofloxacin in Indian star tortoises (*G. elegans*), blood values in free-ranging pancake tortoises (1994), and clinical significance of *Cryptosporidia* in captive and free-ranging chelonians (1997) at the New York Zoological Park. Jes (1989) treated amoebiasis in turtles; drugs were incorporated into a gelatin/food mix to treat 10 species at the Cologne Zoo. Jacobson et al. (1981) treated amoebiasis in red-footed tortoises. Hollamby et al. (2000) documented an epizootic of amoebiasis in a mixed species collection of juvenile tortoises at the Lowry Park Zoo. Beck et al. (1995) compared plasma concentrations of gentamicin injected into the cranial and caudal limb musculature of the eastern box turtle and Calle et al. (1998) initiated a mycoplasma survey of the taxon. Hiller (1984) covered veterinary experience gathered in the breeding of the swamp turtle (*Emys orbicularis*). Marks and Citino (1990) reported on hematology and serum chemistry of the radiated tortoise. Mautino and Page (1993) wrote about the biology and medicine of turtles and tortoises.

Mayeaux (1994) described symptoms of gas-bubble trauma in two species of turtles (*Chelydra serpentina* and *Apalone spinifera*). Mebs (1965) was interested in determining causes for an illness of the eyes frequently found in water turtles. Wenker et al. (1999) diagnosed periarticular hydroxyapatite deposition disease in two red-bellied short-necked turtles.

Rideout et al. (1987) found that viviparous nematodes of the genus *Proatractis* caused mortality of captive tortoises at the Smithsonian National Zoological Park.

Hauser et al. (1977) diagnosed bone metabolism disturbances in Seychelles giant tortoises.

Smith and Coates (1938) from the New York Aquarium diagnosed fibro-epithelial growths in green sea turtles (*Chelonia mydas*). Whitaker and Krum (1999) offered recommendations for medically managing sea turtles in aquaria. Neiffer et al. (1998) from the Pittsburgh Zoo used PC-7 epoxy paste to repair shells in loggerhead sea turtles. Miller et al. (1997) surgically removed a fish hook from a Kemp's ridley sea turtle (*Lepidochelys kempii*) at the Aquarium of the Americas. Bradley et al. (1998) presented hemogram values and morphological characteristics of blood cells in loggerhead sea turtles.

Husbandry

Casares et al. (1995) managed giant tortoises (*Geochelone elephantopus* now *G. nigra* and *Geochelone gigantea*) at the Zürich Zoo. Furrer et al. (2004) compared growth rates of juvenile *G. nigra* at the Zürich Zoo and the Charles Darwin Research Station on the Galápagos Islands. After four years, the zoo tortoises weighed 10 times as much and the carapaces were almost twice as long when compared to the chelonians at the research station: nutrients were reduced and fiber increased for the zoo specimens. Raphael (1980) described sand impaction in a Galápagos tortoise. Wright et al. (1997) experimented with diets for the Galápagos and Aldabran tortoises at the Philadelphia Zoo. At the Rotterdam Zoo, Liesegang et al. (2001) judged the influence of different dietary calcium levels on the digestibility of CA, Mg, and P in captive-born juvenile Galapagos tortoises (*Geochelone nigra*). Pickering (1990) managed the pancake tortoise at the Tulsa Zoological Park. Boyer and Boyer (1994) offered suggestions for tortoise care.

Campbell (1972) developed methods for keeping Central American river turtles (*Dermatemys mawi*) at the Fort Worth Zoo.

From Zoo Atlanta, George (1987) provided an overview of captive maintenance in the genus *Graptemys* and Herman (1991) outlined husbandry of the eastern *Clemmys* group. Stancel et al. (1998) found that calcium and phosphorus supplementation decreases growth but does not cause pyramiding in young red-eared sliders. Zwart et al. (1994) investigated the relationship between intestinal tracts and mineralization of skeletons by referring to normal development of carapaces in young tortoises.

Frolov et al. (1982) described their methods for the captive maintenance of soft-shell turtles at the Moscow Zoo.

Hawksbill sea turtles ply the waters off the coast of Britain and one was collected in the river Severn in the southwest of the country in 1770. According to Dr. Turton, the turtle was placed in his father's fish pond where it survived until winter, likely the first, however brief, captive sea turtle (Coote, 2001). Harwell (1982) and Caillouet et al. (1993) outlined procedures for captive rearing and medical management of Kemp's ridley sea turtles. At the National Aquarium in Baltimore, Stamper and Whitaker (1994) considered the important factors prior to releasing "healthy" sea turtles and Stamper et al. (1997) described single dose pharmacokinetics of ceftazidime in loggerhead sea turtles. Glazebrook and Campbell (1990) surveyed diseases of oceanarium-reared and wild marine turtles in northern Australia. Bels (1987) analyzed growth and Birkenmeier (1972) reared the leatherback sea turtle (*Dermochelys coriacea*) in captivity. Mowbray (1965) outlined procedures for keeping the green sea turtle (*Chelonia mydas*) at Waikiki Aquarium. Owens et al. (1978) determined sex of immature *Chelonia mydas* using radioimmunoassay at the San Diego Zoo. Paulraj et al. (1987) raised the olive ridley sea turtle (*Lepidochelys olivacea*) in artificial sea water. Pritchard (1979) evaluated "Head starting" and other conservation techniques for marine turtles.

Brannian (1984) from the Kansas City Zoo described a soft tissue laparotomy technique in turtles.

General

In 1984, Pedro Trebbau, the director of the Caracas Zoo, Venezuela, published a monumental work on the chelonians of that country with Peter C. H. Pritchard.

Burke (1991) and Curl et al. (1985) discussed zoo captive breeding and recovery programs for the Madagascan ploughshare tortoise (*Geochelone yniphora*). Hirsch (1980) pointed out how Galápagos tortoises were exploited in the past and proposed needed conservation measures and Louis et al. (1990) explored taxonomic relationships among subspecies in zoo populations. Keen (1995) studied seven Aldabran tortoises at the Paignton Zoo. Kirsche (1976) contributed information pertaining to the biology of the star tortoise. Bloxam (1998) recoded distribution, density and abundance of the Madagascar flat-tailed tortoise (*Pyxis planicauda*).

A field team from the Wildlife Conservation Society discovered what is likely the last remaining East Asian giant softshell turtle (possibly *Rafetus swinhoei* or an undescribed species) in Hoan Kiem Lake in the center of Hanoi, Vietnam in March 2003.

Walder (1990) outlined chelonian management practices at the Lowry Park Zoo. Dierenfeld et al. (1999) investigated circulating à-tocopherol and retinol concentrations in free-ranging and zoo turtles and tortoises.

Tuataras

RESOLVED NEVER TO CATCH A TUATERA BY THE TAIL. AS I CONGRATULATED THE MAN UPON HIS ESCAPE, HE SAID , "I AM A BIT LUCKY THIS TIME, FOR I HAVE JUST COME OUT OF THE 'ORSPITAL, WHERE I HAVE BEEN LAID UP WITH BLOOD-POISONING THROUGH THE BITE OF ONE OF THEM PYTHONS."

REVEREND GREGORY BATEMAN (1897)

THAT SPHENODON LIVES TO FOURTEEN YEARS, AND PROBABLY LONGER, IS A FACT OF INTEREST.

MAJOR STANLEY SMYTH FLOWER (1925)

Fig. 16. Tuatara (*Sphenodon punctatus*) at Rotterdam Zoo incorrectly identified as green iguana in the zoo guidebook in 1925. *Provided by Gerard Visser, Rotterdam Zoo.*

History

Hick (1965) solicited comments by specialists of captive tuataras (*Sphenodon*) held at the Auckland, Basel, Chester, Detroit, Dublin, Jersey and San Diego Zoos. Austin (1962) described captive tuatara at the Detroit Zoo. Hoessle (1969) described the tuatara display at the St. Louis Zoo. Cree et al. (1994) highlighted the importance of captive management to the conservation of tuatara in New Zealand.

Behavior

Goetz and Thomas (1994) used annual growth and activity patterns to assess management practices for captive tuataras.

Reproduction

Wood (1967) and Tintinger (1987) bred the tuatara at the Auckland Zoo. Newman (1982) offered suggestions for breeding tuataras in captivity. Brother Island tuatara (*Sphenodon guntheri*) are being bred in large numbers in New Zealand.

Husbandry

Hartley (1963) described observations on tuataras at

the Jersey Wildlife Preservation Trust. Lange (1997) outlined the husbandry and captive maintenance of 10 young tuataras in the Berlin Aquarium.

Captive and Medical Management

Liu and King (1971) described microsporidiosis in the tuatara at the New York Zoological Park. Blanchard (1992) focused on management in the wild and in captivity. Boardman and Sibley (1991) covered captive management, diseases and veterinary care of tuatara, based on experiences at the Auckland Zoo.

General

Newman et al. (1979) used field observations to underscore their recommendations for captive maintenance. See Chapter 9, New Zealand account for additional information and literature citations.

Amphibians

THE BATRACHIANS ARE EXCEEDINGLY INTERESTING CREATURES, AND MOST OF THEM WILL LIVE FOR A VERY LONG TIME IN A PROPERLY ARRANGED VIVARIUM. AS A RULE, THEY ARE MORE EASILY PROVIDED WITH FOOD THAN MANY OF THE REPTILES, AND THEY DO NOT REQUIRE SO MUCH ARTIFICIAL HEAT, WHEN ANY, AS THE LATTER ANIMALS.

REVEREND GREGORY BATEMAN (1897)

TOAD HALL . . . A NAME LIKE THIS SPEAKS FOR ITSELF—A HOUSE FOR AMPHIBIANS, FROM WHICH THERE IS NO ESCAPE, SUCH AS MAY OCCUR IN THE WALLED RETILIARY. SUCH A HOUSE WAS MADE AND USED FOR MANY YEARS BY THE BRITISH HERPETOLOGIST, MR. L. G. PAYNE, AND PROVED TO BE AN EXCELLENT COMMUNITY CENTRE FOR A VARIETY OF FROGS AND TOADS, TREE FROGS AND SALAMANDERS.

ALFRED LEUTSCHER IN *VIVARIUM LIFE. A MANUAL ON AMPHIBIANS, REPTILES AND COLD-WATER FISH*, FIRST PUBLISHED IN 1952

Zoo program designers have traditionally ignored amphibians when constructing exhibits or building collections. Loisel (1912) in his study of the history of menageries found a few references to amphibians in captivity: salamanders at the Royal Menagerie of Versailles; some amphibians in the Menagerie of the Museum of Natural History in Paris in 1839 which had a large center exhibition section called "Aquarium Hall," containing freshwater fishes and amphibians. Longevities included a Japanese salamander for 30 years and a siren for 23 years; 24 amphibians in the Menagerie of the Imperial Cabinet of Natural History in Vienna around 1847; three species of amphibians in

the Kyoto Zoo in Japan in 1908; some amphibians in the Amsterdam Zoo in 1908; the Breslau Zoo in Germany had 79 amphibians of 14 species in 1909; 122 amphibians representing 25 species in 1910 at the Rotterdam Zoo; 10 species of amphibians in the Gizeh Zoo in 1911; some amphibians at the Philadelphia Zoo in 1911; the Hannover Zoo had a collection of 16 amphibians numbering 7 species in 1911; some amphibians in the Schönbrunn Menagerie in 1912; the Frankfurt Zoological Garden maintained a large collection of European amphibians, housed in a large terrarium and conservatory on top of the aquarium; and the Copenhagen Zoo had a small terrarium for amphibians. Aquariums and zoos which contained amphibians in the United Kingdom included the Carnegie Aquarium in Edinburgh and the Blackpool Tower Aquarium, and the zoos in Bristol, Paignton, and Belfast (Schomberg, 1957).

There have been a few papers published over the years but the output is paltry when compared to the number of contributions dealing with reptiles. This deficit is slowly being remediated, based in part on the concern throughout the biological community that amphibians are fast disappearing. Over the past decade, a number of zoos and aquariums have begun to support *in situ* and *ex situ* amphibian projects. When the IUCN/SSC Declining Amphibian Populations Task Force (DAPTF) was established in 1989, the bulk of financial support emanated from the zoo and aquarium community, either through direct donations or through contacts supplied by the staffers. I am proud that my colleagues responded so quickly to this crisis. However, in spite of this increased attention, some groups of amphibians remain virtually unknown, the most dramatic being caecilians.

There are several important papers and books. Maruska (1986) reviewed zoo breeding programs. Nace (1977) outlined procedures used at the Amphibian Facility, University of Michigan. Raphael (1993) presented information on medical management. In 2001, Wright at the Phoenix Zoo and Whitaker at the National Aquarium in Baltimore, published an extraordinary book, *Amphibian Medicine and Captive Husbandry*, which is assuredly the finest source for information on husbandry, diagnosis and treatment of diseases.

Bloxam and Tonge (1995) at Durrell Wildlife Conservation Trust rightly asserted that amphibians were suitable candidates for breeding-release programs. An example is the recovery program for the Mallorcan midwife toad (*Alytes muletensis*) outlined by Konig and Schluter (1991) and Buley and Garcia (1997).

See website and links for the role of zoos and

aquariums in conservation programs http://www.amphibiaweb.org

Salamanders

History

Maruska (1989, 1994) at the Cincinnati Zoo reviewed husbandry and propagation of the Japanese giant salamander (*Andrias japonicus*). A number of studies on reproductive biology, ecology, husbandry and natural history have been published by researchers at the Asa Zoo in Hiroshima, Himeji City Aquarium and other institutions (see Chapter 10, Japan account for additional information and references).

Moore (1994) covered the reproductive behavior and physiology of salamanders by stressing hormones, temperature, light, pheromones, endocrine control, and environmental complexity. Indiviglio wrote a comprehensive book in 1997 called *Newts and Salamanders. Everything About Selection, Care, Nutrition, Diseases, Breeding, and Behavior,* which is an excellent resource.

Behavior

Brodie et al. (1992) investigated the anti-predator mechanisms of three species of salamandrids. Dobroruka (1973) described pellet formation in the Japanese giant salamander.

Reproduction

Kerbert (1904) described the first captive breeding of the Japanese giant salamander at the Amsterdam Zoo. Jordan et al. (1992) and Roberts et al. (1995) duplicated the conditions found in nature to stimulate the Comal Springs salamander (*Eurycea neotenes*) to breed at the Dallas Aquarium. They used a clear plastic tube, filled with rocks, with a continuous water flow from the bottom as these aquatic salamanders were found near spring upwellings. At Berlin Tierpark, Freytag and Petzold (1978) and Rehák at the Prague Zoo (1991) described the natural history and captive breeding in the Vietnamese warty newt (*Paramesotriton deloustali*). Pfaff and Vause (2002) prepared a note on the captive reproduction and growth of the broad-striped dwarf siren (*Pseudobranchus s. striatus*) at the Riverbanks Zoo. Wisniewski and Paull (1986) bred and reared the European fire salamander (*Salamandra s. salamandra*). In 2004, fire salamanders (*Salamandra s. bernardezi*) and fire belly newts (*Cynops cyanurus*) reproduced at the Cincinnati Zoo. The Detroit Zoological Institute, Baltimore Zoo, Metro Toronto Zoo and others have been involved in a broad-based cap-

tive breeding program to conserve the emperor newt (*Tylototriton shanjing*) with production of over 500 metamorphs. Anderson's newt (*Echinotriton andersoni*) reproduced at the Buffalo Zoo in 2005. In 2005, the Galeana false brook salamander (*Pseudoeurycea galeanae*) reproduced at the San Antonio Zoo.

Kramer et al. (1983) sexed aquatic salamanders by laparoscopy at the Cincinnati Zoo. Wright (1994) amputated the tail of a two-toed amphiuma (*Amphiuma means*).

Caecilians

History

Caecilians have been ignored in zoo and aquarium collections so information on captive requirements is lacking. Wright and Minott (1999) identified individual captive Mexican caecilians. The main reference is Wake (1994).

Behavior

Murphy et al. (1977) documented coitus in caecilians (*Typhlonectes compressicauda*) for the first time at the Fort Worth Zoo.

Frogs and Toads

History of Conservation Initiatives

Hugh Quinn (1980) and associates (Harwell and Quinn, 1982; Mays and Peterson, 1996; Quinn and Mengden, 1984) started a propagation initiative for the endangered Houston toad (*Bufo houstonensis*) at the Houston Zoo which included medical management and reintroduction. Burton (1991) and Burton et al. (1995) offered experiences on managing the endangered Wyoming toad (*Bufo hemiophrys baxteri* now *B. baxteri*) and Taylor et al. (1994) reviewed causes of mortality in the captive population at the Cheyenne Mountain Zoo. The Memphis Zoo produced 1,700 tadpoles by artificial fertilization which were released in Wyoming. Martin (1991) recommended captive husbandry as a technique to conserve the Yosemite toad (*Bufo canorus*); satellite colonies were started at the Sacramento Zoo and Chaffee Zoological Gardens, California. Paine et al. (1989) explained the AZA Species Survival Plan recovery program for the Puerto Rican crested toad (*Peltophyrne lemur*) which involves many zoos.

Living specimens of the Mallorcan midwife toad (*Alytes muletensis*) were not discovered in the wild until 1980. In response to a continuing threat to vulnerable small populations, the Durrell Wildlife Conservation Trust and the Barcelona Zoo, Spain, began to study these anurans in captivity and in the field (Tonge and Bloxam, 1989; Roca et al., 1998, Buley and Garcia, 1997; Buley and Gonzalez, 2000).

Zippel (2002) described the breeding program for the Panamanian golden frog (*Atelopus zeteki*) as part of Proyecto Rana Dorada, an *in situ* conservation initiative involving a number of zoos and outside herpetologists.

The Kihansi spray toad (*Nectophrynoides asperginis*) from Tanzania is critically endangered. The Wildlife Conservation Society and zoos in Detroit, Baltimore, Oklahoma City, Toledo, and Buffalo have been involved in a recovery and captive-breeding initiative but even though there has been successful reproduction to the third generation, parasitism of the founders and low survival rate of the offspring have been discouraging (see Lee et al., 2006 for description of program).

Behavior

Based on observations at the Zoological Gardens at Halle, Germany, Hinsche (1928, 1939) recorded antipredator behavior of various anurans directed toward snakes: death shamming, defense, defensive attack and exaggerated postures. Murphy (1976) and Radcliffe et al. (1986) observed pedal luring in leptodactylid frogs.

Husbandry and Reproduction

Because of bright colors, hardiness and diurnal activity, poison-dart frogs have been popular as exhibit amphibians for many years. Dathe and Dedekind (1991) explored the care and breeding of Venezuelan tree climbers (*Colostethus trinitatis*) in the Tierpark Berlin. Grow (1980) described husbandry of the orange striped poison dart frog (*Phyllobates vittatus*) at the Sedgwick County Zoo. Blake (1990) outlined cage design for the Trinidad stream frog (*Colostethus trinitatis*) at the Edinburgh Zoo. Lange (1981) featured the program at the Berlin Aquarium of the green-and-black poison dart frog ((*Dendrobates auratus*), resulting in 32 clutches with 173 frogs successfully reared. Knepper (1993) produced the azure dart-poison frog (*Dendrobates azureus*) at the Chaffee Zoo in California. Fujitani et al. (1998) propagated the Amazonian poison frog (*Dendrobates ventrimaculatus*) at the Ueno Zoo. Preece (1998) described captive management and breeding of poison-dart frogs at the Durrell Wildlife Conservation Trust. Cover and his associates have published many significant papers on the care, maintenance and reproduction of poison dart and hylid frogs

at the National Aquarium in Baltimore (1992, 1994) and Stoskopf et al. (1985) discussed iodine toxicity. In the Cologne Zoo journal *Zeitschrift des Kölner Zoo*, Zimmermann and Zimmermann published many reports on the biology, captive breeding and conservation of poison dart frogs and Malagasy poison frogs (*Mantella*) (see also Zimmermann and Zimmermann, 1994).

In collaboration with outside researchers studying amphibian toxins, zoo workers discovered that the toxins of dendrobatids are the result of dietary preferences; further, captive-reared individuals denied that diet are non-toxic (see Daly et al., 1992, 1994) which would clearly be a significant deficit if reintroduction is to be considered.

Vogt (1974) from the Krefeld Zoo in Germany, Burchfield (1975) from the Gladys Porter Zoo, Smith and Fischer (1975) from the Los Angeles Zoo, and Pawley (1988) from the Brookfield Zoo reproduced the Blomberg toad (*Bufo blombergi*). Endangered Wyoming toads (*Bufo baxteri*) have reproduced at the Houston Zoo on two occasions, the last being in 2005.

Odum et al. (1983) used hormones at the Houston Zoo to induce breeding and rearing of White's tree frog (*Litoria= Pelodryas caerulea*), a species commonly kept in zoos and in the private sector and Bradley and Wright (2000) described captive care and breeding of this tree frog. Miller (1983) prepared notes on a breeding of the red-eyed tree frog (*Agalychnis callidryas*).

Ivanyi (1989) reproduced the black-eared frog (*Leptodactylus melanonotus*) at the Arizona-Sonora Desert Museum. Gibson and Buley (2004) described a novel mode of reproduction in amphibians: foam nests on land, terrestrial tadpoles, and obligatory oophagy in the endangered Montserrat mountain chicken (*Leptodactylus fallax*) at the Jersey Zoo. Johnson (1984) bred the Bell's or ornate horned frog (*Ceratophrys ornata*) without resorting to hormones. Honegger et al. (1985) discussed reproduction using hormonal injections, development and analysis of male call, and breeding of this frog.

Schafer (1981) described her techniques for rearing the Asiatic tree frog (*Rhacophorus leucomystax*) at the San Diego Zoo.

Schmidt and Wicker (1977) observed breeding in the Asian horned frog (*Megophrys nasuta*) at the Frankfurt Zoo.

Birkett et al. (1999) managed and reared the roseate frog (*Geocrinia rosea*) at the Melbourne Zoo.

Weygoldt (1976) offered observations on the reproductive biology of the Surinam toad (*Pipa carvalhoi*). Sclater (1895) and Bartlett (1896, see also Scherren,

1905:212) observed breeding of the Surinam toad (*Pipa surinamensis, P. americana,* now *Pipa pipa*) at the London Zoo. The Ueno Zoo has reproduced Surinam toads (Shibuya, 1978).

Held (1985) maintained, exhibited, and bred the tailed frog (*Ascaphus truei*) at the Washington Park Zoo, Portland, Oregon.

Shiihara and Samejima (1995) propagated Ishikawa's frog (*Rana ishikawae*) at the Nagasakibana Parking Garden, Kagoshima, Japan. Their account included photographs of the enclosure, egg masses, tadpole and metamorph. Gupta (1998) offered observations on ranid reproduction. Yoshimi et al (1996) maintained and bred the Solomon Island leaf frog (*Ceratobatrachus guentheri*) at the Woodland Park Zoo.

Ryboltovsky at the Children's Zoo, Vsevolozhak, Leningradakaya, Russia published interesting papers in 1999 on two poorly known frogs: Tonkin bug-eyed frog (*Theloderma corticale*) and Annamese flying frog (*Rhacophorus annamensis*).

Captive and Medical Management

Stetter et al. (1996) used Isoflurane anesthesia in amphibians and compared five application methods. Upton et al. (1992) discovered testicular myxosporidiasis in the African flat-backed toad (*Bufo maculatus*). Miller et al. (1992) documented an outbreak of disseminated chromoblastomycosis in a colony of ornate-horned frogs and Teare et al. (1991) published on pharmacology of gentamicin in the leopard frog {*Rana pipiens*) at the Smithsonian National Zoological Park. Ackermann and Miller (1992) diagnosed chromomycosis in an African bullfrog (*Pyxicephalus adspersus*). Suedmeyer et al. (1997) discovered chromomycosis in a marine toad (*Bufo marinus*) at the Kansas City Zoo. Also at the Smithsonian National Zoological Park, Nichols was instrumental in documenting respiratory diseases and cutaneous chytridiomycosis (Nichols, 2000, 2003; Nichols et al., 1996, 1998, 2000, 2001). Cutaneous chytridiomycosis has been implicated as a major cause in global amphibian declines.

Mutschmann (1998) found evidence of *Chlamydia psittaci* infections in amphibians, obtained by means of a specific immunofluorescence test (IFT).

McNamara et al. (1999) reported crystalline inclusions associated with vomeronasal organ pathology in red-eyed tree frogs (*Agalychnis callidryas*).

Whitaker and Poyton (1993) identified and recommended treatments for protozoans in dendrobatid frogs. Helman et al. (1998) found parasitic conjunctivitis and lacrimal adenitis in tiger salamanders (*Ambystoma tigrinum mavortium*).

Garner et al. (1995) diagnosed diseases of Solomon Island leaf frogs (*Ceratobatrachus guentheri*) at the Woodland Park Zoo by consulting seventy five necropsy cases. At the Calgary Zoo in Canada, Honeyman et al. (1992) found *Bordetella* septicemia and chlamydiosis in this species.

Burns from the Louisville Zoo (1995) offered suggestions for the humane euthanasia of reptiles, amphibians, and fish.

Stetter and Cook (1994) presented normal and pathological ultrasonographic anatomy of amphibians.

Wright (1996) and Wright and Whitaker (2001) provided an excellent overview of husbandry and medicine and Rundquist (1993) presented suggestions for future veterinary investigations.

General

George Rabb from the Brookfield Zoo has published a number of papers on sound production, fighting, mating behavior, development and general reviews on pipid frogs (Rabb, 1960, 1967, 1973; Rabb and Rabb, 1960, 1963; Rabb and Snedigar, 1960).

From the Duisburg Zoological Garden in Germany, Gewalt (1977) covered catching, transport and keeping of the rare giant or goliath frog (*Conraua goliath*). Gillespie et al. (1988/1989) sexed goliath frogs by laparoscopy at the Cincinnati Zoo.

Hinsche (1941) showed the close reactive relation between organism and floor in various amphibians, i.e., an unsuitable floor/substrate caused problems. Kauffeld (1942) gave tips for raising frog tadpoles.

Japanese Giant Salamander (*Andrias japonica*)

Chapter 2
Evolution of an Idea

Ringed Caecilian (*Siphonops annulatus*)

DESMOND MORRIS (1968)

The idea was exciting but also vexing to zoo managers: how could the world's zoos become significant players in a world of shrinking biodiversity? While the term "biodiversity" had not yet been coined nor its implications understood, the zoo community generated a plan for becoming a force in conservation by using their living collections. There were significant obstacles to overcome. How could zoo workers maximize the utility of their collections, especially since the previous trend had been to accumulate as many species as possible, essentially a "postage stamp" assemblage? How could mortality be managed at acceptable levels? Could groups of breeding animals be assembled and if so, how could their histories be tracked to prevent inbreeding? Could zoo workers be expected to work with each other and outside colleagues to exchange animals and information? Could the competence of animals bred in captivity be assessed to ensure that they bore similarity to their wild counterparts? In an environment of shrinking financial resources, unwelcome political pressures and contrary agendas, could influential conservation and educational programs be expanded beyond zoo boundaries and established *in situ*; how would their efficacy be evaluated? Is reintroduction a viable solution? More specifically, have zoo herpetological programs been effective in any or all of these arenas, especially since amphibians and reptiles are often overlooked by zoo administrators? In my view, the results have been mixed!

To begin the evaluation, it is helpful to look at the history of change in herpetological collections. When

I entered the profession in the mid-1960s, there was little discussion about conservation. Most of our efforts centered on keeping amphibians and reptiles alive; the novel notion of accumulating breeding groups was barely mentioned. In fact, we could rarely sex our animals, much less breed them. When I arrived at the Dallas Zoo to work at the new reptile building (where I was to spend the next thirty years), my boss, the late Director Pierre Fontaine, told me a depressing story. During the initial planning stages of the building and preparation of the budget for assembling the collection, Pierre contacted the late R. Marlin Perkins, Director at the Saint Louis Zoo, to ask what reasonable annual mortality figures could be anticipated. Perkins said that less than 80% would be excellent! There were legitimate reasons for Perkins's gloomy outlook. Virtually all of the collection needed to be assembled within six months to stock the building but established and captive-bred amphibians and reptiles were mostly unavailable; hence, herps from the wild had to be used for display but these specimens were unaccustomed to captivity and the stress of exhibition. Fortunately, our mortality figures for the first year were not unreasonable.

Reptile buildings and accompanying exhibits were often designed by architects and zoo administrators who had little experience with living herpetological collections. During this period, herps were rarely kept in appropriate social situations and little attention was paid to behavioral or environmental cues; social stressors among conspecifics were barely considered. Sadly, a few zoos even had a standing order for "snake dens" to replace specimens lost during each year.

Within the next decade, several significant changes occurred. Veterinary protocols improved considerably and the likelihood of establishing long-term breeding groups was enhanced. Before the mid-1950s, references dealing with medical management were mostly scattered in journals as there were few books dealing with diseases and treatments in a comprehensive manner. To disseminate information, zoo veterinarians interacted with each other describing case histories and successful treatments. As amphibians and reptiles became more popular as pets, veterinarians in the private sector began to see a much greater variety of patients. To treat animals more effectively, these practitioners needed concise information, preferably a single source. In Europe the first significant books about amphibian and reptilian diseases were *Krankheiten der Amphibien* (1961) and *Krankheiten der Reptilien* (1963), both published by Heinz-Hermann Reichenbach-Klinke. In 1973, Fredric Frye published

an important book in English which filled a noticeable gap in the coverage of herpetological medicine and a steady stream of impressive works followed his book. In Table 2.1, influential books by Frye and others have been chosen and listed in chronological order to show the evolution, expansion and sophistication of herpetological medicine. It is evident that significant advances have occurred.

The American Association of Zoo Veterinarians (AAZV) was officially founded in 1960 and has had extensive materials on reptile/amphibian medicine in its proceedings since that time. Prior to that time, a small group of zoo veterinarians met at the American Veterinary Medical Association (AVMA) meetings under the title "Zoo Vets" since 1946. In 1991, the Association of Reptilian and Amphibian Veterinarians (ARAV) was formed and now has over 1300 members. ARAV has been closely associated with AAZV since that time. This society publishes a peer-reviewed quarterly: *The Journal of Herpetological Medicine and Surgery* and hosts an annual meeting. The American College of Zoological Medicine (ACZM) now allows board certification in zoological medicine with a specialty in herpetological medicine. Since its inception, the *Journal of Zoo and Wildlife Medicine* has published numerous manuscripts on herp medicine. The Disease Risk Workshops held by the World Conservation Union (IUCN) and the Conservation Breeding Specialist Group (CBSG) have had a big impact on developing methods of disease risk analysis in animal translocations/reintroductions.

In Europe, Conference Reports of the yearly International Symposiums Concerning Illnesses of Zoo Animals (Internationales Symposium über die Erkrankungen der Zootiere, started in 1959--see: *1. Symposium der Zootierärzte in Berlin-Friedrichsfelde am 21. and 22.3.1959;* Kleintierpraxis, Hannover 1959, 4:121-143) are published which contain papers on herpetological medicine. In 1985 the most important findings were summarized in the book by Rudolf Ippen, Hans-Dieter Schröder & Karl Elze: *Handbuch der Zootierkrankheiten. Band 1: Reptilien.* The European Association of Zoo and Wildlife Veterinarians (founded in 1996) also publishes a number of articles on herpetological medicine in its proceedings.

Our charges were individually marked and sexes determined with greater accuracy so we were able to accumulate proper sex ratios for breeding programs (see Brazaitis, 1969; Laszlo, 1975; Honegger, 1978, 1979 for techniques). Greater attention was paid to providing appropriate thermal gradients, based on the landmark studies of reptilian behavioral thermoregulation by Cowles and Bogert (1944), morphological effects of low temperatures during the embryonic development of the garter snake, *Thamnophis elegans* (Fox et al., 1961), and lists of amphibian and reptilian cloacal temperatures recorded in the field by Brattstrom (1963, 1965) and others; in the past, environmental ambient temperatures had varied minimally in our facilities. Peaker (1969) covered temperature control, activity cycles, seasonal variation, influence of heat and light on reproduction, and determination of a suitable regimen for captive reptiles. Significant contributions on thermoregulation and hibernation by Huey (1982), Avery (1982, 1994), Bartholomew (1982), Gregory (1982), Tryon (1984), Sievert and Hutchison (1988, 1989, 1991) and others helped to refine thermal environments.

Reviews on behavior in *Biology of the Reptilia* (1977) were also helpful: Schoener on competition and the niche, Heatwole on habitat selection, Turner on dynamics of populations, Stamps on social behavior and spacing patterns, Carpenter and Ferguson on stereotyped behavior, and Burghardt on learning. Zoo herpetologists became better at recognizing signs of reproductive behavior and recording these observations in scientific journals. Exciting new possibilities appeared in the literature, such as manipulating clutch and offspring size (Sinervo, 1994) and investigating the evolution of social cooperation in lizards (Dickinson and Koenig, 2003; Sinervo and Clobert, 2003) . Some of the most progressive managers invited ethologists and other researchers from the academic community to work jointly on projects, resulting in better final products (see Sajdak, 1983; Chiszar et al., 1993; Chiszar and Smith, 2005; Garrett, 2005 for reviews). In the United States, researchers such as Gordon Burghardt, Charles Carpenter, David Chiszar, William Cooper Jr., John Daly, Gary Ferguson, William Gehrmann, James Gillingham, Harry Greene, Thomas Jenssen, Hobart Smith, and Paul Weldon, wisely used zoo and aquarium collections to buttress their studies.

One continuing challenge was feeding certain types of snakes, particularly pitvipers. In the beginning, we tried a number of techniques with limited success. Many of our early discussions with colleagues centered on feeding behavior and provision of suitable environments for security. When detailed papers covering the predatory behavior of snakes by Tjard de Cock Buning, Gordon Burghardt, David Chiszar, Harry Greene, Kenneth Kardong, Charles Radcliffe, Hobart Smith, and others began to appear, their findings were incorporated into our methodologies with impressive results. See Murphy and Campbell (1987) for a description of the feeding techniques used to maintain our snakes, developed in large part on the observations of these workers. Improvements in prepared diets for other herps were dramatic, refined by nutritional studies by Mary Allen and Olav Oftedal (see Chapter 8, National Zoological Park account), David Baer (1994), Ellen Dierenfeld (Clum et al., 1996) and Duane Ullrey (1996).

Books on husbandry of reptiles (and to a lesser degree on amphibians) began to be published in greater numbers. Private fanciers developed many innovative techniques although it must be stressed that a number of protocols currently used were generated in zoos. Methods for egg incubation and raising neonates became more refined (see Packard and Phillips, 1994). As an example, eggs laid by reptiles in the early days were routinely discarded as zoo workers felt that hatching them was impossible. The late Thomas Huff was instrumental in advancing the idea of "captive stagnancy" which referred to a condition found in captive snakes where static environmental factors suppressed reproductive behavior; his solution was to change exhibit furniture, potential breeders, temperature and so on to induce courtship and copulation.

The importance and provision of ultraviolet light for herps held in indoor enclosures were highlighted in important papers by Gary Ferguson and associates (1996; Jones et al., 1996), William Gehrmann (1971, 1987, 1994), Carol Townsend and Charles Cole (1985) and Jozsef Laszlo (see Chapter 8, San Antonio Zoo account). R. Andrew Odum (1984) stressed the importance of water quality.

The American Association of Zoological Parks and Aquariums (AAZPA) was founded in 1924. Seventy years later, the name was changed to the American Zoo and Aquarium Association (AZA). Some of the organizational purposes of AZA are (1) to promote discussions and cooperation among zoo and aquarium professionals; (2) to set high standards for its accredited member institutions, now over 180; (3) and to support *in situ* and *ex situ* conservation initiatives. In 1973, the late Ulysses Seal founded the International Species Information System (ISIS), a computerized central database listing animal records worldwide. This database, adopted by AZA, allowed zoo workers to assess histories of animals in captivity and make informed decisions about linking potential breeding partners. In the beginning, amphibians and reptiles were not listed so zoo herpetologists turned to annual inventories of zoo holdings published by Frank Slavens at the Woodland Park Zoo in Seattle (Slavens, 1989). These inventories were critical for early genetic management of herpetological collections. Pérez-Higareda et al. (1985) recognized Slavens for his important contributions by describing a new species of colubrid snake (*Tantilla slavensi*) from Veracruz, México in his honor.

In 1979, a paper by Katherine Ralls, Kristin Brugger, and Jonathan Ballou at the Smithsonian National Zoological Park on inbreeding and juvenile mortality in small populations of ungulates appeared in *Science*. The influence of this paper should not be underestimated; zoo managers were apprised of the deleterious effects caused by inbreeding captive populations. In fact, Directors William Conway (Bronx Zoo), Theodore Reed (National Zoological Park) and Ronald Reuther (Philadelphia Zoo) created "Zoo East Species Trust (ZEST)" to inaugurate cooperative breeding programs among these zoos, based on the caveats reflected in this paper. The AZA approved Species Survival Plans (SSPs) to ensure that cooperative population management plans were in place for threatened or endangered animals and initiatives were put in place for the Chinese alligator, Cuban crocodile, Dumeril's ground boa, rock iguana, Virgin Island boa, Wyoming toad, Puerto Rican crested toad, radiated tortoise and Aruba Island rattlesnake (see Hudson, 1983). In 1996, AZA studbooks included many species: Chinese alligator, common anaconda, annulated boa, Dumeril's ground boa, emerald tree boa, Jamaican boa, Virgin Island boa, bushmaster, king cobra, African slender-snouted crocodile, Cuban crocodile, Morelet's crocodile, Siamese crocodile, Northern Madagascar tomato frog, Malagasy leaf frog, Standing's day gecko, gharial, Fiji banded iguana, Chinese crocodile lizard, Mexican beaded lizard, Asian forest monitor, Komodo monitor, blood python, reticulated python, Arizona ridge-nosed rattlesnake, Aruba Island rattlesnake, eastern diamondback rattlesnake, prehensile-tailed skink, Haitian slider, Louisiana pine snake, Puerto Rican crested toad, Wyoming toad, Tomistoma, Asian brown tortoise, Egyptian tortoise, Galápagos tortoise, Indian star tortoise, pancake tortoise, radiated tortoise, black-breasted leaf turtle, pignosed turtle, side-necked turtle, palm viper, and Philippine pitviper (Read, 1996). To prioritize species for SSP programs, Taxon Advisory Groups (TAGs) were established and currently exist for amphibians, chelonians, crocodilians, snakes and lizards (also mammals and birds).

In 1994, AZA's Wildlife Conservation and Management Committee (WCMC) created Population Management Plans (PMPs) to provide basic population management for various captive populations. PMPs are established for studbook populations that do not require the intensive management and conservation

action of SSPs and these are easier to implement. SSPs and PMPs combine for about 56 studbooks, far behind the mammal initiatives with 203 studbooks. See Wiese and Hutchins (1994) for additional information. The European Association of Zoos and Aquaria (EAZA), Japanese Association of Zoological Gardens and Aquariums (JAZGA), Australasian Regional Association of Zoological Parks & Aquaria (ARAZPA) and others throughout the world fill the same niche as AZA in these regions.

In the 1960s and 1970s at the annual meetings of the American Society of Ichthyologists and Herpetologists (ASIH), Herpetologists' League (HL) and the Society for the Study of Amphibians and Reptiles (SSAR), there were virtually no zoo and aquarium people in attendance, chiefly due, they said, to the fact that they felt unwelcome. Editors and reviewers of peer-reviewed herpetological and biological journals often rejected theoretical and applied papers on amphibians and reptiles from zoo workers, believing that observations made on captive herps were not representative of the biology of their wild counterparts (see Chiszar et al., 1993 for a contrary view). When the Herpetologists' League held an annual meeting in downtown Dallas a few miles from the Dallas Zoo in the late 1960s, I wrote to President Wilmer Tanner inviting the attendees to the Zoo for a tour and social gathering. We were proud of our efforts in opening our new building and assembling a nice collection and we wanted to show colleagues what we had accomplished. Not a single person came to the Zoo! It became clear that zoo herpetology was not taken seriously; we experienced open antagonism or indifference which was bizarre since we were all fascinated with the same group of organisms. At that point, we felt that our somewhat parochial approach had to change in order that zoo herpetology would start being viewed as an important subset of herpetology by our academic and museum colleagues. We tried to open professional ties with other herpetologists by making our collection accessible and developing mutual research projects.

Feeling isolated, zoo workers gravitated toward settings and venues which were more comfortable. When the *International Zoo Yearbook* appeared on the scene in 1960 and its editors actually encouraged studies on captive herpetofauna, zoo colleagues responded with enthusiasm. Important information was disseminated throughout the community. In addition, the International Herpetological Symposium on Captive Propagation and Husbandry was formed and began publishing annual proceedings beginning in 1979. Zoo professionals and serious amateur herpetoculturists attended these meetings. When three years later, the journal *Zoo Biology* was created, papers on captive herpetofauna appeared. In fact, in 1996 when an issue was devoted exclusively to advances in zoo herpetology, a striking feature was that many papers were jointly written by zoo people and specialists outside the zoo field. These important changes led to greater cooperation and interaction between zoo and aquarium personnel.

SSAR began encouraging zoo participation by sponsoring symposia on captive management in 1978, 1991 and 2001, creating a Zoo Liaison committee, including a ZOO VIEW and HERPETOLOGICAL HUSBANDRY column in each issue of *Herpetological Review*, and having a zoo representative on the board of directors. To underscore the fact that zoo associates were recognized for their contributions to herpetology, I am proud to say that a mounted dried fish was ceremoniously given to me at one of those early SSAR business meetings for being the best dressed (I wore a tuxedo) herpetologist in attendance. Unfortunately, it must be said that my keen fashion sense has deteriorated over time.

Beginning in the 1980s, a number of major studies and papers on the importance of genetic management of small populations and the role of zoos in captive management were generated by Jonathan Ballou, Benjamin Beck, William Conway, Robert Fleischer, Nate Flesness, Thomas Foose, Michael Hutchins, Devra Kleiman, Robert Lacy, Katherine Ralls, Oliver Ryder, Ulysses Seal and Robert Wiese (see Table 2.2). At the Dallas Zoo, we saw firsthand the potentially damaging effects of inbreeding. We bred related western diamondback rattlesnakes over six generations to produce an amelanotic strain but some neonates from amelanotic and normally colored snakes had reversed body and ventral scales, anomalous head scales, and some individuals had virtually no scales (Murphy et al., 1987; McCrady et al., 1994).

As zoo herpetologists began to maintain and manage their collections more effectively, offspring were produced in greater numbers. Although this would appear to be a welcome improvement, serious problems developed. The holding capacity within zoos is limited (see Quinn and Quinn, 1993) and there were so many taxa under siege, most could not be accommodated for accelerated breeding programs. When zoo managers met to discuss potential programs, the enormous commitment of time, space and care taking needed to maintain even a single species was overwhelming and the reality of limited resources surfaced. Further, zoo personnel had strong feelings about those species in the greatest need of intervention so heated entreaties often took place during these meetings.

Based on this bittersweet success of more animals produced in captivity, zoo workers began to discuss the feasibility of reintroduction of these threatened species but there were concerns about this approach (see Dodd and Seigel for a review of this issue, 1991). Further, there were often no suitable places for release. R. Howard Hunt from Zoo Atlanta was successful at breeding Morelet's crocodiles for many years but he was unable to find a single location to safely release his rapidly expanding population.

Although attention had been directed toward responsible genetic management, virtually no systematic studies had been done to evaluate whether the captive amphibians and reptiles produced were competent enough to survive if released in the wild. In fact, Chiszar et al. (1993) argued that assessing the competence of captive-produced herps was as important as breeding them. As an example, captive and wild Komodo dragons have significantly different body temperatures (Walsh et al., 1999). Would captives with lower temperatures adjust thermoregulatory patterns upward if released into the wild? Further, if deficits due to the influence of captivity are identified, can these be remediated? How can this be accomplished? Finally, when these propagules are released, what necessary studies will be in place to assess survivability in the wild? These thorny issues have only been marginally addressed but there are some impressive exceptions: Puerto Rican and Wyoming toads, Virgin Island boa, Aruba Island rattlesnake and West Indian iguanas serve as examples of successful field studies and reintroductions in some cases. Since the 1970s, Jersey's Durrell Wildlife Conservation Trust has had several superb breeding programs for threatened taxa (Round Island Boa, Round Island gecko, Jamaican Boa, Malagasy ploughshare and flat-tailed tortoises, Coahuilan box turtle, San Francisco garter snake, Mallorcan toad), directing considerable space and resources to a limited number of species. These initiatives may include an *in situ* component by featuring reintroductions (Round Island boa) and rat-eradication programs (Antigua racer).

Assessing the competence of captive animals is not a trivial issue and there are many examples of allometric or behavioral alterations in a number of organisms induced by captivity or seen in the wild (see Table 2.3 for samples). For example, large constrictors raised in captivity often have abnormally small heads, likely caused by an overabundance of food. Alligators may have abnormally wide heads. I have noticed that long-term captive snakes often are rubbery when fixed in formaldehyde, perhaps caused by deterioration of muscle integrity. Could lack of exercise contribute to this condition? Other negative factors deserve attention. Are muted colors and abnormal patterns sometimes seen in captives irreversible in future generations? The lack of toxins in captive first generation poison dart frogs requires study. Could herps which survive in our captive environments actually be selected against if released whereas those which do poorly in captivity be better equipped to survive in the wild? Are we creating domesticated amphibians and reptiles as Henry Fitch (1980) predicted? To truly understand the effects of captivity, studies on nutrition, behavior, reproduction, susceptibility to disease and so on need to be developed to identify factors which mold our captive animals.

Since zoo managers were dealing with an environmental crisis of massive proportions, it became clear that public awareness was needed, in terms of political and financial support. The stark reality was that virtually all zoo herpetologists had little experience in designing effective educational programs and some, in fact, viewed the zoo visitor as an impediment. One notable departure from this prototype was Dale Marcellini at Smithsonian's National Zoological Park. In a paper on the behavior of zoo visitors in a reptile building written in 1988, Marcellini and Tom Jenssen discovered that the typical visitor spends approximately eight seconds in front of each exhibit, meaning that the animals have to be located and graphics read within this timeframe. This was a disturbing finding and

Fig. 17. In 1988, Dale Marcellini and Thomas Jenssen from Smithsonian National Zoological Park found that zoo visitors spent an average of eight seconds in front of amphibian or reptile displays. Illustration in *Le Jardin des plantes . . .* by P. Bernard in 1842–1843. *Courtesy of Smithsonian Institution Libraries, Washington, DC.*

stimulated the creation of a new interactive exhibit with a series of modules on the biology of amphibians and reptiles called "Reptile Discovery Center." These modules were installed in the National Zoological Park, Zoo Atlanta and the Dallas Zoo to see whether visitor behavior varied in different regions. To determine whether the changes were effective, visitors were tracked and the length of time of their visits recorded before and after installation of the modules. When they exited, an identical set of questions about herps was asked and results compared. There was a significant difference: visitors stayed longer and learned more after the modules were in place. See Marcellini and Murphy (1998) and Doering (1994) for a description of this initiative.

Because humans do not realize that reptiles and amphibians are sometimes treated inhumanely, it is necessary to shock the zoo visitor into comprehension. At the Dallas Zoo, we developed a graphic and disturbing exhibit of the rattlesnake roundups so prevalent in the southwestern United States. The title was "Diamondbacks Can't Scream," referring to the fact that snakes do not vocalize pain. We used photographs of snakes being gassed from dens, dumped in pits, molested by handlers, beheaded by children for a fee, eviscerated and prepared as curios. I was prepared for public outrage because the images were unsettling but to my surprise, the overwhelming response was positive. Unfortunately, these roundups still exist but at least our Zoo took a stand against the carnage.

What is the future of zoo herpetology? The discipline is likely to be very different than my earlier experiences with zoo herpetologists, their buildings, animals and programs, accumulated for over 50 years. There are few new facilities being constructed exclusively for herps so new employees are neither introduced to a herpetologically rich environment nor encouraged to become specialists. The fabulous National Amphibian Conservation Center at the Detroit Zoological Institute is a dramatic exception (see Zippel and Snider, 2001; Zippel, 2002) and plans are underway for a chelonian conservation center. As new facilities are being designed in zoos throughout the world, it seems that amphibians and reptiles are added as an afterthought. Charismatic mammals such as giant pandas, cetaceans, pinnipeds, great apes, elephants, bears, large cats and other highly visible creatures are expensive to acquire and maintain. It is inconceivable that most zoo administrators would be willing to redirect the sizeable monies that even a lone panda garnishes into their herpetological programs. While William Conway (1968, 1973) made the cogent case for

Fig. 18. William Conway suggested that zoo displays depicting the life history of common bullfrog (*Rana catesbeiana*) would be excellent for enlightening the zoo visitor. Illustration from *The Riverside Natural History. Volume III. Lower Vertebrates* in 1888. *Courtesy of Smithsonian Institution Libraries, Washington, DC.*

doing everything needed in a zoo setting to enlighten and educate the visitor by using the common bullfrog as an example, the reality is that bullfrogs do not attract donors or administrators. The fact is that the bullfrog could be the perfect example to highlight the decline of amphibian populations worldwide and the danger of introducing alien species. Conway's point is profound: an animal does not have to be rare, large, expensive or endothermic to be interesting!

Many herp collections are becoming homogenized. Since captive breeding has been successful, the same species are moved from zoo to zoo. One might argue that all of the cooperative breeding programs in place in zoos do not allow for a diverse collection with many new species. The downside of collaborative programs is that I now rarely see something new during my visits and new insights with poorly known taxa are lim-

ited. There is a corollary to this collection-building approach and it is seductive. Some zoo herpetologists devote many hours to AZA-related activities, i.e., preparing studbooks and Taxon Management Accounts (TMAs), administering TAGs and so on, and believing that these activities are research. They are not! Card et al. (1998) published a critical paper which was disturbing. Fifty-two herp departments were contacted to assess involvement with AZA-sponsored programs and formalized research programs during the past decade. Problem # 1: less than half responded; Problem # 2: nearly 80% of the 164 technical papers published were contributed by three institutions; Problem # 3: over 40% of the nontechnical articles were published by one institution; Problem # 4: only four of the over 20 departments having in-house research projects had clearly defined objectives.

In the beginning, SSPs were viewed by the zoo community as the solution for addressing rapidly disappearing indigenous animal populations and most zoo managers were content to simply genetically manage their charges. Zoo directors (and AZA employees) could and did sell the idea to their constituents that meaningful conservation work was being accomplished at their institutions by becoming a refuge for threatened wildlife; the term "Ark" was often used as the symbol for the last safe place to hold these endangered species until they could be reintroduced. When potential breeding programs were planned, assessments and decisions were mostly made by using genetic data which determined relatedness to ensure preservation of the most genetic diversity, that is, preserve as many genes as possible originating from the wild. It was rightly felt that reintroductions would be more likely to be successful if unrelated animals were released. Although there were exceptions, most SSPs did not incorporate much other research on the quality of offspring produced or the likelihood of success if these animals were released.

An *in situ* component was late in coming and it is not clear to me whether pressure from the U.S. Fish and Wildlife Service (USFWS) prompted incorporation of field initiatives or the zoo community came to this conclusion independently, although I suspect the latter. To obtain some permits now, zoo personnel must demonstrate that their programs will enhance survivability in the wild by supporting research or other direct involvement with wild populations. Zoo herpetologists may have visions of being successful researchers and conservation biologists but they are sometimes inexperienced or unprepared to do projects in the field, especially in developing countries. When confronted by the painful realities of extensive habitat destruction, hopeless bureaucracies, nightmares obtaining permits and indifference, as identified in the thoughtful and distressing analysis of conservation biology by Campbell and Frost (1993:53-56), zoo staffers may choose to work at home with their collections; it is simply much easier! But to do significant work in the future with amphibians and reptiles, zoo herpetologists will be expected to broaden their horizons and leave their zoos, hopefully with the bonhomie, blessings and financial support of their superiors.

We must not tarry for the challenges today are awesome, overwhelming and radically different than those we faced 50 years ago. In the early days, amphibians and reptiles were relatively plentiful and available so our efforts were focused on building a diverse collection at a leisurely pace but this is no longer the case. Two recent examples, one *ex situ* and the other *in situ*, will suffice to accentuate this point. Thousands of Asian turtles destined for the food market were confiscated in Hong Kong and distributed to zoos and aquariums, resulting in a major drain on their resources (see Hudson, 2002). Some of my colleagues spent months evaluating chelonian health and dispensing treatments, finding suitable holding areas, arranging transportation and proper documentation, and insuring that these reptiles adapted to captivity. The second example is equally disturbing. John Behler from the Wildlife Conservation Society resurveyed tortoise populations in Madagascar and discovered that hundreds, perhaps thousands, of radiated tortoises had been killed for their livers to make pâté for the Asian food market. Dead tortoises and empty shells were scattered throughout the habitat; virtually no living individuals were located. Populations of the other tortoise species there had collapsed as well, due to human impact. In the future, the zoo and aquarium community will be asked to respond rapidly to these types of crises and we should be prepared to do so immediately. But it is not clear to me how this can be done, especially when facing situations of this magnitude. Reintroduction is not an option today in most cases so the only possibility is to accumulate captive colonies, in spite of the attendant problem of limited resources, with the faint hope that these animals can be returned to the wild in the future. How many of the four spectacular Malagasy tortoise species should be chosen for our captive colonies? Only the rarest one? The biggest one? The prettiest one? As captive managers, we will continually be faced with agonizing decisions. I am not optimistic but it is unconscionable to give up!

Today, zoo administrators are often enamored of and focused on high-tech solutions to protect threatened and endangered species but less attention is being paid to providing environments for our captive colonies which lead to successful reproduction and stable populations over time. There are many examples, mostly with mammals, of these dramatic interventions: artificial insemination, collection and cryo-preservation of gametes or the so-called "frozen zoo," embryo transfer into surrogate mothers, and "hand-rearing" abandoned young. All of these techniques require substantial funding and staff time. Yet, it must be said that these are needed because more conventional approaches have not been fully implemented, that is, providing suitable zoo enclosures and settings so that normal behaviors leading to reproduction are expressed. For example, there is a recent trend to provide an "enriched" setting for zoo animals but it is not clear to me if this "enrichment" is aimed toward the caretakers and visitors or the animals. Further, there are precious few studies in place to determine whether these "enriched" animal subjects actually modify behaviors in a positive way because few baseline data are available. Accordingly, I would hope that future research with zoo herps be directed away from these high-tech manipulations which signal failure and toward understanding the natural behavioral patterns of our captives.

As captive managers, we may be asked to become involved in projects that were unimaginable a few decades ago. Donald K. Nichols, formerly a pathologist at Smithsonian National Zoological Park, published a landmark paper in 2003 describing his experiences in trying to identify a pathogenic fungus years earlier which caused a fatal skin disease in his captive amphibians. Subsequently described as *Batrachochytrium dendrobatidis* (*Bd*) by Nichols and others, this organism has ravaged *in situ* amphibian populations throughout the world. At the Amphibian Conservation Summit during 17–19 September 2005 in Washington DC, an Amphibian Conservation Action Plan (ACAP) was generated to develop captive survival assurance colonies primarily in range countries to combat the effects of *Bd* and climate change so destructive to amphibians. Representative populations have been collected in Panama and Australia and others will be assembled, likely numbering hundreds of species, based on predictive models of threats and to maintain options for reintroduction. This global effort, called Amphibian Survival Alliance, will have multiple functions: (1) rapid-response teams to collect disappearing species; (2) short- and long-term captive management, training and capacity building for captive conservation programs in range countries; (3) research on captive breeding and reproductive science; (4) disease management, and (5) education and outreach. There is no question in my mind that zoo workers will be asked to become major players in this ambitious initiative, especially in training and developing husbandry protocols for captive colonies. Further, it is conceivable that we may be asked to produce thousands of individuals to be used for terminal studies on the etiology and control of *Bd*, a potentially ethical dilemma rarely encountered in our profession.

Another less-likely but thought-provoking example could truly stretch our imaginations as captive managers in the future. Donlan et al. (2005) proposed a controversial conservation plan to restore predators and prey that disappeared 13,000 years ago from Pleistocene North America by using extant representatives as genetic reservoirs of ancestral stock to bring about this new radiation. The impetus for doing this was based on their certitude that virtually all large mammals throughout Asia and Africa are inevitably going to become extinct during this century. They argue that it is better to have some large mammals and other non-mammalian species at risk living on Earth rather than none at all. North America offers the best option for establishing these colonies, comprising both predators and prey, on public lands and private game ranches. Their list includes mostly mammals — African lions, African cheetahs, Asian and African elephants, horses, asses, and Bactrian camels, but also the endangered Bolson tortoise (*Gopherus flavomarginatus*)—a taxon which lived across the Chihuahuan Desert in North America during the Late Pleistocene. Today, these chelonians are restricted to a limited area in northern Mexico. Could zoo herpetologists be asked to produce hundreds of Bolson tortoises, or other herps related to ancestors long gone, to seed protected tracts of land in order to support this alternative conservation strategy? This scenario seems far-fetched and unattainable but it is challenging to think about a plan for building on the past, in essence acting in a new way on the shaping and continuation of biodiversity.

The sad reality is that many zoo herpetologists are concerned but traumatized by the enormity of the problem of disappearing biodiversity. They continue to struggle to find suitable niches, both within their institutions and in the broader conservation and academic community (see Card and Murphy, 2000; Murphy and Card, 1998; Murphy and Chiszar, 1989 for examples). If we are able to do so more effectively, the final beneficiaries will be the amphibians and reptiles entrusted to our care.

Florida Softshell Turtle (*Apalone ferox*)

Chapter 3
Historical Vignettes

Common Viper (*Vipera berus*)

World's First Reptile Building

THE INTRODUCTION OF ELECTRICITY TO THE GARDENS HAS MADE POSSIBLE SOME OF THE GREATEST IMPROVEMENTS THE SOCIETY HAS CARRIED OUT IN RECENT YEARS, AND I DO NOT DOUBT BUT THAT, AS OUR SECOND CENTURY PASSES, IT WILL LEAD TO STILL GREATER ADVANCES.

PETER CHALMERS MITCHELL (1929)
DESCRIBING HOW REPTILES COULD NOW BE
PROPERLY KEPT IN A WARM ENVIRONMENT

In 1829, Edward Turner Bennett wrote an interesting book *The Tower Menagerie: Comprising the Natural History of the Animals Contained in that Establishment, with Anecdotes of Their Characters and History. Illustrated by Portraits of Each, Taken from Life, by William Harvey, and Engraved on Wood by Branston and Wright.* The London Tower Menagerie (1245-1832), with its stone walls and barred dens built specifically for large mammals, was not an ideal facility for reptiles (see Murphy, 2006). At various times over a span of nearly six hundred years according to Bennett, the Indian boa, the anaconda, the rattlesnake, and the alligator lived in the Tower. A female American croco-

Fig. 19. The Anaconda, called *Python Tigris* Var. by Bennett, was likely a Ceylonese python, *Python molurus pimbura. Courtesy of Smithsonian Institution Libraries, Washington, DC.*

dile, *Crocodylus acutus*, recorded as alive in the Gardens in 1831, and the first individual of this species to be received by the Society, may have been the Tower "Alligator" (Coote, 2001). However, the drawing by William Harvey looks like an alligator and is called

Fig. 20. Illustration of American alligator (*Alligator mississippiensis*) living in Royal Menagery, Tower of London from Robert Huish's *The Wonders of the Animal Kingdom . . .* in 1830. Alligators and crocodiles are described as follows: "Living, as it were, in the confines of both land and water, these enormous animals extend their dominion equally over the inhabitants of both elements. Here they enjoy an absolute rule, and dread none of the common dangers which assault other less powerful animals." *Courtesy of Smithsonian Institution Libraries, Washington, DC.*

Crocodilus lucius Cuv., an early name for the alligator. Bennett's specimen was said to be young, measuring not more than three feet in length, did not grow for two years, and was fed raw beef weekly.

It is curious that over 100 rattlesnakes, ranging between four to six feet in length and differing considerably in color and markings, were held in the Tower of London, especially considering that shipping reptiles across the seas must have been a lengthy proposition. These were called *Crotalus horridus* Linn. by Bennett. Coote (2001) provided a likely explanation as to how these snakes arrived in London and their eventual fate. Alfred Cops worked at the Exeter 'Change in the Strand but left this job in 1822 to take charge of the Royal Menagerie at the Tower. Rattlesnakes are somewhat delicate captives and their keeper, Cops, must have been a competent herpetoculturist for he expanded and improved the collection. Cops was an animal supplier and useful contact for visiting American showmen, including one who married his daughter in 1841. Coote speculated that this may have been the conduit for acquiring these pitvipers: "After the tower Menagerie finally closed in 1835, an Englishman from London, called Cops, is known to have arrived in the United States, complete with what was perhaps the first mobile USA reptile show. With such an unusual surname, and his previous dealings with American showmen, surely this must be the same Alfred Cops. If it was then he would certainly have taken his knowledge and husbandry techniques with him, and perhaps even the rattlesnake collection. This would also provide a more realistic opportunity for one

of his American showman friends to become his son-in-law six years later, as described above. It is interesting to speculate if it was this close contact with the Americans that led to the intriguing collection of 100 rattlesnakes at the Tower, mentioned above."

Coote (personal communication, August, 2003) now feels that the speculative evidence that Cops relocated in the USA may not be true. Coote continues "The Zoo's earliest record of receiving one is that of a Timber Rattlesnake in either 1842 or 1843. We can perhaps assume that their keeper Alfred Cops left with them in his possession." Bennett chronicled feeding behavior as follows: "It was long believed, and the notion is still popularly current, that they possessed the power of fascinating their victims, which were thought to be so completely under the influence of their glance as to precipitate themselves of their own accord into the open throat of their enemy; but the truth appears to be that they actually inspire so great a degree of terror that the animals selected for their attacks are commonly rendered incapable of offering such resistance as might otherwise be in their power, or even of attempting to escape from their pursuit."

As a testament to his skill as a caretaker, Cops nearly bred the Tower python, called Indian Boa (*Python Tigris* Daud., now *Python molurus*) by Bennett. In fact, the python laid a clutch of eggs (which did not hatch). Bennett described the event: " The individual figured at the head of the present article is a female; a fact which was proved by the remarkable circumstance of her producing in May last, after having been more than two years in the Menagerie, a cluster of eggs, fourteen or fifteen in number, none of which, however, were hatched, although the mother evinced the greatest anxiety for their preservation, coiling herself around them in the form of a cone, of which her head formed the summit, and guarding them from external injury with truly maternal solicitude."

The Anaconda, called *Python Tigris* Var. by Bennett, were likely Ceylonese pythons, *Python molurus pimbura*. Cops was nearly killed when a python, ready to shed, grabbed his hand while being fed. The snake threw two coils around his neck and Cops was in a bit of trouble until he was rescued by his assistants. Bennett recalled the incident: "His own exertions, however, aided by those of the under keepers, at length disengaged him from his perilous situation; but so determined was the attack of the snake that it could not be compelled to relinquish its hold until two of its teeth had been broken off and left in the thumb."

The Tower collection of animals was presented by the King to the Zoological Society of London in 1831

Fig. 21. Illustration of London Zoo reptile building in 1849. *Provided by Kraig Adler.*

Fig. 22. Snake charmers at the London Zoo from Scherren, H. 1905. *The Zoological Society of London.* Crocodilians occupied the three central pools and snakes and lizards were kept in glass enclosures around the walls. Occasionally, manatees were kept in these exhibits. *Courtesy of Smithsonian Institution Libraries, Washington, DC.*

(Peel, 1903:180) but this is not recorded in the Proceedings of the Society. In May 1849, the London Zoo, which had been founded in 1828, opened the first zoo reptile building, a modified carnivorous mammal facility (Keeling, 1992). The dens, now serving as reptile enclosures, could only be serviced from the front. These unwieldy glass fronts were cranked up by means of pulleys and chains, a hazardous undertaking at best when venomous snakes and speedy lizards were the inhabitants. Clearly, keepers needed to be dexterous, strong and agile. On 13 June 1862, a pair of African pythons mated and the female laid about 100 eggs. She was covered with a blanket for warmth and security but so many people lifted the cloth to see the clutch, she abandoned the lot which, unsurprisingly, did not

hatch. Read the amusing poem "Pity the sorrows of a poor Pythoness," published in Blunt (1976:204).

The second building, designed and constructed specifically for reptiles, was opened in August 1883 (Guillery, 1993) and is still functioning as a tropical bird building today. Ball (1982) presented a description of the facility. The roof was iron with skylights to admit daylight. Amphibians were kept in small cases on a large porch with a double entrance. In the center of the building, three pools were available, the largest containing crocodilians. There were ten large enclosures for constrictors and the same number of smaller ones for venomous snakes and lizards. The smallest exhibits rested on tables on each side of the entrance. The floors of the larger enclosures were constructed of slate with hot water pipes beneath for heat but this arrangement was not effective for providing stable temperatures: too hot in the summer and too cool in the winter. Services passages were behind the exhibits. In 1897, an adjunct tortoise house was built. Hot water pipes running from the reptile house were used for heating and various modifications were tried later, all designed to provide stable temperatures but these changes were only marginally successful. Seventeen years later, a small outdoor enclosure was built for temperate reptiles.

Fig. 23. Tortoise House at Zoological Society of London from H. Scherren's *The Zoological Society of London* in 1905. *Courtesy of Smithsonian Institution Libraries, Washington, DC.*

A new Reptile House, still in use, was designed by curator Joan Beauchamp Procter in 1926-1927. It included a center island to house venomous snakes and other potentially dangerous reptiles and a "Reptiliary" housing iguanas. Several new zoo design features were incorporated. Differential heating was added. The concept of "aquarium principle lighting" was used to illuminate enclosures from above and a darkened visitor space enabled exhibits to be highlighted. Elaborate naturalistic displays enhanced the experience. An improved plan to regulate traffic flow through the exhibit was featured. I loved the last quote in Blunt (1976) about the current 1927 reptile house ". . . it looks quaintly archaic with its neo-classical facade, reminiscent of some provincial railroad station, with rather bored and elegant crocodiles lounging in the pristine waiting rooms."

John Edwards (1996) published a wonderful book *London Zoo from Old Photographs 1852-1914* which includes a description and photograph of the original Reptile House, constructed at the astronomical cost of £9175. There are photographs (some pictured in this volume) of the original Reptile House interior around 1890, the tortoise house ca. 1900, Charles Island (now known as Floreana) tortoise in 1904, tuatara possibly acquired in 1870, and an Abingdon Island tortoise in 1914. Five photographs, taken roughly between 1872-1890, show stiffly seated hirsute male keepers holding a variety of snakes and lizards. The 2nd Baron Walter Rothschild, patron of the Zoo, is shown riding his giant tortoise named Rotumah in front of the Reptile House ca. 1898 (the tortoise, not the Baron, said to have died two years after arriving at the Zoo from sexual over-excitation).

Philip Lutley Sclater was Secretary for the Society (1859-1902) and beginning in 1862, compiled nine editions of an inventory called "List of the vertebrated animals now or lately living in the gardens of the Zoological Society of London," detailing the collection of vertebrates, including amphibians and reptiles. In 1872, his revised list included 14 species of tortoises such as the Galápagos, Indian, radiated, leopard, bowsprit, African spurred, three species of hingebacks, and several forms from the New World. Aquatic turtles were represented as well and included rarities like Hamilton's pond turtle, Central American river turtle, two species of roofed turtle, three species of South American side-necked turtles, alligator snapping turtle, three species of sea turtles and a number of North American emydids. Eight crocodilians were displayed including the Cuban, Nile, slender-snouted, American and mugger crocodiles. Tuataras and many lizard taxa could be found in the collection, such as the moloch, bearded dragon, mastigure (*Uromastyx*), shingleback skink and other skinks, several chameleons, as well as lacertids and iguanids. A dozen boas and pythons, and many colubrids, elapids and vipers were also represented. A nice assemblage of amphibians representing many families,

Fig. 24. Philip Lutley Sclater from *Centenary History of the Zoological Society of London* by P. Chalmers Mitchell in 1929. *Reproduced by permission of Zoological Society of London.*

including giant salamanders, rounded out the collection.

In 1902, Sclater assisted Frederick Hershel Waterhouse, the Society's librarian, in writing the Preface for a catalogue of the library: an author list of ca. 11,000 titles, and a classified list of 629 periodicals and transactions, the whole totaling about 25,000 volumes. He died in 1913 from injuries received in a carriage accident. I would have thought that more likely causes for his death would have been writer's cramp or failing eyesight.

From records between 1870-1924, Major Stanley Smyth Flower selected twenty specimens with the longest lifespan from each group of reptiles at the London Zoo and other institutions (Flower, 1925). As a rule, many reptiles fared poorly in the Society's collection but often these creatures would have arrived at the Zoo in poor condition. Chelonians, snakes and crocodilians survived for longer periods than lizards; chameleons rarely lived one year.

Do Female Pythons Produce Heat To Incubate Their Eggs?

ONE OF THE BASIC PROBLEMS HAS BEEN THAT IT IS DIFFICULT TO OBTAIN THE "ENVIRONMENTAL TEMPERATURE" AND THUS TO DETERMINE THE SOURCE OF ANY HEAT.

HERNDON G. DOWLING (1960)

Even though A. Duméril had noticed an increase in the body temperatures of snakes after feeding (Duméril, 1852), the contention that brooding female pythons produce heat to incubate their eggs was a controversial issue during the nineteenth century (Duméril, 1842, 1858; Krogh, 1916; Lamarre-Picquot, 1835, 1842; Soetbeer, 1898). In 1841, Valenciennes reported his observations on an incubating female python (called *Python bivittatus*) at the Menagerie of the Jardin des Plantes in Paris to the French Academy (Valenciennes, 1841). Controlling temperatures in the snake house at the Menagerie was primitive by today's standards, for snakes were warmed periodically by filling containers beneath the floor with hot water and

Fig. 25. Indian python (*Python molurus*) incubating eggs at Menagerie Jardin des Plantes. Illustration reproduced from S. G. Goodrich's *Johnson's Natural History Volume 2* in 1870. *Courtesy of Smithsonian Institution Libraries, Washington, DC.*

being provided with a cloth. This snake, roughly three meters in length, had been observed copulating with a smaller male several times. On 6 May, she laid 15 eggs beneath a cloth and coiled around the clutch. Prior to oviposition, she was tractable but became defensive after egg deposition. In his attempt to determine whether female pythons produce heat to incubate their eggs, Valenciennes used mercury thermometers to record ambient room temperatures, temperatures beneath the cloth, and temperatures between the coils and on the eggs. He was concerned about the accuracy of his temperature data but nonetheless, recorded a significant increase in temperature in the female at the beginning of egg-laying which gradually diminished until eight of the eggs hatched 56 days later. Although

Duméril had encouraged these experiments on python incubation by Valenciennes, he was not convinced that the results were conclusive; Duméril felt that the water containers beneath the floor, and the heat generated by the decomposing eggs and fecal matter were responsible for the temperature increase (Duméril, 1842). Duméril was convinced that the "cold-blooded nature" of reptiles could not lead to these temperature increases, and the debate raged in the scientific community in France for over two decades. The question remained open: do female pythons brood their eggs or simply guard them?

In 1862, P. L. Sclater, Secretary to the London Zoological Society, described his observations concerning an exceedingly large female African rock python (ca. 7 m) in the Zoo's reptile building. She was housed with a smaller male and breeding was observed several times during June 1861. On 13 January, she laid approximately 100 eggs; she coiled so tightly around the clutch that the eggs were virtually obscured by her coils. Sclater wished to duplicate the experiments by Valenciennes and was fortunate in having better thermometers which were more sensitive and could transmit rapid temperature readings. Since the python's enclosure was warmed with hot-water pipes beneath the floor and temperatures varied considerably, Sclater's solution was to compare temperatures taken from the body surface and folds of both snakes

Fig. 26. Pair of African rock pythons (*Python sebae*) with egg clutch at Zoological Society of London from *Cassell's Popular Natural History* around 1856. Female on bottom right is covered with blanket for warmth and is brooding egg clutch. *From the collections of the Ernst Mayr Library, Museum of Comparative Zoology, Harvard University.*

in the exhibit; the female's temperatures were higher than either the ambient cage temperature or the male's temperatures and validated the earlier observations by Valenciennes. The differences ranged between 2.8-12.4°F for surface temperatures and between 6.8-20.0°F when the thermometer was inserted between the coils. There was a disturbing variable, however, for the clutch began decomposing which likely distorted the temperatures. Nonetheless, Sclater concluded that ". . . it is the normal habit of these highly developed Ophidians, the *Pythones*, to incubate their eggs much as in the superior class of birds."

Nineteen years later, W. A. Forbes, Prosector to the Society, was able to observe brooding behavior in an Indian python (*Python molurus*) at the Zoo. This female was about 4 meters long and lived with two smaller males, one of the same species and another python (*Python bivittatus*). She laid a clutch of around 20 eggs during the nights of 5-6 June 1881 and coiled tightly around the mass. Forbes followed the same experimental model as the one reported by Sclater earlier but his thermometers were more sophisticated. His results were also suspect as part of the clutch spoiled but consistent with his predecessors: female temperatures were higher.

Frances G. Benedict, Director of the Nutrition Laboratory at the Carnegie Institution of Washington, DC, had the opportunity to observe an incubating African rock python (4.6m) at the Smithsonian National Zoological Park and make detailed environmental and body temperature readings during one day (Benedict, 1932). The large cage also contained two males but Benedict later decided to only record female temperatures as the enclosure temperatures fluctuated and the males were moving throughout the cage. On 5 April 1931, this female had laid 20 eggs in the right-hand corner of the exhibit next to the glass and coiled around the clutch. Since she was easily viewed, calm, and did not react defensively, Benedict and his associates were able to survey and record the respiration rate (to assess whether the snake was agitated by human activity), temperature of the air around the snake, temperature of the gravel, temperature between the coils, and the surface of the snake at various points (see Benedict, 1932: Figs. 28-32). His results demonstrated that the female python elevated her body temperature although Petzold (1984) felt that the snake's position against the exhibit glass cooled by the lower temperatures in the visitor space confounded the experiment. Benedict concluded his studies with the following statement: "The disposition of this particular type of snake to incubate its eggs is an astounding

zoological trait. Since in the animal kingdom the birds represent the next highest stage in the development from the reptile, and since birds invariably incubate their eggs, it is possible that the incubating python represents one 'milestone'. . . in the step from the cold-blooded or poikilotherm to the homoiotherm."

Gustav Lederer at the Frankfurt a. M. Zoo described reproduction in the reticulated python (1944) and both subspecies of the Indian python (1956). He found that the temperature of a brooding female near the substrate was several degrees higher than the temperature of the environment. However, at 32.2° C, this temperature is close to the critical minimum later established for the Indian python by Hutchison et al. (1966) to elicit brooding behavior. The environmental conditions were not suited to produce a thermoregulatory response by the snake.

Carl Stemmler-Morath from the Basel Zoo (1956) provided specific data about temperature differences between a brooding snake and its environment as well as information about the spasmodic muscle contractions seen in some species of pythons. This behavior in the literature is also described as jerks, shivering, twitching, or quivering. Stemmler-Morath studied the phenomenon but he was not yet able to recognize the major significance of the "twitches" for metabolic warmth production. He thought that the purpose of these behaviors was to aid air circulation around the clutch.

In 1981, Luttenberger observed that a brooding *Python molurus bivittatus* in the Vienna-Schönbrunn Zoo left the clutch when the temperature was raised but immediately began coiling firmly around it and to twitch when the surrounding temperature was lowered (in Petzold, 1984). The temperature in the clutch proved to be the factor guiding these behaviors (in this case the critical point was 34.1°C which was above the temperature determined by Hutchison et al. [1966] to elicit brood care).

Early researchers were plagued by several problems: lack of experimental controls, inadequate temperature recording instruments, fluctuating environmental temperatures, decomposing egg masses and insufficient data to justify conclusions. At the New York Zoological Society, Victor H. Hutchison, Herndon G. Dowling, and Allen Vinegar published two seminal papers on metabolism, energetics and thermoregulation in brooding female pythons (Hutchison et al., 1966; Vinegar et al., 1970). Their abstract in the first paper reads: "At varying environmental temperatures, measurements of body temperatures and gas exchange of a female Indian python (*Python molurus bivittatus*) show that during the brooding period this animal can regulate its body temperature by physiological means analogous to those in endotherms. Ambient temperatures below 33°C result in spasmodic contractions of the body musculature with a consequent increase in metabolism and body temperature." Their findings were later confirmed by Van Mierop and Barnard (1976, 1978) on a brooding *Python molurus bivittatus* using 6 and 12 channel-telethermometers, especially the significant correlation between temperature difference and rate of contractions.

Not all pythons demonstrate physiological thermoregulation. Petzold (1984) summarized current information as follows: "Genuine brooding behavior" (incubating with physiological thermoregulation manifested by spasmodic body muscular contractions) has been proven for the Indian python (*Python molurus,* ssp. *molurus* and *bivittatus*), blood python (*Python curtus*), amethystine python (*Python = Liasis amethystinus*) and the green tree python (*Morelia = Chondropython viridis*); "Brood care behavior" (brooding) in pythons lacking physiological thermoregulation has been documented in the reticulated python (*Python reticulatus*), African rock python (*Python sebae*), and the Children's python (*Liasis childreni*).

Answers to the questions first raised with the fascinating observations on the breeding of *Python molurus* in the Jardin des Plantes in Paris in the year 1835 have gradually been addressed over 160 years later, mostly by investigating the behavior of captive pythons held in zoos.

First Zoo Snakebite Victim?

C. H. Keeling (1992)

Edward Horatio Girling, head keeper of the snake room in 1852 at the London Zoo, was said to have a propensity for drink and not in moderation (Blunt, 1976). Prior to arriving at the Zoo, he was a guard at the Eastern Counties Railway which almost certainly was not a suitable training venue for handling venomous snakes. After consuming an impressive array of spirits with six fellow Society workers at the Albert Public House at eight o'clock in the morning on 29 October 1852, he went to the Zoo, announced that he was "inspired" and grabbed an Indian cobra a foot behind its head; it bit him on the nose. Application of a tourniquet to a nose

Fig. 27. Edward Horatio Girling, head keeper of snake room in 1852 at London Zoo, was fatally bitten by Cobra di Capello (Adder of the Hood). Illustration reproduced from S. G. Goodrich's *Johnson's Natural History. Volume 2* in 1870. *Courtesy of Smithsonian Institution Libraries, Washington, DC.*

would have been a tricky business! He was taken to University College Hospital where current remedies were tried: artificial respiration and galvanism. Not surprisingly, he expired in an hour rather than being "inspired." The incident elicited a number of suggestions from the public to *The Times* newspaper for the treatment of snakebite. Many respondents suggested liberal quantities of gin and rum but this had already been accomplished. Girling was fortunate that the old remedy of being buried in manure to the neck was not applied. Another helpful citizen stated with certitude that sleep was always fatal and bolstered this fact with a personal experience: two native Indians in the British army had dragged a screaming victim around a verandah for three-and-one-hours to successfully prevent death. The application of a white-hot iron or other fiery instrument for at least an hour was said to be effective but it is unclear whether any victim warmly embraced this treatment. The most ingenious and my favorite remedy: solicit a bite from a second snake to neutralize the effects of the first one. Could Girling have been the first zoo snakebite victim?

Feeding Living Prey to Snakes

MY RULE ABOUT NO LIVING PREY BEING GIVEN EXCEPT WITH SPE-CIAL AND DIRECT AUTHORITY IS FAITHFULLY KEPT, AND PERMIS-SION HAS TO BE GIVEN IN ONLY THE RAREST CASES, THESE GENER-ALLY OF VERY DELICATE OR NEW-BORN SNAKES WHICH ARE GIVEN NEW-BORN MICE, CREATURES STILL BLIND AND ENTIRELY UNCON-SCIOUS OF THEIR SURROUNDINGS.

PETER CHALMERS MITCHELL (1929)

The feeding of live food to snakes was an issue so inflammatory in London over a century ago that it was discussed in Parliament (Blunt, 1976). For many years, live mice, rabbits, birds, frogs and other prey had been fed to snakes at the Zoo but little public outcry ensued; in fact, the feeding demonstrations were enthusiastically viewed by the visitors. In 1869, however, the climate began to change, fueled in part by newspaper campaigns in London to elicit reader responses as to whether the practice should continue. The respondents waxed poetic and a few selected sentences in Blunt (1976) should put the controversy in proper perspective: ". . . that a rabbit should be shut up in a cage with a snake without any chance of his life, deprived of the means of escape allotted to him by nature, and subject to the exquisite torture of terror prolonged by factitious circumstances and enhanced by despair" or ". . . trembling rabbits devoured by a serpent?--a monster reptile maintained and thus feasted for the pleasure of the English--for their children . . ." The superintendent of the Zoo, A. D. Bartlett (1899:182) substituted wild brown mice for tame white mice as a concession to placate women and children visitors. Unfortunately, unless the wild mice were quickly eaten, they sometimes gnawed their way out of the enclosures. The snakes quickly followed and some lived freely for years in the old reptile house. In one instance, a large Egyptian cobra escaped through a mouse hole and was discovered the next day beneath the exhibits and was shot (Ball, 1982). Another cobra escaped into an adjacent cage where it was devoured by a "water-viper"; the latter snake was dispatched and the nearly digested cobra was recovered.

To deal with continuing public pressure, P. Chalmers Mitchell, Secretary of the Society from 1902 to 1935, and curator of mammals Reginald I. Pocock presented dead prey to snakes at the Zoo to determine whether this practice was a viable alternative.

Fig. 28. P. Chalmers Mitchell from *Centenary History of the Zoological Society of London* by P. Chalmers Mitchell. *Reproduced by permission of Zoological Society of London.*

Fig. 29. Reginald I. Pocock from *Centenary History of the Zoological Society of London* by P. Chalmers Mitchell. *Reproduced by permission of Zoological Society of London*

They documented (1907) that dead prey would be taken by many different species of snakes and would often be ingested at night. Many of their observations fit well with our current understanding of snake feeding behavior. When dead prey was used, keepers often wiggled these items to simulate live animals. I was surprised to discover that corn snakes accepted fish. A subsequent study on the acceptance of dead prey by snakes was undertaken by curator Edward George Boulenger in 1915. Blunt (1976) provided an extended account of the controversy surrounding the feeding of live food with a splendid plate of a snake, said to be a boa constrictor but more likely a puff adder, checking out a lovely white rabbit nestled in lettuce with a carrot.

Although offering dead prey is now common in zoos, some reptile buildings remain closed during feeding, due to concern about public sensitivities. When I was employed at the Dallas Zoo, we decided to allow visitors to observe the feeding behaviors of our charges. I steeled myself for controversy and waited for a steady stream of complaints from our patrons. In fact, no complaint was issued to my knowledge and the building was packed with human biomass each week during feeding.

Observations on the Fear of Snakes by Other Vertebrates

It is certainly striking that, of all the mammals, only the apes and monkeys, our nearest relatives, show a dread of snakes. Possibly the common prejudice against these exquisite creatures is a dying inheritance from our far-off ancestors.

Peter Chalmers Mitchell (1929)

In early March 1871, Charles Darwin (1809-1882) published *The Descent of Man and Selection in Relation to Sex*. In Chapter III, he mentioned that ". . . I took a stuffed and coiled-up snake into the monkey-house at the Zoological Gardens, and the excitement thus caused was one of the most curious spectacles which I ever beheld. Three species of Cercopithecus were the most alarmed; they dashed about their cages, and uttered sharp signal cries of danger, which were understood by the other monkeys." Darwin continued ". . . I then placed a live snake in a paper bag, with the mouth loosely closed, in one of the larger compartments. One of the monkeys immediately approached, cautiously opened the bag a little, peeped in, and instantly dashed away. Thus I witnessed what Brehm had described, for monkey after monkey, with head raised high and turned on one side, could not resist taking a momentary peep into the upright bag, at the dreadful object lying quietly at the bottom."

P. Chalmers Mitchell and R. I. Pocock (1907) were interested in detailing the fear of snakes by other animals; they observed responses to snakes by many mammals and birds at the London Zoo and elsewhere.

Two major conclusions were drawn by Mitchell and Pocock. The first was ". . . there is no such thing as a power of fascination possessed by snakes." Smaller birds and mammals are as inquisitive to a slowly advancing snake as to a slowly moving human hand directed toward them. Sudden or noisy movements elicit a flight response.

The second point was that ". . . except in one group, animals have no specific fear of serpents. The vast majority of animals of course frogs, rats and mice, guinea-pigs, rabbits, ruminants, and birds, are totally indifferent to their presence, and even when a snake approaches them directly avoid it, just as they would a stick thrust at them." The excepted group was non-human primates. Corn snakes, a tree boa and small reticulated python were presented to monkeys in the Zoo's collection. Most of the Old and New World monkeys showed varying degrees of threat or escape behaviors, although a few seemed unaffected by the presence of a snake. Baboons, including the Mandrill, ". . . were even more panic-stricken, jumping back in the greatest excitement, climbing as far out of reach as possible and barking." Gibbons were less fearful, which was attributed to their arboreal habits. Chimpanzees were frightened by snakes, except for one baby. These apes fled screaming to the highest reaches of their cages. One chimp was infested with large nematodes which were passed during defecation; when viewed by other chimps, these parasites elicited similar fear responses. Even earthworms caused consternation, perhaps because these invertebrates were cylindrically shaped. Although orangutans were not vocal in the presence of snakes, they slowly moved as far away from them as possible. Mitchell (1929) noticed one exception when a young orang, removed from its mother, behaved differently when presented with a large tame snake: "The orang was not in the least alarmed, but took the snake as a new and attractive toy, playing with it, allowing it to coil over him, and treating it with such friendly roughness that I had to remove it." Lemurs were curious when presented with snakes and advanced toward the reptiles, perhaps to feed upon them. In 1917, a graphic exhibit developed by Mitchell again demonstrated that only some higher primates and a few birds showed fear of snakes (1917–Proc. Zool. Soc. London: 214).

While on holiday at Leith Hall in August 1872, Darwin finished the proofs for his book *The Expressions of the Emotions in Man and Animals*. In Chapter V, he compared the different responses by monkeys to a turtle and snake: "A living fresh-water turtle was placed at my request in the same compartment in the

Zoological Gardens with many monkeys; and they showed unbounded astonishment, as well as some fear. This was displayed by their remaining motionless, staring intently with widely opened eyes, their eyebrows being often moved up and down. They occasionally raised themselves on their hind-legs to get a better view. They often retreated a few feet, and then turning their heads over one shoulder, again stared intently. It was curious to observe how much less afraid they were of the turtle than of a living snake which I had formerly placed in their compartment." When I was at the Dallas Zoo, my colleague, the late Jack Joy hid a newly hatched ornate box turtle in his hand and presented it to an adult male western lowland gorilla. When Jack opened his fist, the gorilla shrieked, jumped backwards and ran to the far end of the enclosure.

Siebold's Giant Salamanders

ALL ATTEMPTS TO MAKE CRYPTOBRANCHUS BREED IN CAPTIVITY HAVE FAILED HITHERTO, OWING NO DOUBT TO THE DIFFICULTY OF OBTAINING IN THE CITY COOL WATER SUCH AS THE ANIMAL IS ACCUSTOMED TO IN ITS MOUNTAIN HOME.

CHUJIRO SASAKI, PROFESSOR IN THE AGRICULTURAL DENDROLOGICAL COLLEGE, TOKYO, WHO COLLECTED 71 SALAMANDERS IN THE SUMMER OF 1880 AND 1881 IN THE INTERIOR OF JAPAN (1887)

A fossil giant salamander had been found in Switzerland in 1726 but Europeans believed the species was extinct. In the 6th year of Bunsei Era (1823) which was in the late Edo Period, Phillipp Franz von Siebold

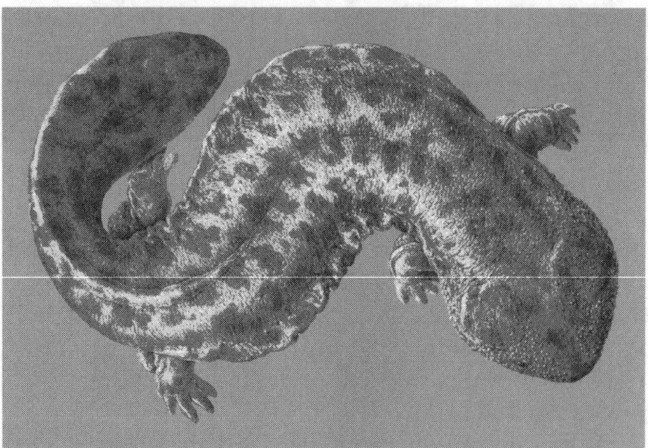

Fig. 30. Japanese giant salamander (*Andrias japonicus*) from a lithograph drawn from life by the Dutch natural history artist, A. Saagmans Mulder (from Philipp Franz de Siebold's *Fauna Japonica,*" *Reptilia part 3, Leiden,* 1838). *Provided by Kraig Adler.*

came to the then closed Japan as a Dutch medical doctor (he was in fact German) and studied Japanese fauna and flora, resulting in his monumental books *Fauna Japonica* and *Flora Japonica*. Siebold was one of only a few foreigners allowed to travel in Japan during the Edo Period. Aside from three months each year when he was permitted to travel under tight security to Edo on "the Captains' trip to greet Shogun" to pay his respects to this Supreme Commander (Shogun was a feudal ruler with absolute power), he stayed within the island of Dejima near Nagasaki. He had special permission to tend to the sick and collect herbal plants. In addition, he taught western medicine to Japanese students at the newly established Narutaki-Gakusha (Falling Water School) in Nagasaki. In 1826, Siebold and his students collected many samples of plants and animals. Among the specimens was a Japanese giant salamander acquired in Suzuka in current Mie Prefecture (there is a photograph of this preserved individual with ruler in Takeyoshi Tochimoto's article written in 2000). This individual was recorded to be 30 cm in length.

Three years later, Siebold left Japan with two live salamanders and arrived in Holland the next year. Kobara (1985) traced Siebold's travel and estimated that he kept one live specimen for four years, three months and 27 days under his care, from acquisition to the arrival at the museum. Siebold acquired the salamanders on 27 March 1826 on the road, and continued on to Edo (now Tokyo), arriving there on 10 April. He began the trip back to Nagasaki on 18 May and got there on 7 July, which indicates that the salamanders were in transit for more than three months. He left Nagasaki for the Netherlands on 29 or 30 December 1829 via Jakarta, departing Jakarta on 5 March 1830 after a month stay. On 7 July he was back in the Netherlands, and the giant salamander arrived at the museum in Leiden on 23 July. This animal died at the Amsterdam Zoo on 3 June 1881. During the voyage, one salamander was bitten to death by the other (picture of this preserved animal with ruler as well). Takeyoshi Tochimoto estimated this individual to be over 70 cm long. The surviving specimen, perhaps bigger than the other, lived for over 51 years at the Amsterdam Zoo. Unfortunately, this specimen has been lost. One interesting aspect of Tochimoto's paper is his investigation to estimate overall longevity in this captive salamander, which could be over 100 years and over 120 cm in length. The nearest rival lived 47 years at the Muroran Aquarium.

It is remarkable that this so-called delicate animal, often intolerant of captive conditions, survived

Fig. 31. Male Japanese giant salamander (*Andrias japonicus*) guards eggs called "Rosenkranzschnüre" or "rose wreath strings" at Natura Artis Magistra. Undated photograph but probably taken around 1902. Reproduction was recorded for the first time by Kerbert (1904) and occurred several times at Artis. *Provided by Eugène Bruins, Natura Artis Magistra Archives.*

the conditions of the feudal era, through many long trips, in primitive accommodations and additionally, established the longevity record for the species. Frequently, giant salamanders are found in Tokyo, sometimes in the sewer system. (Komori Atsushi, "Giant salamanders caught in Tokyo," pp. 202-203, *Animals and Zoos* 22/6, June 1970, in Japanese) reported numerous cases of giant salamanders captured in Tokyo, often in the most polluted aquatic environment imaginable, including three healthy animals captured in one sewage treatment plant between 1960 and 1965. Perhaps one of the common Japanese names for the taxon, *hanzaki*, may be perceptive; the term literally means "torn in half," which hints that the animal is so tough that it keeps living after being split in half.

Three biologists did their PhD theses on this species: L. P. de Bussy, 1904, 112 p., 10 pls.; Daniel de Lange, 1905, 196 p., 4 pls. (de Lange also wrote another book, 1916, 149 p., 8 pls.); P. J. de Rooj (or Rooy), 1906 and 1908, two short papers. In Japan, research on this taxon in earlier years focused on embryology. The Japanese list this salamander as a National Special Natural Treasure.

Herpetological Detective Story

THE RARITY OF THIS HORNED LIZARD MIGHT BE ACCOUNTED FOR IN CONSIDERATION OF ITS RELATIVELY SMOOTH SKIN AND ABSENCE OF HEAD SPINES RENDERING IT AN EASY PREY FOR VARIOUS SNAKES.

RAYMOND L. DITMARS IN *THE REPTILES OF NORTH AMERICA* (1951)

Any sleuth would find this story compelling. In 1890 and 1897, three specimens of Ditmars' horned lizard (*Phrynosoma ditmarsi*) were collected somewhere near the United States-Mexican border but precise locality data were missing. The first, collected in 1890, was deposited in the collection of the American Museum of Natural History in New York. The second and third specimens were donated to Raymond L. Ditmars, who became Curator of Reptiles at the Bronx Zoo on 7 July 1899. Ditmars was able to observe one of these lizards for about a year until it died (Ditmars, 1951). When the saurians died, they were preserved and sent to Leonhard Stejneger at the United States National Museum. Stejneger described the lizard in honor of Ditmars (Stejneger, 1906). Because specific locality data were missing no additional specimens were found alive until 1970, even though herpetologists had searched for additional examples. In an ingenious attempt to pinpoint the lizards' location, two of the preserved lizards' stomachs were examined and insect remnants, plant fragments and small pebbles were recovered. These materials were sent to specialists for identification and a possible locale was postulated, based on known distributions of these animal, vegetable and mineral clues. With this additional information, Ditmars' horned lizard was rediscovered. To pursue this fascinating story, consult Ditmars (1951), Roth (1971, 1997), Lowe et al. (1971), Montanucci (1989) and Sherbrooke et al. (1998).

How to Buy a Zoo!

. . . ZOOS BECAME HIGHLY COMPETITIVE, EACH STRIVING TO EXHIBIT THE LARGEST NUMBER OR THE MOST EXOTIC SPECIES.

THOMAS VELTRE (1996)

One of the most extraordinary zoo documents ever generated was reproduced in the book *Wild Cargo* by Frank Buck and Edward Anthony (1932). The City of Dallas ordered an entire zoo in a letter to Buck, the "Bring-Em-Back-Alive" animal collector on 4 May 1922. Some of the most notable species included tigers, clouded leopards, spotted and black leopards, male elephant, tapirs, hyenas, gibbons, orangutans, zebra stallion, kangaroos, wallabies, cassowaries, pheasants, storks, hornbills, lories, dozens of smaller birds, king cobras, black cobras, 20 and 15 foot pythons, and 6

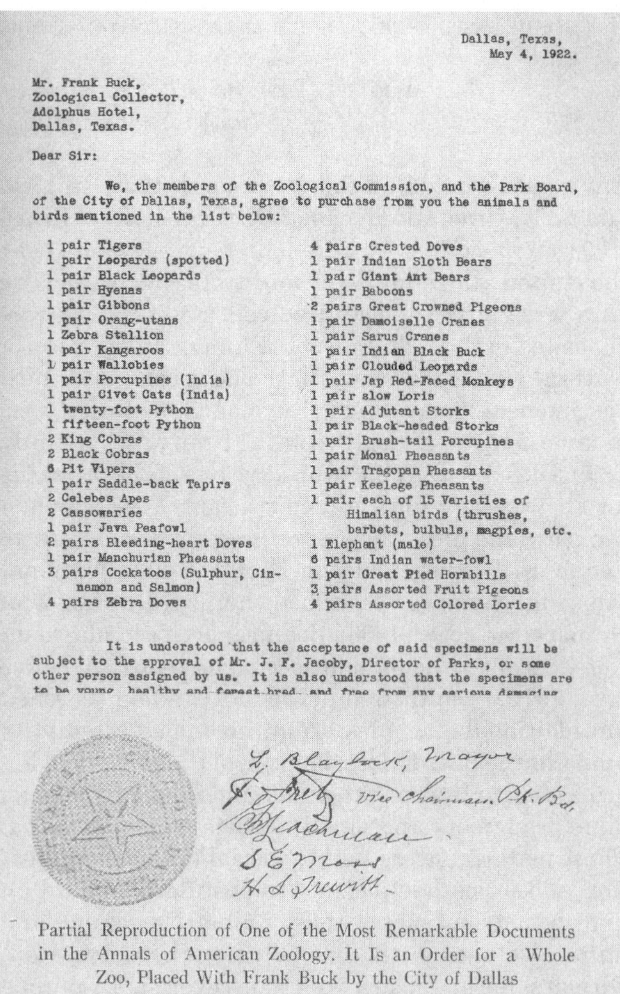

Dallas, Texas,
May 4, 1922.

Mr. Frank Buck,
Zoological Collector,
Adolphus Hotel,
Dallas, Texas.

Dear Sir:

We, the members of the Zoological Commission, and the Park Board, of the City of Dallas, Texas, agree to purchase from you the animals and birds mentioned in the list below:

1 pair Tigers	4 pairs Crested Doves
1 pair Leopards (spotted)	1 pair Indian Sloth Bears
1 pair Black Leopards	1 pair Giant Ant Bears
1 pair Hyenas	1 pair Baboons
1 pair Gibbons	2 pairs Great Crowned Pigeons
1 pair Orang-utans	1 pair Damoiselle Cranes
1 Zebra Stallion	1 pair Sarus Cranes
1 pair Kangaroos	1 pair Indian Black Buck
1 pair Wallobies	1 pair Clouded Leopards
1 pair Porcupines (India)	1 pair Jap Red-Faced Monkeys
1 pair Civet Cats (India)	1 pair slow Loris
1 twenty-foot Python	1 pair Adjutant Storks
1 fifteen-foot Python	1 pair Black-headed Storks
2 King Cobras	1 pair Brush-tail Porcupines
2 Black Cobras	1 pair Monal Pheasants
6 Pit Vipers	1 pair Tragopan Pheasants
1 pair Saddle-back Tapirs	1 pair Keelege Pheasants
2 Celebes Apes	1 pair each of 15 Varieties of
2 Cassowaries	Himalian birds (thrushes,
1 pair Java Peafowl	barbets, bulbuls, magpies, etc.
2 pairs Bleeding-heart Doves	1 Elephant (male)
1 pair Manchurian Pheasants	6 pairs Indian water-fowl
3 pairs Cockatoos (Sulphur, Cin-	1 pair Great Pied Hornbills
namon and Salmon)	3 pairs Assorted Fruit Pigeons
4 pairs Zebra Doves	4 pairs Assorted Colored Lories

It is understood that the acceptance of said specimens will be subject to the approval of Mr. J. F. Jacoby, Director of Parks, or some other person assigned by us. It is also understood that the specimens are to be young, healthy and forest-bred, and free from any serious damaging

L. Blaylock, Mayor
J. Fretz, Vice Chairman Pk. Bd.
...
S E Moss
H L Trewitt

Partial Reproduction of One of the Most Remarkable Documents in the Annals of American Zoology. It Is an Order for a Whole Zoo, Placed With Frank Buck by the City of Dallas

Fig. 32. Letter from City of Dallas to Frank Buck purchasing animals for Dallas Zoo. *Reproduced from* Wild Cargo *by Frank Buck and Edward Anthony (1932) by permission of Barbara Buck Larick.*

pitvipers which totaled over 70 species and 180 individuals. The City insisted that the specimens must be in good health and free from disease. Quarantine would have certainly been interesting, especially in those early days.

Tortoises in Trouble

BUT THE SPREAD OF MANKIND OVER THE FACE OF THE GLOBE, MORE THAN ANY OTHER FACTOR, HAS PLUNGED THE GIANTS INTO IGNOMINIOUS RETREAT. THE OPPORTUNISTIC, INGENIOUS NEW PREDATOR, LITERALLY AS OMNIVOROUS AS A PIG AND MUCH MORE DANGEROUS, PROVED TOO MUCH FOR TORTOISES BEYOND A CERTAIN SIZE.

PETER C. H. PRITCHARD (1996)

The harrowing events that follow are some of the most depressing papers ever published by a zoo worker. Between 1831-1868, fleets of whalers traveled to the Galápagos Islands to procure tortoises for food and oil. Charles Haskins Townsend, Director of the New York Aquarium, scrutinized 79 logbooks and discovered that 151 ships made 189 visits. From these data, he generated a conservative estimate of 13,013 tortoises removed (Townsend, 1925). The last paragraph of his paper summarizes his pessimism: "What a contribution could be made to the world's food supply if the otherwise unimportant islands where, unknown to primitive man, the tortoises reached such an amazing development, could be cleared of the pests introduced by civilized man and the original conditions restored!

Fig. 33. Galápagos tortoises (called *Testudo elephantopus*). Illustration from *The Riverside Natural History. Volume III. Lower Vertebrates* in 1888. *Courtesy of Smithsonian Institution Libraries, Washington, DC.*

Fig. 34. Photograph of Galápagos tortoise being measured at Bronx Zoo (Wildlife Conservation Society) in 1946. *Photograph courtesy of Wildlife Conservation Society, headquartered at Bronx Zoo.*

Fig. 35. Photograph of Galápagos tortoise accepting handout at Bronx Zoo (Wildlife Conservation Society) in 1906. *Photograph courtesy of Wildlife Conservation Society, headquartered at Bronx Zoo.*

This is now unfortunately impossible on the Galapagos. The only remaining hope for the race is the establishment of survivors elsewhere."

In 1928, he was so concerned about potential extinction that he mounted an expedition to collect over 200 tortoises from Albemarle Island which he distributed to zoos to develop captive breeding programs. The next year, he wrote a paper on the extinct Galápagos tortoise that inhabited Charles Island, estimated to have disappeared in 1850 (Broom, 1929). In 1931 and 1937, he published two papers on growth and age in the zoo tortoises, based on data from eight institutions in Florida, Texas, California, Louisiana, Hawaii, Arizona and Sydney, Australia. His other articles on tortoises, description of his life and scientific contributions, and a list of herpetological taxa named in his honor has been published (Grant, 1947). See Pritchard (1996:23) for additional information.

Why Galápagos Land Iguanas From Baltra Island Are Not Extinct

FROM UNPUBLISHED DIARIES OF C. B. "SI" PERKINS WRITTEN DURING HANCOCK EXPEDITION VOYAGE ON VALERO III TO COLLECT ANIMALS FOR THE SAN DIEGO ZOO IN 1933 (*IN* CAMPBELL, 1978)

C. B. "Si" Perkins was the Curator of Reptiles at the San Diego Zoo when a local businessman, G. Allan Hancock, mounted two trips in 1932 and 1933 to the Galápagos Islands to explore the region, as well as study and collect animals for the Zoo (see Banning, 1933 for additional information). His ship, the stately and luxurious Valero III nearly 200 feet in length, was aptly suited for the voyage but Hancock insisted that his fellow passengers, a group of scientists including Perkins, dress for dinner, and more disturbingly, cease

Fig. 36. Galápagos land iguana (*Conolophus subcristatus*) on exhibit at Smithsonian National Zoological Park in Spring 1934. See text for description of interesting intervention by C. B. "Si" Perkins from San Diego Zoo which saved this species found on Baltra Island. *Photograph by E. Hardy, National Zoological Park Photo Archives.*

Fig. 37. C. B. "Si" Perkins and G. Allan Hancock from San Diego Zoo traveled to Galápagos Islands on *Valero III* to collect and translocate land iguanas (*Conolophus subcristatus*) in the early 1930s. *San Diego Zoological Society Archives.*

smoking and drinking. For Perkins, who certainly was not focused on the latest in apparel and, moreover, enjoyed a bit of whisky and strong cigarettes in the evening, the trip was bittersweet. The notation "No cocktails tonight." appeared daily in his diaries (Campbell, 1978).

While on the islands during the first trip, Hancock and Perkins decided to move 40 Galápagos land iguanas (*Conolophus subcristatus*) from Baltra Island, also known as South Seymour Island, to North Seymour Island which contained no iguanas. As Perkins explained in his diaries, their reason: ". . . in a few years come down and see if anything has happened. A good idea, I believe." (Campbell, 1978). On the next trip, they translocated 20 more lizards.

During World War II in the 1940s, Baltra was an American airbase and several thousand military and support personnel were stationed there, in part to guard the Panama Canal. Through a combination of habitat destruction, introduction of feral animals, and direct killing by humans, the Baltra lizards disappeared.

During the subsequent 47 years, there appeared to be virtually no successful reproduction or recruitment on North Seymour so a pair of adults was brought to the Charles Darwin Research Center on Santa Cruz Island to begin a captive colony; additional iguanas were included in this potential breeding group later (Cayot et al., 1994). Fortunately, these iguanas proved to be adaptable captives and began to build nests and lay eggs. Populations of feral cats and dogs were reduced on Baltra. In June 1991, 35 five-year old iguanas were repatriated to Baltra and 24 were released

the next year. This head-starting program was truly an accomplishment deserving praise.

Ironically, if these two intrepid explorers had not done what is now considered to be an unacceptable practice, and one never to be done by conservation biologists without careful studies beforehand, the land iguana population on Baltra would be extinct. Fortunately for Perkins, Hancock relented on the second trip and casual attire, cigarettes and spirits were allowed.

Playful Dragons

AFTER WATCHING THESE GREAT CARNIVORES IN THE WILDERNESS OF ROMANTIC KOMODO, IT WAS PAINFUL TO SEE THE BROKEN-SPIRITED BEASTS THAT BARELY HAD STRENGTH TO DRAG THEMSELVES FROM ONE END OF THEIR CAGE TO THE OTHER. SURELY, IT IS NOT ALL A MATTER OF DIET AND A CHANGE OF CLIMATE. PERHAPS, AS IN THE CASE OF MANY MAMMALS, *VARANUS KOMODOENSIS*, IN ORDER TO SURVIVE, DEMANDS THE FREEDOM OF HIS RUGGED MOUNTAINS.

AN UNFAIR AND UNFLATTERING DESCRIPTION BY
W. DOUGLAS BURDEN (1927) OF DRAGONS IN ZOOS

ON THE OTHER HAND, THE BETTER CONDITIONS FOR THE REPTILES THEMSELVES AND THE FEEDING AND HANDLING OF THEM HAS MADE SOME UNEXPECTED CREATURES DOCILE. NOTABLE INSTANCES ARE THE KOMODO DRAGONS, WHICH ARE AS TAME AS DOGS AND EVEN SEEM TO SHOW AFFECTION.

PETER CHALMERS MITCHELL (1929)

Fig. 38. Komodo dragon (*Varanus komodoensis*) exhibit at Bronx Zoo (Wildlife Conservation Society). Undated photograph but possibly in the 1930s. *Photograph courtesy of Wildlife Conservation Society, headquartered at Bronx Zoo.*

Fig. 39. Komodo dragon (*Varanus komodoensis*) laid eggs in Natura Artis Magistra in 1931. *Provided by Eugène Bruins, Natura Artis Magistra Archives.*

Described scientifically in 1912, Komodo dragons were first placed on exhibit in the New York and Amsterdam Zoos 14 years later. In 1934, a dragon was on display at the Smithsonian National Zoological Park after capture by the Griswold-Harkness expedition. It lived only two years. Between 1934 and 1975, a total of five dragons were exhibited at Smithsonian National Zoological Park, the average life span being five years and the maximum being 12 years. Dragons at other North American and European zoos generally fared poorly as well, perhaps because often the largest specimens were captured (see Jones, 1965 for list). Adult animals often have difficulty adjusting to captivity. Dragons rarely lived beyond five years, and most did not survive the first few months of captivity, although dragons in Europe were kept more successfully than those in the United States. Collecting expeditions to Komodo, and the consequent expense, continued despite high mortality, because the public was excited about viewing these huge, carnivorous lizards. As an example, Lady Broughton (1936) described a trip led by Lord Moyne to secure specimens for display at the London Zoo. World War II brought collecting to an end for many years. As a result, dragons in captivity have not been thoroughly investigated until recent decades when longevity, breeding and rearing improved dramatically.

Monitor lizards are considered to be quite intelligent among lizards (Loop, 1976; Burghardt, 1977). Anecdotal records exist for play in dragons. Hill (1946) mentioned a dragon at the London Zoo pushing a shovel over the stones in his cage ". . . and the more noise he can make with it, the more it seems to please him." Procter (1928) discussed tractability and tameness of adult monitors which included a photograph of

a two-year old child standing next to Sumbawa, a large dragon at the London Zoo. Procter decided to test the ferocity of this lizard at a gathering of zoo supporters: "The dragon, whose name is Sumbawa, walked around a very long table, and without paying attention to the audience ate a large fowl, several eggs, and a pigeon from her hand, allowing itself to be scratched and patted even when swallowing the fowl with enormous gulps, treatment which even dogs will not always permit." Sumbawa responded to the voice of its keeper or curator, but disliked having its tympanum touched. There is a photograph of Keeper Arthur Budd with Sumbawa in 1928 (Zuckerman, 1976). At the Frankfurt a. Main Zoo, a tame Komodo dragon was allowed to follow the keeper through the visitor area (see Lederer, 1942). One of my zoo associates sent an article to me with an old photograph of a large dragon crawling out of its exhibit into the visitor area at the Berlin Aquarium; two persons are standing next to the lizard.

At the Smithsonian National Zoological Park in 1995, biologist Trooper Walsh and keeper Charles Coutris from the herp department asked if I wanted to see a dragon "play"; I was skeptical as animal play is a controversial issue. They said that a young female

Fig. 40. Keeper Roy Jennier next to "Xomo," the first Komodo dragon (*Varanus komodoensis*) at Smithsonian National Zoological Park. This lizard was collected by Griswold-Harkness expedition in 1934, cost $780, and lived two years. Hot water pipes in rockwork provided heat. *National Zoological Park Photo Archives.*

approximately seven feet in length named Kraken, hatched at the Zoo, usually exhibited playlike behavior, such as removing a handkerchief or notebook from a keeper's pocket, scraping the keeper's shoes with her forearm, playing tug of war with a soda can, interacting with empty cardboard boxes, as well as pieces of cloth and scarves. She stood on her hind legs, directed tongue flicks to a keeper's face, rested her head on a keeper's shoulder, and closed her eyes. Kraken carried Frisbees, shoes, action figures for children such as the "Terminator," and other objects around in her mouth but made no attempt to swallow them. She stuck her head into a plastic bucket, raised her anterior trunk so that the container covered her head and walked around the exhibit. She placed her snout inside a shoe, lifted it off the substrate and moved throughout the cage. When a keeper whistled, Kraken turned her head toward the source of the sound. She could discriminate between prey and non-prey; she would gently take a rat offered with tongs and never showed an inclination to bite her caretaker. In fact, Kraken would even "beg" for pizza and never acted aggressively when the pizza slices were offered to her.

After viewing these extraordinary episodes over several days, I called Gordon Burghardt from the University of Tennessee and David Chiszar from the University of Colorado and invited them to join in a study to document these behaviors (see Burghardt et al., 2002 for an initial behavioral inventory and quantitative analysis, definition of animal play and examples described for reptiles). For over 50 years, I have worked with living reptiles and have never seen such complex behavior from a so-called "lower animal;" those days at the Zoo were some of the most exciting in my career!

"Dudley-Duplex" and "Nip-and-Tuck"

From the days of Geoffroy-St.-Hilaire, types of reptilian duplication have been designated by a variety of terms based largely upon the nomenclature used for human monsters.

Bert Cunningham (1937)

Between 1955 and 1974, the San Diego Zoo received several young two-headed California kingsnakes (Shaw and Campbell, 1974). It was unlikely that these snakes, named Dudley-Duplex I, Dudley-Duplex II, and Nip-and-Tuck, were related as they were collected in widely separated localities in San Diego County. The first problem faced by these snakes is emerging from the egg. After slitting the shell with the egg tooth, which head

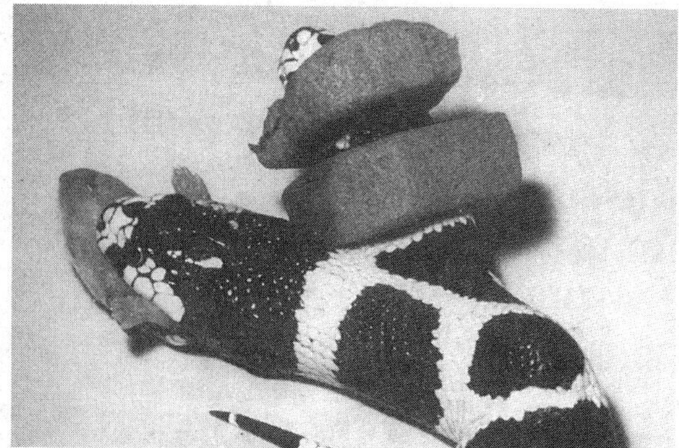

Fig. 41. Two-headed California kingsnake (*Lampropeltis getula californiae*) at San Diego Zoo. *Photograph by Donal Boyer.*

emerges first and do the heads work at cross-purposes? The second problem is locomotion and the snakes at the Zoo often had problems moving, as each head moved in a different direction. The final challenge is feeding as both heads may try to swallow the same prey item. Shaw and Campbell told the story as follows: "Dudley, the right head of Dudley Duplex, the first two-headed snake at the San Diego Zoo, almost always swallowed the mouse even though Duplex, the left head, sometimes got hold of it. In 146 feedings, Dudley ate 139 times, Duplex only 7. Fortunately for Duplex the food wound up in their joint stomach; but if snakes savor their food, Duplex missed a lot of gustatory pleasure." This snake lived at the Zoo for six and one-half years.

In 1995, the Zoo received another dicephalic kingsnake approximately three months old. Both heads attacked mouse prey but often one head struck the other one. The injuries became so severe, unsupervised feeding was discontinued. Curator Donal Boyer and Senior Keeper Brett Baldwin (1997) devised a simple solution to prevent future injury. A foam rubber collar, flattened on the bottom, was slipped over one head during feeding. Placement of the collar was alternated between the heads during feedings and in this way, the concern expressed by Shaw and Campbell about the lack of opportunity by one head to savor a mouse was ameliorated.

Death of a Significant Herpetological Collection: Belle Vue Zoo (1836-1977)

The closure of the zoological gardens involved thousands of animals and hundreds of people, and may one day be the subject of a book of its own. The writer of that book

CLIVE BENNETT AND DAVID BARNABY (1989)

The tragic story of the demise of the Belle Vue Zoological Gardens in Manchester, England, established in the 1830s, is saddening as the operation was important in the history of zoo herpetology (see Keeling, 1983; Murphy 2005). The Zoo was not a conventional zoological garden but was part of an amusement park (Reichenbach, 2002). The decline parallels the history of many inner-city institutions: shrinking financial support and visitation, with increasing vandalism. As an example, windows in the glass roof were replaced five times within two months. The number of species over one 20-year span was extraordinary: 6 salamanders, 15 anurans, a caecilian, 8 crocodilians, 41 chelonians, 94 lizards and 106 snakes. Some notables included Chinese giant salamander, marine and rhinoceros iguanas, false gavial, African dwarf and slender-snouted crocodiles, Galápagos and Aldabran giant tortoises and many unusual snakes. During these exciting days, Matt Kelley was Head Keeper (his brother retired as Curator of Reptiles at the Dublin Zoo). Albert Craythorn followed his father James as Curator of Reptiles. The complete story is chronicled in the fascinating book by Clive Bennett and David Barnaby, published in 1989: *The Reptiles of Belle Vue 1950~1977. A Curator's Viewpoint.* The senior author kept meticulous records and much of the material for this book was derived from his observations. The photographs and floor plans demonstrate how advanced this operation was during this period. Gerald Durrell was employed as a temporary keeper in the Aquarium and traveled to South America to procure reptiles for the Zoo (see his 1964 book *Three Singles to Adventure*). The staff focused on creating naturalistic exhibits with living plants, and American alligators bred in their enclosure for the first time in a European zoo.

There are fascinating historical documents concerning this Zoo: average longevity for 24 chameleons between the years 1898-1901 was slightly over three months and none survived for one year (Flower, 1925), description of the reproduction of a pair of Seba pythons (*Python sebae*) and exhibition of fertilized eggs by the Secretary of the London Zoological Society (Jennison, 1909), a letter by George Jennison to the Secretary of the London Zoological Society describing reproduction and behavior of a pair of pine snakes (Jennison, 1910), and studies on the mating behavior of American alligators (Legge, 1969). When the Zoo closed, the last director Gerald Iles moved to Canada.

World's Rarest Tortoise

LEE DURRELL, BRIAN GROOMBRIDGE, SIMON TONGE, AND QUENTIN BLOXAM (1989)

The future of the tortoises in Madagascar is tenuous at best, but a few committed zoo workers struggle against formidable odds to conserve them. Don Reid has been directing the Durrell Wildlife Conservation Trust's (DWCT) ploughshare (= Angonoka) tortoise (*Geochelone yniphora*) breeding and long-term monitoring project, based at the Ampijoroa Forestry Station for many years and resulting in a number of significant papers (Bourou et al., 2001; Kuchling and Razandrimamilafiniarivo, 1999; Pedrono et al., 2001; Reid et al., 1989, Reid, 1995; Smith et al., 1999). Mallinson (1998) described the collaboration for tortoise conservation between the DWCT and Madagascar which included preparation of colored educational posters for distribution. Curl et al. (1985) investigated current status and distribution, as a component of the recovery program supported in part by DWCT. An account covering the history, biology, conservation partnerships and current status is available (Durbin et al., 1996; Durrell et al., 1989).

During May 1991, lizard ecologist Gary Ferguson and I traveled to Madagascar to study the natural history and social behavior of the panther chameleon (see Ferguson et al., 2004). We met Don Reid in Antananarivo and he described his tortoise project and showed us photographs of the facility over dinner. Gary and I were impressed with his enthusiasm and focus but it was clear that an extended recovery effort was going to be a major challenge.

Young ploughshare tortoises from this captive breeding program in Madagascar have been successfully released (Pedrono and Sarovy, 1998, 2000). In May 1996, this model facility was raided by unscrupulous collectors and 2 adult females and 74 juveniles were stolen. It is disgusting that uncontrolled lust for animals and profit can place this entire endangered group of tortoises at risk and negate the years of work and commitment by the DWCT staff and other conservation partners.

Antiguan Racer Races Toward Extinction?

IRONICALLY, NO OTHER WEST INDIAN SNAKES HAVE HAD THEIR NUMBERS REDUCED, BEEN EXTIRPATED, OR BECOME EXTINCT AS OFTEN AS MEMBERS OF THIS ENIGMATIC GENUS.

ROBERT W. HENDERSON AND
RICHARD A. SAJDAK (1996)

Historically distributed across Antigua and many of the offshore islands, the Antiguan racer (*Alsophis antiguae*) is now confined to Great Bird Island which lies to the north-east of Antigua in the Lesser Antilles (Daltry et al., 2001). Perhaps the rarest snake in the world, the total population in November 1995 was estimated to be only 50-70 individuals, based on capture/mark/recapture studies, and the low numbers were likely caused by introduced rats. There appeared to be no neonate or juvenile recruitment and the sex ratio was heavily skewed toward females. The racer is listed as "Critically Endangered" in the 2000 World Conservation Union (IUCN) *Red Data Book*. Representing less than 1% of the snake's historical range and small in size (0.08km^2), Great Bird Island, a coralline limestone island, can only support approximately 100 snakes. There are significant threats to this small population: invasive predators, killing by humans who visit the island in large numbers, inbreeding depression and/or genetic drift, fluctuating prey numbers, and hurricanes. Without intensive intervention, the future of this ophidian is indeed bleak.

In an effort to stem this downward spiral, the Antiguan Racer Conservation Project (ARCP), a partnership of six national and international organizations, was developed in 1995. The purpose was multidimensional: encourage biological research (feeding ecology, population size and sex ratio, predator-prey relationship, habitat use and activity patterns), promote public education and technology transfer, develop captive colonies for breeding purposes, and initiate ecological restoration. Field work was carried out in October to December 1995 by Fauna and Flora International (FFI) in conjunction with the Environmental Awareness Group of Antigua & Barbuda (EAG), the Forestry Unit (Antigua Ministry of Agriculture) and the Island Resources Foundation (IRF). In December 1995, rats were eradicated from Great Bird Island.

Two male and three female racers were imported to the Jersey's Durrell Wildlife Conservation Trust in February 1996 to establish a captive population. They were kept in large, naturalistic enclosures off-exhibit in the Gaherty Reptile Breeding Centre (see Gibson, 1997 for details). Copulation occurred in July and August 1996, after a simulated rainy season. Temperature, humidity and rainfall regimens duplicated conditions based on data recorded from the island. Both females laid a small clutch of eggs (11 total) two months later. All five of the fertile eggs hatched after 85-95 days under a fluctuating temperature regime. The captives proved to be difficult to maintain successfully and only one female survives as of 2001 (Daltry et al., 2001).

In early 1997, a second population survey to assess the effectiveness of the rat-eradication program was undertaken by an FFI biologist funded by an International Herpetological Society (IHS) donation (Richard Gibson, manuscript in prep.). Preliminary findings were encouraging as there was a noticeable regeneration of vegetation, an increased number of birds, and evidence of an expanding racer population. Juvenile and sub-adult racers were counted and the adult population appears to have increased to around a hundred or so individuals.

A trial reintroduction was tried by moving five male and five female snakes on a small uninhabited island near Great Bird Island in November 1999 and the effort seems to be successful. An additional reintroduction on another rat-free island is planned (Quentin Bloxam, personal communication, 2004).

A Malaysian company has submitted plans to develop Guiana Island, adjacent to Great Bird, and a number of islands in the area for tourism (Richard Gibson and Kevin Buley, personal communication, 2004). This project, which is supported by the Antiguan and Barbudan government, is euphemistically called a "heritage" project: an "Asian Water Village" comprising a 1,000-unit hotel complex, a shopping precinct, a casino and a 36-hole golf course. Bridge and boat access to other islands are also proposed. These plans are in direct conflict with the five years of research into low-impact ecotourism in the form of the proposed "North Sound Marine Park and Wildlife Reserve" encompassing Great Bird and many other islands, the surrounding sea, and contradicts earlier educational involvement with the local community. Could all of these heroic efforts by so many committed conservationists to save the racer be for naught?

In addition to many benefactors outside the zoo community, the Durrell Wildlife Conservation Trust, John Ball Zoo Society and the Columbus Zoo have provided funds to support this project. The future for the recovery of the Antiguan racer depends upon curbing human desire to develop pristine places, restoration of other islands in the region to establish new populations, continued rat-eradication on Great Bird Island and improved success with the captive colony.

Endangered Iguanas and Herpetologist in Danger

As a group, West Indian iguanas are among the most endangered lizards in the world, probably due in large part to their exclusively insular distribution.

West Indian Iguanas: Status Survey and Conservation Action Plan (2000)

The Grand Cayman blue iguana (*Cyclura nubila lewisi*) is one of the most critically endangered of the group of Caribbean rock iguanas and Fred Burton (2000) mentioned that perhaps only a few hundred individuals remained in 2000. Two years later, the population crashed precipitously; now only 15-25 lizards are left. In 2003, five iguanas were hatched at the Indianapolis Zoo. Described in 1940, the lizard was widely distributed in dry habitats over most of the island but is now restricted to a few remnant populations, due to human influences. As is often the case on islands, the introduction of feral animals, combined with habitat destruction and development, places these iguanas at great risk.

Burton, a botanist formerly with the Caymans National Trust, has been developing a captive population of iguanas in the Queen Elizabeth II Botanical Gardens on Grand Cayman, a place comprised of dry forest favored by the lizards. Iguanas have been bred from stock held at the National Trust and small num-

Fig. 42. Adult male Cayman Island blue iguana (*Cyclura nubila lewisi*) on exhibit at Smithsonian National Zoological Park in 2003. *Photograph by Jessie Cohen, Smithsonian National Zoological Park.*

bers released into the Gardens where reproduction has occurred. Although this is a positive step, captive breeding will not be the solution. There have been ongoing discussions to address the problem, one of which is to designate the peninsula known as Barker's Point in the west of the island as a protected area (there is some iguana habitat) and manage the site for ecotourism. Another suggestion is to protect an area in the eastern portion of the island. This region was investigated roughly a decade ago and some iguanas were found but much of the land is owned privately and there are enormous pressures to develop it. Clearly, additional surveys are needed to update the earlier information and locate new sites.

To evaluate the eastern site, Burton and Quentin Bloxam from the Durrell Wildlife Conservation Trust arranged a field trip, hoping that iguanas would still be found. At a zoo meeting in Barcelona, Spain in September 2002, I asked Quentin if his trip had been fruitful. With a strange look on his face, he told me that he had nearly died and the following account is mostly from a newsletter article that he wrote describing his experiences: ". . . unfortunately I managed to get lost in the dry forest. Fred and I went into this area following an old trail, which after about 45 minutes petered out and by this time Fred, who knows the area well had disappeared off in another direction. I continued northwards until I almost reached the far side of the forest block and then returned intending to meet Fred between 11:00 and 11:30 am at a Global Positioning System (GPS) recorded waypoint. Unfortunately my ability to use a GPS was not at the sufficient level of expertise needed. The result was that I missed the rendezvous point and became lost.

This dry forest is incredibly dense, very dry with mid-day temperatures in the high 30s. I was only carrying minimum equipment which included just sufficient water to last me until I reached the intended meeting point. I tried to find my way back to the meeting point for the rest of the afternoon but was unsuccessful. By this time I was becoming increasingly dehydrated which negatively impacted my energy levels and was aware that Fred had organised some kind of search party and indeed was able to see and hear a small spotter plane droning overhead trying to find me.

By late afternoon I decided that I was going to have to spend the night in the forest and found myself a friendly mangrove tree which I curled myself into, simply to get off the ground and avoid the odd mangrove crab and other biting insects that might be around.

At around 3 o'clock in the afternoon I very faintly

picked up voices in the distance. Unfortunately as I subsequently discovered there were three search parties out and they thought my hysterical screeching was actually coming from one of the other parties, so they all ignored me. In the end Fred, who was amongst one of the groups, recognized my typical English wailing and responded.

To give you an idea of the difficulty of the terrain it took me an hour to bring one of the search parties in calling every fifteen seconds or so. To try and wrap up the story, they had to bushwhack a trail out to the road, bundle me into an Ambulance, chuck me into a hospital and then intravenously drip fluid into me for the next five hours. The most galling thing of all is, of course, I did not see any iguanas . . . On a positive note the day before I left, by which time I was on my feet, we surveyed a remote farm and finally found two young iguanas, so clearly there are some breeding animals still left in the wild."

Quentin is not an inexperienced field biologist. He has surveyed tortoises in Madagascar, Antiguan racers and St. Lucia whiptails in the West Indies, boas and lizards on Round Island and has been involved in a multitude of arduous studies of threatened herpetofauna throughout the world. Since he is an old friend, his harrowing experience was particularly unsettling to me.

Chapter 4
Pioneers in Zoo Herpetology

Black Caiman (*Melanosuchus niger*)

HANS GADOW (1901)

My colleague Winston Card and I published two papers on zoo herpetologists. One of them (Murphy and Card, 1998) presented our analysis of the characteristics of our fellow workers in zoos and aquariums and our perception of their unique personalities.

The second contribution (Card and Murphy, 2000) was divided into two parts: (1) lists of persons in chronological order under their institutional affiliations in the United States; (2) biographies of deceased and retired staffers who have had an impact on our discipline. This chapter, focusing on deceased zoo herpetologists, updates and expands upon earlier information presented in that publication. In addition, the histories and accomplishments of deceased foreign professionals have been added to show how our profession is global in scope.

Leopold Joseph Franz Johann Fitzinger
(1802-1884)

TRANSLATED FROM GÜNTHER SCHULTSCHIK (2001)

KRAIG ADLER (1989)

Leopold Fitzinger, Curator at the Natural History Cabinet in Vienna and later Director of the zoos at Budapest and Munich, wrote the first genuine "history" of the

65

Fig. 43. Leopold Fitzinger. *Provided by Kraig Adler.*

Fig. 44. Cover of *Die Kaiserliche Menagerie zu Schönbrunn. Eine populäre Schilderung sämmtlicher Thiere derselben* by Leopold Fitzinger in 1875. *Courtesy of Widener Library at Harvard University, provided by James Hanken and Marie McLaughlin (Museum of Comparative Zoology).*

Fig. 45. Illustration of Der indische Helm Chameleon (*Chameleon coromandelicus*, now *Chamaeleo zeylanicus*) in Leopold Fitzinger's *Bilder-Atlas* published in 1867. This picture atlas was produced for popular consumption. *From the collections of the Ernst Mayr Library, Museum of Comparative Zoology, Harvard University.*

Viennese zoo in the conference (session) reports of the Mathematical-Natural Science Class of the Imperial Academy of Sciences in 1853. This publication is translated as *Attempted History of the Zoological Gardens of the Austria Imperial Court*, where he discussed Galápagos tortoises in captivity in Europe. This detailed history covered the collections of the Austro-Hungarian royalty, often with detailed descriptions of mammal, bird, amphibian and reptile specimens received and their longevities.

Aside from several smaller publications about Schönbrunn animals, mostly for the Austrian Academy of Sciences, in 1875, he published a "guide" addressed to the visitors of the zoo which is translated as *The Imperial Menagerie at Schönbrunn. A popular description of all animals kept there.*

Although Fitzinger is known for his herpetological contributions, he covered some mammals at the zoo as well. In 1830, he wrote a paper on illness and the death of a giraffe. In 1879, he published a paper on cavies or guinea pigs. After retirement, he was director of a private zoo in Munich and later guided the Budapest Zoo through its three-year construction phase.

See Adler (1989) and Schultschik (2001) for a biographical sketch of Fitzinger.

Johann von Fischer (1850-1901)

Johann von Fischer from Vienna is considered to be the father of terrarium science as he provided detailed descriptions on the maintenance and behavior of captive amphibians and reptiles, many of which were published in the journal *Der zoologische Garten*. He had an enormous private collection of chelonians including rarities like the bowsprit tortoise (*Chersina angulata*) and bog turtle (*Clemmys muhlenbergi*). J. von Fischer was director of the old Düsseldorf Zoo between 1880 to 1890. After he left Düsseldorf, he traveled to France and became Director of the "Laboratoire d'Erpétologie" in Montpelier. He sold southern European and North African reptiles to interested terrarium practitioners and presumably died in Montpelier.

During 1872-1874, he discussed the habits, maintenance and care of turtles in captivity, addressing treatment and prevention of illnesses. He wrote papers on the reptiles and amphibians, as well as mammals, of the St. Petersburg area (1874). He designed a new heatable terrarium for reptiles (1879). The next year, he recorded observations on the ringed agama (*Oplurus torquatus* now *O. cuvieri*) in captivity. In 1882, he was the first to document copulatory behavior and reproduction in common chameleons (*Chamaeleo vulgaris* now *C. chamaeleon,* see Murphy 2005 for translation of his papers and list of publications). He attempted to hatch nearly 800 eggs; he was successful

Fig. 46. In 1882, Johann von Fischer was the first to document copulatory behavior and reproduction in chameleons (*Chamaeleo vulgaris*, now *C. chamaeleon*). He attempted to hatch nearly 800 eggs; he was successful with a few using an incubation medium of camel dung, leaves, peat soil and sand. Illustration from *The Riverside Natural History. Volume III. Lower Vertebrates* in 1888. *Courtesy of Smithsonian Institution Libraries, Washington, DC.*

with a few using an incubation medium of camel dung, leaves, peat soil and sand. The same year, he recorded observations on the Cape spiny lizard (*Uromastix capensis auct.*), skink (*Trachydosaurus asper* now *T. rugosus asper*), iguana (*Iguana tuberculata,* now *Iguana iguana*), Mexican vine snake (*Oxybelis aeneus*) and reproduction in the cylindrical lizard (*Gongylus ocellatus* now *Chalcides ocellatus*). Over the next six years, his output included publications on the Australian tree toad, (*Pelotryas coeruleus = Hyla cyanea*), panther toad (*Bufo pantherinus = B. mauritanicus*), southern smooth snake (*Coronella girundica* now *C. girondica*), pygmy lateral fold lizard (*Ablepharus pannonicus*), salamander (*Salamandrina perspicillata*), care of European newts, common skink (*Scincus officinalis* now *S. scincus*), spotted lizard, (*Eremias pardalis* now *Mesalina guttulata*), Dabb's mastigure (*Uromastix acanthinurus*), worm lizard (*Trogonophis Wiegmanni* now *T. wiegmanni*), and relationships of lacertid lizards. Published in 1884, his book on the terrarium, its plantings and population stands as a major landmark in the science of maintaining herps in captivity.

Arthur E. Brown (1850-1910)

In the field of science the name of Arthur Erwin Brown is an important and enduring one. His score and more of publications in zoology were chiefly in herpetology, but he also wrote papers on bears, anthropoids, and monkeys. These were based in large part upon specimens that were exhibited in the Zoological Gardens. He was also deeply interested in evolution, a topic that profoundly altered the thinking of most naturalists of his day and the various facets of which were the basis of many and prolonged learned arguments.

Roger Conant (1957)

Arthur E. Brown, the second Superintendent of the Philadelphia Zoological Garden from April 1876 until his untimely death, was a consummate herpetologist. Most of his studies were in systematics, taxonomy and zoogeography. In 1878, he published a guide to the zoo. He studied garter snakes and documented his findings in several papers. Brown described Marcy's checkered garter snake, short-tailed snake, Trans-Pecos rat snake, and gray-banded kingsnake. Other papers cov-

Fig. 47. Arthur Erwin Brown. *The Zoological Society of Philadelphia, provided by Karl Krantz.*

ered snakes from tropical America living in the collection of the Zoo, a review of the genera and species of American snakes north of Mexico, reptiles and batrachians from Sumatra, Borneo and the Loo Choo Islands (Ryuku today), generic types of Nearctic reptiles and amphibians, Texas reptiles, post-glacial Nearctic centers of dispersal for reptiles, and theories of evolution since Darwin. Expanded descriptions of many of his publications can be found in the Philadelphia Zoological Garden account, Chapter 8.

Joan Beauchamp Procter (1897-1931)

Mr. Boulenger and I were able to give her full information about the arrangements in the great combined Aquarium, Reptile House and Insectarium in Berlin, and with negative hints of what to avoid from our knowledge of other houses for reptiles—hints which she was able to amplify from her experience in our own old Reptile House—in many respects the worst that was ever built . . . but from the beginning to the end it was her house.

Peter Chalmers Mitchell (1929)

Joan Beauchamp Procter was plagued with poor health throughout her life (MacBride, 1931-1932). Originally, she had planned to attend Cambridge University after schooling at St. Paul's School for Girls but her physical ailments prevented this from happening. George A. Boulenger from the British Museum offered her a post in 1917 until she left six years later to work for his son Edward George Boulenger, whom she followed as the next Curator of Reptiles at the London Zoo. She lobbied incessantly for construction of a new reptile building and her constant entreaties to the Council bore fruit. Procter was involved in the planning of the new 1927 reptile building with architect E. G. Dawber and her bust stands in the lobby. Her compassion for animals was well known. As an example, a very large python was declining, due to a septic abscess. The snake was restrained by fourteen handlers as she lanced the abscess and irrigated it with lysol. She remained committed to her job until the end; although bed-ridden, she called the Zoo regularly and issued directives to the staff.

She published many papers in herpetology: tailless batrachians from East Africa, variation of the scapula in the batrachian groups Aglossa and Arcifera, new and rare reptiles from South America, new and rare reptiles and batrachians from Australia, unrecorded characters seen in living snakes, description of a new tree-frog, nests of some African frogs, albino grass snake, Komodo dragon at the Zoo, and the Namib

Fig. 48. Joan Beauchamp Procter from *Centenary History of the Zoological Society of London* by P. Chalmers Mitchell in 1929. *Reproduced by permission of Zoological Society of London.*

Fig. 49. Photograph by F. W. Wood of Komodo dragon (*Varanus komodoensis*) named Sumbawa standing next to two-year-old child from Joan Procter's article "Dragons that are alive to-day," published in *Wonders of Animal Life* in 1928–1929. *Provided by Kraig Adler.*

sand gecko. While she was at the Museum, she published a classic paper on the development of the African pancake tortoise (*Malacochersus tornieri*), then known as *Testudo loveridgei* (Procter, 1922). The elder Boulenger thought that its flexible shell was not

present during its early years; he had mistakenly ascribed the young of another species with a rigid shell to this taxon. Procter rightly demonstrated that young pancake tortoises have flexible shells as well as adults, where the bony plates do not disappear but become thinner and more pliable as the tortoise ages. See the London Zoological Society account, Chapter 5 for additional information on her papers.

Raymond L. Ditmars (1876-1942)

Ray was speaking only the simple truth when he told his friends that he was having to feel his way along. In those days there were no books available to the amateur snake-fancier telling him the habitat, habits and temperament of the various reptiles. All he could learn from books were facts about their scientific classification. For clues to the care and treatment of his pets, Ray had to rely on his own judgment, and the kind of intuition developed, he always insisted, by one who has a strong and sympathetic interest in a subject . . . To the young man who found himself suddenly in charge of it, the Reptile House seemed too good to be true. It was the intention of the Zoological Park's founders that this zoo, unlike many others at the time, should place the same emphasis on reptiles as on birds and mammals and exhibit a representative series of them from all parts of the world. Funds for their purchase were adequate, and the building to accommodate them was to be in keeping with the importance of the exhibit.

L. N. Wood (1944)

Raymond L. Ditmars, curator of reptiles (also curator of mammals for many years) at the Bronx Zoo, was a popular author of many books on reptiles and amphibians which helped spread the word about their habits, distribution and diversity to a broad readership. It is

Fig. 50. Raymond L. Ditmars perusing spider in insect cage at Bronx Zoo (Wildlife Conservation Society) during September 1934. *Photograph courtesy of Wildlife Conservation Society, headquartered at Bronx Zoo.*

Fig. 51. Curator Raymond L. Ditmars (far right) force-feeding large constrictor at New York Zoological Society on 9 April 1913. *Photograph courtesy of Wildlife Conservation Society, headquartered at Bronx Zoo.*

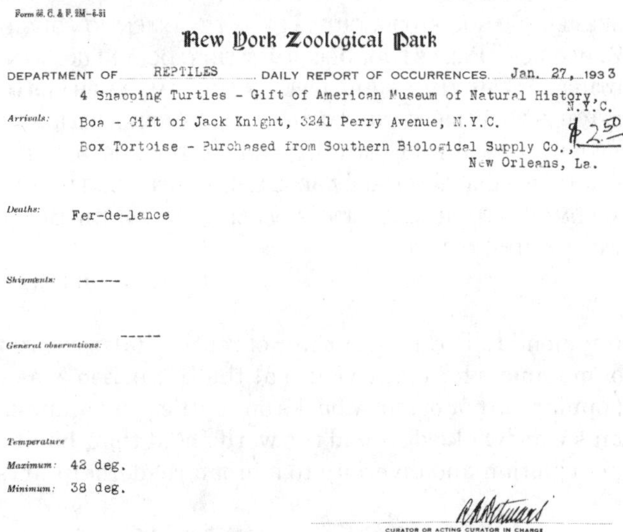

Form 66. C. & P. 2M—6-31

New York Zoological Park

DEPARTMENT OF ___REPTILES___ ___ DAILY REPORT OF OCCURRENCES. ___ Jan. 27, 1933

Arrivals: 4 Snapping Turtles – Gift of American Museum of Natural History, N.Y.C.

Boa – Gift of Jack Knight, 3241 Perry Avenue, N.Y.C.

Box Tortoise – Purchased from Southern Biological Supply Co., $2.50
                                                New Orleans, La.

Deaths: Fer-de-lance

Shipments: – – – – –

General observations: – – – – –

Temperature
Maximum: 42 deg.
Minimum: 38 deg.

CURATOR OR ACTING CURATOR IN CHARGE

Fig. 52. Daily Report by Raymond L. Ditmars on 27 January 1933. *Provided by Peter Brazaitis.*

not necessary here to give a detailed account of his career as L. N. Wood (1944) published a book entitled *Raymond L. Ditmars. His Exciting Career with Reptiles, Animals and Insects.* Wood's narrative began with Ditmars' early childhood, development of interest in natural history, and subsequent career at the Zoo. Consult Card and Murphy (2000) and Chapter 8, Wildlife Conservation Society/Bronx Zoo account for additional information.

Fig. 53. Raymond L. Ditmars extracting venom from water moccasin or cottonmouth (*Agkistrodon piscivorus*) at Bronx Zoo (Wildlife Conservation Society). Undated photograph. *Photograph courtesy of Wildlife Conservation Society, headquartered at Bronx Zoo.*

Stanley Smyth Flower (1871-1946)

BEFORE HE WAS ABLE TO READ OR WRITE, HE COULD NAME MOST OF THE ANIMALS, LIVING OR EXTINCT, EITHER IN THE MUSEUM OR IN THE ZOOLOGICAL GARDENS.

OBITUARY IN *HERPETOLOGICA* (1946)

Major Stanley Smyth Flower was a renaissance man in biology, publishing books and papers on a variety of subjects. His father, William, was the President of the Zoological Society of London (later Director of the British Museum of Natural History) when Stanley was a youngster. The younger Flower was the Director of the Giza Zoological Gardens in Cairo, Egypt from 1898 until 1924 so many of his studies center on birds and mammals from that region. Although his tenure was interrupted by World War I, this Zoo became known throughout the world as a progressive and important institution because of his commitment to its improvement. In 1902, a tropical house was constructed which had " . . . numerous reptiles, Soft River Turtles, Carolina Box-Tortoise, Flat-backed Terrapin, Long-necked Tortoise, and a representative collection of Egyptian lizards, chameleons and snakes." The collection num-

Fig. 54. Stanley Smyth Flower. *Photograph provided by Kraig Adler.*

bered over 40 species of amphibians and reptiles (Flower, 1903). While he was director, he compiled a list of the zoos of the world and regularly updated this compilation; these publications were the only comprehensive source for this information at that time. Around 1920, he was made an Officer of the Order of the British Empire for his many accomplishments at the Zoo. After retiring from Giza, he returned to the London Zoological Society where he served on the Council (1924-1936) and was Vice President (1927-1929).

His publication "Notes on the Fauna of the White Nile and its Tributaries," published in 1900, shows how broad his background was in biology for there were references to mammals, birds, reptiles, batrachians, fishes, mollusks, insects, millipedes, centipedes, arachnids and crustaceans. His herpetological output was impressive as well. In 1896, he studied the herpetofauna in the Malay Peninsula and made a list of species. Three years later, he published two checklists reflecting collections of reptiles and batrachians made in the Malay Peninsula and Siam. In 1921, he covered the exhibition of living specimens of *Testudo leithii* (now *T. kleinmanni*) and *T. ibera.* Several papers dealt with Egyptian and African reptiles: an extensive treatment of the Sinai desert cobra *Walterinnesia* (1923), *Testudo ibera* (now *T. graeca*) hatched between 1907-1921 at Giza (1924), and African spurred tortoise reproduction at Giza (1928). Another paper on loss of memory accompanying metamorphosis in amphibians was published in 1927.

When Flower traveled to Karachi, Pakistan, during May 1913, he visited the "Mugger Pir," at Pir Mangho, the tomb of Haji Mangho (Flower, 1914). Haji Mangho was a holy hermit who was thought to have resided in this area in the 13th century. During earlier times, hundreds of mugger crocodiles, some very large, were kept in a natural swamp at the base of the tomb ". . . but as they used to stray, attack people, or steal children, and probably owing to the increased value of land in the oasis, the swamp was drained and the crocodiles confined to a tank surrounded by a high wall." Flower saw 24 adult crocodiles on 27 May, rang-

Fig. 55. Photograph of Tomb of Haji Mangho from *Ministry of Public Works, Egypt. Zoological Service. Report on a Zoological Mission to India in 1913* by Stanley Smyth Flower. *Courtesy of Smithsonian Institution Libraries, Washington, DC.*

Fig. 56. Photograph of "Mugger Pir" from *Ministry of Public Works, Egypt. Zoological Service. Report on a Zoological Mission to India in 1913* by Stanley Smyth Flower. *Courtesy of Smithsonian Institution Libraries, Washington, DC.*

ing in size between 6 feet (1.82m) to about 9 feet (2.74m) as well as juveniles. Two goats were fed to the reptiles and Flower vividly described the event: "There appeared to be no method in the feeding. It was no orderly meal but a wild scramble, crocodiles snapping, fighting, and rolling about with much splashing of dirty water.

An incident which helped one to realize what great strength of a crocodile's jaws was an animal, not more than seven or eight feet in length, seized the severed head of a goat and while holding it high above the water crushing it to pieces, with a horrid sound of breaking bones, as easily as a man might break a hen's egg between his hands."

Between 1925 and 1937, he wrote four papers on the lifespan in vertebrate animals, using data from zoos, museums, universities and private collections worldwide: batrachians, amphibians, and two studies on reptiles. The longest life span recorded was an American alligator which lived for 46 years.

In 1928, Flower recommended methods for transporting mammals, birds, fishes, amphibians and reptiles. A list of his herpetological papers is available in *Herpetologica* 3:85-88, 1946.

Edward George Boulenger (1888-1946)

PETER CHALMERS MITCHELL (1929)

Fig. 57. Edward G. Boulenger from *Centenary History of the Zoological Society of London* by P. Chalmers Mitchell in 1929. *Reproduced by permission of Zoological Society of London.*

Edward George Boulenger was the son of George Albert Boulenger from the British Museum. The younger was appointed as the first Curator of Reptiles, Amphibia and Fish in 1911 at the London Zoo. He received the princely sum of £150 yearly. Seems as though curatorial salaries haven't improved much since then! Three years later, he was elevated to Director of the Aquarium and his chief assistant Joan Beauchamp Procter became Curator of Reptiles in 1923.

Boulenger had a productive research career, describing new tree frogs from Sierra Leone and Trinidad, vertically movable maxillary bone in *Xenodon*, a new horned lizard, and a new giant salamander. A number of papers dealt with herps on exhibit: male specimens of the midwife toad carrying eggs, young matamata, a blue specimen of the edible frog, an Australian pygopodid lizard, a melanistic specimen of the green lizard, spines of an insectivore found in the excrement of a boa, a living tiger salamander, a photograph of the giant saddle-backed tortoise thought to be female *Geochelone abingdonii*, a new land tortoise (*Testudo loveridgii*), and the mudpuppy. Other publications covered metamorphosis of the Mexican axolotl (*Amblystoma tigrinum* now *Ambystoma*), lizards of the genus *Chalcides* and notes on snake feeding behavior. He published several books: *Reptiles and Batrachians*

(1914); *The Zoo and Aquarium Book* (1932); and *World Natural History* (1938). Boulenger was even placed in charge of the rat eradication program at the Zoo and published his findings (1919. Reports on methods of rat destruction, p. 227-244. LZS Sci. Proc.), proving yet again how herpetological curators are remarkably flexible. See the London Zoological Society account, Chapter 5 for additional information.

Grace Olive Wiley (1883-1948)

Condensed Death, Squirming and Crawling About Museum, Will Add Interest to State Fair. Deadly Snakes are Just Pets to This Woman.
Saint Paul Daily News- 30 August 1925

Incidentally, she does a lot of her reading with a snake coiled around her neck taking an evening's doze.
The Minneapolis Star - 18 February 1937

Grace Olive Wiley, a librarian at the Minneapolis Public Library, started an exhibit of reptiles at that institution in the 1920s. She had a penchant for publicity, much to the delight of reporters and editors, and hundreds of newspaper articles were written about her love of reptiles and interest in taming them, including venomous ones (Table 4.1). Many of these features were accompanied by photographs showing Wiley holding rattlesnakes, Gila monsters, king cobras, large alligators and a bewildering array of potentially dangerous creatures. Wiley gave many demonstrations at sportsmen's shows, nature study groups and civic organizations which included tame king cobras, alligator feeding, tug-of-war between two garter snakes, 2-headed snapping turtle and feeding minnows from her fingers to water moccasins.

In early days, it was highly unusual for women to hold curatorial positions in zoos and the situation has not changed dramatically since that time. Hence, it was remarkable that Wiley was hired at the Brookfield Zoo outside of Chicago in the 1930s (see Murphy and Jacques, 2005). Her stay was tumultuous since she believed that venomous snakes could be tamed and handled with bare hands, a practice that did not endear her to her superiors. In fact, she was repeatedly instructed to discontinue these escapades but she refused to do so. After a series of 19 snake escapes, some of which included one bandy-bandy, three Egyptian cobras, and several venomous sand snakes, the upcoming liability insurance payment for the Zoo was increased to the point it exceeded her annual salary; her boss Acting Director Robert Bean lost patience and

Fig. 58. Although there is no information attached to this photograph of Grace Olive Wiley and her elapid friend, it appears to be the service area of the reptile building at Brookfield Zoo. It was episodes such as this that led to her dismissal as reptile curator in 1935.

fired her in September 1935. Hardly repentant, Wiley did not accept the termination gracefully for she complained to a *Time* magazine reporter on 30 September: "I hate to say it and I know some persons who don't like snakes are very nice persons but Mr. Bean was frightened and frightened persons will exaggerate. I do not feel I was guilty of carelessness. I just forgot, simply forgot, to close the door to the cobra's cage after I cleaned it. I couldn't do everything at once. All the other snakes that got away were harmless except Bandy-Bandy and I'm sure he went down the drain pipe. 'The cobra,' she added affectionately, 'just found the coziest place it could in the whole reptile house. If most persons were half as nice as snakes, this world would be a better place.'" Her pet, "Bandy-Bandy," was found by keeper Emil Rokosky when he thrust his hand into a bag of dried leaves which he planned to use for cage decorations.

After the Zoo debacle, she started a private roadside reptile exhibit called "Grace Wiley —Reptiles" in Cypress, California where she allowed venomous

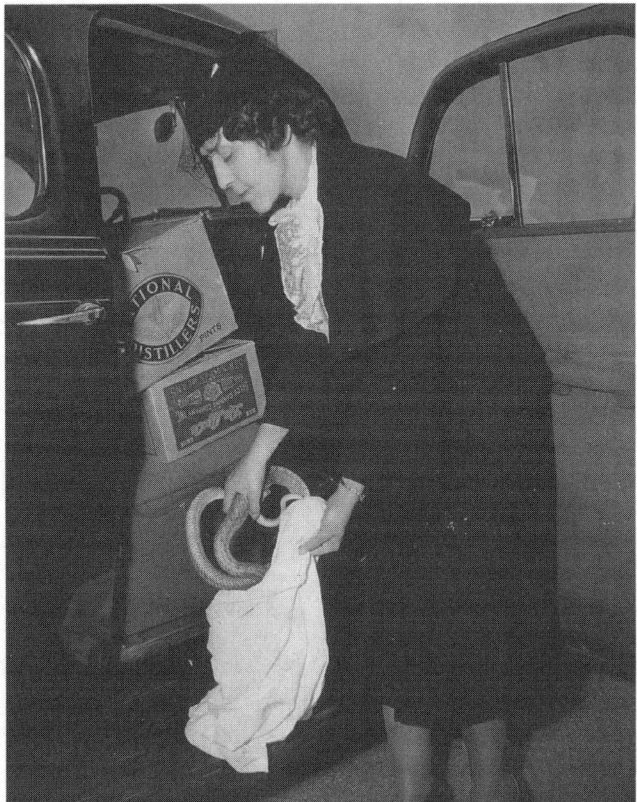

Fig. 59. Grace Olive Wiley removing king cobra from bag for film "Trade Winds" in 1938. *Photograph provided by George Rabb, Chicago Zoological Society.*

snakes, including king cobras, tiger snakes, gaboon vipers, kraits, copperheads, and rattlesnakes, to be handled by children and crawl through crowds of visitors for a small fee. A number of movie studios used her charges, such as king cobras, for films like *Trade Winds* with Fredric March and Joan Bennett in 1938. Jule Mannix wrote a book called *Married to Adventure* in 1954 where she described Wiley's fatal snakebite in detail in the chapter "Grace Wiley, Reptiles—Death of a Remarkable Person." Her husband, Dan Mannix, was a photographer for *True* magazine and he wanted to get some pictures of Wiley with an Indian cobra spreading its hood. She used a newly imported snake since her tame specimens did not spread their hoods. Being uncomfortable being photographed wearing eyeglasses (she was near-sighted), she removed them for the photograph. When working with wild cobras, she would allow them to strike the palm of her hand and stroke their heads; in this way, the snakes eventually calmed until they could be handled. With this new cobra and her vision impaired, she misjudged the snake's strike. The cobra was able to seize

her middle finger and chew to inject venom. Because anti-snakebite serum was expensive and she needed so many types, she had virtually none on hand. She was rushed to Long Beach Hospital in an ambulance but died ninety minutes after being bitten (see Murphy and Jacques, 2006).

Two of her admirers, Jouette and Eugenia Heaton, wrote a poem in her honor in 1941.

THE HERPETOLOGIST

She strokes the cobras tenderly that they
May feel no other touch save kindliness,
They raise their hooded heads for her caress
And sense her voice vibrations as they sway,
With confidence she turns one to display
Gargantuan length of graceful deadliness,
She needs no words, deft fingers can express
Her wishes to her novel protégée.
She moves among the cages of her pets
Explaining habitats that she may stir
A scientific interest, not quick threats
Ophidian species usually incur.

She is so calm and yet so debonair
It seems in apropos to say BEWARE!

C. B. "Si" Perkins (1888-1955)

A TALL, SLENDER, SLIGHTLY STOOPED MAN WITH A NEATLY TRIMMED MUSTACHE AND A HABIT OF PEERING OWLISHLY OVER HIS READING GLASSES, HE WAS ONE OF THE EARLIEST AND MOST RESPECTED STAFF MEMBERS OF THE SAN DIEGO ZOO, WHERE HE COMPILED A STILL UNEQUALED RECORD OF ECCENTRICITIES.

SHELDON CAMPBELL (1978)

C.B. "Si" Perkins, from the San Diego Zoo was reluctant to reveal his real name: Clarence Basil Perkins. His nickname "Si" was derived from a cartoon character. Laurence Klauber, who wrote *Rattlesnakes: Their Habits, Life Histories, and Influence on Mankind*, referred to him repeatedly in his opus and many behavioral observations in that book were done at the Zoo with Perkins and Charles Shaw, his curatorial successor. Klauber named the leaf-nosed snake after Perkins, *Phyllorhynchus decurtatus perkinsi*. Papers by Perkins include several compilations in *Copeia* on longevity of amphibians and reptiles in captivity, notes on captive-bred snakes, incubation and hatching snake eggs, frequency of shedding in injured snakes, and hybrid rattlesnakes. See Card and Murphy (2000) and Chapter 8, San Diego Zoological Society account for additional information.

Fig. 60. C. B. "Si" Perkins (right). *San Diego Zoological Society Archives.*

Gustav Lederer (1892-1962)

The crowning achievement of his life was the construction of the aquarium, now called "Exotarium." His understanding for exhibit facilities, the impressive air conditioning system and his ability to fit the technology to the needs of the animals—to make sure of fresh air supply and temperature variations in the terraria made the Exotarium an impressive show piece.

Translated from Cristoph Scherpner (2001)

On 18 March 1944, the Frankfurt a. Main Zoo was destroyed by Allied bombing. The Aquarium was redesigned by curator Gustav Lederer, an employee since 1913. The facility was reopened in 1957 and renamed "Exotarium." There were many unique features. A hall of climatic landscapes was designed, ranging from a polar environment with seals and penguins to a tropical riverbank with birds, reptiles and fishes. Fourteen large aquatic exhibits were arranged geographically: Southern-Pacific reef, Black Forest River, and Ama-

Fig. 61. Plaque commemorating Gustav Lederer at Frankfurt a. M. Zoo in Germany. *Photograph provided by Christian Schmidt.*

zon River biotope serve as examples. The reptile section incorporated innovative design and aesthetic features such as movable glass roofs and living plants and the crocodilian enclosure included a simulated tropical rainstorm. Lederer personally collected cage furniture for his exhibits. A tame Komodo dragon was allowed to follow the keeper through the visitor area of the old destroyed aquarium.

In 1931, he published a paper on the ethology of the sailfin lizard (*Hydrosaurus amboinensis*) and another on whether cold-blooded animals recognize their keepers. Ten years later, Lederer outlined care of the Chinese alligator. The next year, he wrote papers on the Komodo dragon and reproduction and development of the yellow anaconda. In 1944, he published a lengthy paper on food supply, development, mating and brood care of the reticulated python. Five years later, his study of the viperine water snake was published. The next year, Lederer described a ratsnake hybrid. In 1956, he chronicled the reproductive biology and development of Indian and Burmese pythons.

Lederer was interested in entomology, with a special emphasis on butterflies, and gained great prestige in this field. The Frankfurt Zoo had a long tradition in keeping insects and opened an insect house in 1903.

In 1922, Lederer established a section for pest knowledge. His most important publications included "Introduction to Pest Science" and "The Natural History of Day Butterflies." He also edited the *Entomological Periodical* for many years. In 1953, the University of Frankfurt awarded him an honorary doctorate (Scherpner, 2001).

Moody Lentz (1901-1970) and Richard Marlin Perkins (1905-1986)

Two pioneers in zoo herpetology started their careers in the reptile building, which opened in 1927, at the Saint Louis Zoo: R. Marlin Perkins and his assistant Moody Lentz. Perkins was assuredly the most well-known zoo personality in the United States with his television shows *Zooparade* and *Mutual of Omaha's Wild Kingdom*. The AZA issues a "Marlin Perkins Award for Excellence" to a zoo worker chosen for significant contributions to the profession.

Perkins wrote a paper with Roger Conant on mite control in 1931 and Lentz published a paper 40 years later detailing the use of Shell No-Pest Strips to control mites at the Zoo.

Although I mentioned this experience earlier (Card and Murphy, 2000), it bears repeating as it was a defining moment of my life: "While a high school student, JBM was introduced to Perkins by Gene Hartz. JBM timidly knocked on his office door, entered cau-

Fig. 62. Moody Lentz. *Undated photograph provided by St. Louis Naturalist Club, courtesy Ronald Goellner.*

Fig. 63. R. Marlin Perkins, probably in the 1930s. *Photograph provided by St. Louis Zoo, courtesy Ronald Goellner.*

Fig. 64. R. Marlin Perkins. *Photograph provided by Kraig Adler.*

tiously and spent an unforgettable thirty minutes with him. Perkins told JBM that getting into the zoo business was the greatest mistake of his life. His frustrations were exacerbated by indifferent politicians, zoo workers, and personal friends who could not understand his passion for animals. Further, zoo finances were always unstable and his bosses were jealous of his TV fame. Finally, Perkins advised the impressionable lad that he were stupid enough to embark on a career in zoo herpetology, fate would deal a cruel and unrelenting blow to his entire life. Perhaps he was right?"

Werner Krause (1905-1970)

TION, NEVERTHELESS, IN COMBINING THE FUNCTIONS OF A CENTRE OF PRODUCTION FOR THE BASIC PHARMACEUTICAL RAW MATERIAL, SNAKE VENOM, OF A SCIENTIFIC RESEARCH ESTABLISHMENT AND AT THE SAME TIME OF A PUBLIC EXHIBITION CENTRE OF GREAT EDUCATIONAL VALUE, IT REPRESENTS SOMETHING QUITE UNIQUE IN CENTRAL EUROPE . . . THE NAME OF WERNER KRAUSE IS INSEPARABLY ASSOCIATED WITH THE SNAKE FARM.

WOLF-EBERHARD ENGELMANN
AND FRITZ JÜRGEN OBST (1982)

Werner Krause started working at the zoological park Tierpark Berlin shortly after the terrarium building opened and was known as "Snake Farmer Krause." In 1956, the snake house (farm) opened, housing a large number of venomous snakes, including many from his private collection. He was Headkeeper (Leader) of the Terrarium until 1970. Before Krause came to the Tierpark, he was involved in the construction of the Stockholm Aquarium beginning in 1929 and through the recommendation of King Gustav, was able to study several semesters of biology in Uppsala. Four years later, Krause built the aquarium in Oslo. After returning to Germany, he established the department for poisonous snake research by order of the Dessau Hygienic Institute. After World War II, he built a snake farm in Tomau near Dessau; later he went to the serum factory in Bernburg. He came from there to the Tierpark where he developed an exhibit truly unusual for a zoo, a combination venom extraction laboratory and public exhibit. The venom extracted was made available for medical research (Leipzig University) and

in hospitals. Aspects of his life and professional career are not widely known as he rarely published (see Scheffel, 2001; Pederzani, 2000). He died in an automobile accident in 1970.

Charles E. Shaw (1918-1971)

ROGER CONANT (1997)

Charles E. Shaw from the San Diego Zoological Society followed "Si" Perkins as herpetological curator. He published papers on chuckwallas, eggs and young of some African, United States and Mexican lizards and other reptiles, longevity of snakes in North American collections, and husbandry of sea snakes. Based on observations at the Zoo, he documented reproduction in the Cuban and rhinoceros iguana, and Galápagos tortoise. His studies on the male combat "dance" of some crotalid snakes, American colubrid snakes and additional remarks on combat in other colubrid and elapid snakes serve as classic behavioral reports. Prior to Shaw's papers, combat was confused with courtship. There is evidence that ancient Greeks knew the difference and probably the Romans as well. After the fall of Rome, this insight was lost, only to be resurrected centuries later in Shaw's publications. He wrote

Fig. 65. "Snake Farmer" Werner Krause and wife Pia extracting venom from water moccasin or cottonmouth (*Agkistrodon piscivorus*) at Tierpark Berlin Snake Farm on 25 November 1959. *Photograph by Gerhard Budich, provided by Falk Dathe and Gisela Petzold.*

Fig. 66. Charles E. Shaw (right). *San Diego Zoological Society Archives.*

Fig. 67. Photograph of late herpetological curator Charles E. Shaw and "Maggie" taken in 1964. *Courtesy of San Diego Zoological Society Archives.*

many articles in the zoo publication "ZOONOOZ." In 1974, Shaw and Sheldon Campbell, a Zoo board member and later president, published the book *Snakes of the American West.*

Shaw did not allow veterinarians to treat his reptiles and none were turned over to pathologists on death for necropsies; rather, specimens were ground up in a huge meat grinder and then flushed down the drain (Marvin Jones, personal communication).

See Card and Murphy (2000) and San Diego Zoological Society account, Chapter 8 for additional information.

Carl F. Kauffeld (1911-1974)

KAUFFELD WAS ABLE TO DEVELOP ONE OF THE LARGEST COLLECTIONS OF LIVING REPTILES IN THE UNITED STATES, ONE THAT ONCE BOASTED HAVING ON EXHIBIT ALL TAXA OF NORTH AMERICAN RATTLESNAKES, WHICH WERE ANIMALS OF LIFELONG INTEREST TO HIM.

KRAIG ADLER (1989)

As ideas, protocols and technologies improved for keeping reptiles in zoos during the mid-20th century, there is no question that one of the most important persons to contribute to our discipline was Carl F. Kauffeld at the Staten Island Zoo. His observational skills, honed as he cared for his charges, were so acute that many of the practices that are commonly in use today were pioneered by him years earlier. His publications covered a range of topics: growth and feeding of newborn Price's and green rock rattlesnakes using food lizards frozen in ice cube trays and thawed later, feeding captive snakes, treatment of mouthrot in snakes, removal

of abnormal snake eggs by sectioning, mites and ticks in captive snakes with remarks on cage sanitation, manipulation of odor as an aid in feeding captive snakes, the effect of altitude, ultraviolet light, and humidity on captive reptiles, shedding, and the importance of environmental factors such as temperature. For snakes, he stressed the need for security, maintaining them singly, and offering pre-killed food. In my experience, some snakes tend to thrive better in groups.

But his influence goes beyond captive management and husbandry. His inspirational writings of collecting adventures throughout the United States motivated scores of his readers to follow in his footsteps. His favorite sites such as Okeetee preserve in South Caro-

Fig. 68. Carl F. Kauffeld (left), Robert T. Zappalorti (center) and Jim Bockowski (right) place eastern diamondback rattlesnake (*Crotalus adamanteus*) collected at Duck Pond, Okeetee, Jasper County, South Carolina in trunk to transport to Staten Island Zoo on 7 April 1965. *Photograph by Manny Rubio, provided by Robert Zappalorti.*

Fig. 69. Carl F. Kauffeld collecting Gila monster (*Heloderma suspectum*) in southern Arizona around 1947 for Staten Island Zoo. Credit: photograph by Roy Pinney, provided by Robert Zappalorti.

lina, Ajo Road and Ramsey Canyon in Arizona, and Lake Okeechobee and Payne's Prairie in Florida were often overrun with collectors. Since he was so explicit in setting the stages for his accounts, he often provided pinpoint localities which spurred many to exploit these places.

The reptile building was (and is) an older facility which was part of a larger complex housing mammals and aquatic life. Many of the enclosures being relatively small, Kauffeld decided to specialize in snakes, especially rattlesnakes. He assembled an exhibit of all U.S. varieties for the first time in any zoo. An impressive list of longevity records, especially for snakes, began to accumulate, clear testimony to his care and attention.

Kauffeld arrived at the Zoo as Curator of Reptiles in 1936 and began augmenting the reptile collection. As his reputation broadened, a number of zoo workers who later gained prominence, came to the Island to work with him: Carl Alimonti, Lou Pistoia, John Werler, Bob Zappalorti, and Dave Zucconi. He was able to add the title of Zoo Director to his résumé in 1963 and retired a decade later.

Fig. 70. Carl F. Kauffeld (holding snake head) pictured in "News Bulletin of the Staten Island Zoological Society" in March 1937. Each issue cost 5 cents. *Staten Island Zoological Society, provided by Ken Kawata.*

Many of Kauffeld's reflections were initially published in the quarterly zoo publication *Animaland* and later incorporated into his two books: *Snakes and Snake Hunting* and *Snakes: The Keeper and the Kept.* The following vignette shows his powers of observation and ability to tell interesting stories about reptiles.

In 1960, he published "Reptile Wing personality sketches" in *Animaland* 27(4). Kauffeld provided three

sketches. The first was an account of a fatal dominance struggle between two large male alligators. One amusing anecdote described a keeper who had the unpleasant habit of leaping out of the gator enclosure by bounding off the crocodilian's dorsum after cleaning the pool; the keeper refused to discontinue this practice until threatened with dismissal. The second account was a description of a very nasty tegu lizard, cleverly foisted on Kauffeld by James Oliver, Bronx Zoo herpetology curator. The last was a description of a pair of exceedingly passive green mambas that were so sensitive to their dead rodent prey offered with tongs that the mouse whiskers had to be removed during feeding so they wouldn't tickle the snakes; if this were not done, the snakes were disinclined to feed. In my experience, it seems as though there is an explanation: if the snakes described by Kauffeld had been offered prey frontally, they might have been reacting defensively as normally mambas and other venomous snakes prefer to strike prey when the lateral aspect is presented.

In the late 1960s, I planned to travel to Arizona to collect ridge-nosed rattlesnakes for the Dallas Zoo collection. I wrote a long letter to Kauffeld asking advice about these pitvipers. Although we had never met, I was sure that he would gladly help a fellow zoo worker. To my dismay, he promptly answered that I was not worthy to either see them in the field or collect specimens for our zoo.

Louis Pistoia (1917-1976)

Take your tongue off the glass at once and put it back into your mouth where it belongs!

Louis Pistoia screaming at a zoo visitor who flicked and pressed her tongue against the exhibit glass to entice a snake to strike during one of my visits.

Are you ready for *this*!

This enthusiastic demand, shouted in my face dozens of times by Pistoia, was reinforced by poking and grabbing as he pulled me from cage to cage to show his enormous collection behind the scenes in the early 1960s.

Curator Louis Pistoia at the Columbus Zoological Gardens in Ohio was convinced that zoo reptile buildings were hardly imaginative and he was determined to design a facility that was radically different when zoo administrators approved his request for a new one. Many times, Lou told me that his design was revolutionary and if I paid attention to his every word, I could

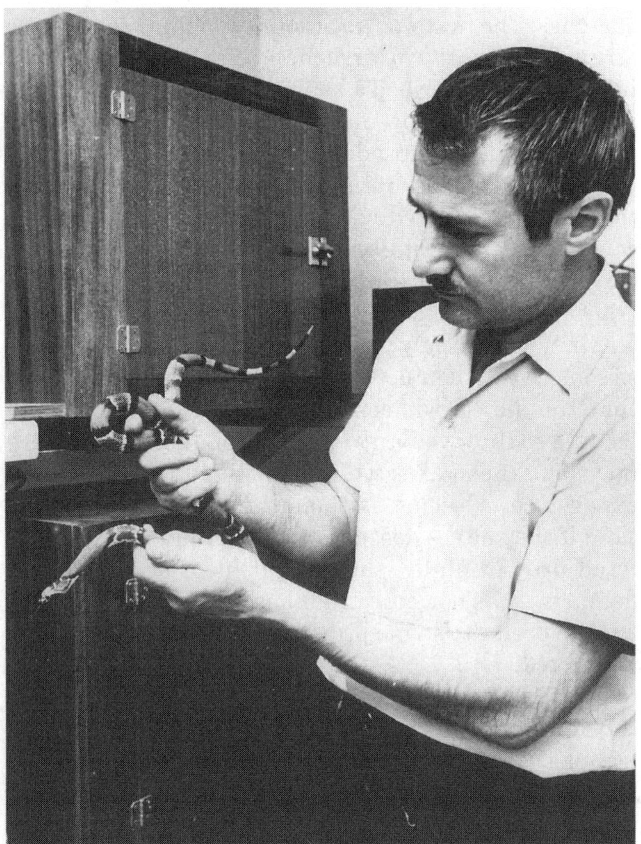

Fig. 71. Louis Pistoia. *Undated photograph provided by Kraig Adler.*

learn something about zoo architecture. He was certain that the exhibits should be in the center of the building and the public area outside, in part to provide a thermal barrier for the animals to outside temperatures. He also believed that exhibits should be sterile to facilitate cleaning and disinfection. His solution was to build a circular building to fit his philosophy. The Reptilia-Amphibia Hall opened on 13 April 1968.

Lou had amassed an enormous collection, most of which was held behind the scenes, but I was not allowed to see his charges until my credibility had been established. In fact, even the director did not have a key to the building and had to obtain permission to enter the rear section. When he finally deigned to allow entrance to his hallowed inter-sanctum, the most extraordinary array of snakes, many venomous, greeted my eyes. Many of the species were so rare and delicate that I am certain that most herpetologists had never seen living examples. As he showed these creatures to me, Lou insisted that this occasion would be the most instructive and important landmark in my life.

It is a shame that Lou's name is not widely known but this almost certainly is due to the fact that he rarely published his observations. One day, he told me two extraordinary stories. The first was that a Javan wart snake (*Acrochordus javanicus*) lived for over four years in his collection without feeding. The second was equally intriguing. A former keeper was stationed in Viet Nam during the war and collected a baby bamboo pitviper (*Trimeresurus stejnegeri*) which was sent to Pistoia. The snake lived in isolation in a small terrarium and when it matured, gave birth to a living neonate. When I suggested that both observations were worthy of publication, Lou said that the fact that he had seen both episodes was quite enough and the information did not need to be shared with unworthy colleagues. Had he been anywhere near as prolific a writer as Carl Kauffeld at the Staten Island Zoo where he started his career, his reputation as a gifted animal keeper would have been assured.

I first met Lou during the late 1950s while an undergraduate college student in Cincinnati. Our initial meetings did not go well as he felt that my youth and herpetological inexperience were liabilities not easily overcome. Over the years, I made many pilgrimages to Columbus, and as Lou slowly accepted our developing relationship, he began to share his husbandry secrets. He insisted that captive snakes die from overfeeding and should not be offered food until ravenous. No one but himself was allowed to feed his snakes. Since Lou believed that reptiles and amphibians did not respond well to any medical treatment, the zoo veterinarian was not allowed to touch any of his animals. The snake mite caused massive problems in zoo collections for years and a number of chemical treatments had been tried to control outbreaks. Not believing in chemicals, Lou covered all of his snake cages with white towels every night since mites tended to migrate to the cloth; every morning, he would soak the towels in hot water to kill the pests; it worked but was hardly a time-saving procedure!

When I visited his new building a few months after opening, I had the audacity to suggest that the exhibits might be enhanced with a little cage furniture, since there were only water bowls and sand or gravel substrates. His arboreal reptiles having no places to climb, I mentioned that a few branches in each cage might be a good idea. He asserted that I was only a naive neophyte and further, I had a great deal to learn before I was competent to question his professional judgment. This was a typical interchange between us but somehow we remained friends; he even offered a keeper job later. When I pointed out that I was already head of the herpetological department at the Dallas Zoo, he said that the opportunity to work under him was more important.

Lou's collection included a number of potentially dangerous elapid snakes but there were neither shift cages nor warning labels. When I asked how these snakes were maintained, he said that procedures varied, depending on species. When the large king cobra's cage was cleaned, Lou crawled into it, tied a handkerchief around the snake's eyes as it reared in a defensive posture, and commenced cleaning. In the cage with many common mambas (*Dendroaspis angusticeps*), Lou cleaned around them but watched to make sure that the snakes did not begin moving their heads from side-to-side, a sure sign that they were becoming agitated. The most difficult snake to control was the Gould's tree cobra (*Pseudohaje goldii*); Lou said that it was the only snake taxon which made him nervous. When I asked about the lack of labels, he said that his keepers were required to remember the location of every venomous snake to keep alert so no warnings were needed. This practice was not enthusiastically endorsed by his staff!

James A. Oliver (1914-1981)

On November 20, 1954, in the laboratory above the Reptile House, Jim Oliver opened the first meeting of a new herpetological society. While none of the professsional biographers mention his association with herp societies, those associations may well be among his most memorable contributions. Jim Oliver was much more than a founder of **NYHS**—he was mentor, guide, inspiration, and friend to our members. The guiding priniciples of **NYHS** — EDUCATION AND CONSERVATION—were established under his genial guidance. All of us who became herpphiles at an early age know the feeling of being an oddball in one's own neighborhood and the real need to share our knowledge and interest. Dr. Oliver's establishment of the Metropolitan **NYHS** (now **NYHS**) opened a niche many were waiting for. **NYHS** was certainly not the first herp society but it may be the oldest in continuous operation. It is difficult to express in mere words the feeling of being able to easily approach and converse with someone of Dr. Oliver's stature. I wonder how many other Curators were as interested in and accessible to herpphiles of all ages.

Nat Sloan describing James Oliver's role in creating the New York Herpetological Society (1978)

Early in his career, James A. Oliver was a college professor. Amazingly, he was the director at the three

Fig. 72. James A. Oliver holding newly hatched king cobra (*Ophiophagus hannah*) at Bronx Zoo (Wildlife Conservation Society). Undated photograph, but probably around 1956. *Photograph courtesy of Wildlife Conservation Society, headquartered at Bronx Zoo.*

paramount institutions focused on natural history in New York: Bronx Zoo, American Museum of Natural History and the New York Aquarium. In 1951, he became curator of reptiles at the zoo where he took charge of the renovation of the reptile house. Seven years later, he was installed as director of the museum; his zoo associates regretted his departure but wished him well in his new post. Fairfield Osborn, president of the New York Zoological Society, praised Oliver in an article in the *New York Times* on 3 June 1959: "He's a first-class zoologist and administrator. These are not the usual qualities in combination. We're delighted he has this opportunity."

His breadth of research included a preliminary analysis of the herpetofauna of Sonora, check list of the snakes of the genus *Leptophis*, relationships and zoogeography of the snake genus *Thalerophis,* and the anoline lizards of Bimini, Bahamas. He published two books: *The Natural History of North American Amphibians and Reptiles* and *Snakes in Fact and Fiction.* Oliver also wrote a number of articles for the zoo publication *Animal Kingdom* (now *Wildlife Conservation*). Consult Card and Murphy (2000) and Chapter 8, Wildlife Conservation Society/Bronx Zoo account for additional information.

Ensil Ross Allen (1908-1981)

He was an extremely likable person, handsome, and with a splendid physique and a ready smile. He was an athlete

Ensil Ross Allen was a showman who appeared in movies, television, newspapers and magazines but he also was involved with the Boys Scouts of America, local civic groups, and the U.S. military for which he taught herpetology and outdoor survival courses.

He created three publications under the banner of his Ross Allen's Reptile Institute, located in Silver Springs, Florida: "Special Publications," "Publications of the Research Division," and "Bulletins." The "Bulletins" were as inexpensive as 15 cents each, $1.25 post paid and covered topics ranging from snake biology and husbandry to disease treatment for captive specimens. One of the "Bulletins" (#79) described the development of his Rattlesnake Control Stations in Florida and Georgia over many years. These stations were created to control rattlesnakes in inhabited areas. Allen also created the Research Division in 1949 and hired Wilfred T. Neill (1922-2001) as director. Allen and Neill published a large number of short papers in *Herpetologica* and *Copeia*, some of which are listed in the Ross Allen's Reptile Institute account, Chapter 8 to give a picture of the breadth of research conducted at the Institute. Twelve papers on the herpetofauna of British Honduras were produced. In addition to herpetological papers, several extensive treatments on Florida's Seminole Indians were developed and printed by the Institute, beginning in 1952. This productive

collaboration lasted until a disagreement severed the relationship in 1960. Ross was interested in spreading the word about herpetology so he published articles in *All-Pets*, *Pet Shop Management*, and *Florida Wildlife* magazines.

While in high school, I first met Ross Allen, accompanied by my long-suffering parents. When he discovered that I was interested in herpetology, he spent hours regaling me with stories about his experiences with reptiles, occasionally interrupted when he wrestled an alligator or did his venomous snake show. When I was in college in the late 1950s, I visited George Campbell, author of *Jaws, Too!* who lived in Grosse Point, Michigan. He had the second largest collection of crocodilians in the world, topped only by the Berlin Aquarium. All of his charges were housed in his basement, called "Grosse Point Underground Zoo," and virtually all were less than six feet in length. When the reptiles predictably grew to unmanageable lengths, the entire collection was transferred to the Reptile Institute to fill in the gaps as Ross wanted to have every taxon on display.

Years later, my college roommate called excitedly and said that he had just met Allen. Allen was in declining health, and running his facility had become difficult. When Ross discovered that we were friends, he told him to offer the job of director to me. Since I was comfortable in Dallas, I declined the offer although I was flattered that he still remembered me. As I had bought an array of living reptiles during each visit, he probably considered me his best customer.

Hans-Günter Petzold (1931-1982)

Hans-Günter Petzold, first as Curator of Lower Vertebrate Animals and later also as Deputy Director at the Tierpark Berlin, published a number of herpetological

Fig 73. Undated photograph of Ensil Ross Allen enjoying a leisurely swim with companion. *Provided by Kraig Adler.*

Fig. 74. Hans-Günter Petzold. Photograph by Günther Barkowsky on 9 November 1973. *Provided by Falk Dathe and Gisela Petzold.*

papers, mostly on behavior and reproductive biology. His publications often addressed the biology of reptiles and amphibians from Cuba and Vietnam in his zoo collection. Petzold began employment at the Tierpark on 1 April 1955. Beginning in 1956, the collection was kept in an old, small but well-kept facility. Petzold had done research in Cuba and Vietnam; this interest was represented in the collection by a number of Cuban reptiles, including the ground iguana (*Cyclura n. nubila*), boa (*Epicrates angulifer*), dwarf boa (*Tropidophis maculatus*), Antillean terrapin (*Trachemys decussata*), water snake (*Tretanorhinus variabilis*), and curly-tailed lizards (*Leiocephalus carinatus*).

Petzold published three books on herpetological topics. In 1971, he described systematics, morphology, natural history, reproduction, behavior, and representatives of the family Anguidae. Twelve years later, he investigated the same topics in anacondas (*Eunectes*). His magnum opus was published in 1982 in the zoo journal *Milu, Berlin* and 1984 by BINA, Berlin and is translated as *Tasks and Problems Connected with Re-*

search into the Life Expressions of the Lower Amniotic Animals (Reptiles). This scientific volume, published in the former German Democratic Republic (East Germany), contains references on the Marxist theory of class struggle mostly in the introductory chapter. Without these necessary remarks, common in the East-German scientific community, it was almost impossible to publish a book. If today the introduction appears to be biased in terms of ideology, the scientific argumentation however remains robust. For example, his discussion on chameleon behavior and reproductive biology includes an extensive description of the functional morphology of the chameleon tongue, based in part on observations made in captivity.

In the book, he described observations and experiments that resulted from his herpetological research: reproduction, post-embryonic ontogeny, behavior and nutrition. It often focuses on the importance of captive reptiles to answer broader biological questions. The bibliography is extensive and lists a number of relatively obscure German and other foreign-language references, mostly unknown to the English-speaking world, but some of which will interest current researchers. This book is important to contemporary scientists because it reveals a number of important and unique discoveries in zoo biology, herpetology and animal behavior. The information is still current, the historical references are fascinating and the bibliography represents a unique collection of references. (See Heichler and Murphy [2007] for a translation of this book.) As an example of the scope of his interests, Petzold also published a book on guppies. His many papers and books on herpetology and obituaries by his colleagues are listed in Chapter 7, Tierpark Berlin account.

When I visited Petzold in 1969, the Zoo was in the eastern sector of Berlin, divided by the Wall. After an hour waiting at the border, we traveled to the Tierpark where Petzold gave me a tour of the herpetological facility, which included a number of Cuban reptiles. This was exciting as I had not seen living examples of most forms. After the tour, we sat at the Tierpark restaurant and agreed that we would not discuss politics; rather he told me about his plans for his books, other herpetological projects, and his collecting experiences in Cuba. We maintained a wonderful professional relationship until his death, exchanging ideas, literature and animals.

Werner Schröder (1907-1985)

Werner Schröder was a vivarium scientist of the old school who always lent an ear to the beginner and was

AVAILABLE WHENEVER POSSIBLE NOT ONLY WITH ADVICE BUT ALSO WITH DEED. FOR HIS MERITS IN THE RECONSTRUCTION OF THE BERLIN AQUARIUM HE WAS AWARDED THE FEDERAL CROSS OF MERIT.

TRANSLATED FROM WERNER RIECK (2001)

The overseer of the Berlin Aquarium, Werner Schröder (see Rieck, 2001 for his biography), watched as his beloved building was destroyed on 22-23 November 1943, during an Allied bombing run in World War II. All of the floors except the cellar were leveled. Few survivors remained from the prized collection; the few that survived the bombs later perished during the winter as there was no heat available. Many animals were killed (over 750 species). Two false gavials, some giant tortoises, two pythons, several alligators, some hawksbill sea turtles and four giant salamanders survived. The Komodo dragon, although alive, later succumbed to injuries. Dead crocodiles, snakes, lizards and fishes were scattered through the wreckage. The large naturalistic crocodilian exhibit, covered with glass panes, was destroyed; in fact, every glass-fronted exhibit was gone. This crocodilian enclosure was the first walk-through display in any zoo. Damage to the bird and mammal collection was equally dramatic and horrifying. The zoo staff valiantly tried to save as many of the survivors as possible. Since food was at a premium; the staff was forced to eat their lifeless charges; they cut the tails off of dead crocodilians to cook in the zoo kitchen. Schröder was involved in cleaning up the Aquarium after the war. When he was asked to justify rebuilding the Aquarium over building a new facility, he argued that the former edifice was magnificent and could never be replaced. The designs of the original exhibits were modified to accommodate different types of animals in the rebuilt aquarium. The new Aquarium was completed on the original site in 1950 but there were many problems due to faulty building materials.

In 1969, I visited the Zoo and Schröder was my guide during a tour of the Aquarium. Several memories remain. The first was I saw my first Lake Tanganyikan rock-dwelling cichlids which excited me greatly. The second was the impressive crocodilian collection. The third was the only live specimen of the Philippine water monitor (*Varanus salvator nuchalis*) I have ever seen. The last was more humbling. He asked if I would like to come up to his apartment in the Aquarium for refreshments. When I entered his living room, I couldn't help but notice a ping pong table in the center. During my college days, I had avoided studying by playing the game constantly so I was confident that the match would end quickly, especially since Werner was considerably older than I. It did but I was unmercifully thrashed; all of our rematches ended similarly!

Zdenek Vogel (1913-1986)

VOGEL PLAYED AN IMPORTANT ROLE IN THE POPULARIZATION OF AMPHIBIANS AND REPTILES, THROUGH HIS LECTURES IN CZECHOSLOVAKIA AND OTHER COUNTRIES AND PARTICULARLY THROUGH HIS SEVERAL BOOKS. BEGINNING IN 1930, HE TRAVELED WIDELY IN EUROPE, ASIA, AFRICA AND SOUTH AMERICA.

KRAIG ADLER (1989)

Zdenek Vogel was born on 11 November 1913 in Kutna Hora (central Bohemia, at that time part of the Austro-Hungarian Monarchy). In 1947, he was successful in creating a facility devoted to amphibians and reptiles called "Herpetological Station in Suchdol," a town on the outskirts of Prague. In the early 1950s, he was employed at the Prague Zoo but only stayed about one year. His station was administered by the Prague Zoo as a special entity but was located in Vogel´s private house in Suchdol. Vogel was the director of this station. In the early days, the station was part of the Zoo but later became his private facility; Vogel lived then as an independent writer.

He closely cooperated with the National Museum in Prague where he was the chief of the Ichthyological and Herpetological section of the Society of Friends of the National Museum and later was head of the Terraristic Section of the Society. Vogel donated a number of preserved specimens to the Museum. He played a significant role in forming the Czechoslovak herpe-

Fig. 75. Werner Schröder on September 1952. *Berlin Aquarium. Courtesy of Jürgen Lange and Christina Unger.*

Fig. 76. Zdenek Vogel. *Photograph provided by Kraig Adler.*

tological society (1973) and was an important participant in many annual conferences of Czechoslovak herpetoculturists. Vogel developed important connections abroad, e.g., Stuttgart State Natural Sciences Museum in Germany and the Cuban Academy of Sciences. He traveled extensively to some African and Mediterranean countries, Cuba, and the Orinoco region in South America.

Vogel imported a number of exotic amphibians and reptiles into Czechoslovakia, including many rare taxa, which led to major advances in husbandry and breeding. He was an advocate of the so called "hygienic method" of keeping herps by using simple, practical enclosures. In 1948, a special glass reptile house was built and he designed wooden cages with sliding glass fronts and mesh tops to insure air circulation. There were sizable metal tanks for crocodilians and large constrictors. His collection of reptiles must have been enormous, based on the many photographs and descriptions in his books. He demonstrated that his captive charges were not only to be kept as a hobby but were useful for scientific research and contributed to conservation efforts.

Vogel attended almost all significant herpetological meetings where he gave many presentations accompanied with a number of slides. He published a great number of popular articles as well as several scientific papers and monographs on boids. Vogel published twenty books (the first in 1946) in Czech, German and English, e.g.: *Aus dem Leben der Reptilien* (Artia, Praha, 1954), "Reptile Life" (Artia, Praha, 1956), *Reptiles and Amphibians. Their Care and Behaviour*

(Viking Press, New York, 1964), *Terrarien Taschenatlas* (Artia, Praha, 1966), *Riesenschlangen aus aller Welt* (Ziemsen Verlag, Wittenberg, 1968), and *Die Riesenschlangen, Die Nattern* (in Grzimek's *Tierleben*, Band VI, 1971). Among his Czech titles are, e.g., *Na lovu vzacných krokodylu* (*On the Hunt for Rare Crocodiles*) (Orbis, Praha, 1967), which was devoted to his Cuban travels and maintenance of the herps found there. Another was "Džungle na tisíci ostrovech (Jungle on the Thousands of Islands)" (Academia, Praha, 1973) about his experiences in Venezuela.

The big-headed turtle (*Platysternon megacephalum vogeli*) was named in his honor. He died on 16 November 1986 in Prague.

Jozsef Laszlo (1935-1987)

VIPERS SHOULD COVER THE WORLD LIKE SHERWIN-WILLIAMS PAINT.

LASZLO DISCRIBING TO ME WHY HIS COLLECTION AT THE SAN ANTONIO ZOO WAS FILLED WITH SO MANY OLD WORLD VIPERS.

YOU'VE GOT TO KEEP THEM COOL, MON!

LASZLO EXPLAINING HOW DINOSAURS SHOULD BE KEPT IN CAPTIVITY IF THEY HAD STILL BEEN AVAILABLE.

Herpetology flourished at the San Antonio Zoological Gardens where Jozsef (Joe) Laszlo was curator. Through Laszlo's efforts, the Zoo became recognized as a place where the tradition of captive management reached new heights with studies on critical husbandry parameters necessary to keep reptiles and amphibians in good health. Since he was such an important force from the late 1970s until his untimely death in 1987, his career and influence demand expanded treatment. Laszlo was certainly a pioneer in focusing on husbandry issues, and the International Herpetological Symposium (IIIS) presents an annual award in Joe's name to a person who significantly contributes to captive management. At IHS or other herpetological meetings, Joe would be seated in a comfortable chair, surrounded by many persons, as he explained his new findings. Since he retained his native Hungarian accent and struggled a bit with English, these sessions were often hilarious and occasionally filled with malapropisms. When he died, his loss cast an unbelievable gloom in the zoo community as he was truly beloved.

Joe's office was in the center section of the reptile building and his desk was stacked with piles of reprints, correspondence and handwritten notes reflecting his observations. He had two passions: the vagaries of human sexuality and discovering necessary elements

Fig. 77. Jozsef (Joe) Laszlo holding newly hatched water monitors (*Varanus salvator*) at San Antonio Zoo in 1981. *Photograph by John Tashjian.*

Fig. 78. Undated photograph but probably taken around the late-1960s of Jozsef (Joe) Laszlo (center in dark coveralls) at Houston Zoo. *Provided by Kraig Adler.*

to keep herps alive in captivity. The former will not be pursued here but the latter deserves mention. Laszlo was convinced that the interface between temperature and light was critical for success and he poured his energies into investigations to duplicate conditions found in the wild. Since the off-exhibit area was so small, he was unable to build a spacious hibernaculum to cool his temperate reptiles so he installed a soft drink cooler instead. Banks of fluorescent lights were suspended everywhere with an array of different bulbs which he was testing. Racks of shoe boxes and terrariums held a bewildering assortment of snakes to be used for his trials. Many of the enclosures contained living plants. Joe loved technology so any new light bulb, humidifier, cooling chip or heating device would be acquired. The written instructions and specifications enclosed with the products were never ad-

equate so he composed lengthy letters to manufacturers asking mostly unanswerable questions. There were occasional glitches. When he placed a home air conditioner in the service area to cool his animals, he neglected to include an exhaust vent to the outside so the temperature remained the same which frustrated him greatly.

Since Joe was rotund and not particularly graceful, squeezing through tiny openings or handling venomous snakes was a major effort fraught with danger. The staff had placed a large eastern diamondback rattlesnake in a clear plastic tube for examination. Joe was gazing at the snake's head at the front of the tube when one of his keepers released the tail. It took a few moments for Joe to realize that snake was moving toward his eyeball; a tornado of movement followed as Joe clumsily tried to avoid the snake, all the while cursing the chap who had released the rattler.

When Joe's daughter and my son were born around the same time, we agreed to call one another when each said his or her first words. Some months later, Joe called and said his daughter had uttered her first words. When I asked what they were, Joe said "Burger King!" He was my first herpetological friend to die and I still miss him.

Curt Heinrich Dathe (1910-1991)

IN ESSENCE, HE WAS REGARDED AS ONE OF THE DISTINGUISHED OF ALL ZOO PERSONS. AS HE AGED, HE CONTINUED TO BE ACTIVE, AND HIS SEVENTY-FIFTH BIRTHDAY CELEBRATION WAS MARKED BY A VAST OUTPOURING OF PRAISE FOR WHAT HE HAD ACCOMPLISHED DURING HIS BUSY LIFE.

ROGER CONANT (1997)

Heinrich Dathe was the scientific assistant to the director at the Leipzig Zoo from 1 October 1936 where he was responsible for maintaining animal records. He was drafted into the Army on 3 September 1939, released from a prisoner of war camp, and returned to the Leipzig Zoo. The well known Karl Max Schneider who ran the Zoo was his mentor. Dathe was the first Director of the Tierpark Berlin in 1955. He published over one thousand papers on zoo biology, including a number dealing with herpetology. In addition to his herpetological papers, he wrote about the Tierpark as well: *Vom Tierpark Berlin* (*About the Berlin Animal Park=Tierpark*) and a small guidebook called *Jahrbuch für den Tierfreunde* (*Yearbook for Animal Lovers*). He was editor of the journals *Der zoologische Garten* (*N.F.*) (from 1953 up to his death in 1991), *Beiträge zur Vogelkunde* (a journal of ornithology / from 1952

Fig. 79. Heinrich Dathe. Credit: photograph by Klaus Rudloff on 27 June 1984. *Provided by Falk Dathe and Gisela Petzold.*

up to his death), *Milu* (journal of Berlin Tierpark / from 1960 up to his death), and *Nyctalus (NF)* (journal of bats / from 1978 up to 1990).

His herpetological papers cover a wide range of topics: swimming of slow-worms, birth of a panther (=tokay) gecko (*Gekko gecko*), effect of Wofatox on amphibians and reptiles, tenacity of a giant snake in clinging to life, snake conservation, tail regeneration in caimans, battle between a smooth snake and a sand lizard, herpetological sketches from Cuba and Southeast Asia, Nile crocodile distribution, "snake charming" in Rangun, crocodile farms in Cuba, and ancient lizard sculptures.

His son Falk Dathe is the current Reptile and Amphibian Curator at the Tierpark.

Heini Hediger (1908–1992)

To recapitulate Heini Hediger's major contributions to the science of animal behavior, he made the study of captive wild animals professionally legitimate; he helped us see captivity not as a situation of deprivation but of modification of the circumstances of natural behavior, often allowing increased variability in the expression of natural behavior; he practiced a comparative approach to animal behavior in nature and in captivity that is both theoretically and empirically useful; and he demonstrated that for captive wild animals, humans invariably play a variety of roles in regard to their behavior, and these roles are of practical and scientific value in managing and understanding wild animals in captivity. And, finally, he helped

stimulate study and ongoing debate on the subjective mental capability of animals.

George Rabb (1993)

Heini Hediger, Director of the Basel Zoo in Switzerland at the start of his career and Zürich Zoo at the end, was a pioneer in zoo biology who wrote several books on psychology, psychopathology and captive management which influenced zoo workers for generations. In fact, his writings should be required reading for the serious zoo professional. He had a strong interest in herpetology. His doctoral dissertation "Beiträge zur Herpetologie und Zoogeographie Neu Britanniens und einiger umliegender Gebiete (Contribution to the herpetology and zoogeography of New Britain and some adjacent regions)" included information on the habits of species collected there and the relationship to the environment. His specimens were deposited in the Basel Natural History Museum. For his contributions, the venomous snake

Fig. 80. Heini Hediger on 6 December 1988. *Photograph provided by Christian Schmidt.*

Parapistocalamus hedigeri was named in his honor by Jean Roux in 1934.

During the period between 1932 and 1943, he published many papers in herpetology on a variety of topics: problem of the "flying" snakes, peculiar biotope of *Lacerta muralis muralis,* reptiles and amphibians collected by Dr. A. Bühler in the Admiralty Archipelago and several neighboring islands, herpetology and zoogeography of New Britain, poisonous snakes and snake poison, snakes of Central Europe, two papers on his herpetological observations in Morocco, unusual reptiles and amphibians of the Solomon Islands, and a feeding mud snake (*Farancia*). Hediger's *Die Schlangen Mitteleuropas,* published in 1936, was for many years up to the end of World War II, the only available modern guide for Central Europe until the appearance of *Kriechtiere und Lurche* by Robert Mertens in 1952.

In 1964, he published *Wild Animals in Captivity.* In it, Hediger discussed many topics related to captive and wild herps: territoriality in alpine salamanders, thermoregulation, housing, cage furniture, flight distance and security, space, quarantine, feeding techniques and space-time rhythm in snakes (sloughing, feeding, digestion, sloughing again). He pointed out that snakes often refuse to feed during the shedding period and varanids regurgitate pellets after feeding. Hediger noticed that snakes have varied responses to different caretakers, ranging from passivity to aggression. He explained that he had seen no instance of reptiles viewing humans as companions (but see Komodo dragon vignette in Chapter 3). Hediger mounted a passionate plea for careful record keeping; his message was that no zoo worker could be perceived as professional if accurate records were not kept. Based on fecal examinations of wild and captive snakes, Hediger found many shed fangs; reptile keepers must be careful not to be impaled when removing scats from cages (1958).

Four years later in his book *The Psychology and Behaviour of Animals in Zoos and Circuses*, Hediger discussed "fascination" (hypnosis) of prey when encountering snakes. His point was that prey animals are not hypnotized but rather may be attracted to "fascinatory organs"; a prey's response is not to escape but rather to investigate the snake's puzzling appearance. Examples of these "organs" include caudal luring, tongue extension, nasal appendages, and anterior trunk movement. Toads can deceive predatory snakes as well by assuming exaggerated postures or becoming immobile. He listed "expression phenomena": acoustic, optic, olfactory and internal, including herp examples.

In the book *Man and Animal in the Zoo; Zoo Biology* published the next year, his readable account of the inner-working of zoos included a disturbing story of the times that Hediger visited other zoos, only to find that antivenin stock had deteriorated, expired or was the wrong type; in one case, there wasn't even a clean hypodermic needle available to administer the serum!

His contributions to the zoo field are covered in detail by Rübel and Honegger (2001).

Gerald Durrell (1925-1995)

GERALD HAD ALWAYS SEEN THE DEVELOPMENT OF HIS ZOO AS TAKING PLACE IN TWO PHASES. THE FIRST WAS THE ESTABLISHMENT OF A BASIC STRUCTURE, WITH A STOCK OF ANIMALS SUFFICIENT TO ATTRACT A PAYING PUBLIC IN NUMBERS LARGE ENOUGH TO GUARANTEE THE CONTINUANCE OF THE EMBRYONIC INSTITUTION. THE SECOND PHASE WOULD BE THE ZOO'S EVOLUTION INTO A PUBLIC BODY AND SERIOUS SCIENTIFIC ESTABLISHMENT WHOSE PRIMARY FUNCTION WAS THE CAPTIVE BREEDING OF SPECIES THREATENED WITH EXTINCTION. AT THE REQUEST OF JULIAN HUXLEY HE HAD DELAYED PUSHING THIS THROUGH IN ORDER TO GIVE PRIORITY TO THE FOUNDATION OF THE WORLD WILDLIFE FUND, WHICH CAME INTO BEING IN 1961. BUT NOW, WITH JERSEY ZOO ON A STABLE, ALBEIT STILL IMPOVERISHED FOOTING, GERALD DECIDED THE TIME HAD COME TO MOVE TO THE SECOND PHASE OF THE ORGANISATION'S DEVELOPMENT AND TRANSFORM HIS SMALL LOCAL ZOO INTO A WORLD-CLASS CHARITABLE SCIENTIFIC TRUST.

> DOUGLAS BOTTING (1999) IN HIS BOOK *GERALD DURRELL. THE AUTHORIZED BIOGRAPHY*

The founder of Jersey Wildlife Preservation Trust (JWPT is now Durrell Wildlife Conservation Trust [DWCT], named for him by his widow Lee), and noted author Gerald Durrell was a visionary far ahead of his time. With a lifelong interest in animals and their protection, he started the Trust on the Channel Islands off the coast of Great Britain to conserve endangered fauna, especially insular taxa. His approach was simple but effective. Arrangements were made with wildlife officials and governmental agencies in those countries where endangered species occurred, including management plans with both an *in situ* and *ex situ* component. When animals were taken to Jersey for breeding, they remained the property of the country of origin and were often reintroduced after captive breeding had increased numbers to a safe level. In this way, the captive population in Jersey was a hedge against extinction of the wild population. In addition, especially on islands such as Round Island, feral mammals and

Fig. 81. Gerald Durrell with female Parson's chameleon. This picture was taken during the same filming trip in 1981 for 'The Ark on the Move,' probably in October, in Berenty, Madagascar. *Durrell Wildlife.*

alien vegetation were eliminated before reintroduction was attempted. Few zoos have embraced such effective conservation enterprises and fewer still have supported them with time and money. The Durrell Institute of Conservation and Ecology at the University of Kent at Canterbury, UK, and Durrell Wildlife Conservation Trust promote important conservation initiatives.

The facility is rather small so large numbers of animals are not kept; the aim of the staff is to exhibit and breed a select number of endangered species: ploughshare (*Geochelone yniphora*), flat-tailed (*Pyxis planicauda*) and radiated tortoises, Round Island boa (*Casarea dussumieri*), gecko (*Phelsuma guentheri*), and skink (*Leiolopisma telfairii*), Mallorcan midwife toad (*Alytes muletensis*), Montserrat mountain chicken (*Leptodactylus fallax*), Coahuilan box turtle (*Terrapene coahuila*), San Francisco garter snake (*Thamnophis sirtalis tetrataenia*), Antiguan racer (*Alsophis antiguae*), Jamaican and Puerto Rican boas (*Epicrates subflavus, E. inornatus*), St. Lucian whiptail (*Cnemidophorus vanzoi*) and ground iguanas (*Cyclura collei*). Mammals and birds are included in the collection. Some of these include a reintroduction component.

In his writings, Durrell stressed the need for strategic planning for species conservation and the need to change from captive breeding, research and education in zoos to field projects with a conservation focus. A single example will demonstrate how a broad-based zoo initiative involving the Fort Worth Zoo, San Diego Zoo, DWCT, and approximately 15 other zoos has been developed by arranging a partnership with the Hope Zoological Gardens, Kingston, Jamaica. Approximately one hundred captive specimens of one of the most critically endangered reptile species, the Jamaican iguana *Cyclura collei* are being maintained at the Hope Zoo where studies are underway on reintroduction, nutrition, social behavior, daily activity patterns and thermoregulation (Gibson, 1993). DWCT and other zoos, particularly the Ft. Worth Zoo and the San Diego Zoo, are supporting this program in Jamaica by building holding enclosures for head-started captives to be released, paying for upkeep for these young lizards until release, preparing educational materials to explain the purpose for protecting this unique taxon, and contributing monies for field researchers' salaries. This program fills the criteria proposed by Durrell: partnerships with other zoos, building an *in situ* component, and supporting institutions within the country of origin.

When I visited the Trust in 1989, I was impressed by the care and attention given to each animal and the overall health of the collection. This was my first opportunity to see Round Island boas, skinks and geckos, ranging from juveniles to adults.

Miguel Alvarez del Toro (1917-1996)

DURING HIS LIFETIME, ALVAREZ WAS THE LEADER IN ESTABLISHING WILDLIFE PRESERVES IN CHIAPAS, AND HE WON MANY HONORS FROM CONSERVATION ORGANIZATIONS IN SEVERAL DIFFERENT COUNTRIES. HE WAS ALSO THE RECIPIENT OF TWO HONORARY DOCTORAL DEGREES FROM UNIVERSITIES IN CHIAPAS.

ROGER CONANT (1997)

Miguel Alvarez del Toro was the premier zoo herpetologist and conservationist in México. He was born in Colima and moved to Mexico City in 1939, to be a taxidermist and preparator at the Natural History Museum. While there, he also collected specimens for the Philadelphia Academy of Sciences. In 1942, he was asked by the governor of the state of Chiapas to assist in developing a zoo linked with the new Natural History Museum and Tropical Vivarium. He accepted the challenge, was elevated to director of this renamed Natural History Institute, and remained its head for over 50 years.

Don Miguel, as he was affectionately known, was a very productive herpetologist who published a number of books and papers. The World Wildlife Fund and the Mexican government invited him to conduct research on the biology and feasibility of captive breeding of the Morelet's crocodile. His findings, coupled with his observations on the other Mexican crocodilians, were contained in his magnum opus, published

Fig. 82. Miguel Alvarez del Toro (in black jacket) receiving award from Chicago Zoological Society in January 1989 for his work in conservation. *Audiovisual Services, Chicago Zoological Park, provided by George Rabb.*

in 1974: *Los Crocodylia de Mexico*. Photographs comparing egg sizes, neonates, and adults of the spectacled caiman (*Caiman crocodilus chiapasius*), American and Morelet's crocodile, with additional images of mating pairs, habitats, nesting sites, and gastroliths, are included. There were three editions (1960, 1973, 1982) of his book *Los Reptiles de Chiapas*, an excellent treatment of the reptiles of Chiapas. The final edition included many color photographs.

At the Zoo, the American crocodile reproduced for the first time in captivity in México and detailed observations were made on parental care in caimans in 1969.

Miguel Alvarez del Toro received many awards throughout the years for his work as an ardent conservationist, author of several books on the spiders, mammals and birds of Chiapas, and consummate zoo professional. As a testimony to the high esteem accorded him by his associates, several amphibians and reptiles have been named in his honor: beaded lizard (*Heloderma horridum alvarezi*), tropical night lizard (*Lepidophyma alvarezi*), salamander (*Nototriton alvarezdeltoroi*), colubrid snake (*Coniophanes alvarezi*) and anole (*Anolis alvarezdeltoroi*).

Today the institution, located in Tuxtla Gutiérrez, is known as Zoologico Regional Miguel Alvarez Del Toro (ZOOMAT) and specializes in exhibiting and breeding the fauna from Chiapas. When I visited some years ago, the beauty of the setting and the exhibits made an indelible impression.

Thomas A. Huff (1948-1998)

THERE APPEARS TO BE A STRONG TENDENCY ON THE PART OF RESEARCHERS AND REPTILE BREEDERS TO INTERPRET SUCCESSFUL PROPAGATION EFFORTS AS IMPLYING THE PRESENCE OF NECESSARY EXCITATORY STIMULI. INDEED, THE WORDS "STIMULUS" AND "EXCITATION" ARE FREQUENTLY USED INTERCHANGEABLY. WHILE WE CERTAINLY DO NOT DENY THE EXISTENCE OR THE IMPORTANCE OF EXCITATORY STIMULATION, WE WANT TO POINT OUT THAT SOME (PERHAPS MANY) STIMULI ACT BY REMOVING INHIBITION RATHER THAN BY GENERATING EXCITATION.

DAVID CHISZAR, HOBART SMITH, AND JAMES MURPHY UNDERSCORING THE IMPORTANCE OF HUFF'S INSIGHT ON THE 'CAPTIVE STAGNANCY' SYNDROME IN CAPTIVE SNAKES (1992)

Thomas Huff started his career as a keeper at the San Francisco Zoo. Beginning in 1975, he was Director of the Reptile Breeding Foundation in Picton, Ontario, Canada, and began accumulating sizeable captive colonies of insular boas of the genus *Epicrates* for the purpose of observing and recording reproductive patterns. His efforts were successful as nine taxa gave birth, resulting in several hundred young. For many years, Tom also served on the Board of Directors of the Institute for Herpetological Research in Stanford, California, a nonprofit foundation started by Richard Ross.

Several endangered forms produced at the Foundation deserve mention: Jamaican, Puerto Rican, and Cuban boas. Huff developed a number of important husbandry techniques to maintain his captives in good health and was particularly adept at raising neonates, which can be notoriously difficult to feed. The Foundation was a private enterprise closed to the public

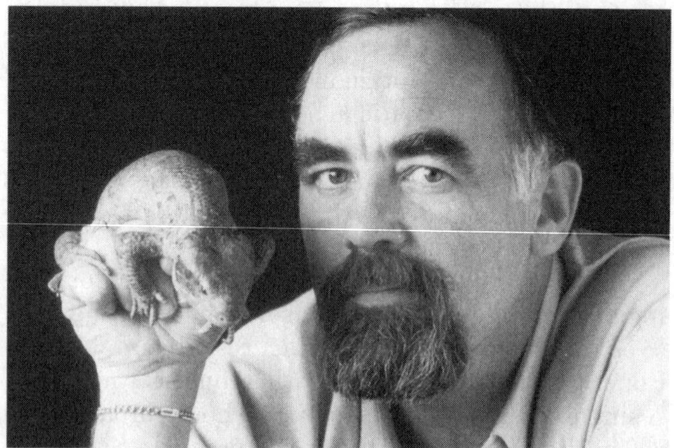

Fig. 83. Thomas A. Huff holding Solomon Island prehensile-tailed skink. *Photograph provided by Jeremy Huff.*

and Tom felt that this factor was critical for success since the snakes were rarely disturbed.

Perhaps his most important contribution was the realization that captive snakes require a changing environment to stimulate reproductive behavior. Huff changed cage furniture, temperature, humidity, feeding times, diurnal cycles and other stimuli to vary routines. He allowed males to engage in combat. His discovery that snakes could recognize other conspecifics as individuals and had mating preferences for some but not others led him to change combinations of breeders. Tom argued that it was not a specific parameter which led to success, i.e., a certain temperature range or humidity level, but rather that **any** change was the important element. He coined the term "captive stagnancy" to refer to the condition in captive snakes which are lethargic, inactive and disinclined to breed.

Some years after his Foundation had been operating successfully, Tom told me that he was the victim of his success. After producing hundreds of snakes, he discovered that there were few colleagues willing to accept his surplus specimens. This can be the danger of any successful program, particularly when large reptiles are involved. Unfortunately, the Foundation no longer exists.

Sean McKeown (1944-2002)

The world of herpetology recently suffered a great loss upon the death of Sean McKeown. His aim was to make a difference by sharing his love of herpetology—the wonders of the many species, the helpful tips on care, the history of some famous creatures, the commitment to conserving reptiles and amphibians.

Honolulu Zoo Director Emeritus Paul Breese (2002)

For two decades beginning in 1975, Sean McKeown was the herpetological curator at the Honolulu Zoo and later at the Chaffee Zoo in Fresno, California. Sean followed Ron Tremper at Fresno. When I visited the Chaffee Zoo years ago, Sean proudly demonstrated the computerized environmental chambers for herps (see Baker for description, 1986). Each unit duplicated the environmental parameters encountered by its inhabitants in the wild, such as day length, humidity, temperature and so on. Sean explained that a recent power outage had erased the settings; since there was not a backup, all had to be reinserted.

McKeown published many papers on geckos, especially day geckos of the genus *Phelsuma* (1982, 1984, 1989, 1992). His book *The General Care and Mainte-*

Fig. 84. Sean McKeown at Black River, Mauritius, on 21 June 1981. Male panther chameleon (*Furcifer pardalis*) on his shoulder contributed to production of fertile eggs and offspring at Chaffee Zoo. *Photograph provided by Wendy McKeown.*

nance of Day Geckos is an excellent guide covering this diverse group so highly prized by fanciers (1993).

McKeown's first book, published in 1978, was on Hawaii herpetofauna: *Hawaiian Reptiles and Amphibians*. His second book, written 18 years later and titled *Field Guide to the Reptiles and Amphibians in the Hawaiian Islands*, documented the establishment of alien species such as red-eared sliders, Jackson's chameleons and Madagascar day geckos. He was concerned about the possibility of accidental releases of the brown tree snake from Guam into Hawaii (1992) and was a member of the Brown Tree Snake Control Group, an active cluster of conservationists who implemented detection and removal plans.

He espoused the ecosystem approach for managing and displaying reptiles and amphibians in zoos (1985). In 1989, he documented the first captive breeding of the Madagascar ground boa in North America from long-term captive adults.

Tortoises were a strong interest, with papers on managing and breeding them in captivity (1991) and observations on the reproductive biology of the Asian brown and Angonoka tortoises in the Honolulu Zoo (1982). See Chun and Beaman (2003) for additional information on Sean's career.

Roger Conant (1909-2003)

Most herpetologists and herpetoculturists in North America, including teenagers and our most senior scien-

TISTS, HAVE BEEN INFLUENCED IN ONE WAY OR ANOTHER BY ROGER CONANT. MORE THAN ANY OTHER AMERICAN HERPETOLOGIST ACTIVE TODAY, HE HAS HAD AN EVEN BROADER IMPACT ON THE PUBLIC AT LARGE, THROUGH HIS MANY BOOKS, RADIO AND TELEVISION PROGRAMS, HIS CONSULTANTSHIP TO THE BOY SCOUTS OF AMERICA, AND MANY OTHER EDUCATIONAL ACTIVITIES.

KRAIG ADLER (1994)

Adler (1994, 2004) described Roger Conant's life and contributions to herpetology in detail so I will not duplicate his tribute to him here. Conant's herpetological output was so astounding and he has set such a high standard that it is virtually impossible for any zoo worker to duplicate his accomplishments. He was the model zoo professional.

Conant started his career at the Toledo Zoo but the majority of time was spent in Philadelphia where he was Curator of Reptiles (1935-1973) and later Director (1967-1973). He designed the new reptile building with naturalistic displays shortly before his retirement and the exhibits are named in his honor.

Besides a number of seminal works describing natricine and other colubrid snakes, and pitvipers, he published many papers on a variety of amphibians and reptiles (see Chapter 8, Toledo Zoo and Philadelphia Zoological Garden accounts for list). His books have become classics. In 1939, his *What Snake is That? A Field Guide to the Snakes of the United States East of the Rocky Mountains, by Roger Conant . . . and William Bridges . . . with 108 Drawings by Edmond Malnate* was created. Malnate, from the Philadelphia Academy of Sciences, was Conant's friend for years. In fact, they often drove together to work.

In 1958, seven years after *The Reptiles of Ohio* was produced, at the urging of Roger Tory Peterson,

Fig. 85. Undated photograph of Roger Conant and orangutan. *The Zoological Society of Philadelphia, provided by Karl Krantz.*

the creator of the Field Guide series for birds, Conant's *A Field Guide to the Reptiles and Amphibians of the United States and Canada East of the 100th Meridian* was published. This book immediately set the standard as the premier model for herpetological field guides. It is important to stress that this publication (and subsequent editions) has the largest distribution of any herpetological book ever and is regularly used by naturalists and amateurs as well as herpetologists who encompass a variety of disciplines.

The same year, his *Reptile Study. Reptile Study Merit Badge Pamphlet, Boy Scouts of America* (Revised Edition)" was distributed to countless thousands of youngsters. Many gained an appreciation of reptiles by reading this publication and enjoying the illustrations of these lovely creatures. This is precisely the sort of community outreach program that zoo workers seek today, yet Conant figured it out years ago! This publication was rewritten by Conant in 1972 and revised four years later.

The late Howard K. Gloyd and Conant finished *Snakes of the* Agkistrodon *Complex. A Monographic Review* in 1990, which took over 60 years to complete. In the 1970s, Gloyd visited the Dallas Zoo almost yearly with his grandchild to see the rattlesnake collection. At the Zoo, he saw for the first time living examples of some species he had written about in his monograph, especially montane forms from Mexico. We could always count on his enthusiasm each time he discovered a new one! During these visits he told us about his affection and admiration for Conant.

In 1997, a fascinating autobiography with certainly one of the most clever titles and dust jackets of any herpetological book appeared: A *Field Guide to the Life and Times of Roger Conant*. In it, Conant described his collecting adventures, research initiatives, zoo career and retirement activities. His portraits of zoo and herpetological personalities are delightful. This is must reading for anyone interested in the growth of zoo herpetology. Four color photographs show taxa named in honor of Conant: Conant's false brook salamander *Pseudoeurycea conanti*; spotted dusky salamander *Desmognathus conanti;* Florida cottonmouth *Agkistrodon piscivorus conanti*; and Conant's milk snake *Lampropeltis triangulum conanti.*

Personal Reflections: In the late 1960s, the American Zoo and Aquarium Association held a regional meeting in the Dallas-Fort Worth area. Since the focus was herpetology, I invited Roger and Edward H. Taylor from the University of Kansas to stay at my home and attend the conference. I thought that both would enjoy visiting the collections at the two zoos and interacting with the delegates.

Early each morning, Roger and Ed would be sitting in my living room, listening to Mozart, eating graham crackers, drinking coffee, and recalling their careers in herpetology and adventures in the field. Every evening, colleagues and friends came to my home to listen to these two great herpetologists reminisce about their herpetological experiences and their favorite places, especially Mexico. Roger was very careful about overextending himself and would retire at around ten o'clock each evening. On the other hand, Ed, who was an octogenarian, stayed up until two or three in the morning. Frankly, I was younger than either but awoke exhausted each morning. It was a remarkable week.

When I decided to retire in 1996, I asked Roger what dangers might lie ahead. He said that it is critical to always have a future project in mind to counteract boredom. His second rule was to ensure that the preceding day and the following day are different from the present day. He lived by this philosophy. As a nonagenarian, he finished a monograph on Mexican garter snakes, based on data acquired on earlier field trips. In 2003, as proof of his continuing productivity, he sent me an inscribed copy of this recently completed monograph published by the American Museum of Natural History where he was a research associate for over 50 years.

In 2001, I asked Roger to write the foreword and review drafts of chapters for this book. We kept in close contact by telephone but it was a bittersweet experience for me. Although he was always supportive, alert, and conversant, his voice was weaker, his eyesight failing, and his hearing less acute each time we communicated. Toward the end of his life, he easily tired and so our conversations became shorter and different. As older people are prone to do, he spent much time talking about his deteriorating health. He often said that he needed to end our discussions after a few minutes and put his "old bones" to rest.

It will be unusual if our profession ever produces another Roger Conant. He was my confidant, mentor, supporter, and friend—I miss him.

John E. Werler (1922-2004)

SPEAKING OF HERPETOLOGY CHRIS WEMMER OF THE SMITHSONIAN NATIONAL ZOOLOGICAL PARK WONDERED (1991), "HOW COME HERPETOLOGISTS TEND TO BE MORE COMMITTED BIOLOGISTS THAN BIRD AND MAMMAL CURATORS?" HE ALSO OBSERVED, "THEIR INTELLECTUAL COMMITMENT IS MANIFESTED BY HIGHER INDIVIDUAL PUBLICATION RATES, MORE COLLABORATION WITH NON-ZOO BIOLOGISTS, MORE FIELD EXPERIENCE, AND PROBABLY HIGHER PER CAPITA MEMBERSHIP IN PROFESSIONAL SOCIETIES." JOHN WERLER TYPIFIES THIS PARTICULAR BREED OF ZOO PROFESSIONALS. DR. WEMMER THEN REVEALED A LIST OF 21 SCHOLARS WORLD-WIDE TITLED "A GALLERY OF ZOO RESEARCHERS AND ZOO BIOLOGISTS." INCLUDED IN THIS IMPRESSIVE ROSTER WERE HIGHLY RESPECTED EUROPEAN SCHOLARS, JOHN WERLER AND HIS MENTOR CARL KAUFFELD.

KEN KAWATA AND JAMES B. MURPHY (2004)

John E. Werler was born on 11 June 1922 in Oldenburg, Germany, and emigrated to Weehawken, New Jersey in 1926. When he was 18, he started his career as a keeper in the zoo field at the Staten Island Zoo, New York City. His boss was Carl Kauffeld, the zoo's reptile curator, a mentor who frequently would spend an hour or two explaining the basics of taxonomy, the secrets of captive reptile maintenance, and proper identification. Since the zoo had a large number of rattlesnakes, one of Werler's jobs was to assist Kauffeld with individually marking rattlesnakes by pinning each snake while Kauffeld clipped subcaudal scales; Werler suffered a snakebite which was serious enough to keep him in a local hospital for five days.

In 1946 John became curator of reptiles at the San Antonio Zoo and was promoted to assistant director in 1952. He moved to the Houston Zoo and served as general curator from 1954 to 1963, when he became the director.

As director, he remained active in the reptile collection. As part of the ceremonial opening of the zoo's Reptile House in 1964, Werler demonstrated proper snake handling with a cobra. Some of the most unusual boas and pythons ever exhibited in a reptile

Fig. 86. John E. Werler and his rattlesnake friend while he was curator at San Antonio Zoo. A variation of this photograph was used on cover of his bulletin on the venomous snakes found in Texas. *Photo courtesy of John E. Werler.*

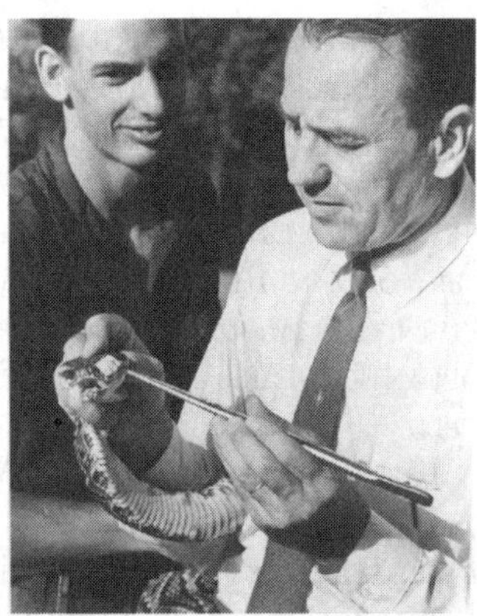

Fig. 87. Houston Zoo director John E. Werler (right) with reptile curator Tommy Logan during the mid-1960s. *Photo courtesy of John E. Werler.*

building were displayed in Houston, starting around 1965: all three Madagascan boas, Angolan dwarf python, Calabar python, white-lipped python, black-headed python, Bismarck ringed python, Panamanian dwarf boa, all of the Pacific boas, Argentinian boa, Boelen's python, and many insular Caribbean forms. The venomous snake assemblage was equally impressive: many Asian and Neotropical pitvipers, Hind's viper, Berg adder, many horned adder, night adders, Levant vipers, various Indo-Pacific elapids and several coral snakes. Although not as extensive, colubrid snakes were represented as well. A group of yellow-bellied sea snakes lived for years in the Aquarium. The reptile building was large and traditional with numerous exhibits on the outer walls. Two central islands housed smaller reptiles, mostly snakes. On one end was a giant tortoise enclosure and the other end had an exhibit of small lizards. Many of the smaller taxa were kept in rear sections off-exhibit where husbandry protocols were developed and refined.

Werler's scientific publications included reproduction of reptiles, taxonomy, and a monumental book on Texas snakes, coauthored with James R. Dixon in 2000 (see Chap. 8, San Antonio Zoo and Houston Zoo accounts for titles).

Werler was in charge of an operation in Houston which attracted an astounding number of zoo herpetologists over the years: Tim Jones, John Banks, Joe Laszlo, Ardell Mitchell, Tommy Logan, Gregory Mengden, the late John McLain, Bern Tryon, Hugh Quinn, R. Andrew Odum and Karl Peterson. Unique in the zoo world, 10 of these men have held or currently hold curatorial posts; virtually no other zoo can duplicate placement of so many former employees on this administrative level.

Werler served as AAZPA president for the period 1975 to 1976. His wise counsel and forward-thinking were critical skills in facing formidable challenges buffeting the zoo industry: shrinking finances, criticism from animal welfare groups, federal regulations and international treaties controlling wild animal trade, the need to maintain self-sustaining populations of captive wild animals, and the problems in finding homes for surplus animals.

Werler had a wicked sense of humor and enjoyed playing practical jokes. To quote Murphy and Card (1998): "The director at a major zoo was inspecting the reptile house with the staff. As he leaned over the railing to examine an exhibit closely, he failed to notice some foreign material on the rail. Earlier, a zoo visitor had placed his child on top of the railing whereupon the child promptly defecated. When the director checked his shirt, there was a prominent horizontal stripe emblazoned on the garment. He called each member of the staff in turn over to the exhibit and demanded that each bend over the railing to view the exhibit closely for signs of uncleanliness. As a result, each had a similar design on his shirt. Although the zoo had a standard uniform policy, matching brown stripes had not been envisioned." That director was John Werler.

Werler had an incredibly strong handshake and I lived in fear every time he proffered his hand. He would challenge his zoo employees to try and crush his hand —they never succeeded, as far as I know. Many staffers were imposing physical specimens, but they left these contests with heads lowered in ignominious defeat!

Joe Laszlo, a keeper in the reptile department, was a sensitive person easily intimidated by criticism. In the office of the reptile building, Werler's large personal library filled many shelves. Since he had been collecting for many years, a number of rare and irreplaceable books and monographs were in evidence. These works were available to his herpetological staff as John felt that it was vital that they develop a sense of herpetological history. One day, he opened his prized copy of Joseph Fayrer's *The Thanatophidia of India* and discovered that large portions of the text had been underlined in permanent ink, with notations in Hungarian written in the margins. When he checked other books, some of these were disfigured as well. Since Laszlo was the only person in the department who spoke Hungarian, clearly Joe had done the dastardly deed; he was afraid that he might forget important points contained in these tomes unless they were highlighted and annotated. Werler confronted Laszlo but the encounter did not go well. Following the meeting, Joe's enthusiasm and commitment toward his job plummeted precipitously; he even talked of suicide. During one of my visits, Werler pulled me aside to solicit my advice about how to deal with Joe's depressive state. Against my better judgment, I suggested that John forgive Joe, for it was clear that Laszlo could rebound only after he was forgiven. To his great credit, Werler did so and Joe recovered his bearings.

In Werler's honor, Roger Conant described the Tabasco watersnake *Natrix (Nerodia) rhombifera werleri* in 1953. Two amphibians from Veracruz, Mexico were named after Werler: Werler's false brook salamander (*Pseudoeurycea werleri*) and the leptodactylid frog *Eleutherodactylus werleri* [now *E. laticeps*]. In 2004, a paper by Jonathan Campbell and Werler appeared in *Southwestern Naturalist* describing a new species of lizard (*Diploglossus ingridae*) named after his beloved late wife, Ingrid. Werler died of cancer on 21 March 2004. Consult Card and Murphy (2000) and Kawata and Murphy (2004) for his biography.

Common Wall Lizard (*Podarcis muralis*)

Chapter 5
Zoological Gardens of Great Britain

Thorny Devil (*Moloch horridus*)

Zoological Society of London (1828)

History and Mission: The Zoological Society's history of herpetology, generated in the zoological garden in Regent's Park where the reptile building is located, is so rich and diverse that many topics are covered in individual accounts below, written by a number of zoo historians. When one reviews the extraordinary scientific output generated from this Zoological Society over many years, issues hotly debated in the 1960-70s by zoo workers were, in fact, addressed in London decades earlier.

Facility and Collection: According to *International Zoo Yearbook (*1998:Volume 28), composition of the collection was 160 taxa of reptiles with 529 specimens and 19 taxa of amphibians numbering 238.

Staff and Scientific Achievements: Edward George Boulenger was the first Curator of Reptiles in 1911. Joan Beauchamp Procter was the next Curator of Reptiles. She was involved in the planning of the 1927 building. Burgess Barnett was the next person to oversee the operation, followed by Jack Lester. R. A.

Fig. 88. London Zoo keeper with boa constrictor (*Boa constrictor*) ca. 1872. *Photograph by Frederick York, provided by John Edwards.*

Lanworn occupied the newly created position of Herpetologist, beginning in 1964. The Assistant Director of Science and Director of the Aquarium, H. G. Vevers, served as Acting Curator of the Reptile House during the 1970s. Angus d'A. Bellairs was honorary advisor to the staff. The Overseer of Reptiles, Dave Ball, retired from the Zoo. Simon Tonge was Senior Curator between 1994 and 2000 and is now Executive Director at the Paignton Zoological and Botanical Gardens. Before 2003, the reptile curator was responsible for the aquarium but this is no longer the case. Former herpetological curator Heather Hall is now only responsible for aquarium operations. The current Curator of Herpetology is Richard Gibson.

Publications: Not only did the Zoo staff pioneer many technological achievements for keeping captive reptiles and amphibians but also developed four important serials. The *Proceedings of the Zoological Society of London*, begun in 1833 and later changed to *Journal of Zoology*, was the vehicle for publishing scientific papers, many in herpetology. The annual *International Zoo Yearbook*, launched in 1960, is the premier publication for research by zoo and aquarium personnel worldwide to the present day. In 1864 the first volume of *The Record of Zoological Literature* was edited by Albert C. L. G. Günther and this series remains a valuable bibliographic reference over 135 years later. The Zoological Society of London Symposia series was started in 1960 and many of the volumes are devoted to topics of interest to zoo professionals: captive and medical management, conservation, and other relevant issues.

Personal Reflections: When I visited the Zoo in Regent's Park in 1969, Dave Ball spent the entire day showing the collection to me. Upon my return to the Dallas Zoo where I was employed, some of my nonherpetological colleagues asked if I had been impressed with the giant pandas and other rare mammals in London. Since I had been so engrossed with the European newts in naturalistic enclosures on the roof of the Reptile Building, I realized that I had totally forgotten the famous pandas. But I pointed out to my disbelieving listeners that newts were really neater than pandas anyway!

Historical Overview

Ball, D. J. 1982. A brief history of the Reptile Department at the Zoological Society of London, p. 41-62. In D.L. Marcellini (ed.), 6th International Herpetological Symposium on Captive Propagation and Husbandry. International Herpetological Symposium, Thurmont MD. [Overseer of Reptiles at the Zoo in 1982, Dave Ball, wrote this immensely enjoyable account of the rich history of his Department. Of particular interest is the section on feeding live food in 1849 and the public outcry which ensued.]

Bartlett, A. D. 1899. Wild Animals in Captivity Being an Account of the Habits, Food, Management and Treatment of the Beasts and Birds at the 'Zoo' with Reminiscences and Anecdotes by A. D. Bartlett/ Compiled and Edited by Edward Bartlett. Chapman and Hall, London. [book covered the habits of tortoises, including a story of a curator at a museum who was sitting quietly alone at night when a loud commotion commenced. One of the tortoises, varnished and properly labeled, revived and knocked the other rightly preserved chelonians hither and yon. In the next section, the habits of lizards such as the stump-tailed skink, thorny devil, iguana and chameleon are treated.]

Bartlett, A. D. 1900. Bartlett's Life Among Wild Beasts at the 'Zoo' . . . being a Continuation of Wild Animals in Captivity. The Habits, Food, Management and Treatment of the Beasts and Birds at the 'Zoo' with Reminiscences and Anecdotes/ Compiled and Edited by Edward Bartlett. Chapman and Hall, London. [This may be one of the longest titles in zoo literature. Chapter 1 covered the habits of snakes, including reproduction, venom, cannibalism, anecdotes and breeding of the Surinam toad. The last observations were made with A. Thompson, Head Keeper and Keepers Tyrrell and Tennant.]

Beddard, F. E. 1905. Natural History in Zoological Gardens Being Some Account of Vertebrated Animals, With Special Reference to Those Usually To Be Seen in the Zoological Society's Gardens in London and Similar Institutions. Archibald Constable, London. [Chapters 8 and 9 covered reptiles, amphibians and fishes.]

Bellairs, A. d' A., and D. J. Ball. 1976. Reptiles. Symp. Zool. Soc. London 40:119-132.

Bennett, E. T. 1829. The Tower Menagerie: Comprising the Natural History of the Animals Contained in that Establishment, with Anecdotes of Their Characters and History. Illustrated by Portraits of Each, Taken from Life, by William Harvey, and Engraved on Wood by Branston and Wright. Printed for R. Jennings, London.

Bennett, E. T. 1835. The Gardens and Menagerie of the Zoological Society Delineated/Published With the Sanction of the Council, Under the Superintendence of the Secretary and Vice-Secretary of the Society . . . Printed by Maurice, Clark . . . for the Proprietors, Published by Thomas Tegg [etc.], London [2 volumes: Quadrupeds, Birds].

Berridge, W. S. 1926. Marvels of Reptile Life. Thornton Butterworth, London. [anecdotes and photographs of herps at Zoo].

Berridge, W. S. 1935. All About Reptiles and Batrachians. George G. Harrap, London, Bombay, Sydney [anecdotes and photographs of herps at Zoo, with section on reptiles as pets].

Blatchford, D. R. 1986. History of reptile keeping at London Zoo. Herptile 11(3):95-101.

Blunt, W. 1976. The Ark in the Park, the Zoo in the Nineteenth Century. Hamilton: Tryon Gallery, London. [chapter 20, "Snake troubles," detailed the history of herpetology at the London Zoo. There are wonderful plates of the first two reptile buildings (p. 220, 221).]

Cassell's Popular Natural History. ca. 1856. [Illustration is of African rock pythons (*Python sebae*) at Zoological Society of London. Female on bottom right is covered with blanket for warmth and is brooding egg clutch. ca. 1856 4:51.]

Chambers, P. 2006. A Sheltered Life. The Unexpected History of the Giant Tortoise. Oxford University Press, Oxford, New York [history of giant tortoises at London Zoo].

Flower, S. S. 1929. Reptiles. In *List of the Vertebrated Animals Exhibited in the Gardens of the Zoological Society of London, 1828-1927*. Centenary Edition in Three Volumes. Zoological Society of London, London 1929; III: 1-272. [The third part on fishes was written by E. G. Boulenger. The first part on reptiles was mostly based on G. A. Boulenger's *Catalogue of Reptiles in the British Museum* in seven volumes published between 1885-1896. The lizard families were arranged according to the classification by C. L. Camp in his *Classification of the Lizards* published in 1923.]

Hopley, C. G. 1882. Snakes: Curiosities and Wonders of Serpent Life. Griffith & Farran, London. [In the last chapter titled Notes from the Zoological Gardens, Catherine Cooper Hopley discussed some of the snakes at the London Zoo, including the king cobra, rhinoceros viper, cottonmouth and Neotropical rattlesnake.]

Huish, R. 1830. The Wonders of the Animal Kingdom Exhibiting Delineations of the Most Distinguished Wild Animals in the Various Menageries of the Country. Thomas Kelly, London [description of the alligator, housed in the Royal Menagerie of London (p. 51-52)].

Keeling, C. H. 1985. Where the Crane Danced: More About Zoological Gardens of the Past. Clam Publications, Guilford UK. [expansive discussion of the Zoo collection.].

Keeling, C. H. 1992. Here, There and Regent's Park. Clam Publications, Guilford UK. [Important historical events which occurred elsewhere in Britain are juxtaposed with happenings at the London Zoo on the same day, gleaned from the Zoo's daily reports. Herpetological episodes are included.]

Keeling, C. H. 1992. A Short History of British Reptile Keeping. Clam Publications, Guilford UK. [Zoo's reptile collection was augmented by donations from a number of private benefactors. The history of the program there, the world's first reptile building, is fascinating reading. Two other Keeling publications are of interest. The first *Wonderful Year* covered the relatively high mortality rate of reptiles in the Zoo in 1938. The other *Year of Janus* documented very high reptile death numbers in 1900. In a letter dated 24 January 2001 to me, Keeling suggested that this high rate might have been intentional as Lord Rothschild was continually depositing his animals at the Zoo, although he really wanted them for his museum.]

Loisel, G. 1907-1908. The zoological gardens and establishments of Great Britain, Belgium, and the Netherlands. [Translated and abridged from the *Rapport sur une mission scientifique dans les jardins et établissements zoologiques publics et privés du Royaume-Uni, de la Belgique et des Pays-Bas*, par M. Gustave Loisel. Extrait des Nouvelles Archives des Missions Scientifiques, 19. Paris, 1907.] Washington DC, Smithsonian Inst. Rep. 1907, 1908, (407-448).

Loisel, G. 1912. Histoire des ménageries de l'antiquité à nos jours (History of Menageries from Antiquity to Present Times). O. Doin et fils, Paris.

London Zoo (London, England).1929. List of the Vertebrated Animals Exhibited in the Gardens of the Zoological Society of London, 1828-1927. Printed for the Society, London. [Centenary edition in three volumes (10th edition) included chapters on reptiles by S. S. Flower, amphibians by Malcolm A. Smith and fishes by E. G. Boulenger.]

Mitchell, P. C. 1929. Centenary History of the Zoological Society of London. London Zoological Society, London.

Peel, C. V. A. 1903. The Zoological Gardens of Europe. Their History and Chief Features. F. E. Robinson,

London. [Reptile building housed the heloderm lizard where Sir Joseph Fayrer demonstrated its venomous character, the largest python the author had ever seen, as well as a Chinese alligator and giant tortoises placed by Hon. Walter Rothschild (p. 192).]

Roberts, W. J. 1912. Zoo Folk. T. Werner Laurie, Clifford's Inn, London. [Quaint narrative tour of the Reptile Building at the London Zoo (chapter 2) is punctuated with a few anthropomorphisms but is delightfully written. The description of the Tortoise House (chapter 3) with the Charles Island tortoises, those domes of silence, is charming.]

Scherren, H. 1905. The Zoological Society of London. Cassell, London.

Sclater, P. L. 1884. Guide to the Gardens of the Zoological Society of London. Bradbury, Agnew, London.

Sclater, P. L. 1896. List of the vertebrated animals now or lately living in the Gardens of the Zoological Society of London. 9th ed. Printed for the Society, London.

Smith, M. A. 1929. Amphibia. In List of the Vertebrated Animals Exhibited in the Gardens of the Zoological Society of London, 1828-1927. Centenary Edition in Three Volumes. Zoological Society of London, London, 1929; III: 275-309. [Malcolm Arthur Smith (1875-1958) served as Curator of Reptiles at the London Zoo for a short time around 1937. He is known for his work on the herpetological fauna of British India, Malay Peninsula and Britain, as well as sea snakes. Smith wrote articles on herpetological topics for the magazine *Zoo Life*, published by the Zoological Society of London. See Adler (1989: 86) for his biography.]

Street, P. 1965. Animals in Captivity. Faber and Faber, London. (Chapter 7, "Reptiles in captivity," chronicled the history of herpetology at the London Zoo. Subjects covered included the various buildings which housed reptiles over the years, exhibition, feeding live food, departmental procedures, snakebite and antivenin production, and an overview of reptile biology.]

Toovey, J. W. 1976. 150 years of building at London Zoo. Symp. Zool. Soc. London 40:179-202. [This architectural history of the London Zoo is interesting reading. The 1882-83 Reptile Building is pictured on p. 182. This symposium volume covers the Zoological Society of London from 1826-1976.]

Zuckerman, L. 1976. Golden Days. Historical Photographs of the London Zoo. Gerald Duckworth, London. [Photographs of green sea turtle, Jackson's chameleon, alligator enclosure, cane toad, four line snake with neonates, Keepers Dexter and Wilson holding snakes, Keeper Arthur Budd extracting venom from Russell's viper, and Komodo dragon "Sumbawa" with Budd in 1928.]

Morphology, Systematics & Taxonomy

Boulenger, E. G. 1911. On a new tree-frog from Trinidad, living in the Society's Gardens. Proc. Zool. Soc. London 1911:1082-1083 [description of *Hyla goughi*, named for Lewis H. Gough, now known as *Hyla minuta*. Color plate].

Boulenger, E. G. 1915. On two new tree-frogs from Sierra Leone, recently living in the Society's Gardens. Proc. Zool. Soc. London 1915:243 [description of *Rappia aylmeri*, now known as *Hyperolius fusciventris*, and *Rappia chlorostea*, now known as *Hyperolius chlorosteus*].

Boulenger, E. G. 1915. On a colubrid snake (*Xenodon*) with a vertically movable maxillary bone. Proc. Zool. Soc. London 1915:83-85.

Boulenger, E. G. 1916. On a new lizard of the genus *Phrynosoma*, recently living in the Society's Gardens. Proc. Zool. Soc. London 1916:537. [description of *Phrynosoma brevicornis* with photographs, now known as *Phrynosoma cornutum*.].

Boulenger, E. G. 1924. On a new giant salamander, living in the Society's Gardens. Proc. Zool. Soc. London 1924:173-174. [Description is of *Megalobatrachus sligoi*, named after the Marquess of Sligo who procured the specimen. This taxon is now known as *Andrias davidianus*.]

Boulenger, G. A. 1888. Description of a new land-tortoise from South Africa, from a specimen living in the Society's Gardens. Proc. Zool. Soc. London 1888:251 [description of *Homopus femoralis*].

Gray, J. E. 1860. Description of a new species of *Emys* lately living in the Gardens of the Zoological Society. Proc. Zool. Soc. London 1860:232-233 [description of *Emys fuliginosus*].

Procter, J. B. 1920. On a collection of tailless batrachians from East Africa made by Mr. A. Loveridge in the years 1914-1919. Proc. Zool. Soc. London 1920:411-420.

Procter, J. B. 1921. On the variation of the scapula in the batrachian groups Aglossa and Arcifera. Proc. Zool. Soc. London 1921:197-214.

Procter, J. B. 1923. On new and rare reptiles from South America. Proc. Zool. Soc. London 1923:1061-1067.

Procter, J. B. 1923. On new and rare reptiles and batrachians from the Australian region. Proc. Zool. Soc. London 1923:1069-1079.

Procter, J. B. 1924. Unrecorded characters seen in living snakes, and description of a new tree-frog. Proc. Zool. Soc. London 1924:1125-1129. [Descriptions are of albino cobra, neck inflation of false water cobra, tongue extension in arboreal snakes, and a new tree frog (*Hyla blandsuttoni*), named in honor of Sir John Bland-Sutton. This taxon is now *Litoria aurea*.]

Husbandry

Ashby, G. J. 1962. Breeding locusts for food at London Zoo. Inter. Zoo Yearb. 4:128-130.

Ball, D. 1969. Housing reptiles. J. Inst. Anim. Techs 20:137-154.

Ball, D. 1970. Terrapins, p. 79-93. In M. Knight (ed.), Tortoises and How to Keep Them. 2nd ed. Brockhampton Press, Leicester UK.

Ball, D. J. 1974. Handling and restraint of reptiles. Inter. Zoo Yearb. 14:138-140.

Ball, D. J., and A. d'A. Bellairs. 1972. Reptiles. The UFAW Handbook on the Care and Management of Laboratory Animals. 4th ed. Williams and Wilkins, Baltimore. p. 490-510.

Medical Management

Griffith, A. S. 1928. Tuberculosis in captive wild animals. J. Hyg. 28:198. [Cases that were reported from the London Zoo from 1924-1933. Griffith isolated *Mycobacterium marinum* from 71% of cases. See Griffith (1939) in Proc. Royal Soc. Medicine 32:1405.]

Scott, H. H. 1925. Congenital malformations of the kidney of reptiles, birds, and mammals. Proc. Zool. Soc. London 1925:1259-1270. [H. Harold Scott was pathologist to the Zoo.]

General

Balmford, A. 2000. Separating fact from artifact in analyzes of zoo visitor preferences. Conserv. Biol. 14:1193-1195. [Two of the three most popular exhibits were the aquarium and reptile house at the Zoo.]

Bartlett, A. D. 1894. On a single case of one snake swallowing another in the Society's Reptile House. Proc. Zool. Soc. London 1894:669-670.

Bateman, G. C. 1897. The Vivarium, Being a Practical Guide to the Construction, Arrangement, and Management of Vivaria, Containing Full Information as to all Reptiles Suitable as Pets, How and Where to Obtain Them, and How to Keep Them in Health. L. Upcott Gill, London. [Keeper Tyrrell from the Zoo assisted Bateman with the development of this book.]

Boulenger, E. G. 1911. Exhibition of living male specimens of the midwife toad (*Alytes obstetricans*) carrying the eggs. Proc. Zool. Soc. London 1911:696.

Boulenger, E. G. 1913. Exhibition of a young specimen of the matamata terrapin (*Chelys fimbriata*). Proc. Zool. Soc. London 1913:1097.

Boulenger, E. G. 1913. Exhibition of a blue specimen of the edible frog (*Rana esculenta*), and of the remarkable Australian lizard, *Pygopus lepidopus*. Proc. Zool. Soc. London 1913:151.

Boulenger, E. G. 1913. Exhibition of a living melanistic specimen of the green lizard (*Lacerta viridis*). Proc. Zool. Soc. London 1913:546.

Boulenger, E. G. 1913. Exhibition of the spines of an insectivore found in the excrement of a boa. Proc. Zool. Soc. London 1913:152 [Madagascar].

Boulenger, E. G. 1913. Exhibition of a living specimen of the salamander (*Amblystoma tigrinum*). Proc. Zool. Soc. London 1913:2-3.

Boulenger, E. G. 1913. Experiments on the metamorphosis of the Mexican axolotl (*Amblystoma tigrinum*), conducted in the Society's Gardens. Proc. Zool. Soc. London 1913:403-413.

Boulenger, E. G. 1914. Exhibition of photograph of the giant saddle-backed tortoise. Proc. Zool. Soc. London 1914:220-222 [thought to be female *Geochelone abingdonii*].

Boulenger, E. G. 1914. Reptiles and Batrachians. J. M. Dent & Sons, London; E. P. Dutton, New York.

Boulenger, E. G. 1920. Exhibition of, and remarks upon, a remarkable new land tortoise (*Testudo loveridgii*). Proc. Zool. Soc. London 1920:190.

Boulenger, E. G. 1920. Exhibition of, and remarks upon, living specimens of *Necturus*. Proc. Zool. Soc. London 1920:659.

Boulenger, E. G. 1920. On some lizards of the genus *Chalcides*. Proc. Zool. Soc. London 1920:77-83.

Boulenger, E. G. 1932. The Zoo and Aquarium Book. Duckworth, London.

Boulenger, E. G. 1938. World Natural History. C. Schribner's Sons, New York [first published November 1937 with imprint by B. T. Batsford, London].

Brightwell, L. R. 1947. The Zoo Aquarium. Zoo Life 2(2):35-37 [references to reptile collection].

Brightwell, L. R. 1947. House to house at the Zoo. The Reptile House. Zoo Life 2(3):88-91. [view of 119 years of reptile keeping at the Zoo.].

Cansdale, G. S. 1955. Reptiles of West Africa. Penguin

Books, London. [George Soper Cansdale spent 14 years in the Gold Coast as a member of the Colonial Forest Service where he collected small mammals and reptiles for museums and zoos. He published a number of books: *Animals of West Africa* (1946), *Animals and Man* (1952), *George Cansdale's Zoo Book* (1953), *Belinda, the Bush Baby* (1953) with his wife Sheila and works on Ghana string figures, West African lilies and orchids. Cansdale was in charge of the London Zoo for nearly five years until his post was abolished in 1953. He was a lecturer, broadcaster, writer and television personality. This book is filled with black-and-white photographs.]

Cansdale, G. S. 1961. West Africa Snakes. Longmans, Green, London [book illustrated by John Norris Wood and reprinted in 1965].

Cornish, C. J. 1894. Wild Animals in Captivity or, Orpheus at the Zoo and Other Papers. Macmillan, New York [description of lizards and crocodilians at the reptile building (p. 263-269)].

Cott, H. B. 1932. The Zoological Society's expedition to the Zambesi, 1927: No. 4. On the ecology of tree-frogs in the Lower Zambesi Valley, with special reference to predatory habits considered in relation to the theory of warning colours and mimicry. Proc. Zool. Soc. London 1932:471-541. [Hugh B. Cott collected frogs for the Zoo on this trip.]

Davidson, W. E. 1910. Land tortoises of the Seychelles. Letter sent to the Secretary of the London Zoological Society. [Interesting letter (1 June 1910) discussed longevity, size, reproductive behavior, growth and development, and location of specimens sent to the Zoo.]

Huxley, J. 1936. Most babies never see their mother. Zoo Life 1(5):22-24. [Julian Huxley, Advisory Editor and Secretary of the Zoological Society of London (1935-1943), published his wireless talks on animal life in this magazine which was the official organ of the Society. The magazine began in June 1936. This article described the reproductive biology of amphibians and reptiles. Captain Charles Pitman, former game warden in Uganda and author of *A Guide to the Snakes of Uganda,* also published in this magazine.]

Lanworn, R. A. 1972. The Book of Reptiles. Hamyln, London.

Mcfarlane, R. G., and B. Barnett. 1934. The haemostatic possibilities of snake venom. Lancet 1934(Nov. 3):985-987.

Morris, R., and D. Morris. 1965. Men & Snakes. McGraw-Hill, New York, San Francisco.

Procter, J. B. 1920. Exhibition of, and remarks upon, a living specimen of the tailed batrachian *Spelerpes fuscus* Bonaparte. Proc. Zool. Soc. London1920:437.

Procter, J. B. 1925. Notes on the nests of some African frogs. Proc. Zool. Soc. London 1925:909-910 [account covered *Chiromantis, Arthroleptis* and *Anhydrophryne*].

Procter, J. B. 1926. A note on an albino grass snake. Proc. Zool. Soc. London 1926:1095.

Procter, J. B. 1928. On the remarkable gecko *Palmatogecko rangei* Andersson. Proc. Zool. Soc. London 1928:917-922 [color plate].

Regan, C. T. (ed.). 1946. Natural History. Ward, Lock & Co., Ltd, London. [Sections on amphibians and reptiles were written by Edward George Boulenger.]

Durrell Wildlife Conservation Trust (1959)

History: The Durrell Wildlife Conservation Trust was formerly the Jersey Wildlife Preservation Trust and is located in Jersey, Channel Islands.

Facility and Collection: Prior to 1976, the Jersey Zoo reptile house was a smallish edifice not particularly effective for either displaying or keeping amphibians or reptiles but the situation changed dramatically with the addition of the Gaherty Reptile Breeding Centre (now re-named the Gaherty Amphibian and Reptile Conservation Centre), a facility with the majority of breeding and research taking place behind the scenes. The Gaherty building was different from most other reptile houses of the period in that it emphasized these off-view facilities for breeding and research programs rather than many public displays.

The aim of the staff is to work with a select number of endangered species: ploughshare, flat-tailed and radiated tortoises, Round Island boa, gecko and skink, Mallorcan midwife toad, Montserrat mountain chicken (*Leptodactylus fallax*), agile frog (*Rana dalmatina*), Coahuilan box turtle, San Francisco garter snake, Antiguan racer, Jamaican and Puerto Rican boas, and ground iguanas. Some of these include a reintroduction component. The Mallorcan midwife toad project is one of longest running programs.

According to *International Zoo Yearbook* (1998:Volume 28), composition of the collection was 36 taxa of reptiles with 357 specimens and 6 taxa of amphibians numbering over 220.

Staff and Scientific Achievements: Quentin Bloxam

has been in charge of the herpetological program for many years. Simon Tonge was his assistant between 1980 and 1991. Richard Gibson was Herpetology Department Head until 2001 but has moved to the London Zoo. For two years, Kevin Buley replaced Gibson but has relocated to the Chester Zoo. Gerardo Garcia is the incumbent.

Publication: The scientific journal *Dodo* was inaugurated in 1977 and includes many papers by DWCT staff.

Historical Overview

Botting, D. 1999. Gerald Durrell. The Authorized Biography. Carroll & Graf, New York. [Durrell published 38 books and many other articles dealing with animals.]

Conservation

Behler, J. L., Q. M. C. Bloxam, E. R Rakotovao, and H. J. A. R Randriamahazo. 1993. New localities for *Pyxis planicauda* in West-Central Madagascar. Chelonian Conserv. Biol. 1(1):49-51 [contribution of Tsimbazaza Zoo, DWCT and Wildlife Conservation Society, New York].

Bloxam, Q. M., and S. J. Tonge. 1995. Amphibians: Suitable candidates for breeding-release programs. Biodiv. Conserv. 4:636-644.

Buley, K. R., and G. Garcia. 1997. The Recovery Programme for the Mallorcan midwife toad *Alytes muletensis*--an update. Dodo, J. Jersey Wildl. Preserv. Trust 33:86-90.

Curl, D. A., J. C. Scoones, and M. K. Guy. 1985. The Madagascar tortoise *Geochelone yniphora*: current status and distribution. Biol. Conserv. 34:35-54 [recovery program supported in part by DWCT].

Durrell, L. 1998. Strategic planning for species conservation by Jersey Wildlife Preservation Trust. Dodo, J. Jersey Wildl. Preserv. Trust 34:176-177.

Durrell, L., and J. J. C. Mallinson. 1998. The impact of an institutional review: a change of emphasis toward field conservation programmes. Inter. Zoo Yearb. 36:1-8. [Authors stress the change from captive breeding, research and education in zoos to field projects with a conservation focus.]

Escalona, T., and J. E. Fa. 1998. Survival of nests of the terecay turtle (*Podocnemis unifilis*) in the Nichare-Tawado Rivers, Venezuela. J. Zool. (London) 244:303-312.

Mallinson, J. J. C. 1998. Collaboration for conservation between the Jersey Wildlife Preservation Trust and countries where species are endangered. Inter. Zoo Yearb. 27:176-191. [Account covered this model program which included preparation of colored educational posters for Jamaican boa, plowshare tortoise and midwife toad for distribution in country of origin. Field observations included plowshare tortoise, various endemic reptiles and amphibians in Madagascar, Round Island boa and skink, Mallorcan midwife toad and Jamaican boa.]

McHenry, T., W. Brady, D. Candland, J. Echeverria, J. Ferguson, J. Jensen, F. Koontz, A. Jolly, V. Mars, A. Model, M. Pearl, and J. Tuten. 1999. Wildlife Preservation Trust International's conservation perspective and strategic directions. Dodo, J. Jersey Wildl. Preserv. Trust 35:113-116.

Rakotombololona, W. F. 1998. Study of the distribution and density of the Madagascar flat-tailed tortoise *Pyxis planicauda* in the dry deciduous forest of Menabe. Dodo, J. Jersey Wildl. Preserv. Trust 34:172-173.

Rakotoniaina, L. J. 1998. The role of community-based activities in the conservation of endangered animals in Madagascar. Dodo, J. Jersey Wildl. Preserv. Trust 34:173-174.

Reproduction

Bloxam, Q. 1983. A preliminary report on the captive management and reproduction of the Round Island boa, *Casarea dussumieri*, p. 115-117. In P. J. Tolson (ed.), 7th International Herpetological Symposium on Captive Propagation and Husbandry. International Herpetological Symposium, Thurmont MD. [DWCT developed a plan to eradicate the introduced mammals on Round Island to restore native vegetation. Several endangered species, including this taxon, have benefited but the other boa (*Bolyeria*) is apparently extinct.]

Bloxam, Q. M. C., and S. J. Tonge. 1986. Breeding programmes for reptiles and snails at Jersey Zoo; an appraisal. Inter. Zoo Yearb. 24/25:49-56.

Durrell, G., and L. M. Durrell. 1980. Breeding Mascarene wildlife in captivity. Inter. Zoo Yearb. 20:112-119 [*Phelsuma guentheri* featured but other herps mentioned].

Kuchling, G., and O.C. Razandrimamilafiniarivo. 1999. The use of ultrasound scanning to study the relationship of vitellogenesis, mating, egg production and follicular atresia in captive ploughshare tortoises *Geochelone yniphora*. Dodo, J. Jersey Wildl. Preserv. Trust 35:109-115 [Tortoise and Turtle Breeding Centre, Durrell Wildlife Conservation Trust].

Medical Management

Arnold, L. 1980. Pathology report. Dodo, J. Jersey Wildl. Preserv. Trust 17:96-104.

Arnold, L. 1981. Pathology report. Dodo, J. Jersey Wildl. Preserv. Trust 18:86-94.

Blampied, N. le Q., and A. F. Allchurch. 1978. Veterinary report. Dodo, J. Jersey Wildl. Preserv. Trust 15:102-106 [mouthrot].

Cooper, J. E., C. J. Dutton, and A. F. Allchurch. 1998. Reference collections: Their importance and relevance to modern zoo management and conservation biology. Dodo, J. Jersey Wildl. Preserv. Trust 34:159-166.

Tagg, J. 1987. Pathology report. Dodo, J. Jersey Wildl. Preserv. Trust 24:128-137.

Husbandry

Allchurch, A. F. 1986. The Nutrition Handbook of the Jersey Wildlife Preservation Trust. A Collection of the Diets in Current Use at the Jersey Wildlife Preservation Trust [publisher and place not given].

Brice, S. 1995. A review of feeding and lighting requirements for captive herbivorous lizards: *Cyclura* and *Iguana*. Dodo, J. Jersey Wildl. Preserv. Trust 31:120-139.

Hartley, J. 1963. Notes on tuataras (*Sphenodon punctatus*) in captivity. Inter. Zoo Yearb. 5:170-171 [pair received at JWPT on 6 August 1963].

Hick, U. 1965. Die Tuatara, ein Relikt aus längst vergangener Zeit (The tuatara, a relic of the ancient past). Freunde Köln. Zoo 8(3):75-82 [descriptions by specialists of captive specimens held at Auckland, Basel, Chester, Detroit, Dublin, Jersey and San Diego Zoos].

Reid, D. 1987. Rearing juvenile spur-thighed tortoises *Testudo graeca*. Reptiles: Proceedings of the 1986 U.K. Herpetological Societies. p. 55-60.

Watson, G. 1969. Notes on the care of Mastigure lizards *Uromastix acanthinurus* at Jersey Zoo. Inter. Zoo Yearb. 9:49-50.

Wheler, C. L., and J. E. Fa. 1995. Enclosure utilization and activity of Round Island geckos (*Phelsuma guentheri*). Zoo Biol. 14:361-369. [Study stressed environmental enrichment at JWPT by using cage furniture, various thermal levels and alternate focal heat.]

General

Clemons, J., and M. Lambert. 1996. Herpetology in Jersey; a report of the 1996 visit to Jersey organised by the conservation committee. Brit. Herpetol. Soc. Bull. 57:33-40.

Durrell, G. M. 1961. The Whispering Land. R. Hart-Davis, London [collection and preservation of zoological specimens, and zoology of Argentina].

Esson, M., and K. Cowan. 1998. Cross-curricular activities: A richness of opportunities for zoo educators. Dodo, J. Jersey Wildl. Preserv. Trust 34:115-124.

Feistner, A. T. C. 1998. Jersey Wildlife Preservation Trust scientific output 1997. Dodo, J. Jersey Wildl. Preserv. Trust 34:125-129.

Feistner, A. T. C. 1999. Durrell Wildlife Conservation Trust scientific output 1998. Dodo, J. Jersey Wildl. Preserv. Trust 35:176-180.

Phillpot, P. 1996. Visitor viewing behaviour in the Gaherty Reptile Breeding Centre, Jersey Wildlife Preservation Trust: a preliminary study. Dodo, J. Jersey Wildl. Preserv. Trust 32:193-202.

Price, E. C., and A. T. C. Feistner. 1999. Durrell Wildlife Conservation Trust scientific output 1998. Dodo, J. Jersey Wildl. Preserv. Trust 35:176-180.

Waugh, D. 1988. Training in zoo biology, captive breeding, and conservation. Zoo Biol. 7:269-280 [description of DWCT's International Training Program].

Westley, F., U. Seal, and C. C. M. Clark. 1999. The Population and Habitat Viability facilitators' course: A retrospective. Dodo, J. Jersey Wildl. Preserv. Trust 35:124-133.

Edinburgh Zoo (1913)

According to *International Zoo Yearbook* (1998:Volume 28), composition of the collection was 31 taxa of reptiles with 216 specimens and 10 taxa of amphibians numbering 240.

Historical Overview

Gillespie, T. H. 1934. Is It Cruel? A Study of the Condition of Captive and Performing Animals. H. Jenkins, London [arguments and examples of animal behavior and welfare from the Zoological Park at Edinburgh, Scotland].

Gillespie, T. H. 1964. The story of the Edinburgh Zoo: The Royal Zoological Society of Scotland and the Scottish National Zoological Park. An Account of their Origin and Progress. M. Slains; Old Castle, Slains, Aberdeenshire, Scotland. [Thomas Haining Gillespie was first Director-Secretary of the Society who described an anaconda which fasted for at least 14 months. A 13-foot king cobra normally fed on grass snakes but refused a live adder.]

General

Gillespie, T. H. 1934. The Way of the Serpent (A Popular Account of the Habits of Snakes). R. M. McBride and Co., New York. [There are many references to snakes in zoos and captivity, including a chapter on the reaction of other animals to snakes. Gillespie was Director at Zoo at time of publication.]

Skelton, T. M., N. K. Waran, and R. J. Young. 1996. Assessment of motivation in the lizard, *Chalcides ocellatus*. Anim. Welfare 5:63-69.

Chester Zoo (1931)

Personal Reflections: When I visited in the late 1960s, the Chester Zoo in Chester, England, had a varied collection of amphibians and reptiles. The founder and Director, George S. Mottershead, was my host and he was particularly proud of the North American representatives we had sent to the Zoo from Dallas several years earlier. Since that time, the staff at the Zoo has been involved in a number of conservation initiatives and has focused less on a large and varied collection. Kevin Buley is Curator of Lower Vertebrates and Invertebrates.

General

Cowley, D. L. 1981. Herptiles on view at Chester Zoo, 13 July 1981. Herptile 6(4):27-29.

Conservation

Anon. 2000. Sand lizards return again to Wales. Zoo Life, Issue 3. Autumn: 10. [Presents a portrait of the Chester Zoo recovery and reintroduction program for the endangered sand lizard *Lacerta agilis* which included captive breeding, resulting in 27 young.]

Medical Management

Lyon, D. G. 1971. A survey of parasite problems, treatment and control at Chester Zoo, p. 147-152. In Ippen, R., and H.-D Schröder, (eds.). Erkrankungen der Zootiere. Verhandlungsberichte des 23. Internationalen Symposiums über die Erkrankungen der Zootiere von 2. bis 6. Juni 1971 in Helsinki. Akademie-Verlag, Berlin.

Zoo Miscellany

General

Croudace, C. 1989. Cooperative, reptile management in the U.K. and Europe: History and Proposals. Reptile 14:31-45. [Charlotte Croudace is Scientific Officer at the Bristol Zoo.]

Keeling, C. H. 1985. Where the Crane Danced: More About Zoological Gardens of the Past. Clam Publications, Guilford UK. [This is an expansive discussion of the London Zoo collection; Crystal Palace (1871-1936) which had axolotls and "The Royal Exhibition of Working Ants" in 1884 and crocodiles, vipers and salamanders 26 years later; Doncaster Zoological Garden (1956-1960) which had an impressive number of amphibians and reptiles on exhibit, including some unusual taxa such as moloch, lace monitor, beaded dragon, montpellier snake, egg-eating snake, and tegu; Aberdeen Zoological Garden (1962-1973) with an Exhibition House displaying reptiles and amphibians.].

Keeling, C. H. 1992. A Short History of British Reptile Keeping. Clam Publications, Guilford UK. [Clinton H. Keeling must have enjoyed this project as he searched through old literature to document captive reptile keeping. The first herper, born in 1573, was William Laud, Archbishop of Canterbury, who kept a spur-thighed tortoise, acquired around 1625. What was surprising was that a chameleon, usually difficult to keep in captivity, was obtained in 1820 by Sir Robert Heron. After that, several British zoos kept a variety of reptiles, ranging from pythons to venomous snakes.]

Keeling, C. H. 2001. Zoological gardens of Great Britain, p. 49-74. In V. N. Kisling Jr., (ed.). Zoo and Aquarium History. Ancient Animal Collections to Zoological Gardens. CRC Press; Boca Raton, London, New York, Washington DC [number of references to herpetofauna in zoo collections].

Schomberg, G. 1957. British Zoos. A Study of Animals in Captivity. Allan Wingate, London. [Chapter 9 covered the herpetological collections at the Dublin (1831), Edinburgh, Chester, Paignton (1923), Belle Vue, and Dudley (1937) Zoos. Aquariums which contain amphibians and reptiles included the Carnegie Aquarium in Edinburgh, Bristol, Paignton (1923), Belfast (1933), and Blackpool Tower Aquarium (1875-1963). In 1972, he published a book on zoo design.]

Webb, C. S. 1954. The Odyssey of an Animal Collector. Longmans, Green, New York, London, Toronto. [At time of publication, Cecil S. Webb was Superintendent of the Dublin Zoo. He wrote about his experiences while collecting a variety of herps throughout the world for zoos.]

Anaconda (*Eunectes murinus*)

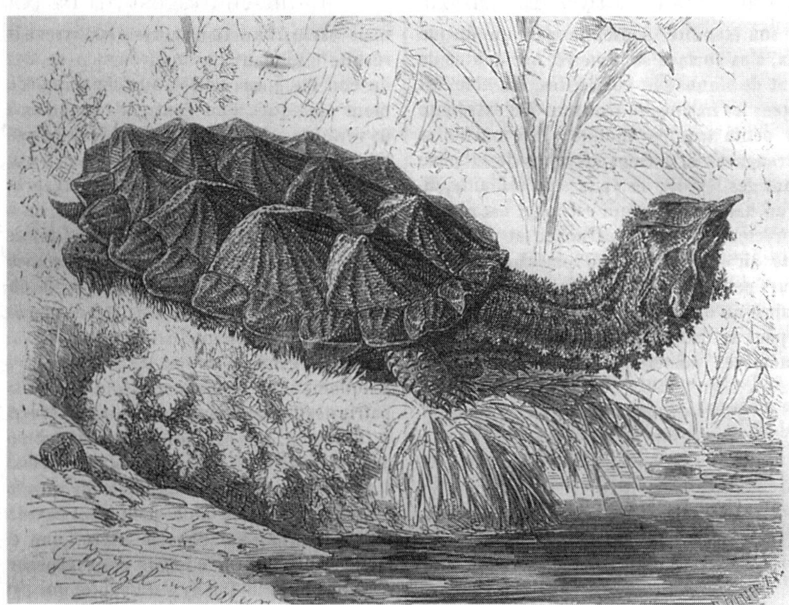

Mata Mata Turtle (*Chelus fimbriata*)

(Austria)
Menagerie Schönbrunn/Tiergarten Schönbrunn (1752)

History and Mission: The following account is mostly gleaned from Petzold (1984) and Schwammer (2001). In 19th century Vienna, the Schönbrunn Menagerie consisted of a small collection of amphibians and reptiles: European land tortoises from 1816 on, a specially built grotto for olms (1802-1808) as a gift from Baron von Zois zu Laibach, Galápagos tortoise (1835-1851), and cobras, saw-scaled vipers and rattlesnakes (around 1850). There were several branches in Vienna. The first, founded in 1800, the Menagerie in the Imperial-Royal Natural Produce Cabinet, was a precursor to zoological museums and was oriented toward research but a section was destroyed by fire nearly 50 years later. The collection at that time included 72 reptiles and 24 amphibians, augmented by merchants, foreign diplomats and expeditions. Starting in 1808, the collection included fire salamanders, slow worms, grass snakes, smooth snakes and geckos. Ten years later, chameleons from North Africa were added to the collection and eggs were laid in 1845 but did not hatch.

By 1820, a number of agamids, gekkonids and turtles were inventoried. Leopold Joseph Franz Johann Fitzinger, who later was Director of the Munich and Budapest Zoos, listed 74 reptiles and 22 amphibians. The second branch, the Menagerie in the Imperial Hofburg [Castle] Garden in Vienna, established in1820, was more oriented toward exhibition and included Galápagos and radiated tortoises, turtles, and crocodilians such as the American crocodile.

Facility and Collection: The aquarium and terrarium were destroyed in 1945 during World War II so the fishes and reptiles were moved to the monkey house. In October 1959, a new aquarium and terrarium were opened, including an improved crocodilian exhibit. Karl Schopper, Head Keeper, designed the new facility. Forty-one years later, this facility (Crocodile Pavilion and Aquarium-Terrarium House) was renovated and enlarged. The Schönbrunn Menagerie contained 217 reptiles and amphibians from 49 species on 1 January 1912 which were kept on thick brown felt carpets to show off their colors.

According to *International Zoo Yearbook* (1998:Volume 28), composition of the collection was nearly 50 taxa of reptiles with over 300 specimens and 7 taxa of amphibians numbering 60.

Staff and Scientific Achievements: Franz Luttenberger was in charge of the department from 1969 to 1998 and was interested in herpetoculture. Ekkehard Wolff is the current curator and is also interested in herpetoculture.

Historical Overview

Ash, M. G., and L. Dietrich. 2002. Menagerie des Kaisers ~ Zoo der Wiener. 250 Jahre Tiergarten Schönbrunn (The Emperor's Menagerie ~ Zoo of the Viennese. 250 Years Schönbrunn Zoo). Pichler Verlag, published by order of the Schönbrunn Zoo, Inc.

Fitzinger, L. J. F. J. 1853. Versuch einer Geschichte der Menagerien des Österreichisch-Kaiserlichen Hofes. Wien, 1853. Aus dem März - und Aprilhefte des Jahrganges 1853 des Sitzungsberichtes der mathem.-naturw. Classe der kais. Akademie der Wissenschaften, besonders abgedruckt. (Attempted History of the Zoological Gardens of the Austria Imperial Court. Vienna, 1853. From the March and April Issues, 1853, of the Conference Reports of the Mathematics-Natural History Class of the Imperial Academy of Sciences Printed Specially) Reptiles, pp. 109, 110, 135-147, 152, 188, & 189. [Galápagos tortoises in captivity in Europe are discussed. See Adler (1989) for a biographical sketch of Fitzinger.]

Loisel, G. 1912. Histoire des ménageries de l'antiquité à nos jours (History of Menageries from Antiquity to Present Times). O. Doin et fils, Paris, France.

Petzold, H.-G. 1984. Aufgaben und Probleme bei der Erforschung der Lebensäusserungen der niederen Amnioten (Reptilien) (Tasks and Problems Connected with Research into the Life Expressions of the Lower Amniotic Animals [Reptiles]). Bina, Berlin.

Schwammer, G. V. 2001. Tiergarten Schönbrunn. Schönbrunn Zoological Garden, p. 1216-1220. In C. E. Bell (ed.), Encyclopedia of the World's Zoos. Fitzroy Dearborn Publishers, Chicago, London.

Werner, F. 1892. Das Vivarium in Wien (The Vivarium in Vienna). Zool. Gart., Frankfurt a. M. 33:22-26.

Werner, F. 1897. Neues aus dem Wiener Vivarium (The latest from the Vienna Vivarium). Zool. Gart., Frankfurt a. M. 38:257-263.

Werner, F. 1899. Des Wiener Vivariums Ende (The end of the Vienna Vivarium). Zool. Gart., Frankfurt a. M. 40:33-38.

Werner, F. 1901. Noch einmal das Vivarium in Wien (Once more the Vienna Vivarium). Zool. Gart., Frankfurt a. M. 42:1-5.

Wolff, E. 1993. Terrarienbau und - gestaltung: Schaumglas - ein anderes Material für die Landschaftsgestaltung (Construction and equipment of terraria - foamed glass - another material for the formation of landscapes). Sauria 15(1):17-19.

(Belgium)
Royal Zoological Society of Antwerp (1843)

History: The Antwerp Zoo has exhibited reptiles since 1900. The reptile building was located on the top floor of the Aquarium. This structure was severely damaged by bombing during the final days of World War II and was rebuilt in 1972.

Facility and Collection: The building is a long rectangle with a one-way zigzag traffic flow. At the entrance, there is a walk-through area which displays lizards, snakes, birds and mammals. Fiberglass rockwork festooned with live plants offers a naturalistic setting for the visitor. Boa constrictors, several Indo-Australian pythons, some colubrids, water dragons,

Fig. 89. Title page from *Promenade au jardin zoologique d'Anvers par Eugéne Gens* by Eugéne Gens in 1861. *Courtesy of Smithsonian Institution Libraries, Washington, DC.*

Fig. 90. Reptile display on right at Antwerp Zoo from *Promenade au jardin zoologique d'Anvers par Eugéne Gens* by Eugéne Gens in 1861. *Courtesy of Smithsonian Institution Libraries, Washington, DC.*

lacertids, basilisks and other lizards may be seen and at the end of the walkway, an enclosure houses Aldabran tortoises.

After passing through this room, one encounters Komodo dragon and aquatic turtle displays. In the central wing of the building are a number of exhibits of various sizes which display a variety of medium-sized snakes and lizards such as small boas and pythons, several colubrids, Gila monster, cordylids, and agamids. Live plants are used throughout this section and painted dioramas depicting appropriate habitat settings provide the visitor with a sense of place.

At the end, a large crocodilian exhibit utilizes live plantings, an expanse of soil substrate and a large pool to depict a river bank along and forest stream. A simu-

Fig. 91. Plate of "Caiman a Tète de Brochet," translated as "Caiman with the snout of a pike" from *Promenade au jardin zoologique d'Anvers par Eugéne Gens* by Eugéne Gens in 1861. *Courtesy of Smithsonian Institution Libraries, Washington, DC.*

lated thunderstorm with oncoming darkness, thunder, lightning and rain create a convincing downpour at periodic intervals.

There are several innovative features incorporated in this building. An open cage uses a stream of cold air which flows from the public area toward the cage, thus providing a thermal barrier for the small python living there. An insect gun, using a blast of air, blows insects at regular intervals into an enclosure where insectivorous lizards await their meals. In the beginning, movie projectors were used throughout the building and films were projected either on free-standing screens in the public area or behind exhibit animals on the back wall of their cages. Now, television monitors are used.

According to *International Zoo Yearbook* (1998:Volume 28), composition of the collection was 82 taxa of reptiles with 381 specimens and 8 taxa of amphibians numbering 130.

Staff and Scientific Achievements: A. Paul Van den Sande from the Zoo has published papers on captive management and exhibition of reptiles. The Curator of Reptiles and Sea Mammals is P. Jouk.

Publication: *Acta Zoologica et Pathologica Antverpiensia* is a superb journal focusing on zoo biology, published in association with the Université de Liège, Belgium. There are papers on captive management and exhibition of amphibians and reptiles in zoos.

Personal Reflections: When I met with the herpetological staff in 1969 for several hours, their topic of conversation for most of the time centered on amphibians and reptiles of the southwestern United States, especially the Trans-Pecos region in Texas. As we became more comfortable with one another, they finally asked if I were as wealthy as all other Americans and did I own a Lincoln or Cadillac stretch limousine. When I replied that the bank and I jointly owned a Volkswagen Beetle, they could hardly contain their glee as this was the car owned by most of them.

Historical Overview

Gens, E. 1861. Promenade au jardin zoologique d'Anvers par Eugéne Gens (Promenade of the zoological garden of Antwerp by Eugéne Gens). J.-E. Buschmann [brief discussion on pythons and crocodilians].

Husbandry

Bels, V. L., and A. P. Van den Sande (eds.). 1984. Maintenance and reproduction of reptiles in captivity. Volume I. Maintenance and reproduction. Acta Zool. Path. Antverpiensia No. 78. [Volume included 24 papers by a number of prominent specialists covering reproduction, husbandry and ethology.]

Behavior

Bels, V. L. 1986. Analysis of the display-action-pattern of *Anolis chlorocyanus* (Sauria: Iguanidae). Copeia 1986:963-970.

Medical Management

Bels, V. L., and A. P. Van den Sande (eds.). 1986. Maintenance and reproduction of reptiles in captivity. Volume II. Diseases. Acta Zool. Path. Antverpiensia No. 79. [Volume included 10 papers on disease and pathology by a number of prominent specialists.]

Vuysteke, C. 1955. Quelques nématodes de la collection d'helminthes récoltés à la Société royale de zoologie d'Anvers (Some nematodes in the collection of helminths gathered at the Royal Society of Zoology of Antwerp). Bull. Soc. r. Zool. Anvers No. 5:5-27.

Reproduction

Bels, V. L., and C. Gans. 1986. Conclusions: Perspectives on breeding reptiles in captivity. Acta Zool. Path. Antverpiensia 79:110-112.

Bels, V. L., and A. P. Van den Sande. 1986. Breeding the Australian carpet python *Morelia spilotes variegata* at Antwerp Zoo. Inter. Zoo Yearb. 24/25:231-238. [Account covered reproductive behavior, housing effects, cage design and furnishings, temperature, male and maternal behavior, feeding techniques and diet.]

General

Bels, V. 1987. A behavioural approach to the maintenance of *Anolis* lizards in captivity, p. 47-57. In P.W. Scott and A. G. Greenwood (eds.), Exotic Animals in the Eighties. Proceedings from the 25th Anniversary Symposium of the British Veterinary Zoological Society, 18th-20th April 1986. British Veterinary Zoological Society [place of publication not given].

Bels, V. 1987. Analysis of the growth of *Dermochelys coriacea* (Reptilia: Testudines) in captivity, p. 157. In P.W. Scott and A. G. Greenwood (eds.), Exotic Animals in the Eighties. Proceedings from the 25th Anniversary Symposium of the British Veterinary Zoological Society, 18th-20th April 1986. British Veterinary Zoological Society [place of publication

not given].

Bels, V. 1989. Analysis of the psychophysiological problems of reptiles in captivity. Herpetopathologica 1:11-18.

Gans, C., and A. P. Van den Sande. 1976. The exhibition of reptiles: Concepts and possibilities. Acta Zool. Path. Antverpiensia No. 66:3-51. [Excellent contribution covered the how and why of exhibiting reptiles by using the renovated reptile building of the Zoo as a model. Topics included enclosure design and size, interpretative graphics, animal movement within exhibits, service areas, handling venomous snakes, and methods of stimulating public interest. This paper should be required reading for the zoo curator.]

Gijzen, A., and H. Wermuth. 1958. Schildkröten-pflege in öffentlichen Schau-aquarien nach biologischen Gesichtspunkten (The care of turtles in public exhibit aquaria in accordance with biological principles). Bull. Soc. r. Zool. Anvers No. 6:1-65. [At time of publication, Agatha Gijzen was associated with the Zoo and Heinz Wermuth was at the Berlin Museum of Zoology. This comprehensive work, accompanied by 52 plates and a chart, is an important treatment of turtles in captivity.]

Leloup, P. 1980. Liber amicorum Walter van den Bergh. Tielt. [Chapter "Prophylaxie Elementaire des Accidents dans Elevage de *Bothrops atrox*" covered venomous snakebite maintenance and first aid procedures. This book honored Walter van den Bergh, former Director of the Zoo, on occasion of his 70th birthday.]

(France)
Menagerie du Jardin des Plantes (1793)
(Paris)

History and Mission: Menageries were important as precursors to natural history museums and veterinary schools, and they had an important role in experimental zoology. Crocodiles, chameleons, common snakes, geckos, salamanders, East Indian and Indian tortoises were held at the Royal Menagerie of Versailles (1665), considered to be the first "zoological garden" as plants and animals were combined. In fact, chameleons were scattered throughout several menageries. Some living tortoises and the arrival of a crocodile at the Vineuil Menagerie in 1783 caused a stir. The Menagerie of the Museum of Natural History in Paris was created in 1793 and was the first of the national menageries. Its rich history as a center for experimental zoology is detailed extensively by Gustave Loisel. Constant Duméril was in charge of the Terrarium, a separate section in the old monkey house, which housed 80 reptiles of 24 species, amphibians, fishes and insects from 1839. Travelers were given written instructions to collect specimens for the collection. In 1870-1874, an architect Emile Blanchard designed a reptile menagerie with a pavilion. The facility was 30 meters long, with two halls which contained smaller squamates and chelonians. Two large center exhibition halls were called Crocodile Hall and Aquarium Hall; the latter contained freshwater fishes and amphibians. Bronze sculptures were placed throughout the menagerie: "The Snake Charmer" and "The Crocodile Hunter" by Arthur Bourgeois in front of the reptile pavilion and "Eve" by Guitton near the outdoor crocodile pool. In 1910, the collection numbered 216 reptiles and 237 amphibians. Longevities included a 60-year-old American alligator, 35-year--old Australian turtle, 22-year-old snapping turtle, three giant tortoises over 20 years, Madagascan boa over 21 years, 14-year-old reticulated python and a "molure" python two years older, Japanese salamander for 30 years and a siren for 23 years.

Facility, Staff and Collection: In the fall of 2002, I revisited the reptile building and spent a lovely afternoon looking at the architecture and collection. One unusual feature is that the rear walls of enclosures on the outer wall are glass. As a result, these elevated wooden displays contain living plants which thrive in natural light. There is a large aquatic exhibit which houses an impressive collection of crocodilians. At one end of the building, large tortoises are on display. Madagascan reptiles are well represented which is not surprising since there have been many seminal studies in the region by French herpetologists such as E.-R. Brygoo and Charles Blanc. Several extensive graphic displays with lifelike models of reptiles are situated throughout the building. Some of the larger enclosures are being remodeled. In the newer Vivarium building located near the reptile building, smaller amphibians and reptiles are represented in naturalistic settings.

According to *International Zoo Yearbook* (1998: Volume 28), composition of the collection was 56 taxa of reptiles with over 150 specimens.

Françoise Perrin is the current curator.

Historical Overview

Bernard, P. 1842-1843. Le Jardin des plantes: description complète, historique et pittoresque du Muséum d'histoire naturelle, de la ménagerie, des serres, des galeries de minéralogie et d'anatomie, et de la vallée suisse: moeurs et instincts des animaux, botanique, anatomie comparée: minéralogie, géologie, zoologie (The Jardin des plantes: A complete, historical and picturesque description of the Museum of Natural History, of the menagerie, the green-houses, the galleries of mineralogy and anatomy and of the Swiss Valley: mores and instincts of the animals, botany, comparative anatomy: mineralogy, geology, zoology). L. Curmer, Paris. [Beginning on page 153, Pierre Bernard (1810-1876) discussed the habits and characteristics of reptiles living in the menagerie and included several excellent plates of a scorpion, lizard and snake. He defended the need for a living animal collection as a valuable vehicle for research and education.]

Duméril, A. H. A. 1854-55. Notice historique sur la ménagerie des reptiles du Muséum d'histoire naturelle et observations qui y ont été recueillies (Historical notice on the menagerie of reptiles of the Museum of Natural History and observations collected there). Archives du Muséum d'histoire naturelle 7:193-320. [In four papers, Auguste-Henri-André Duméril recorded the longevities of specimens in the Menagerie. See Adler (1989:43) for his biography.]

Duméril, A. H. A. 1858-61. Deuxième notice (Second notice). Archives du Muséum d'histoire naturelle (Archives of the Museum of Natural History) 10:429-460 [longevities of specimens in Menagerie].

Duméril, A. H. A. 1862. Troisième notice (Third notice). Nouvelles archives du Muséum d'histoire naturelle (New archives of the Museum of Natural History) 1:31-50 [longevities of specimens in Menagerie].

Duméril, A. H. A. 1869-70. Quatrième notice (Fourth notice). Nouvelles archives du Muséum d'histoire naturelle (New archives of the Museum of Natural History) 5:47-60. [longevities of specimens in Menagerie.].

Lacepède, B.-G.-É. 1801. La Ménagerie du Muséum national d'histoire naturelle; ou, Les animaux vivants, peints d'après nature, sur vélin, par le citoyen Maréchal, et gravés au Jardin de plantes, avec l'agrément de l'Administration, par le citoyen Miger. Avec une note descriptive et historique pour chaque animal, par les citoyens Lacépède et Cuvier (The Menagerie of the National Museum of Natural History or its Living Animals Painted from Nature on Velin by Citizen Maréchal, a painter of the Museum and Engraved in the "Jardin des Plantes" with Agreement of the Administration by

Fig. 92. Portrait of Georges-Louis Leclerc Buffon in M. Boitard's *Le Jardin des Plantes. Description et Moeurs des Mammifères* in 1845. *Courtesy of Smithsonian Institution Libraries, Washington, DC.*

Fig. 93. Portrait of Baron Georges Cuvier in M. Boitard's *Le Jardin des Plantes. Description et Moeurs des Mammifères* in 1845. *Courtesy of Smithsonian Institution Libraries, Washington, DC.*

Citizen Miger, Engraver and Member of the Royal Academy of Paintings, with a Description and Historical Note by the Citizens Lacepède and Cuvier). Miger, Paris. [Bernard-Germain-Étienne de la Ville-sur-Illon, Compte de Lacepède (1756-1825) also published *Histoire Naturelle des Quadrupèdes Ovipares et des Serpens* in 1788-1789. See Adler 1989:14 for biography.]

Lacepède, B.-G.-É. 1801. Discours sur les établissements publics destinés à renfermer des animaux vivants, et connus sous le nom de ménageries; réimpr. dans Oeuvres de L. (1883) 1:106-111. (Lecture about the public installations intended to confine living animals, and known by the name of menageries; reprinted in L.'s works [1833] Part 1, p. 106-111). [important historical paper detailing the mission of zoos, design recommendations, and integration of science].

Loisel, G. 1912. Histoire des ménageries de l'antiquité à nos jours (History of Menageries from Antiquity to Present Times). O. Doin et fils, Paris. [Loisel mentioned that the Zoological Garden of Lyon contained a few large American alligators, several reptiles and a few small tortoises in 1907.]

Mullan, B., and G. Marvin. 1999. Zoo Culture. Second Edition. University of Illinois Press, Urbana and Chicago [two photographs of reptile cases and crocodilian enclosure].

Murphy, J. B., and G. Iliff. 2004. Count de Lacepède: Renaissance Zoo Man. Herpetol. Rev. 35:220-223.

Medical Management

Nouvel, J., G. Chauvier, and L. Stazielle. 1961. Rapport sur la mortalité enregistrée à la ménagerie du Jardin des Plantes pendant l'année 1959 (Mortality reports of the menagerie of the Jardin des Plantes). Bull. Mus. Hist. nat. Paris (2) 33:63-78. These authors published a number of annual mortality reports.]

General

Duméril, A. H. A. 1867. Métamorphoses des batraciens urodèles à branchies externes du Mexique, dits axolotls, observés à la ménagerie des reptiles du Múséum d'histoire naturelle (Metamorphoses of the urodele batrachians with external gills of Mexico, called axolotls, observed in the menagerie of reptiles of the Museum of Natural History). Ann. Sci. nat. 7: 229-354.

Vaillant, L.-L. 1898. Múséum d'histoire naturelle Guide à la ménagerie des Reptiles. Laboratoire d'Herpétologie, Paris. [In addition to his regular duties in the museum, Léon-Louis Vaillant (1834-1914) was in charge of the reptile menagerie and aquarium. On the front cover of this guide are drawings of Lacepède, C. Duméril, A. Duméril and Vaillant. There are also nice drawings of the building's façade and floorplan of the menagerie.]

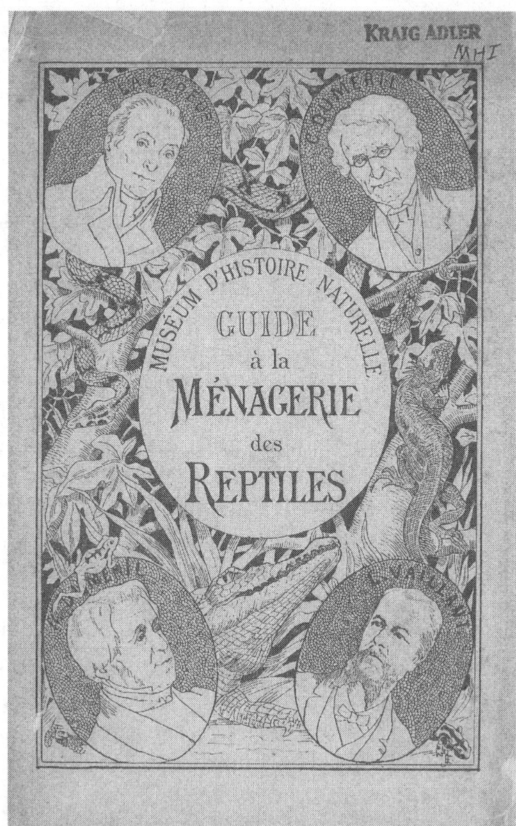

Fig. 94. Cover of Guidebook for reptile building at Menagerie Jardin des Plantes in Paris by Léon-Louis Vaillant in 1898. On the front cover of this guide are drawings of Lacepède, C. Duméril, A. Duméril and Vaillant. *Provided by Kraig Adler.*

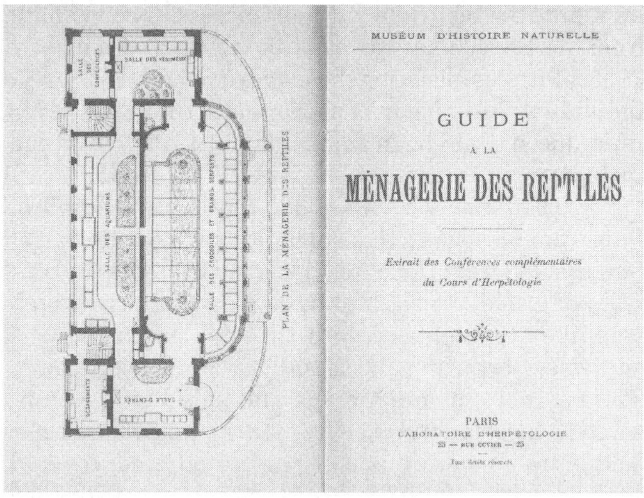

Fig. 95. Title page and floor plan of the Menagerie Jardin des Plantes in Paris by Léon-Louis Vaillant in 1898. There are nice drawings of the building's façade. *Provided by Kraig Adler.*

(The Netherlands)
Natura Artis Magistra (1838)
(Amsterdam)

History and Mission: Coenraad Jacob Temminck (1778-1858), herpetologist and Director of the Rijksmuseum van Natuurlijke Historie in Leiden, and G. F. Westerman unsuccessfully attempted to establish a zoo in Amsterdam in the early 1830s. In 1838, J. J. Wijsmuller, G. F. Westerman and J. W. H. Werleman were able to build a zoo, Natura Artis Magistra. The zoological society attracted about 400 members within a few weeks. The Zoo was a center for comparative anatomy and descriptive zoology. The stated mission was "to found a zoological society to enlarge public knowledge of natural history by collecting and exhibiting a collection of live animals and zoological objects."

In 1852, King Willem III of the Netherlands gave the predicate Royal to the Zoo, together with some pheasants from his own collection. In 1868, Charles Darwin queried the Zoo staff about peacocks.

Fig. 96. Illustration of snake hall (Schlagen-Zaal) at Natura Artis Magistra in Amsterdam from *De Dierentuin van het Koninklijk Zoölogisch Genootschap te Amsterdam* by Hermann Schlegel in 1872. *Courtesy of Smithsonian Institution Libraries, Washington, DC.*

The objectives of the Royal Zoological Society Natura Artis Magistra, usually abbreviated to Artis Zoo Amsterdam, are to offer recreational and educational opportunities to the public in the field of natural history and the environment. The Planetarium, living animals and plants, fossil and mineral collections, and preserved plant and animal materials, are used to meet these objectives.

Artis Zoo occupies a 14-hectare park, located in the old center of the city. Some of its buildings are prized as historical monuments. In addition to its zoological collection, Artis Zoo houses a botanical garden with more than 200 species of trees, which, together with the buildings, create an atmosphere typical of a 19th century park. A zoological museum (owned by the University of Amsterdam), a geological museum, and a planetarium with a permanent exhibition on astronomy and space exploration are also part of the zoo complex.

Facility and Collection: The collection started with preserved animals privately mounted by Meindert Draak, and some living mammals were added. One year after the establishment of the zoo, the traveling menagerie from Cornelis van Aken was purchased. The collection contained a lion, four tigers, an elephant, other mammals, birds and reptiles, among them a python and a caiman. Temperature control was erratic so the giant snakes were kept warm by using hot water. Unbelievably, a small python was worth about $100 US dollars in those early days, roughly the annual salary for an office worker at the zoo.

In 1829, a Japanese giant salamander was imported by Franz von Siebold for the Rijksmuseum voor Natuurlijke Historie in Leiden, the first one in Europe (see Chapter 3). In 1840, this animal was moved to the snake-room of Artis Zoo, where it died in 1881. This specimen died at an age of at least 52 years which is still the longevity record. During its time at the Zoo, it was fed frogs or small fish, about 20 during each feeding.

One of the world's oldest reptile buildings began its operation at this Zoo in 1852. The reptile collection was worldwide in scope and had a sizeable number of giant reptiles in 1864, including giant constrictors, turtles and crocodilians. A boa constrictor gave birth to 26 young in 1868. In 1885, tuataras were added to the collection for the first time. There were pythons from Java, boa constrictors and other large snakes, tortoises, lizards, alligators and a Temminck's snapper (alligator snapping turtle) at the Reptile House in 1903.

In 1903, the most spectacular herpetological success in the history of the Zoo occurred: the Japanese giant salamander was propagated. After a month of increased activity by the two adults, about 500 eggs were found on 18 September 1902 but were infertile. The next year, eggs were deposited on almost the same date (19 September). The female had to be removed because the male was aggressive. This male protected the eggs successfully, until the last one hatched on 26 November. The larvae were 30 mm in length. Drs. Kerbert, De Bussy and De Lange studied reproduction and embryonic development and Japanese scientists came to Amsterdam to study the reproduction of their spectacular salamander. In 1905, the salamanders reproduced again. A few years later, more eggs were laid, but this last time without success. One neonate from the 1903 clutch stayed at the zoo and died in July 1955, almost duplicating the longevity of the 1829-1881 individual. After a gap of many years, the taxon never reproduced again in an aquarium until 1979, this time in Japan.

In 1910, the Zoo housed 380 reptiles and amphibians from 98 species in a splendid new herpetological facility, 38 meters long and designed by architect B. J. Ouëndag. This house contained a central area which was like a greenhouse with two side pavilions. One side was a terrarium and the other was an insectarium. The pavilion for amphibians and small snakes contained a large humid cage and some small glass cages. The bottoms were made of perforated zinc, covered with a layer of damp earth with moss, ferns and other plants, some rocks of peat and in one corner, a tiny pond. The tops consisted wholly of glass or were partly grilled to allow air circulation. This building is still in use but unfortunately the monumental towers and the

Fig. 97. Postcard of Artis Aquarium at Natura Artis Magistra in 1910. *Provided by Eugène Bruins, Natura Artis Magistra Archives.*

Fig. 98. Artis reptile building with impressive towers and classic windows at Natura Artis Magistra, constructed in 1910. This building still exists but the towers and windows are gone. *Provided by Eugène Bruins, Natura Artis Magistra Archives.*

Fig. 99. Great Hall of Artis Aquarium after renovation in 1995-1997. *Provided by Eugène Bruins, Natura Artis Magistra Archives.*

classic windows have disappeared. In 1912, spectacled caimans bred; the eggs had been incubated in the terrarium. Three years later, an American alligator hatched; its history was documented until it died in 1955. In 1926, the Komodo dragon was added to the collection. Five years later, 12 eggs were laid in a hole and protected by the female. They however never hatched. The collection also contained king cobras which were fed grass snakes (and once an anaconda), other venomous snakes, giant tortoises, Gila monsters and many more species.

During World War II Artis Zoo barely suffered damage, unlike many other European zoos. During a bombing episode of the city in 1941, some venomous snakes had to be killed as concern for public safety became an issue. In 1963 a crocodile managed to escape some distance to the pelican pond but was quickly recaptured. In 1975, the reptile house was enlarged. Seven years later, the reptile collection was moved to the insect pavilion. In 1998, an outdoor enclosure for giant tortoises was added to the reptile house.

Aquarium

In 1879, the building of the aquarium was started; 1740 wooden poles were placed into the ground. The Aquarium, containing 640,000 liters of seawater and 225,000 liters of fresh water, was designed by a father and son architectural team named Salm. It was built according to the circulation system of W. Alford Lloyd. The Aquarium was opened on 2 December 1882 and at that time it was the biggest and most modern aquarium in the world. Total costs for the building were about

$200,000 US dollars. As a comparison, the renovation in 1995 cost 6 million US dollars. An amphibian room was added to the aquarium in 1957, renovated in 1973 and closed during the renovation in 1995. Several aquatic turtle species were kept in the aquarium.

In the period 1953-1982 the recirculation system was redesigned in order to have more variety of water systems and better water filtration was introduced in the aquarium. With the extensive renovation of 1995-1997, the historic Great Hall was restored to its former glory, with salt water tanks on the left and fresh water tanks on the right hand sides. Four giant new "mammoth-aquariums" were added as new elements. These ecosystem tanks house the Amazonian flooded forest, the tropical coral reef and an Amsterdam canal. The water recirculation system now contains five different types of seawater and two fresh water systems.

In 2002, the composition of the collection was 80 taxa of reptiles with nearly 235 specimens and 20 taxa of amphibians numbering over 100.

Staff and Scientific Achievements: Curator F. de Graaf not only published many articles on aquarium subjects but also in applied herpetoculture in hobby magazines like *Het Aquarium*. In 1999 Eugène Bruins became curator of lower vertebrates and invertebrates of the Zoo. Unlike his predecessors who were mainly aquarium experts, he specialized in herpetology and entomology.

In the reptile house B. Bollee worked as one of the keepers. He became so fascinated with reptiles that he started making concrete sculptures, including the two large white dinosaurs in front of the aquarium.

Fig. 100. In 1963, a crocodile escaped from Artis reptile house and was recaptured in pelican pond at Natura Artis Magistra. Keepers F. Buurman and J. Struyk are trying to put it in boat. *Provided by Eugène Bruins, Natura Artis Magistra Archives.*

Conservation: Bruins is an active member in the EAZA Amphibian and Reptile TAG, which resulted in a proposal for the European studbook for the frilled lizard and a proposal for the agamid section of the Regional Collection Plan of the TAG. Besides these initiatives, Artis Zoo employees, in concert with Zürich Zoo, expanded the scope of the blue poison dart frog studbook (*Dendrobates azureus*) from the United Kingdom to Europe.

Publications: Natura Artis Magistra founded the Artis library, with G. F. Westerman in charge. The periodical *Contributions to Zoology* (*Bijdragen tot de Dierkunde*) was founded in 1848. The library became integrated into the University of Amsterdam in 1939.

Personal Reflections: When I visited the Zoo in 1969 I was surprised to see an adult African goliath frog in a very small aquarium in quarantine. The anuran, with a damaged snout, was unable to leap. Since these frogs are often injured by crashing into the walls of their enclosure, this solution seemed to be ingenious and successful. Others were placed in the middle of the amphibian room in a very big aqua-terrarium and looked healthier than many other goliath frogs I have seen elsewhere.

Historical Overview

Bruggen, A. C., van. 2003. Kerbert and the Japanese giant salamander: Early scientific achievements in the Amsterdam Aquarium. Inter. Zoo News 50: 481-486.

Loisel, G. 1912. Histoire des ménageries de l'antiquité à nos jours (History of Menageries from Antiquity to Present Times). O. Doin et fils, Paris, France.

Nieuwendijk, J. G. 1970. Zoo was Artis, zo is Artis. Bussy, Amsterdam.

Peel, C. V. A. 1903. The Zoological Gardens of Europe. Their History and Chief Features. F. E. Robinson, London [brief description (p. 41) of pythons from Java, boa constrictors and other large snakes, tortoises, lizards, alligators and a Temminck's snapper (alligator snapping turtle) at the reptile house].

Portielje, A. F. J., and S. Abramsz. 1922. Het Artisboek (The Book of Artis), part 1 & 2. P. van Belkum Az., Zutphen. [Part 1 describes the insectarium on p. 172-222 and the reptile house on p. 223-295. Part 2 covers the aquarium on p. 211-288.]

Rombouts, J. E. 1888. Artis; kijkjes in den dierentuin (Artis; Taking a Look at the Zoo). Tj. van Holkema, Amsterdam. [Book covered reptiles on p. 50-60 and axolotl on p. 320-325.]

Schlegel, H. 1872. De Dierentuin van het Koninklijk Zoölogisch Genootschap te Amsterdam (The Zoological Gardens of the Royal Zoological Society Natura Artis Magistra in Amsterdam). Amsterdam [beautiful plates].

Strehlow, H. 2001. Zoological gardens of western Europe, p. 75-116. In V. N. Kisling Jr. (ed.), Zoo and Aquarium History. Ancient Animal Collections to Zoological Gardens. CRC Press, Boca Raton, London, New York, Washington DC [portraits of many European zoos, including Amsterdam, some of which had herpetological collections].

Tochimoto, T. September 2000. Siebold's salamanders. Newsletter Himeji City Aquarium No. 37:2-3. [In Japanese, this is a wonderful study.]

Vosmaer, A. B. 1804. Description of an Exquisite Collection of Rare Animals, Consisting of Quadrupeds, Birds and Serpents, of East and West Indies, Found Formerly Living at the Menageries Belonging to His Highness My Lord the Prince of Orange Nassau, by the Late M. A. Vosmaer, in His Life Advisor of His Most Serene Highness...with Figures Designed and Illuminated as per Nature. Amsterdam [first edition under a reduced title, dated 1767].

Vries, L. de. 1981. Het boek van Artis (The book of Artis). A.W. Bruna & Zoon, Utrecht.

General

Bruins, E. 1999. Encyclopedia of Terrarium. Rebo International b. v., The Netherlands. [Provides extensive coverage of insects, arachnids, amphibians and reptiles. This is one of the most valuable and comprehensive treatments available on invertebrate care. Topics include legislative issues, captive management, and feeding. There are many color photographs.]

Kerbert, C. 1888. Het aquarium en zijne bewoners (The aquarium and its inhabitants). Bijdragen tot de Dierkunde (Contributions to Zoology) 12:1-98.

Natura Artis Magistra. 1848-. Bijdragen tot de Dierkunde (Scientific Magazine of the Royal Zoological Society Natura Artis Magistra).

Natura Artis Magistra. 1955-. Artis. Magazine of the Foundation to sustain the Zoo of the Royal Zoological Society Natura Artis Magistra [6 magazines/year].

Petzold, H.-G. 1984. Aufgaben und Probleme bei der Erforschung der Lebensäusserungen der Niederen Amnioten (Reptilien) (Tasks and Problems Connected with Research into the Life Expressions of the Lower Amniotic Animals (Reptiles)). This book (Nr. 38) is in the series "Berliner Tierpark-Buch," published by Bina in Berlin [excellent overview of history of Japanese giant salamanders in European zoos, including Amsterdam].

Smit, P., 1988. Artis, een Amsterdamse tuin (Artis, a garden in Amsterdam). Rodopi b.v., Amsterdam, the Netherlands.

Morphology, Systematics & Taxonomy

Groot, T. V. M., E. Bruins and J. A. J. Breeuwer, 2003. Molecular genetic evidence for parthenogenesis in the Burmese python, *Python molurus bivittatus*. Heredity 90 (2):130-135.

Schlegel, H. 1851. Description d'une nouvelle espèce du genre Eryx, Eryx REINHARDT. Bijdragen tot de dierkunde, 3: pp:1-3.

Werken, H. van de. 1969-1971. Artis encyclopedie (Artis encyclopedia 6 parts, 936 pp). Ploegsma, Amsterdam.

Husbandry

Veltman, J., and W. Wilhelm, 1991. Husbandry and display of the jewel wasp *Ampulex compressa* and its potential value in destroying cockroaches. Inter. Zoo Yearb. 30:118-126.

Medical Management

Janse, M., D. Overbosch, P. Kager, G. Carels, and P. S. J. Klaver. 1998. The management for a poisonous snake bite. Mém. Inst. Océano. P. Ricard, p. 123-132.

Addendum: Since 1898, Artis Zoo has contained the world's oldest Insectarium. In the beginning, only pinned insects were displayed but this collection grew quickly and was housed in its own building in 1853. The first living insects were honey bees, using a beehive designed by Dzierzon. C. Kerbert was the first custodian of the aquarium until he became director in 1890. After a proposal of Rudolf A. Polak was accepted by director Kerbert, Polak started an Insectarium in 1898, containing mostly wooden enclosures with caterpillars. Since zoo visitors were ecstatic about insects, better insectaria and more space were needed. In 1899, the Insectarium became part of the reptile department. Nicer and more permanent insect displays constructed of metal and glass replaced the wooden ones. Each cage, a glass case sometimes without a cover and with the opening at the bottom, rested on a zinc box. These boxes were pierced with holes and covered with earth or sand, and supported some bottles with wide openings to contain fresh plants for the caterpillars. The glazed part at the top was often replaced by a grill and supported small boxes containing preserved insect specimens of the same species as those living below. The collection contained mostly Dutch insects. In 1903, a walking stick (*Carausius morosus*) was added to the collection, the first one in the Netherlands. It competed for popularity with the hippo and the giraffe, certainly a happening to warm any entomologist's heart. Newspapers reported its death three months after arrival, but many eggs had been laid so a stable population was ensured. Soon thereafter, Polak became the world's supplier for walking sticks.

Fig. 101. These insectariums as constructed in 1899 at Natura Artis Magistra. *Provided by Eugène Bruins, Natura Artis Magistra Archives.*

A few years later, leaf insects, tarantulas, scorpions and centipedes were added to the collection. In 1910, the insects were moved to the new reptile facility. Sixteen years later, the Insectarium was closed to await renovation. In 1980, the Insectarium was moved into a separate building, together with sea lions.

Rotterdam Zoo (1857)

History and Mission: The Rotterdam Zoo is also known as Diergaarde Blijdorp. This is an old zoo, founded in 1857, and bombed during World War II. Curator Gerard Visser told me that a number of mammals escaped from their enclosures during the War: sea lions in the canals, zebras in the city center, chimps locked in telephone booths, as well as reptiles. The new zoo began in 1940.

Facility and Collection: The staff at the Rotterdam Zoo opened a state-of-the-art reptile house in May 1906. The walls were made of Falconnnier glass bricks and the glass roof covered three rooms with murals, smaller naturalistic aquatic and terrestrial exhibits, and large enclosures for crocodilians, boas and pythons. Many of the specimens were brought back by sailors traveling to the Dutch overseas colonies in the East Indies (now Indonesia) and West Indies (Netherlands Antilles) since Rotterdam was a major seaport. Mortality was high. Green iguanas had an average life span of one month but the exhibits were always stocked, since the sailors provided a steady supply of replacements. One of the old guidebooks showed a picture of a tuatara, labeled incorrectly as a green iguana. In 1910, 270 reptiles comprising 78 species and 122 amphibians representing 25 species occupied the facility. During World War II, the collection suffered since food was scarce and many of the enclosures were not heated properly. The few survivors included pairs of Ameri-

Fig. 102. Reptile building at Rotterdam Zoo in 1911. *Provided by Gerard Visser, Rotterdam Zoo.*

can alligators and African slender-snouted crocodiles. One of two teahouses was modified into a reptile house in 1953. Nineteen years later, the reptiles were transferred to Rivièrahal (Rivièra Hall) and additional aquariums and terrariums were added. In this new setting with high humidity and tropical plants, reproduction dramatically increased: Fiji Island and rhinoceros iguanas; ridge-tailed, mangrove, lace, white-throated, and yellow monitors; leopard, Madagascar spider and Bell's hinged tortoises; East Indian and Philippine sailfin lizards, and false water cobras with the total reproductive output comprising over 100 taxa. Incubating eggs and neonates were displayed. A new wing for crocodilians was added in 1976 with space for the alligators and slender-snouted crocodiles, by now over 50 years of age. Spectacled caimans and water monitors were also on display. Three years later, an Amazonian wing was added to Rivièrahal where red-footed tortoises, South American giant river turtles, and yellow-spotted giant river turtles were exhibited. In July 2001, a new exhibit called Oceanium was added displaying Caribbean herpetofauna: Cuban anoles and boas; rhinoceros and Roatan black iguanas; Aruba Island rattlesnakes and several leptodactylid frogs.

According to *International Zoo Yearbook* (1998:Volume 28), composition of the collection was 100 taxa of reptiles with 830 specimens and 13 taxa of amphibians numbering 90.

Staff and Scientific Achievements: Johann Büttikofer (1850-1927), who published on herpetological topics, was appointed director in 1887 (see Visser, 2003 for additional information). The teahouse was curated by Joop van der Werff until he left the facility in 1974. His boss was headkeeper Pierre van Leeuwenn since there were no formal curatorial positions. The first biologist with formal academic training was Hans van Roon, who published papers in herpetology; his successor was Han Assink, who was also responsible for fishes and birds. When Gerard Visser arrived at the Zoo in 1972, he was trained as an aquarist and reptile keeper by van der Werff and van Roon. There is no question that Visser, current curator, has brought an international reputation to the zoo, especially for his work in conservation. He is co-chair of the European Aquarium and Zoo Association (EAZA) Amphibian and Reptile Taxon Advisory Group (AR TAG), and European Endangered Species Program (EEP) Coordinator and studbook keeper for the Komodo dragon. Assistant headkeeper Henk Zwartepoorte (also chairman of the Dutch Turtle Society), is currently coordinator of the EEP for the

Egyptian tortoise, and Turtle Conservation Officer at the Zoo to deal with the Asian turtle crisis. Recently, Judith van der Koore has been appointed general headkeeper.

Conservation: The Zoo supports the "Sea Turtle Tracking Project" of the Caribbean Conservation Corporation/Sea Turtle Survival League, based in Gainesville, Florida; (see: http://www.cccturtle.org/ and also http://www.rotterdamzoo.nl/).

The "Antiguan Racer Conservation Project" to save the Antiguan Racer *Alsophis antiguae,* coordinated by the Antiguan Forestry Unit, Fauna and Flora International and the Durrell Wildlife Conservation Trust as well as other organizations, is also sponsored. (See also http://www.fauna-flora.org/).

The Zoo is involved in the Turtle Conservation and Ecology Project in Cuc Phuong National Park, Vietnam, where additional enclosures have been built and funds provided for the care of confiscated Asian turtles before they are returned to protected sites.

Personal Reflections: Several years ago, approximately 10,000 Asian chelonians were confiscated in Hong Kong, to be used for food and medicinal purposes in China. Many were seriously injured and some had been caught with fish hooks. Gerard has been a major force in finding homes for them in European zoos and other facilities. He arranged medical evaluations and care, interacted daily with colleagues to keep them informed, and coordinated shipping. In this way, he demonstrated how resourceful and important one individual in the zoo community can be when faced with a crisis of this magnitude.

Historical Overview

Loisel, G. 1907. Rapport sur une mission scientifique dans les jardins et établissements zoologiques publics et privés du Royaume-Uni, de la Belgique, des Pays-Bas, de l'Allemagne, de l'Autriche-Hongrie, de la Suisse et du Danemark. Nouv. arch. miss. sci. litt., Paris, 14 (124); 15 (125-282). [Gustave Loisel published several treatments on zoos with herpetological exhibits in Europe, Canada and the United States. These works give a good view of the state of the art in the early 1900s with photographs of the Rotterdam Zoo Reptile Building and exhibits, constructed at the beginning of the 20th century.]

Loisel, G. 1907-1908. The zoological gardens and establishments of Great Britain, Belgium, and the Netherlands. [Translated and abridged from the Rapport sur une mission scientifique dans les jardins et établissements zoologiques publics et privés du Royaume-Uni, de la Belgique et des Pays-Bas, par M. Gustave Loisel. Extrait des Nouvelles Archives des Missions Scientifiques, 19. Paris, 1907.] Washington DC, Smithsonian Inst. Rep. 1907, 1908, (407-448).

Loisel, G. 1912. Histoire des ménageries de l'antiquité à nos jours (History of Menageries from Antiquity to Present Times). O. Doin et fils, Paris, France.

Visser, G. 2001. Diergaarde Blijdorp. Rotterdam Zoo, p. 353-358. In C. E. Bell (ed.), Encyclopedia of the World's Zoos. Fitzroy Dearborn Publishers, Chicago, London.

Visser, G. 2003. Herpetology at the Rotterdam Zoo. Herpetol. Rev. 34:11-16.

Morphology, Systematics & Taxonomy

Horn, H.-G., and G. J. Visser. 1988. Freilandbeobachtungen und einige morphometrische Angaben zu *Varanus giganteus* (Gray, 1845) (Sauria: Varanidae) (Observations in the wild and some morphometric data concerning *Varanus giganteus* [Gray, 1845] [Sauria: Varanidae]). Salamandra 24:102-118. [Hans-Georg Horn has had an avocational interest in varanids for years, which has led to a significant output of books and scientific papers on this genus. In fact, his home is filled with varanid enclosures, housing some rare and seldom seen taxa. This account discussed morphometrics, ecology and ethology.]

Husbandry

Polder, J. J. W. 1969. Feeding king cobras *Ophiophagus hannah* at Rotterdam Zoo. Inter. Zoo Yearb. 9:56 [account described use of frozen food snakes].

Medical Management

Bemmel, A. C. V. van, J. C. Peters, and P. Zwart. 1960. Report on births and deaths occurring in the Gardens of the Royal Rotterdam Zoo during the year 1958. Tijdschr. Diergeneesk. (Netherlands Journal of Veterinary Science) 85:1203-1213.

Bemmel, A. C. V. van, P. Zwart, and J. C. Peters. 1962. Report on births and deaths occurring in the Gardens of the Royal Rotterdam Zoo during the years 1959 and 1960. Tijdschr. Diergeneesk. (Netherlands Journal of Veterinary Science) 87:826-836.

Borst, G. H. A., C. Vroege, F. G. Poelma, P. Zwart, W. J. Strik, and J. C. Peters. 1972. Pathological findings on animals in the Royal Zoological Gardens of the Rotterdam Zoo during the years 1963, 1964 and 1965. Acta Zool. Path. Antverpiensia No. 56:3-20 [birds, mammals, reptiles].

Zwart, P., F. G. Poelma, W. J. Strik, J. C. Peters, and J. J. W. Polder. 1968. Report on births and deaths occurring in the Gardens of the Royal Rotterdam Zoo during the years 1961 and 1962. Tijdschr. Diergeneesk. (Netherlands Journal of Veterinary Science) 93:348-362.

General

Visser, G. 1992. Monitors and the Rotterdam Zoo. Vivarium 4(3):19-22.

(Switzerland)
Zoologischer Garten Zürich (1929))

History and Mission: The Zürich Zoo enjoys an impressive reputation in the areas of conservation and the output of publications in ethology, reproductive biology and zoo philosophy is a model which should be emulated by zoo workers worldwide.

Facility and Collection: In 1929, a large building was constructed in a park of 10 ha which included an aquarium and terrarium besides other animal buildings. The collection has been properly utilized for many years since a number of significant papers on behavior, conservation and reproductive biology have been published.

Staff and Scientific Achievements: Heini Hediger (1908-1992), late Director of the Zoo, was a pioneer in zoo biology who wrote several books on psychology and captive management of wild animals. René Honegger was the Curator of Herpetology between 1960 and 1999, and was in charge of the Exotarium since 1972. He was responsible for many innovations in terrarium displays, captive husbandry and management, and zoo philosophy. Further, he has been a major player in national and international conservation issues relating to herps. René compiled the first Red Data Book on amphibians and reptiles; he convinced members of the Council of Europe to become involved in their conservation. His career serves to demonstrate that a zoo worker can be effective in the conservation arena; many of his seminal publications deal with conservation subjects. He retired in October 1999.

Personal Reflections: When I first met Hediger, Honegger and Christian Schmidt (now the Director of the Frankfurt Zoo) in 1969 at the Zoo, René showed me the fine collection, housed in naturalistic exhibits with living plants. One of my most memorable herpetological experiences occurred that day. He reached into an enclosure and retrieved an adult New Caledonian giant gecko (*Rhacodactylus leachianus*) which he placed on his forearm. The lizard rested quietly as René continued his pace throughout the building. Although I had heard of these impressive lizards, nothing had prepared me for the sheer delight of seeing this massive saurian in the flesh for the first time.

Historical Overview

Hediger, H. 1964. Wild Animals in Captivity, Dover Publications, New York. [German edition was *Wildtiere in Gefangenschaft*, published in 1950 and the English version was translated by G. Sircom. At the time of publication, Hediger was Director of the Basle Zoo, Switzerland.]

Hediger, H. 1990. Ein Leben mit Tieren im Zoo und in aller Welt. Werd Verlag, Zürich. [Hediger's biography contains a wealth of information and anecdotes on herpetology.]

Honegger, R. E. 1971. Die Freianlage für Riesenschildkröten im Zürcher Zoo (The open air [free] facility for giant tortoises at the Zurich Zoo). Anthos 10(3):14-16.

Honegger, R. E. 1993. Heini Hediger (1908-1992). Copeia 1993:584-585.

Honegger, R. E. 2001. Heini Hediger (1908-1992). In W. Rieck, G. T. Hallmann, and W. Bischoff (eds.), Die Geschichte der Herpetologie und Terrarienkunde im deutschsprachigen Raum (History of Herpetology and Terrarium Science in German-Speaking Areas) - Mertensiella 12:473-474. Deutschen Gesellschaft für Herpetologie und Terrarienkunde e.V. (DGHT).

Klages, J. 1968. Born in the Zoo. Viking Press, New York. [Text and captions by Heini Hediger; there are two plates, one in color, of hatchling Greek tortoises.]

Maple, T. L. 1992. In memoriam. Prof. Dr. Heini Hediger (1908-1992). Zoo Biol. 11:369-372.

Rabb, G. 1993. Heini Hediger - A pioneer in the science of animal behavior. Zool. Gart. (N.F.), Jena 63:163-167.

Röthlin, O., and K. Müller. 2000. Zoo Zürich-Chronik eines Tiergartens. NZZ Verlag, Zürich.

Schmidt, C. R. 1993. Obituary. Professor Dr. Dr. h. c. Heini Hediger, 1908-1992. Inter. Zoo News No. 241. 39:2-5.

Schmidt, C. R., and R. E. Honegger. 1968. Bibliographie von Prof. Dr. Dr. Heini Hediger (Bibliography of

Prof. Dr. Dr. Heini Hediger). Zool. Gart. (N.F.), Leipzig 36:5-11.

Schmidt, C. R., and R. E. Honegger. 1993. Bibliographie von Prof. Dr. Dr. Heini Hediger II (Bibliography of Prof. Dr. Dr. Heini Hediger II). Zool. Gart. (N.F.), Jena 3:159-162.

Husbandry

Honegger, R. E. 1969. Notes on some amphibians and reptiles at Zurich Zoo. Inter. Zoo Yearb. 9:24-28.

Honegger, R. E. 1997. Aussichten-Einsichten; Gedanken zur Gestaltung von Grossterrarien für Grossechsen und Riesenschlangen (Views and insights: Thoughts on the construction of large terraria for large lizards and giant snakes). Z. Köln. Zoo 40(2):71-75. [Honegger felt that those persons designing vivaria should utilize natural materials, including live plants. Plastic foliage should never be considered as its use sends the wrong message to the zoo visitor.]

Behavior

Hediger, H. 1963. Camouflaged still-fishing. Strange, wormlike lure draws turtle's prey. Natural History 72(6):18-21.

Hediger, H. 1963. Die Alligator-Schnappschildkröte, die vollkommenste Todesfalle des Tierreiches (The snapping turtle, the most perfect lethal trap of the Animal Kingdom). Das Tier (The Animal) 3(11):4-6.

Hediger, H. 1968. The Psychology and Behaviour of Animals in Zoos and Circuses. Dover, New York. [Translation of *Skizzen zu einer Tierpsychologie im Zoo und im Zircus* was published in 1955 by Butterworths Scientific Publications. Starting on p. 128, a series of photographs of male combat in captive red rattlesnakes at the San Diego Zoo was provided.]

Reproduction

Casares, M., A. Rübel, M. Döbeli, R. E. Honegger, and E. Isenbügel. 1994. Non-invasive assessment of reproductive patterns in tortoises. Verh. ber. Erkrg. Zootiere 36: 81-87.

Hediger, H. 1965. Environmental factors influencing the reproduction of zoo animals, p. 319-354. In F. A. Beach (ed.), Sex and Behavior. Wiley & Sons, New York, London, Sydney.

Honegger, R. E. 1971. Beitrag zur Fortpflanzungsbiologie einiger tropischer Reptilien (Paper on the reproductive biology of several tropical reptiles). Z. Köln. Zoo 13:175-179.

Honegger, R. E. 1978. Geschlechtsbestimmung bei Reptilien (Determination of gender in reptiles). Salamandra 14:69-79 [discussion of sexing techniques for reptiles].

Honegger, R. E. 1982. Breeding crocodiles in captivity, a retrospect. Proc. 5th Working Meeting Croc. Spec. Group SSC/ IUCN. Florida State Mus., Gainesville, 12-16 Aug 1980.

Honegger, R. E. 1986. Zur Pflege und langjährigen Nachzucht von *Siebenrockiella crassicollis* (Gray, 1831) (Concerning the care and long-term reproduction of *Siebenrockiella crassicollis* (Gray, 1831)). Salamandra 22:1-10. [Discussion of reproduction and husbandry; one female lived at Zürich Zoo for nearly 60 years.]

Medical Management

Dollinger, P., O. Pagan, T. Jermann, R. Baumgartner, and R. E. Honegger. 1997. Husbandry and pathology of land tortoises. Verh. ber. Erkrg. Zootiere 38.

Honegger, R. E., and J. Furrer. 1975. Einige bemerkenswerte Todesfälle bei Reptilien (Several noteworthy cases of death in reptiles). Salamandra 11:179-181.

Conservation

Honegger, R. E. 1965. Galapagos Seychellen retour . . . oder das grosse Sterben der Schildkröten (Galapagos Seychelles return . . . or the great dying of the tortoises). Schweizer Spiegel (Swiss Mirror) III:39-45.

Honegger, R. E. 1966. Beobachtungen an eingeführten Säugetieren auf Galapagos (Observations concerning mammals imported to the Galapagos Islands). Natur und Museum 96:20-27.

Honegger, R. E. 1967. Beobachtungen an den Riesenschildkroten (*Testudo gigantea*) der Inseln im Indischen Ozean (Observations of the giant tortoises [*Testudo gigantea*] of the islands of the Indian Ocean). Salamandra 3:101-121 [many habitat photographs and discussion of pressures on tortoise population].

Honegger, R. E. 1968. Red Data Book Vol. 3: Amphibia/ Reptilia. International Union for Conservation of Nature and Natural Resources, Survival Service Commission, Morges, Switzerland. 1st ed. [2nd ed. published in 1970].

Honegger, R. E. 1969. Bedrohte Amphibien und Reptilien (Endangered amphibians and reptiles). Zool. Garten (N.F.) 36 Hediger-Festschrift:173-185 [discussion of conservation problems, caused by mostly human pressures].

Honegger, R. E. 1972. Breeding and maintaining reptiles in captivity, p. 1-12. In R. D. Martin (ed.), Breeding Endangered Species in Captivity. Academic Press, London.

Honegger, R. E. 1972. Amphibians and reptiles appearing in the Red Data Book, vol. III, listed by country or, in the case of Island species, under Oceanic regions. IUCN Bull. Suppl. 3 (7):1-8.

Honegger, R. E. 1972. The status of four threatened crocodilian species of Asia. Proc. 1st Working Meeting Croc. Specialists, IUCN Publ. Ser. Suppl. Paper 32:44-50.

Honegger, R. E. 1974. The reptile trade. Inter. Zoo Yearb. 14:47-52.

Honegger, R. E. 1975. The public aquarium and terrarium as a consumer of wildlife. Inter. Zoo Yearb. 15:269-271.

Honegger, R. E. 1979. Some aspects of captive propagation of amphibians and reptiles with the aim of conservation, p. 121-134. In R. A. Hahn (ser. ed.), 2nd International Herpetological Symposium on Captive Propagation and Husbandry. Zoological Consortium, Inc., Thurmont MD.

Honegger, R. 1980-81. List of amphibians and reptiles either known or thought to have become extinct since 1600. Biol. Conserv. 19:141-158.

Honegger, R. E. 1987. CITES Identification manual III. Dermochelyidae, Chelidae. Genève. [Honegger published nine sections in this series, mostly with Urs Woy: Testudinidae (1980); Trionychidae (1982); Dermochelyidae, Chelidae (1983); Emydidae (1985); Pelomedusidae (1986); Rhynchocephalidae (1987); Cryptobranchidae (1987); Turtle-shell and tortoise-shell products (1987); all but this one were published in Lausanne.]

Honegger, R. E. 1989. Crocodile utilisation and public education: Difficulties in explaining conservation. Crocodiles. Proc. Meeting Croc. Spec. Group SSC/IUCN. Quito. IUCN Publ. N.S.:183-184.

Honegger, R. E. 1992. Undesirable trends in captive management and conservation of reptiles. Bull. Chicago Herpetol. Soc. 27(19):207-210. [Article was reprinted in Inter. Zoo News 40(5):11-17 (1993). A German version was published the next year: Albinos und Hybriden oder die Gier nach dem Abartigen in Naturmuseum 124(8):258-270.]

General

Grzimek, B. 1975. Grzimek's Animal Life Encyclopedia, In B. Grzimek, H. Hediger, K. Klemmer, O. Kuhn, and H. Wermuth (eds.), Reptiles. New York.

Hediger, H. 1932. Zum Problem der "fliegenden" Schlangen (Regarding the problem of the "flying" snakes). Rev. Suisse Zool. 39:239-246.

Hediger, H. 1932. Ein eigenartiges Biotop von *Lacerta muralis muralis* (Laurenti) (A peculiar biotope of *Lacerta muralis muralis* [Laurenti]). Bl. Aquar. Terrkd. 43:316-317.

Hediger, H. 1933. Über die von Herrn Dr. A. Bühler auf der Admiralitätsgruppe und einigen benachbarten Inseln gesammelten Reptilien und Amphibien (Concerning the reptiles and amphibians collected by Dr. A. Bühler in the Admirality Archipelago and several neighboring islands). Ver. naturf. Ges. Basel 44. 2. Teil,:1-25.

Hediger, H. 1934. Beitrage zur Herpetologie und Zoogeographie Neu-Britanniens und einiger umliegender Gebiete (Contributions to the herpetology and zoogeography of New Britain). Zool. Jb. Syst. 65:441-582.

Hediger, H. 1934. Giftschlangen und Schlangengift (Poisonous snakes and snake poison). Ciba-Z. 1:265-269.

Hediger, H. 1936. Die Schlangen Mitteleuropas (The Snakes of Central Europe). Wiss. Abt. Ges. Chem. Industrie Basel (Section for the Society of Chemical Industries in Basel [Switzerland].).

Hediger, H. 1936. Les Serpents de l'Europe Centrale (The Snakes of Central Europe). Dép. scient. Soc. chem. Industrie Bâle.

Hediger, H. 1936. Herpetologische Beobachtungen in Marokko (Herpetological observations in Morocco). Ver. naturf. Ges. Basel 46:1-49.

Hediger, H. 1937. Herpetologische Beobachtungen in Marokko II. Zur Herpetofauna der Umgebung von Ouezzan und Tanger (Herpetological observations in Morocco II. Regarding the herpetofauna in the environs of Ouezzan and Tangier). Ver. naturf. Ges. Basel 48:183-192.

Hediger, H. 1937. Seltsame Reptilien und Amphibien der Salomon-Inseln (Unusual reptiles and amphibians of the Solomon Islands). Natur. u. Volk 67:590.

Hediger, H. 1941. Eine fressende *Farancia* (A feeding *Farancia*). Zool. Gart. (N.F.), Leipzig 13:251-255.

Hediger, H. 1963. Sind Schlangen eigentlich gefährlich? (Are snakes actually dangerous?). Das Tier 3(1):20-22.

Hediger, H. 1969 (published one year later due to publication delays). Man and Animal in the Zoo; Zoo Biology. Routledge & K. Paul, London [originally published as *Mensch und Tier im Zoo in 1965*].

Heusser, H., and R.E. Honegger. 1955. Die Verbreitung der Amphibien am mittleren Zimmerberg (The distribution of amphibians on the Middle Zimmerberg). Vierteljahrsschr. Natf. Ges. Zürich (Quarterly of the Society for Natural History Research of Zurich) 100:282-290.

Honegger, R.E. 1958. Umweltveränderungen und ihre Wirkung auf die Amphibien (Changes in the environment and their effects on amphibians). Leben Umwelt (Life/Environment) :149-154.

Honegger, R.E. 1960. Beobachtungen an einigen Ostamerikanischen Schildkröten im Freiland und in Gefangenschaft (Observations on some Eastern American turtles, both living free and in captivity). Z. Vivaristik 6:54-65.

Honegger, R.E. 1963. Bei den letzten Riesenschildkröten auf den Galapagos-Inseln (Among the last giant tortoises on the Galapagos Islands). Atlantis Jan.:11-20.

Honegger, R.E. 1964. Beobachtungen an der Spaltenschildkröte *Malacochersus tornieri* in Ost-Afrika (Observations on Tornier's tortoises *Malacochersus tornieri* in East Africa). Natur Mus., Frankf. 12:462-470.

Honegger, R.E. 1967. The green turtle (*Chelonia mydas japonica* Thunberg) in the Seychelles Islands. Brit. J. Herpetol. 4:8-11.

Honegger, R.E. 1972. Zoo Breeding and Crocodile Bank. Proc. 1st Working Meeting Croc. Specialists, IUCN Publ. Ser. Suppl. Paper 32:86-97.

Honegger, R. E. 1972. Die Reptilien-Bestände auf den Galapagos-Inseln (Populations of reptiles on the Galapagos Islands). Natur und Museum 102: 437-454.

Honegger, R.E. 1975. The crocodilian situation in European zoos. Inter. Zoo Yearb. 15:277-283.

Honegger, R.E. 1979. Marking amphibians and reptiles for future identification. Inter. Zoo Yearb. 19:14-22.

Honegger, R.E. 1981. Breeding endangered species of amphibians and reptiles: Some critical remarks and suggestions. Brit. J. Herpetol. 6:113-118. [Presents discussion of need for cooperation between amateurs and professionals to expand breeding projects. Honegger provided a list of species bred consecutively in zoos during 1978-1979.]

Honegger, R.E. 1998. Zürichs Riesenschildkröten (Zurich's giant tortoises). Reptilia 13(3):62-68.

Honegger, R.E., and H. Hunt. 1990. Breeding crocodiles in zoological gardens outside the species range, with some data on the general situations in European zoos, p. 200-228. Crocodiles: Proc. 10th Working Meeting of the Crocodile Specialist Group, Gainesville, Florida. IUCN. The World Conservation Union Publ. N.S., Gland, Switzerland.

Honegger, R.E., and C. R. Schmidt. 1964. Herpetologisches aus dem Züricher Zoo (Herpetological material from the Zurich Zoo). Aquar.-u. Terra.-Z. 1964(17).

Honegger, R., and F. W. Zeigler. 1991. The 1989/1990 crocodile surveys in European and American zoos and aquaria. Inter. Zoo Yearb. 30:153-157.

(Czech Republic)
Zoologicka Zahrada Praha (1931)

History and Mission: The Prague Zoo (Zoologicka Zahrada Praha) is located in the Czech Republic. In August, 2002, the Zoo was badly damaged by flooding when the Vltava River was swollen by heavy rains.

Facility and Collection: At this time, there is not a special "Reptile house" but an exhibit was built in 1995. Reptiles are featured in various places within the Zoo: indoor/outdoor crocodile exhibit, building for large tortoises, displays in the feline building which includes a giant enclosure for iguanas and some exhibits in the elephant building. Others are kept off exhibit. In 2004, a new pavillon called "Indonesian Jungle" was opened where Komodo dragons and Malaysian giant turtles are displayed.

There have been many impressive breedings: caiman lizard, green tree and mangrove monitors, Cayman Islands and rhinoceros iguanas, prehensile-tailed skink, blood and Indian pythons and pancake tortoise. Giant tortoises are well represented with nine Aldabran and three Galápagos specimens.

Fig. 103. Desert exhibit at Prague Zoo, Czech Republic. There is not a reptile building but an exhibit was built in 1995. *Photograph provided by Ivan Rehák.*

Fig. 104. Giant tortoise exhibit at Prague Zoo, Czech Republic. *Photograph provided by Ivan Rehák.*

According to "International Zoo Yearbook" (1998:Volume 28), composition of the collection was 41 taxa of reptiles with 149 specimens and 3 taxa of amphibians numbering 14.

Staff and Scientific Achievements: Ivan Rehák has been supervisor of herpetology since 1991 but the Zoo does not have formalized curatorial positions. Before 1991, he was also associated with the Zoo as a member of "Ethology Research Group," a cooperative laboratory of the former Czechoslovak Academy of Sciences where he was employed and which is located at the Zoo. He has developed a linkage with Charles University where he is a lecturer for two courses: "Batrachology and herpetology" and "Ecology of amphibians and reptiles." Rehák serves as President of the Czech Herpetological Society and is involved in conservation activities. From 1993, he has been chair of the Amphibia and Reptilia Taxon Advisory Group of the European Association of Zoological Gardens and Aquaria (EAZA) where he maintains studbooks for Cuban boas and ground iguanas.

This association has a number of species management initiatives called EEPs. Rehák participates in the Gila monster, beaded lizard, Chinese alligator, Madagascan tree boa, and several Asian turtle programs. In 1992, Rehák published many accounts on amphibians, reptiles and herpetological history in the publication "Fauna ÈSFR. Svazek 25. Obojzivelníci, Amphibia, Reptilia. V. Baruš and O. Oliva (eds.), Academia, Praha." Petr Volensky is the current curator.

Personal Reflections: At the EAZA meeting in Barcelona in September 2002, Ivan described the catastrophe which impacted the herpetological collection, due to the rapidly rising floodwaters. The entire collection had to be moved to higher ground, including giant tortoises and large constrictors. Virtually all of his records, computers and personal effects were destroyed. In spite of this devastating loss, he was optimistic that the Zoo would be rebuilt in the near future.

Historical Overview

Rehák, I. 1991. Herpetology and herpetoculture in Czechoslovakia, p. 74-84. In A. W. Zulich (ed.), 14th International Herpetological Symposium on Captive Propagation and Husbandry. International Herpetological Symposium.

Morphology, Systematics & Taxonomy

Rehák, I. 1983. Changes in body measures during the growth of the newts *Triturus vulgaris*, T. *alpestris* and *T. cristatus* (Amphibia: Urodela). Vìst. ès. Spoleè. zool. (Praha) 47:51-67.

Rehák, I. 1984. A study on *Paramesotriton deloustali* in captivity, with description of the egg, the larva, the juvenile and the adult (Amphibia: Caudata: Salamandridae). Vìst. ès. Spoleè. zool. (Praha) 48:118-131.

Rehák, I. 1985. *Coluber rubriceps thracius* ssp. n. from Bulgaria (Reptilia: Squamata: Colubridae). Vìst. ès. Spoleè. zool. (Praha) 49:276-280.

Rehák, I. 1986. The ontogenetic development in *Paramesotriton deloustali* (Caudata: Salamandridae), p. 239-242. In (Z. Roèek, ed.) Studies in Herpetology. Charles University, Prague.

Rehák, I. 1986. Taxonomic evaluation of *Coluber rubriceps* (Venzmer, 1919) from Bulgaria, p. 289-292. In (Z. Roèek, ed.). Studies in Herpetology. Charles University, Prague.

Rehák, I. 1987. Color change in the snake *Tropidophis feicki* (Reptilia: Squamata: Tropidophiidae). Vìst. ès. Spoleè. zool. (Praha) 51:300-303.

Rehák, I. 1992. K výskytu èolka karpatského, *Triturus montandoni*, na Moravì a jeho køízení s èolkem obecným, *T. vulgaris* (On the occurrence of the Montandon's newt, *Triturus montandoni,* in Moravia and its hybridization with the smooth newt, *T. vulgaris*). Abstr. XI. Konf. Herpetol. sekce ÈSZS ÈSAV, Libìchov 1992, p. 11. [in Czech, English summary.].

Rehák, I. 1993. Morfometrická charakteristika, taxonomická determinace a rùst galapázské zelvy v Zoo Praha (Morphometrics, taxonomic determination and growth of Galapago tortoise in Zoo Prague). Gazella (Praha) 20:73-78. [in Czech, English summary].

Rehák, I., and D. J. Osborn. 1988. Notes on the distribution of reptiles and amphibians in Egypt. Vìst. ès. Spoleè. zool. (Praha) 52:271-277.

Husbandry

In 1958, a journal named *Akvárium terárium (Praha)*, now published monthly, was inaugerated. This publication features articles on captive maintenance. Another journal *Ziva (Praha)* also focuses on papers dealing with herpetoculture. Rehák has published many papers in both, based on the collection of the Zoo. Species covered include Montandon´s and smooth newts, several poison dart frogs, black-spined toad, banded gecko, Chinese water dragon, inland bearded dragon, Cuban iguana, Chinese softshell, Malayan snail-eating and Asian box turtles, banana, eyelash, Oaxacan, Ambergris Cay and Feick's dwarf boas, several insular *Epicrates*, rosy boa, New Guinea viper boa, boa constrictor, Burmese, Children's, and blood pythons, Great Plains, Japanese, red-tailed, corn, green and Trans-Pecos ratsnakes, eastern garter snake, California, Sinaloan, scarlet and grey-banded kingsnakes, western hognose snake, and bluntnose viper.

Rehák, I. 1995. Biologie a chov agamy *Physignathus* cf. *cocincinus* z Tonkinu (Biology and captive care of the agamid lizard *Physignathus* cf. *cocincinus*). Gazella (Praha) 22:91-98 [in Czech, English summary].

Rehák, I. 1997. Poznámky k oboím zelvám *Geochelone gigantea* na Seychellských ostrovech (Notes to the giant tortoises *Geochelone gigantea* in Seychelles islands). Gazella (Praha) 24:155-176. [in Czech, English summary.].

Rehák, I., and P. Velenský. 1997. Biologie varanù *Varanus prasinus*, *V. rudicollis* a *V. salvadorii* v lidské péèi (Biology of the varanids *Varanus prasinus*, *V. rudicollis* and *V. salvadorii* in captivity). Gazella (Praha) 24:108-138 [in Czech, English summary].

Behavior

Rehák, I. 1990. Poznámky k agonistickému chování hadích samcù (Notes on the agonistic behavior in snake males). Gazella (Praha) 17:115-137 [in Czech, English summary].

Conservation

Kovács, T., Z. Korsos, I. Rehák, K. Corbett, and P. S. Miller (eds.). 2002. Population and Habitat Viability Assessment for the Hungarian meadow viper (*Vipera ursinii rakosiensis*). Workshop Report: IUCN/SSC Conservation Breeding Specialist Group, Apple Valley MN.

Rehák, I. 2001. Batrachologický poklad vietnamské fauny - mlok *Paramesotriton deloustali* (Batrachological treasure of Vietnamese fauna - the Tam Dao newt, *Paramesotriton deloustali*. Gazella (Praha) 28:95-128 [in Czech, English summary].

(Russia)

Many of the titles below were originally published in the Russian language but have been translated into English. One problem encountered by Russian authors is their names may be changed during translation and editing into English. Two names have been standardized here, even though they may be different in the original publications: Kudryavtsev (Kudrjavtsev in English), and Vassiliev (Vasiliev or variations in English).

Moscow Zoo (1864)

History and Mission: The Zoo is supported by the city but is viewed by the citizenry as a national institution. In the past, it has had partial control over the dispersal of monies to other zoos in Russia and acted as a clearinghouse for inventories and conservation efforts. It was started by a Board of Trustees from a nearby university for research and educational purposes, with public attendance as an added responsibility.

Facility and Collection: Beginning in 1864, a few specimens were exhibited in various pavilions, mainly in the Aquarium. Occasionally the number of reptiles reached 10 species and 50 specimens. The reptile department was founded in 1926 and its official opening was one year later. The collection was housed in the Terrarium, located on the second and third floors inside an artificial mountain; the exhibits and interior were similar to the Exotarium in Frankfurt. At that time, the reptile collection grew to 70 species and remained at that level for many years. During World War II, known as the Great Patriotic War in Russia, the collection was nearly eliminated, with anti-aircraft artillery placed on the Terrarium roof.

In 1989, the collection was moved to a larger, two-story building adjacent to the Zoo, constructed about the time of Napoleon. Originally a music school, it was modified with public exhibits on the ground floor and around the outside of the structure. On the upper floor

Fig. 105. Terrarium-Aquarium building in Moscow Zoo, Russia, in 1877. *Provided by Sergei Kudryavtsev, Moscow Zoo Archives.*

Fig. 106. Artificial mountain housing herpetological collection at Moscow Zoo in 1927. *Provided by Sergei Kudryavtsev, Moscow Zoo Archives.*

Fig. 107. Current reptile building at Moscow Zoo. *Photograph provided by Sergei Kudryavtsev.*

are offices and sections for breeding and research colonies. Many amphibians have been bred at the Zoo. There is also a breeding farm outside of Moscow, which includes a building with some herps. The 1980s were the most active and fruitful period. The collection expanded and in the late 1990s, consisted of 226 species and 846 specimens, comprising both rare and common species.

Staff and Scientific Achievements: Many distinguished biologists worked in the Terrarium between 1930-1950. Alexey Sergeiyev was an honorary member of Moscow State University, specializing in the evolution of reptiles. He died at age 31 in Stalin's labor camps. Roman Khecin-Lourie was a famous geneticist. Ilja Darevsky, who worked at the Zoo in 1962, was the Head of Laboratory for Ornithology and Herpetology of Zoological Institute of Russian Academy of Science. Valentina Orlova is Curator of the Herpetological Department of Zoological Museum of Moscow State University from 1962. In 1946-1980, the Zoo's herpetological curator was Zoja Kovaleva. At that time, the reptile collection consisted mainly of common species and no scientific research was conducted.

Beginning in 1980, Vladimir Frolov curated the collection when it was in the artificial mountain but in 1985, he was elevated to Assistant Director for zoological and veterinary affairs. When Sergei Kudryavstev took over as curator, Sergei Mamet became his assistant curator two years later. The Department of Research was founded in 1980. In concert with the Institute for Nature Conservation, Zoological Institute of Russian Academy of Science, and zoological organizations of various Soviet Republics, research has centered on the ecology and reproductive biology of rare amphibian and reptile species of the former Soviet Union. Research programs were directed at the reproductive biology of nearly all rare species of vipers and ratsnakes there as well as many rare pythons and boas. The Central Asian cobra was bred for the first time in captivity, and successful reproduction in the Caspian monitor occurred. Over 170 scientific and popular articles as well as several books were published. Numerous field surveys in Central Asia, Caucasus, Far East and other regions in the former Soviet Union were conducted. In the late 1980s and early 1990s, staff members participated in herpetological studies in Ryukyu Archipelago (Japan), East, West and South Africa, Vietnam, Cambodia, Indonesia (Java, Sumatra, Irian Jaya), in the United Sates and other regions.

Personal Reflections: Many years ago, Sergei Kudryavstev and Sergei Mamet visited us in Dallas on their way to see the herps of southwestern Texas. During their visit, we discussed their herpetological programs, facility and collection, which was and remains global in scope. It was clear that many exciting initiatives were in place in Moscow but conditions have changed. There are serious financial problems and it is unlikely that a new reptile building will be constructed in the foreseeable future. The collection is decreasing, as well as the research and scientific activity. In spite of these difficulties, the main research interest during the last several years continues to be reproductive biology and in developing methods for long-term captive husbandry of rare and problematic species of poisonous snakes found throughout the world.

Publications: Beginning in 1990 (volume 2), the journal *Scientific Research in Zoological Parks, Moscow* has been an important venue for the staff at the Zoo which has published many papers on captive reptiles and amphibians therein, mostly based on observations at that institution. These include studies on behavior and/or reproduction in Caucasian parsley frog; rock, Indonesian scrub, Boelen's and other species of Asian and Asian-Pacific pythons; Central-Asian and red-footed tortoises; Chinese, Central Asian, Egyptian and other cobras; four-lined, Russian, keeled, Persian, tropical and other species of ratsnakes; Honduran milksnake; pigmy rattlesnake; eastern box turtle; desert and Caspian monitors; and many rare colubrids and vipers such as Caucasian, Latifii's and Radde's viper. D. B. Vassiliev has authored many papers on medical management: helminths and treatment with antihelminth drugs, surgical techniques, infectious diseases, anesthesia and immobilization, and ultrasonography.

The *Russian Journal of Herpetology* is another publication containing many studies by the Zoo staff. Research has centered on husbandry, breeding and reproduction in the Chinese crocodile lizard, Papuan python, Mexican burrowing python, Asian ratsnake, and long-nosed, Caucasian and Field's vipers.

Historical Overview

Kudryavtsev, S. V., and S. V. Mamet. 2003. History of Moscow Zoo's herp department. Herpetol. Rev. 34:292-294.

Solski, L. 2001. Zoological gardens of Central-Eastern Europe and Russia, p. 117-146. In V. N. Kisling Jr. (ed.), Zoo and Aquarium History. Ancient Animal Collections to Zoological Gardens. CRC Press; Boca Raton, FL; London; New York; Washington DC [portraits of the Moscow and St. Petersburg Zoos].

Morphology, Systematics & Taxonomy

Frolov, V. Å. 1989. New morphometric features of sexual dimorphism in Central Asian tortoise *Agrionemys horsfieldi*. Voprosy gerpetologii, Kiev, ð. 264–267.

Husbandry

Bozhanskyi, À. Ò., and S. V. Kudryavtsev. 1982. Special aspects in captive maintenance of rare vipers of USSR. Razvedenie I sozdanie novikh populyatsii redkikh I tsennykh vidiv zhivotnykh, Àshkhabad, p.175-177.

Frolov, V. Å., and Y. P. Tsvetkova. 1985. Artificial hibernation of Central Asian and Mediterranean Tortoises. Voprosy herpetologii, Òàshkent, p. 215-216.

Frolov, V. E., and Y. P. Tsvetkova. 1986. Raising young freshwater and land turtles in captivity, p. 89-94. In A. F. Kovshar' (ed.), Sokhranim dikikh zhivotnykh (Let Us Preserve Wild Animals). Almar-Ata: Kaynar.

Frolov, V. Å., S. V. Kudryavtsev, and A.V. Êîrolev. 1982. Experience in working with rare species in Reptile Department of Moscow Zoo. Razvedenie I sozdanie novikh populyatsii redkikh I tsennykh vidiv zhivotnykh, Àshkhabad, pð.162-166.

Kudryavtsev, S. V. 1986. The Malayan viper; laboratory species, p. 131-133. In V. E. Sokolov, and E. E. Syroyechkovsky (eds.), Tezisy vsesoyuznogo soveshaniya po problemam zookultury (All-Union Conference of Zoo Animal Maintenance). Moscow: Nauka.

Captive Management

Kudryavtsev, S.V. 1987. Malayan pit viper (*Calloselasma rhodostoma*) - a laboratory species. Òàzisy I Vsesiyuznogo soveschaniya po problemam zookul'tury, Ìîscow, II, ð. 131–133.

Ìàkeev, V. Ì., and V. E. Frolov. 1986. On perspectives for establishing captive population of tortoises in the USSR. Pervoe Vsesoyuznoe soveschanie po problemam zookul'tury, Ìîscow, p. 134-136.

Ìàkeev, V. Ì., A. T. Bozhanskyi, K. A. Kispoev, and S. V. Kudryavtsev. 1989. Guidelines for collecting, captive maintenance and milking snakes for venom production. Ìîscow, All-Union Research Institute for Nature Conservation and Reserves.

Medical Management

Dyagilets, Å. Y., V. I. Karabak, and D. B. Vassiliev. 2001. Bacteriological flora and its place in pathogenesis of main complex symptoms in captive reptiles. Materials of IX-th Moskovskogo mezhdunarodnogo veterinarnogo kongressa (Moscow, 12-14.04.2001), p. 121-123.

Ganina, L. V., and D. B. Vassiliev. 2001. Body inclusion diseases of snakes. Materials of IX-th Moskovskogo mezhdunarodnogo veterinarnogo kongressa (Moscow, 12-14.04.2001), ð. 124-125.

Gorokhov, V. V., L. N. Romanenko, G. A. Kozlov, À. E. Îîskvin, D. B. Vassiliev, and A. N. Volichev. 1999. Tongue worms – parasites of man and animal. Trudy Vserossiiskogo Instituta Gelmintologii. No. 35, p. 45-56.

Khutoryansky, A. A. 1968. Cause of death of (*Vipera lebetina turanica*) Chernov at Moscow Zoo, p. 72-74. The Herpetology of Central Asia. Tashkent: FAN Uzbek, SSR.

Kudrjavtsev, S. V. 1986. Diagnostic algorithms for some pathological syndromes in reptiles, p. 313-316 [in Russian with English summary]. Verh. int. Symp. Erkrank. Zootiere No. 28.

Kudryavtsev, S. V. 1987. Surgical removal of infected eggs from glued snake clutches, p. 305-310. Verh. int. Symp. Erkrank. Zootiere No. 29.

Kudryavtsev, S. V. 1987. Methods for cure of some reptile diseases in Moscow Zoo, p. 315-317. Verh. int. Symp. Erkrank. Zootiere No. 29.

Kudryavtsev, S. V., and V. E. Frolov. 1985. First attempt in Moscow Zoo for high-continuity vitaminisation of reptiles, p. 399-404. Verh. int. Symp. Erkrank. Zootiere No. 27.

Kudryavtsev, S. V., and A. M. Òimerina. 2000. Paramyxoviral infection in snakes. Ìaterialy 8-go Mezhdunarodnogo kongressa po problemam veterinarnoi mediciny melkikh domashnikh zhivotnykh. Ìîscow, p. 152-153.

Kudryavtsev, S., V. Frolov, and A. Korolev. 1985. Hydrops in amphibians. p. 481-484. Verh. int. Symp. Erkrank. Zootiere No. 27.

Skorokhodov, V. À., E. Y. Dyagilets, and D. B. Vassiliev. 2001. Special features of applying antibacterial drugs for treating reptiles. Materials of IX-th Moskovskogo mezhdunarodnogo veterinarnogo kongressa (Moscow, 12-14.04.2001), ð. 97.

Vassiliev, D. B. 2000. Clinical and laboratory methods for diagnostics of internal diseases of reptiles. Ìaterialy III Mezhdunarodnoi konferentsii Aktual'nye problemy veterinarnoi mediciny melkikh domashnikh zhivotnikh na Severnom Kakaze, Persianovsky, ð. 28-31.

Vassiliev, D. B. 2000. Clinical and laboratory methods for diagnostics of internal diseases of reptiles. Aktual'nye problemy veterinarnoi meditsiny, Sankt-Pitersburg, ð. 41-46.

Vassiliev, D. B. 2001. Reptiles: Practical anesthesia and surgery. Materials of IX-th Moskovskogo mezhdunarodnogo veterinarnogo kongressa (Moscow, 12 - 14.04.2001), ð. 100-120.

Vassiliev, D. B., and O. V. Balakina. 2000. Haematological studies in infectious and parasitic diseases of rare python *Morelia boeleni*. Ìaterialy 8-go Mezhdunarodnogo kongressa po problemam veterinarnoi mediciny melkikh domashnikh zhivotnykh. Ìîscow, ð. 153-156.

Vassiliev, D. B., and A. M. Òimerina. 2000. În some peculiarities of immobilization and anesthesia of reptiles. Ìaterialy 8-go Mezhdunarodnogo kongressa po problemam veterinarnoi mediciny melkikh domashnikh zhivotnykh. Ìîscow, ð. 163-166.

Vassiliev, D. B., A. A. Åvglevskyi, and I. V. Timerin. 1999. Viral diseases of reptiles. Puti povysheniya produktivnosti, vosproizvoditel'noi sposobnosti, profilaktiki I lecheniya sel'khoz zhivotnykh. Kursk, Russia, p. 65-66.

Reproduction

Bozhansky, A. T., and S. V. Kudryavtsev. 1986. An experience of Caucasian viper breeding in captivity. Vestnik Zool. 1986:78-81.

Frolov, V. E. 1981. Reproduction in four gecko species in the Moscow Zoological Gardens, p. 138-139. In I. S. Darevsky, N. B. Ananeva, Z. S. Barkagan, L. Y. Borkin, T. M. Sokolova, and N. N. Shcherbak (eds.), Voprosy herpetology. Leningrad: Nauka.

Kudryavtsev, S. V., and V. E. Frolov. 1985. Experience obtained from use of oxytocin to stimulate oviposition in snakes, p. 405-407. Verh. int. Symp. Erkrank. Zootiere No. 27.

Kudryavtsev, S. V., and S. V. Mamet. 1989. Breeding reptiles at the Moscow Zoo, p. 29-32. In V. V. Spitsin (ed.), Achievements in Zoos in Breeding Rare and Endangered Species of Animals. Proceedings of the International Conference. Ministry of Culture of the USSR, Moscow.

Kudryavtsev, S. V., and S. V. Mamet. 1989. Research and breeding of reptiles at the Moscow Zoo. Inter. Zoo Yearb. 28:199-204.

Kudryavtsev, S. V., and S. V. Ìàmåt. 1989. Breeding of reptiles at Moscow Zoo. Dostizheniya zooparkov v izuchenii biologii razvedeniya redkikh I ischezayuschikh vidov zhivitnykh, Îiscow, p. 23-24.

Kudryavtsev, S. V., and S. V. Mamet. 1993. Keeping and breeding in captivity snakes of Russia and adjacent countries (within the former USSR). Part I. Snake 25:39-53.

Kudryavtsev, S. V., S. V. Mamet, and M. Proutkina. 1993. Keeping and breeding in captivity snakes of Russia and adjacent countries (within the former USSR). Part II. Snake 25:121-130.

Ìamåt, S. V., and S. V. Kudryavtsev. 1992. Snakes of Russia: breeding in captivity. Proceedings of International Boidae Group meeting (Sweden).

Îdinchenko V. I. 1992. Notes about of breeding of five rare species of Boid snakes at Moscow Zoo. Proceedings of International Boidae Group meeting (Sweden).

Shubravy, O. I., I. A. Serbinova, V. K. Uteshev, and B. F. Goncharov. 1986. Maintenance and reproduction in captivity of rare and vanishing species of amphibian fauna in SSSR, p. 68-73. In A. F. Kovshar' (ed.), Sokhranim dikikh zhivotnykh (Let Us Preserve Wild Animals). Almar-Ata: Kaynar.

Uteshev, V. K., O. L. Shubravy, I. A. Serbinova, and B. F. Goncharov. 1986. Breeding of rare and endangered amphibian species in captivity, p. 731-733. In Z. Roèek (ed.), Studies in Herpetology. Charles University, Prague, Czech Republic.

Conservation

Goncharov, B. F., O. I. Shubravy, I. A. Serbinova, and V. K. Uteshev. 1989. The USSR programme for breeding amphibians, including rare and endangered species. Inter. Zoo Yearb. 28:10-21 [data on 43 taxa].

General

Berdyeva, Z. S., V. F. Îrlova, and V. F. Frolov. 1981. Two new finds of *Coluber spinalis* in the Soviet Far East and in Eastern Kazakhstan. Fauna I ekologia amphibii I reptili palearkticheskoi Azii, Leningrad, p. 28.

Bozhanskyi, A. T., and S. V. Kudryavtsev. 1986. Ecological observations of the rare vipers of the Caucasus, p. 495-498. In Z. Rocek (ed.), Studies in Herpetology. Charles University, Prague, Czech Republic.

Chebyshev, N. V., I. A. Val'tseva, V. N. Krylov, and S. V. Kudryavtsev. 1997. Venomous Animals of the Lands and Seas. Îiscow Medical Academy named after I. M. Sechenov, Îiscow [text book, pp. 60].

Koroljov, A. V. 1986. Some data on the larvae of *Mertensiella caucasica*. Studies in Herpetology, Proceedings of the European Herpetological Meeting, Prague, p. 281-283.

Kudryavtsev, S. V. 1983. A verified case of envenomation by Central Asian cobra (*Naja oxiana* Eichw., 1831). Terepavtichesky arhiv, Moscow 55:113-114.

Kudryavtsev, S. V., and A. T. Bozhanskyi. 1988. A case of hybridization of two species of vipers of the genus *Agkistrodon* in Moscow Zoo's terrarium. Vestnik Zool. 1988:69-71.

Kudryavtsev, S. V., and S. V. Mamet. 1989. Finding of *Dinodon rufozonatum* (Colubridae, Squamata) in the Promorski Territory. Zoological Journal, Ìoscow 11:153-154.

Kudryavtsev, S. V., and S. V. Mamet. 1989. Terrestrial snakes of the Soviet Union. Snake 21:29-35.

Kudryavtsev, S. V., and S. V. Ìamåt. 1998. Venomous snakes: Simply, on the essence. Univesitet Knizhnyi Dom, Îiscow.

Kudryavtsev, S. V., V. E. Frolov, and A. V. Korolev. 1991. The Terrarium and Its Inhabitants (A List of Species and Their Maintenance in Captivity, A Handbook). Lesnaya promyschlennost, Moscow.

Ìakeev, V. Ì., and S. V. Kudryavtsev. 1982. On perspectives for raising venomous snakes for venom production. Razvedenie I sozdanie novikh populyatsii redkikh I tsennykh vidiv zhivotnykh, Àshkhabad, ð.181-183.

Makeev, V., A. Bozhanskyi, and V. Frolov. 1986. Distribution of the Central Asian tortoise (*Agrionemys horsfieldi* Gray, 1844) in the South of the Turkmen SSR. Studies in Herpetology, Proceedings of the European Herpetological Meeting, Prague, p. 711-712.

Petukhov, E. B., N. P. Aleksandrova, A. V. Savushkin, N. N. Kvitko, and D. B. Vassiliev. 1995. Effect of native venom of red cobra (*Naja pallida*) on morphological and rheological properties of erythrocytes. Bull. Exp. Biol. Med. 120:1055-1057.

Sopyev, Î. S., V. M. Ìakeev, S. V. Kudryavtsev, and A. N. Ìakarov. 1987. Case of intoxication by grey monitor (*Varanus griseus*) bite. Izvestya Akademii Nauk Turkmenskoi SSR, Ashkhabad, 1, p. 29.

Storozhilova, À. N., M. D. Smirnov, A. B. Dobrovolsky, S. V. Kudryavtsev, and V. N. Òitov. 1989. Isolation and parameters of protein activator C from venom of copperhead. Biblioteka experimental'noi biologii I meditsiny, Moscow 7, p. 57-59.

Vassiliev, D. B. 1999. Turtles: Husbandry, Diseases and Treatment in Captivity. Àkvarium, Îiscow.

Vassiliev, D. B., and A. Sokolov. 1999. Turtles, Lizards, Snakes: Husbandry, Care and Treatment in Captivity. Àkvarium, Ìîscow.

Vassiliev, D. B., S. V. Kudryavtsev, and O. V. Shumakov. 1997. Manual for methods for handling poisonous snakes at zoos, for prophilaxy, and treating snake-bites --Antivenom Index. Ìîscow Zoo, 83 pp.

Leningrad Zoo (1865)

The staff at the Leningrad Zoo (St. Petersburg), known as Leningradskii Zoopark, has published on captive management. Currently, Yuri Lukin is Chief of the Dept. of Herpetology and Eugeny Kamelin is the Herpetologist. According to *International Zoo Yearbook* (1998: Volume 28), composition of the collection was 76 taxa of reptiles with 252 specimens and 3 taxa of amphibians numbering 8.

Publications:

Igolkina, V. A. 1981. Some peculiarities in feeding activity in artificial conditions, p. 59-60. Problems in Herpetology: Abstracts of 5th Herpetological Conference. Leningrad: Nauka.

Igolkina, V. A. 1986. Incubation of reptile eggs and some peculiarities of early ontogenesis of snakes from breeding data at Leningrad Zoo, p. 73-81. Keeping and Breeding Wild Animals. Alma-Ata: Kaynar.

Igolkina, V. A., and W. A. Tscherlin. 1977. Über die Aufzucht von Schlangen unter den Terrarien bedingungen des Leningrader Zoos (Concerning the breeding of snakes under terrarium conditions in the Leningrad Zoo). Aquavarien Terrarien, L.I.B. Urania Verlag 12:426-427.

Tula Exotarium (1987)

History and Mission: This facility, a foundation initially supported by local town authorities, the Zoological Institute of the Russian Academy of Sciences and the Moscow Zoo, was founded in September of 1987. Located ca. 200 km south of Moscow and specializing in amphibians and reptiles, its purpose is to study reproductive biology, taxonomy, ecology and principles of captive management. For nine years, the Exotarium was self-supporting but between 1993-1995, growth and development ceased. Since 1996, the Exotarium has been supported by the State: salaries, animal food, electricity and other utilities. Expenses for the animal collection, including transportation and customs fees, collecting expeditions, reconstruction work, new equipment, and new laboratories are borne by the staff. Even with these supplemental efforts, financial support remains tenuous.

Fig. 108. Sergei Ryabov, Director of Tula Exotarium in Russia, next to dinosaur statue near entrance. *Photograph provided by Sergei Ryabov.*

Fig. 109. Exhibit at Tula Exotarium housing Chinese water dragons (*Physignathus cocincinus*). *Photograph provided by Sergei Ryabov.*

Facility and Collection: The Exotarium consists of one building with a floor area of over 600² m and roughly 1 ha of surrounding land. When donated by local town authorities, the building required major refurbishing over many months. At the entrance, a 3-meter copper statue of a dinosaur greets the visitor who then encounters 50 terrariums and aquariums, where beside reptiles and amphibians, fish, birds, mammals, arthropods and mollusks may be seen during the tour. In addition, 24 small terrariums are used for temporary displays focused thematically: in the year 2001, which according to the Oriental calendar was the year of the Snake, beautiful snakes from all continents worldwide were featured and the visitor was asked to vote for the most stunning example.

Amphibians and reptiles maintained in the off-exhibit laboratory are used for research and breeding. These include Asian arboreal snakes (*Boiga, Rhynchophis, Dinodon, Ahaetula,* some *Elaphe, Philodryas* and African *Toxicodryas*); terrestrial Asian snakes (*Elaphe* sensu lato; *Spalerosophis, Coluber, Rhabdophis, Oligodon, Lycodon* and others); kingsnakes (*Lampropeltis*) and other North American colubrids; Euro-Asian ratsnakes; Asian vipers (*Trimeresurus, Protobothrops, Ovophis, Tropidolaemus* and *Azemiops*); European and Caucasian vipers (*Vipera*); pythons and boas (*Morelia, Corallus, Sanzinia, Candoia, Python, Liasis, Leiopython, Antaresia, Boa, Acranthophis, Epicrates, Eunectes, Eryx* and others); geckos and eublephares (*Gekko, Rhacodactylus, Uroplatus, Gehyra, Ptychozoon, Eublepharis, Hemitheconyx, Coleonyx, Goniurosaurus* and others); Agamas and skinks (*Pogona, Physignathus, Tiliqua, Corucia, Tribolonotus, Eumeces* and others); chelonians; and amphibians (*Polypedates, Theloderma, Ichtyophis*),

Two additional laboratories for hibernation of Palearctic species of southern latitudes (with the temperature regimen +9-12°C) and for hibernation of snakes of temperate latitudes and Alpine species (with temperature +3-5°C) are available.

In 2002, the collection included 408 species, subspecies and forms of reptiles and 30 species of amphibians.

Staff and Scientific Achievements: The Exotarium staff is unusual as they are specialists responsible for specific tasks, rather than generalists as in many zoos. The staff is led by Sergei Ryabov, founder of the Exotarium. Other workers from the beginning to present include Vladimir Vander (Assistant Director during 1988-1990); Vladimir Dmitriev (Assistant Director of science during 1987-1992); Yelena Bortuleva (Head of the International Department during 1988-1994, customs broker since 1998 until present); Sergei Prohorchik (herpetologist working with colubrids during 1988-1994); Raisa Prohorchick (herpetologist working with lizards during 1987-1994); Sergei Tereshkin (craftsman responsible for most construction and her-

Fig. 110. Timo Paasikunnas from Helsinki Zoo visits laboratory at Tula Exotarium to examine Asian colubrid snakes. *Photograph provided by Sergei Ryabov.*

petologist from 1987 until present); and Yuri Kaverkin (gecko specialist during 1989-1991).

Nikolai Orlov, from the Zoological Institute of the Russian Academy of Sciences (St. Petersburg) has always been involved with this organization. Presently, the staff includes Tatjana Moiseeva (Assistant Administrative Director since 1992); Konstantin Shiryaev (venomous snake specialist since 1999); Eugeny Astreiko (Assistant Director specializing in boas and pythons since 1997); Svetlana Popovskaya, (Curator specializing in snake neonatal care since 1998); Ilja Korshunov (specialist with colubrid snakes since 1995); Anna Lebedeva (specialist with geckos since 1995); Svetlana Krasnova (Administrator and press-secretary since 1998); Oksana Tishenko (Curator of the International Department since 1997), and others. Refer to the list of publications below to see examples of many reproductive events, many for the first time in captivity.

Personal Reflections: The research done at Tula with staff and scientists from the Zoological Institute is important. Nearly 140 taxa bred at this facility in 2001 which is a remarkable accomplishment. Furthermore, the large reptile and amphibian collection serves as an important public exhibition for visitors, and is known as a significant research operation with ongoing studies on reproduction and conservation of rare species. These include optimum regimens of hibernation and the use of molecular techniques. It is a shame that adequate financial support remains elusive; the total is approximately US $100,000 annually which is much too low for an operation of this magnitude.

Publications: In the Russian journal *Scientific Researches in Zoological Parks* a number of papers were published describing breeding accomplishments: Stuart's milk snake and other kingsnakes, rare Southeast Asian ratsnakes, white-lipped python, six species of *Boiga,* rough-scaled sand boa, prairie kingsnake, Jalisco milksnake, and Ashy pitviper (*Trimeresurus puniceus*), western hog-nosed snake, two species of geckos (*Paroedura, Goniurosaurus murphy*), bearded dragons, sex determination in young colubrid snakes, and an extensive study on the green tree python. Many other reproductive studies are in press: Trans-Pecos rat snake, Euro-Asian group of ratsnakes, Lotiev's viper, black mangrove snake, (*Toxicodryas blandingi*), Central American ratsnake (*Elaphe flavirufa*), Chinese bamboo ratsnake, green-eyed gecko (*Gehyra marginata*), New Caledonian bumpy gecko (*Rhacodactylus auriculatus*), and scarlet kingsnake. In the publication "The Problems of Herpetology, Seventh Herpetological Conference." Kiev, 1989, there are papers on hybrid offspring from smooth and Chinese soft-shelled turtles, breeding the Oriental ratsnake (*Ptyas mukosus nigriceps*), reproductive biology of the Japanese rat snake (*Elaphe climacophora)* in captivity and in nature, and breeding the Caucasian rat snake (*Elaphe hohenackeri*). A new book by N. Orlov and S. Ryabov on snakes of the Far East, Eastern Siberia and Mongolia is currently in press.

Historical Overview

Ryabov, S. A. 2001. The Exotarium – Snake Centre in Russia. Litteratura Serpentium 21:40-46.

Ryabov, S. 2002. The Tula Exotarium. Exso (Kiev, Ukraine) 1:45-47.

Morphology, Systematics & Taxonomy

Lebedeva, A. M. 2000. Comparative data on two forms of *Gekko gecko,* p. 163-164. In: The Problems of Herpetology, Proceedings of the 1st Meeting of the Nikolsky Herpetological Society, 4-7 December 2000, Pushchino-na-Oke.

Orlov, N. L., and S. À. Ryabov. 2002. New species of the genus *Boiga* (Serpentes: Colubridae: Colubrinae) from Tanahjampea Island and description of "black form" of *Boiga cynodon* complex from Sumatra (Indonesia). Russian J. Herpetol. 9:69-80.

Ryabov, S. A. 1998. The unique case of hybridization of Persian ratsnake and leopard snake *Elaphe persica*

and *E. situla*. Scientific Researches in Zoological Parks 10:296-298.

Ryabov, S. A. 2000. The reproductive biology and some taxonomic researches of *Spalerosophis* genus. Scientific Researches in Zoological Parks 13:177-181.

Reproduction

Orlov, N. L. 1982. The breeding of snakes in the terrarium. Zh. Priroda Mosk. 10:72-80.

Orlov, N. L., and V. E. Dmitriev. 1982. Some aspects of snake breeding in the terrarium, p. 166-169. The Breeding and Creating of Rare and Valuable Animal Populations. Ashchabad: Obschestvo okhrany prirody Turkm. SSR.

Orlov, N., and S. Ryabov, 1991. Diversity of forms, maintenance and breeding of Palearctic *Vipera, Agkistrodon* and *Elaphe*, p. 1-5. In A.W. Zulich (ed.), 14th International Symposium on Captive Propagation and Husbandry. International Herpetological Symposium.

Orlov, N., S. Ryabov, and K.-D. Schulz. 1999. Eine seltene Natter aus Nordvietnam, *Rhynchophis boulengeri* Mocquard, 1897 (Squamata: Serpentes: Colubridae) (A rare snake out of North Vietnam, *Rhynchophis boulengeri* Mocquard, 1897 [Squamata: Serpentes: Colubridae]). Sauria 21:3-8.

Orlov, N. L., S. A. Ryabov, K. A. Shiryaev, and Nguen Van Sang. 2001. On the biology of pit vipers of *Protobothrops* genus (Serpentes: Colubroidea: Viperidae: Crotalinae). Russian J. Herpetol. 8:159-164.

Panteleev, D., S. Ryabov, N. Orlov, and K. Shiryaev. 2002. Unique data on reproductive biology of pitviper from Sumatra *Trimeresurus sumatranus*. Russian J. Herpetol. 9:243-254.

General

Gumprecht A., F. Tillack, N. L. Orlov, A. Captain, and S. A. Ryabov. 2004. Asian Pitvipers. GeitjeBooks, Berlin.

Gumprecht A., Ryabov, S. A., N. L. Orlov, D. J. Panteleev, and K. Tepedelen. 2002. Die Bambusottern der Gattung *Trimeresurus* Lacepede Teil VII: Anmerkungen zur Biologie, Haltung und Nachzucht von *Trimeresurus sumatranus* (Raffles, 1822) Sauria 25:37-44

Ryabov, S. A. 1998. Green tree python: results on breeding and perspective programme on studying and conservation of the rare species of the world fauna. Scientific Researches in Zoological Parks 10:219-225.

Ryabov, S. A., N. L. Orlov, D. J. Panteleev, and K. A. Shiryaev. 2002. *Trimeresurus hageni, Trimeresurus puniceus* and *Trimeresurus sumatranus* (Ophidia: Viperidae: Crotalinae). The data on reproductive biology methods of captive breeding in laboratory conditions. Russian J. Herpetol. 9:243-254.

Miscellaneous Russian Titles

Cherlin, V. A. 1983. Ways of adaptation of reptiles to environmental temperature. Zh. obshch. Biol. 44:753-764.

Pestinsky, B. V. 1939. Materials for venomous snakes in biology: their capture and keeping, p. 4-62. The Works of Uzbek Zoo. Tashkent: Gostechizdat.

Zinyakova, M. P. 1964. The reproduction and rate of growth of *Vipera lebetina* in captivity. Dokl. Akad. Nauk Uzbek. SSR 7: 63-65. 106.

Zinyakova, M. P. 1966. The feeding of *Vipera lebetina* in captivity, p. 14-19. The Problems of Herpetology and Toxicology of Snake Toxins. Tashkent: Akademiya Nauka Uzbek SSR.105.

Zinyakova, M. P., et al. 1972. The keeping of Central Asian venomous snakes in captivity, p. 64-122. Tashkent: FAN Uzbek. SSR.

European Common Toad (*Bufo bufo*), European Green
Toad (*Bufo viridis*), and Natterjack Toad (*Bufo calamita*)

Chapter 7
The German Tradition

Water Moccasin (*Agkistrodon piscivorus*)

Introduction

The art of maintaining rare and unusual amphibians and reptiles in elaborate terrariums and aquariums has been a long tradition in Europe, especially in Germany. It was not unusual for professional academic and museum herpetologists to have sizeable collections at home. As an example, no less a luminary than the venerable Robert Mertens from the Senckenberg Museum in Frankfurt/Main had outdoor tortoise pits and a greenhouse with such rarities as a pair of tuatara, New Caledonian geckos, Borneo earless monitor, geometric tortoise, Fly River turtle and emerald tree monitor. Over the years, over 1500 taxa were brought to his greenhouse and numbered 250 species and subspecies in 1967 (see Bull. Maryland Herpetol. Soc. 1969, vol 5, no. 1 for his justification for keeping living collections). In the United States, professionals do not normally maintain large personal collections, although many did so earlier in their lives before getting their positions. When one views the history of amphibian and reptile keeping, Germany stands as one of its premier centers. To underscore this fact, one needs only to review a recent catalogue by the antiquarian book dealer Chimaira in Frankfurt/Main, listing hundreds of titles in German and English dealing with captive herpetofauna. Although most of the books summarized below were not written by zoo professionals, these publications stand as important references outlining and summarizing relevant information for zoo workers.

Johann Matthaeus Bechstein (1757-1822) wrote the first book on captive care of domestic animals and pets in 1797. This intriguing volume was called *Naturgeschichte; oder, Anleitung zur Kenntniss und Wartung der Säugethiere, Amphibien, Fische, Insecten und Würmer, welche man in der Stube halten kann* (*Natural History, or, Guide to the Knowledge and Care of Mammals, Amphibians, Fish, Insects and Worms Which Can Be Kept in the Home*) and stressed animal behavior. He also published a multivolume set on zoology and wrote extensively on the care of birds and mammals. See Rieck (2001) and Heichler and Murphy (2004) for his biography.

In 1884, Johann von Fischer from Vienna published *Das Terrarium, seine Bepflanzung und Bevölkerung* (*The Terrarium, Its Plantings and Population*). This important book was the first to cover care of amphibians and reptiles in a detailed way. There are sections on aquarium and terrarium design and construction, and recommendations of plants suitable for these enclosures. One interesting feature are the lists of amphibians and reptiles which can be kept together and specific suggestions for the types of terraria/aquaria (with suitable plants) to be used for them. See Murphy (2005) for his biography.

In 1908, Paul Krefft wrote a seminal guide to terrarium science called *Das Terrarium* which covered terrarium design and construction, plants suitable for terrariums, food and feeding, husbandry of amphibians and reptiles, and climatological data illustrated with three maps of the world.

Wilhelm Klingelhöffer (1871-1953) was a German medical doctor who compiled this extraordinary treatment of reptiles and amphibians in captivity, called *Terrarienkunde*, edited by Christoph Scherpner. The work was the bible for European zoo workers and herpetoculturists. The first part was "Allgemeines und Technik" (General and Technique) and covered lighting, housing, terrarium design and outside enclosures, food and feeding, capture and transporting, and diseases. The second part was "Lurche" (Amphibians). The third was "Echsen" (Lizards) and the last was "Schildkröten, Panzerechsen, Schlangen, Reptilienzucht--Sachregister" (Turtles, Crocodilians, Snakes, Reptile Breeding--Index). If one wishes to get a feel for herpetocultural activity in Europe in the 1950s, this is the book to acquire. The exquisite complexity of the terrariums and aquariums, with live plants, naturalistic cage furniture, and the latest in environmental technology, can be appreciated as one views the many photographs. There are images of European zoo displays, heavily festooned with living plants in a natural setting, which contrast sharply with some of the rather sterile and simplistic exhibits in many zoos in the United States at that time. When I first encountered this book many years ago, I had not even seen pictures of many of the taxa represented, much less living examples. In fact, many still remain rare in living collections. See Adler (1989:102-3) and Scherpner (2001) for his biography.

Between 1957 and 1962, Heinz Reichenbach-Klinke published *Krankheiten der Amphibien* (*Diseases of Amphibians*) and *Krankheiten der Reptilien* (*Diseases of Reptiles*), later translated and revised by E. Elkan. The new work in English, *Principal Diseases of Lower Vertebrates*, published in 1965, was an important reference for maintaining captive herpetofauna and included recommendations for the treatment of diseases.

Günther Nietzke (1911-1998) published two volumes (1969, 1972) covering terrarium design, husbandry, diseases and treatments, and descriptions of amphibians and reptiles in captivity. This important work was *Die Terrarientiere: Bau, technische Einrichtung und Bepflanzung der Terrarien: Haltung, Fütterung und Pflege der Terrarientiere in zwei Bänden* (*Terrarium Animals: Construction, Technical Equipment, and Planning of Terraria: Care and Feeding of Terrarium Animals in Two Volumes*). See Schmidt (2001) for his biography.

Hans-Günter Petzold (1931-1982) wrote a significant book in 1984 on the importance of living collections which lead to scientific discoveries using captive herpetofauna. This work, largely unknown outside of German speaking areas, was entitled *Aufgaben und Probleme bei der Erforschung der Lebensäusserungen der Niederen Amnioten (Reptilien)* (*Tasks and Problems Connected with Research into the Life Manifestations of the Lower Amniotic Animals (Reptiles)*. See Klemmer (2001) and Chapter 4 for his biography.

Elke Zimmermann, mostly known for her work with captive amphibians, published a splendid book entitled *Das Züchten von Terrarientieren* in 1983. Three years later, the work was translated into English: *Breeding Terrarium Animals*.

In 1984, Fritz Jürgen Obst, Klaus Richter and Udo Jacob published *Lexikon der Terraristik und Herpetologie*. Four years later, the book was translated into English: *The Completely Illustrated Atlas of Reptiles and Amphibians for the Terrarium*. This enormous book is an invaluable reference.

One year later, Rudolf Ippen, Hans-Dieter Schröder and Karl Elze published *Handbuch der Zootierkrankheiten. Band 1: Reptilien.* (*Handbook of Diseases of Zoo Animals. Vol. 1: Reptiles*). During the same year, Ewald Isenbügel and Werner Frank produced *Heimtierkrankheiten* (*Diseases of Domestic Animals*).

In 2001, an exciting book *Die Geschichte der Herpetologie und Terrarienkunde im deutschsprachigen Raum* (*History of Herpetology and Terrarium Science in German-Speaking Areas*) was published which nicely traces the history of amphibians and reptiles in captivity in Germany. It is number 12 in the series Mertensiella by Deutschen Gesellschaft für Herpetologie und Terrarienkunde e.V. (DGHT). There are descriptions and photographs of zoos and aquariums, photographs and biographies of zoo workers, exhibitions, publications and an array of materials relevant to zoo history. Some of the most ornate and spectacular aquariums and terrariums imaginable are pictured.

Zoo professionals published in a number of periodicals: *Blätter für Aquarien- und Terrarienkunde* 1890–1901; *Wochenschrift für Aquarien- und Terrarienkunde* 1904–1950; *Lacerta* 1908–1929; *Aquarien, Terrarien* 1954–1990; *Aqua-Terra* 1964–1973; *Das Aquarium mit Aqua-Terra* 1967–1978; *Aquarien-Magazin* 1967–1978; *Die Aquarien- und Terrarienzeitschrift (DATZ)* since 1948; *Salamandra* since 1965 and *herpetofauna* since 1979.

The Past To Now

Loisel (1912) and Petzold (1984) offered brief remarks about herpetological collections in Germany. In the early 19th century, small collections of amphibians and reptiles were often maintained in German aquariums and natural history museums. It was only later in the 20th century that terrariums designed for amphibians and reptiles were built and collections became more varied. At the Frankfurt am Main Zoo, 11 species of amphibians were kept in 1860. The large Hamburg Aquarium exhibited a Japanese giant salamander in 1864 and Aldabran and Galápagos tortoises in the 1890s. In 1866, the Hanover Zoo only had 17 reptiles. The "Laboratoire d' Herpetologie" in Montpellier, France began supplying European zoos with amphibians and reptiles in the 1880s, enlarging collections significantly. Johann von Fischer (1850–1901) was director of the old Düsseldorf Zoo between 1880 and 1890 and may be considered as "the old master" of terrarium science in Germany. He was the first to combine a scientific approach with terrarium keeping, based in part on his observations using his extensive private collection of turtles and other reptiles.

Leopold Josef Fitzinger (1802–1884) was born in Vienna and was a gifted student of natural history, chemistry, anatomy, physiology, and botany (Schultschik, 2001). Fitzinger described a number of new species. After retirement, he went to Munich to head a private zoo, and finally became director of the Budapest Zoo.

During the early 20th century, the Breslau Zoo had 79 amphibians of 14 species in 1909. The zoological gardens of Hamburg maintained grass snakes and lizards in a terrarium, as well as a few other houses for reptiles; it closed in the 1930s but had a huge collection of herps throughout its history. Glass cages contained some large crocodiles, lizards, tortoises, snakes and salamanders (Peel, 1903:103). The Hanover Zoo had a collection of 66 reptiles numbering 36 species and 16 amphibians of 7 species on 1 January 1911.

The following portrait is based on the description by Hubert Bosch and Joseph Boscheinen (2001). In 1876 the Fauna Society opened a zoological garden in the center of Düsseldorf with an initial inventory of about 200 animals but 27 years later, its fiscal state was so perilous that only a private donation prevented bankruptcy. The zoo was transformed into a foundation in 1905 and given to the City of Düsseldorf. There was a large mammal and bird collection with an inventory of 1,238 animals, 60 of which were reptiles and amphibians. The Zoo nearly ceased to exist during the first World War. In 1914, G. Aulmann was Director of the Löbbecke Museum and began administering the Zoo seven years later. In 1927, he planned a large aquarium and insect collection in the zoo. In 1930, the new Museum was opened on land owned by the zoo on Brehmstrasse.

As director of the Löbbecke Museum and commissioned as zoo director, Horst Sieloff established new aquariums and open-air terrariums in 1932/33. The Zoo was destroyed by Allied bombing raids on 2 November 1944. There existed along with the larger Düsseldorf Zoo, a small, private "miniature zoo" in the East Park in Düsseldorf. Beginning in 1929, the Society for Terrarium and Aquarium Science ("Iris") established a terrarium and aquarium exhibit facility to combat unemployment on about 1000m² of a garden plot since amphibians and reptiles were under-represented in the Zoo. The subsequent Director of the Düsseldorf Zoo, R. Weber, provided scientific advice.

An animal sculptor Pallenberg from Düsseldorf occasionally provided this small zoo with animals brought back from his travels. In 1947 the remnants which had survived the war were absorbed into the reopened Löbbecke Museum, now located in an edifice known as "Zoobunker" on Brehmstrasse, opposite the former zoo. The next year, the aquarium and terrarium were added.

Seven years later, a new name "Aquazoo" was created. An expanded terrarium installation was opened in 1971 in this "Zoobunker," fortifying the breadth of herpetology in the former Löbbecke Museum and Aquarium.

In 1975, the City of Düsseldorf solicited competitive bids for construction of a new aquarium in the North Park. The cornerstone for the new building was laid in 1983, and it was dedicated on 10 July 1987 with the name "Löbbecke Museum and Aquazoo." Ekkehard Wolff published papers on general herpetoculture. When he left in the late 1980s to work at the Munich Zoo, Hubert Bosch assumed his position and has studied iguanids, boids, monitors, and amphibian and reptile parasites. In 1998, the collection included over 70 varieties and over 200 reptiles and 25 species and 150 individual amphibians. In 2001, the Düsseldorf Aquazoo, consisted of 62 reptile species (about 240 individuals) and 26 species of amphibians (about 120 individuals) with many impressive captive breedings.

Günther Praedicow is employed at the Thüringer Zoopark in Erfurt where the collection in 1998 numbered over 60 types of reptiles with over 200 individuals and nearly 50 specimens of amphibians, comprising 9 varieties. The history of Vivarium Darmstadt (1961) was described by Ackermann (1976, 1979 [1980]), Koch-Isenburg (1971) and Hallmann (2001). The first Vivarium was developed by Heinz Ackermann (1921-1986), Senior Magistrate Director of the City of Darmstadt. The facility was opened as a "school pavilion" for teaching biology on July 21, 1956 in the dilapidated greenhouse of the city nursery in the Orangerie Garden. Approximately six years later, the Vivarium on the Schnampelweg was renovated and modernized. The rededication of the four-acre facility took place on August 28, 1965. The Zoo published *Vivarium Darmstadt* which has included a number of articles on amphibians and reptiles by Heinz Ackermann, Hanns Feustel, Hans-Georg Horn, Konrad Klemmer, Ludwig Koch-Isenburg, Dietrich Mebs, Wilhelm Schweinfurth, Ludwig Trutnau, and Hartmut Wilke. The Chinese crocodile lizard and Macklot's python have been bred here. In 1998, the collection consisted of 53 species and over 200 reptiles and 16 species and over 90 amphibians. Zoo Dresden had 45 varieties and nearly 160 reptile specimens and 138 individual amphibians comprising 13 taxa in 1998.

(Berlin)
Zoologischer Garten (1844)
and
Aquarium Unter den Linden (1869)

History and Mission: Alfred Edmund Brehm (1829-1884), who coined the term "Vivarium," was Director of the Hamburg Zoo from 1863 onward and founder of the Aquarium Unter den Linden (Berlin's main street) in 1869. This three-story facility was located in the center of town, not at the Zoo. There was a large aviary, some mammal exhibits (gorilla and orangutan), and a geological grotto made from natural rock. By then, Otto Hermes, a pharmacist, had developed artificial sea water so a sizeable number of salt water organisms could be exhibited. The first Chinese alligator in Europe was on display as well as other reptiles. In 1862, a Japanese giant salamander was on exhibit. The Aquarium closed in 1910. The new

Aquarium, a large, beautiful facility built at the Zoo in 1913, housed an outstanding collection of reptiles, amphibians, fishes and invertebrates. Although known as an ornithologist, Oskar Heinroth (1871-1945) designed and was Director of this edifice. The overseer of the Aquarium, Werner Schröder, was involved in cleaning up the Aquarium after World War II.

Facility and Collection: The Aquarium was rebuilt on the original site in 1950. This facility was modernized and enhanced, and after six years of construction, was reopened in August 1983. The facade of the 1913 building was reproduced. Animals were exhibited in naturalistic enclosures. Quarantine and holding space was increased and the physical plant improved and upgraded.

According to *International Zoo Yearbook* (1998:Volume 28), composition of the collection was 75 species of reptiles with 387 specimens and 33 species of amphibians numbering over 600.

Staff and Scientific Achievements: Jürgen Lange followed Schröder as the head of the Aquarium in 1978 and is interested in Komodo dragons and tuataras. Three years earlier, he published a book with Karl Probst on marine aquariums. Dietmar Jarofke is the veterinarian in charge of fishes, amphibians and reptiles. Lange is now Director of the Zoo.

Publication: *Bongo*, started in 1977.

Historical Overview

Anders, K. 1981. Die Wirbeltiere des grossen Tiergartens von Berlin (The vertebrates of the large Berlin Zoo). Sitzungberichte Ges. naturf. Freunde Berlin (N.F.) (Conference Reports of the Berlin Friends of Nature (N. F.) 20-21:100-106.

Hallmann, G., and W. Bischoff. 2001. Alfred Edmund Brehm (1829-1884). In W. Rieck, G. T. Hallmann, and W. Bischoff (eds.), Die Geschichte der Herpetologie und Terrarienkunde im deutschsprachigen Raum (History of Herpetology and Terrarium Science in German-Speaking Areas) - Mertensiella 12:429-430. Deutschen Gesellschaft für Herpetologie und Terrarienkunde e.V. (DGHT).

Heck, L. 1954. Animals. My Adventure. Methuen, London. [In the chapter "Zoo in Flames," Director Lutz Heck described the stunning destruction of the Zoo during World War II. Originally published in German as *Tiere-Mein Abenteuer* in 1952.]

Hoage, R. J., and W. A. Deiss (eds.). 1996. New Worlds,

New Animals. From Menagerie to Zoological Park in the Nineteenth Century. The Johns Hopkins University Press, Baltimore and London [expanded account of the zoos and aquariums of Berlin].

Klös, H.-G. 1969. Von der Menagerie zum Tierparadies. 125 Jahre Berlin Zoo (From Menagerie to Animal Paradise. 125 Years of Zoo Berlin). Haude & Spenersche, Berlin [an arresting account of the history of the Berlin Zoo by Director Heinz-Georg Klös (1956-1991), detailing the immense destruction to the Zoo during World War II].

Klös, H.-G. 1972. Werner Schröder zur Vollendung seines 65. Lebensjahres (To Werner Schröder on occasion of his 65th birthday). Zool. Gart. (N.F.), Leipzig 42:328-330.

Klös, H.-G., and J. Lange. 1985. Modernes Aquarium im alten Gewande - das umgebaute Aquarium des Zoologischen Gartens Berlin (seine Geschichte und Modernisierung) (A modern aquarium behind an old facade - the rebuilt Aquarium of the Berlin Zoological Garden [The Aquarium's history and its modernization]). Zool. Gart. (N.F.), Jena 55:273-297. [The account included floor plan and many photographs of visitor spaces, exhibits, and service area.]

Klös, H.-G., and J. Lange. 1986. The modernization of the Aquarium at Berlin Zoo. Inter. Zoo Yearb. 24/25:322-332 [original building opened in 1913].

Randow, H. 1958. Zoo Hunt in Ceylon. Doubleday & Co., Inc., Garden City NY. [Heinz Randow built the Aquarium in the Wuppertal Zoo and later was in charge of the operation. This book described his herpetological collecting adventures, including the capture of a 25 ft. python for the Berlin Aquarium.]

Husbandry

Heinroth, O. 1937. Erfahrungen aus der Seewasser-Abteilung des Berliner Aquariums (Experiences made in the saltwater section of the Berlin Aquarium). Zool. Gart. (N.F.), Leipzig 9:278-285.

Medical Management

Jarofke, D., and F. Fiolka. 1997. Bestandssanierung nach dem Ausbruch einer atypischen Amöbendysenterie im Zooaquarium Berlin (Stock rehabilitation after an outbreak of atypical amoebic dysentery in the Aquarium of Berlin Zoo). Erkrg. Zootiere 38:43-45.

Jarofke, D., and H.-J. Herrmann. 1997. Amphibien: Biologie – Haltung – Krankheiten – Bioindikation (Amphibians: Biology-Care-Diseases-Bio-indications). Ferdinand Enke Verlag, Stuttgart.

Jarofke, D., and J. Lange. 1993. Tierärztliche Heimtierpraxis. Band 3. Reptilien: Krankheiten und Haltung (Veterinary Care of Animals Kept in the Home. Vol. 3: Reptiles: Diseases and Care). Verlag Paul Parey, Berlin & Hamburg.

Klös, H.-G., and E. M. Lang. 1982. Handbook of Zoo Medicine: Diseases and Treatment of Wild Animals in Zoos, Game Parks, Circuses, and Private Collections as well as their Therapy. Van Nostrand Rheinhold, New York. [Translation is of *Zootierkrankheiten: Krankheiten von Wildtieren im Zoo, Wildpark, Zirkus und in Privathand sowie ihre Therapie.* Werner Frank discusses amphibians and reptiles.]

General

Brehm, A. E. 1890-1893. Brehms Tierleben: allgemeine Kunde des Tierreiches (Brehm's Life of Animals: General Zoology). Bibliographisches Institut (Bibliographic Institute), Leipzig, Wien (Leipzig, Vienna). [Ten-volume compilation set the standard for a comprehensive treatment of animals. The reptile section (vol. 4) was revised by Franz Werner in 1912-1913.]

Lange, J. 1989. Observations on the Komodo monitors *Varanus komodoensis* in the Zoo-Aquarium Berlin. Inter. Zoo Yearb. 28:151-153.

Lange, J. 1991. Brückenechsen im Zoo-Aquarium Berlin (Tuatara in the Berlin Zoo Aquarium). D. Aquar. Terrar.-Z. 44:366-368.

Lange, J. 1991. Drei Komodowarane im Zoo-Aquarium Berlin (Three Komodo monitors in the Berlin Zoo Aquarium). D. Aquar. Terrar.-Z. 44:656-659.

Lange, J. 1996. Das Zoo-Aquarium Berlin bürgert Seeschildkröten im Mittelmeer aus (The Berlin Zoo expatriates sea turtles in the Mediterranean). Bongo 26:83-88.

Lange, J. 1996. Zoo-Aquarium Berlin bürgert Seeschildkröten im Mittelmeer aus (Berlin Zoo Aquarium expatriates sea turtles in the Mediterranean). D. Aquar. Terrar.-Z. 49:39-41.

Schubert, G. 1879. Ein interessanter Molch im Berlin Aquarium (An interesting newt in the Berlin Aquarium). Zool. Gart., Frankfurt a. M. 20:1-3.

Zoologischer Garten Der Stadt Frankfurt (Frankfurt-Am-Main) (1858)

History and Mission: From the beginning, the Zoo has had an impressive tradition of supporting conservation and scientific endeavors. In 1859, the Zoo founded the publication *Der zoologische Garten,* which ceased publication around 1913. It was later rejuvenated and became an exceedingly important scientific journal which deals to this day with contributions related to zoo biology. This association with the journal lasted until 1913. Bernhard Grzimek was director from 1945-1974 and instituted internationally known conservation programs. He was a prolific author and editor.

Facility and Collection: The Zoo received its first Japanese giant salamander in 1863. By 1912, the Zoological Garden maintained a large collection of European reptiles and amphibians, aquatic turtles, tortoises, monitors, small crocodilians, venomous snakes, and smaller lizards, housed in a large terrarium and conservatory on top of the aquarium. The Aquarium was renovated in 1928 and displayed a Chinese alligator and Komodo dragon. On 18 March 1944, the Zoo was destroyed by Allied bombing. The Aquarium was redesigned by curator Gustav Lederer, reopened in 1957 and renamed "Exotarium." The reptile section incorporated living plants with movable glass roofs and the crocodilian enclosure included a simulated tropical rainstorm. A tame Komodo dragon was allowed to follow the keeper through the visitor area. The Zoo is known for breeding the blue rock lizard from Baja California, Klemmer's day gecko and Malayan water monitor.

According to *International Zoo Yearbook* (1998:Volume 28), composition of the collection was 67 taxa of reptiles with 365 specimens and nearly 30 taxa of amphibians numbering over 200.

Staff and Scientific Achievements: During the 1960-1970s, Dieter Backhaus was curator of the Exotarium and his interest was venomous snakes. After he left the Zoo, he worked for the German animal magazine *"Das Tier."* Hartmut Wilke followed Backhaus and his interests were Chinese crocodile lizards, turtles and tortoises. Rudolf Wicker is currently curator and is known for his work on developing guidelines for the humane transport of amphibians and reptiles. His interests include turtles, tortoises and lizards (especially iguanids and varanoids).

Publications: *Grzimek's Animal Life Encyclopedia:* This series, comprising 13 volumes, covered lower animals, insects, mollusks and echinoderms, fishes, amphibians, reptiles, birds and mammals, certainly a remarkable achievement. The original German edition was *Tierleben.* Bernhard Grzimek, editor-in-chief, was Professor, Justus Liebig University of Giessen and Trustee, Tanzania and Uganda National Parks, East Africa at the time of publication. He was a major player in conservation initiatives in Africa and author of *Serengeti Shall Not Die,* written with his son Michael in 1961. Grzimek solicited contributions from prominent biologists for each volume. Every volume was of high quality with excellent photographs, nicely rendered plates and distribution maps. The series is now revised and enlarged. Grzimek was editor-in-chief of three other volumes: *Grzimek's Encyclopedia of Ecology, Grzimek's Encyclopedia of Evolution,* and *Grzimek's Encyclopedia of Ethology.* All of these were published by Van Nostrand and Reinhold, New York in 1976-77.

Fig. 111. Exotarium at Frankfurt a. M. Zoo. *Photograph provided by Christian Schmidt.*

Fig. 112. Interior of Exotarium at Frankfurt a. M. Zoo. *Photograph provided by Christian Schmidt.*

Personal Reflections: In 1969, I visited the Zoo with some American colleagues. Grzimek hosted a wonderful dinner for his staff and us in a remodeled castle, then serving as a luxury hotel. Each diner had three wine glasses; after each sip of wine, a waiter immediately refilled the glass which made for a merry evening! The next day, Grzimek toured the zoo with us and it is the only time that I have seen a director surrounded by visitors seeking autographs. This was due to his popular television program "A Place for Animals" which aired for over two decades. In the Exotarium, I saw my first green tree monitors and the largest giant salamander imaginable. Later, Curator Dieter Backhaus and I toured the city to see what herp species were available in local pet shops. In Germany during this period, there were a number of specialty shops which catered to the serious herpetoculturist and the range of taxa was impressive. I was quite surprised to discover that venomous snakes such as bamboo pitvipers as well as Asian long-nosed snakes and other opistoglyphs were being offered for sale. As there is a tradition of keeping elaborate terrariums with living plants in Europe, the range of tropical plants offered for sale was stunning.

Historical Overview

Grzimek, B. 1959. Exotarium at Frankfurt Zoo. Inter. Zoo Yearb. 1:31-33.

Husbandry

Wilke, H. 1981. Die Situation von *Cyclura cornuta cornuta* Bonnaterre in Zoologischen Gärten, ihre natürlichen Lebensbedingungen und der Versuch einer Übertragung auf die Tiere des Zoo Frankfurt a. M. (The situation of *Cyclura cornuta cornuta* Bonnaterre in the Zoo, its natural conditions of life, and an attempt of a transfer to the animals of the Zoo at Frankfurt a. M.). Zool. Gart. (N.F.), Jena 51:177-190. [Average longevity in zoos is 2.4 years, due in part to enclosures which are too small. The Frankfurt Zoo enclosure was much larger (13 square meters) which contributed to greater activity and reproduction.]

Medical Management

Backhaus, D. 1965. Gegen Strongyloides-Infektionen bei Reptilien (Against Strongyloides infections in reptiles). Salamandra 1:27-28.

Schildger, B. 1994. Endoscopic examination of the urogenital tract of reptiles. Proc. Amer. Assoc. Zoo Vet. :60-61.

Reproduction

Eidenmüller, B., and R. Wicker. 1992. Über einige Nachzucht von *Heloderma suspectum* (Cope, 1869) (Concerning some breeding of *Heloderma suspectum* (Cope, 1869)). Salamandra 28:106-111. [Account covered the first captive breeding in Germany.]

Schildger, B.-J., and R. Wicker. 1989. Sex determination and clinical examination in reptiles using endoscopy. Herpetol. Rev. 20(1):9-10.

Behavior

Lederer, G. 1931. Ein weiterer Beitrag zur Ethologie der Segelechse (*Hydrosaurus amboinensis* Schloss.) (A further contribution to the ethology of the sailfin lizard [*Hydrosaurus amboinensis* Schloss.]). Zool. Gart. (N.F.), Leipzig 4:277-279.

General

Backhaus, D. 1963. Zur Behandlung und Pflege von Schlangen, p. 223-253. Die Giftschlangen der Erde (The Venomous Snakes of the World), Behringwerk-Mitt. Spec. Suppl.

Backhaus, D. 1972. Unterschiede in der Giftwirkung von Sandrasselottern (*Echis carinatus*) und Kettenvipern (*Vipera russelii*) (Differences in the poison effects of saw-scaled vipers [*Echis carinatus*] and Russell's vipers [*Vipera russelii*]). Salamandra 8:177-178 [discussion of geographical variation of venom toxicity and antivenin efficacy].

Grzimek, B. 1963. Twenty Animals, One Man. A. Deutsch, London. [This book, translated from the German edition (*20 Tiere und 1 Mensch*), covers animal behavior, folklore, juvenile fiction and biography.]

Grzimek, B. 1966. Wild Animal White Man. Some Wildlife in Europe, Soviet Russia and North America. Andrew Deutsch and Thames and Hudson, London [Chapter 17 is titled "What is to become of 300,000 pet tortoises?" and discusses habits, care, exploitation and conservation of the Greek tortoise.]

Grzimek, B. 1974. Grzimek's Animal Life Encyclopedia, G. E. Freytag, B. Grzimek, O. Kuhn, and E. Thenius (eds.), Fishes 2/Amphibians. Van Nostrand and Reinhold, New York. [The amphibian volume included appendices on systematic classification, supplementary readings, animal dictionary in French-English- German-Russian, and a conversion table of metric to US and British systems.]

Grzimek, B. 1975. Grzimek's Animal Life Encyclopedia, B. Grzimek, H. Hediger, K. Klemmer, O. Kuhn, and H. Wermuth (eds.), Reptiles. Van Nostrand and Reinhold, New York. [The reptile volume included appendices similar to the amphibian volume.]

Seitz, A. 1897. Mitteilungen aus dem Zoologischen Garten zu Frankfurt a. M. (News from the Zoo in Frankfurt a. M.). Zool. Gart., Frankfurt a. M. 38:289-291.

Weigl, R. 1999. Visits to some Latin American zoos: Part 1 - Peru and Colombia. Inter. Zoo News 46(8):485-492. [At time of publication, Richard Weigl was a keeper at the Zoo who recorded the herpetological collections in Latin America.]

Zoologischer Garten Köln (1860) (Cologne)

History and Mission: In the 19th century, the Cologne Zoological Garden essentially evolved from a menagerie to an important German zoo. During World War II, the Zoo was bombed by the Allies and completely destroyed. There has been a sustained building program since that time and the Zoo has become an important regional conservation center.

Facility and Collection: The Aquarium opened on 29 April 1971 and includes a large collection of fishes, amphibians and reptiles, as well as an insectarium. The terrariums are located in a tropical hall with lush vegetation. Dwarf caimans were bred for the first time worldwide in this facility. The REGENWALD is the new tropical house specializing in the fauna of the tropical rainforests of Southeast Asia.

According to *International Zoo Yearbook* (1998:Volume 28), composition of the collection was 83 taxa of reptiles with 526 specimens and 21 taxa of amphibians numbering 188.

Staff and Scientific Achievements: Harald Jes was in charge of the herpetological collection between 1969–1994 and specialized in crocodiles and boids. Under his guidance, many unusual reptiles were successfully bred. After Jes retired, Hans-Werner Herrmann was Head of the Aquarium and Curator for Fishes, Amphibians, Reptiles and Invertebrates and in charge of Zoo conservation projects. His interests include anurans, varanoids, viperids, and herpetological conservation. Hans-Werner and his staff developed two impressive field projects. Supported by the German branch of the American Linen Supply Company (ALSCO), the ALSCO Biodiversity Conservation Initiative in Cameroon was started in 1998 at Mt. Nlonako, a mountain rainforest area in the vicinity of Nkongsamba, Southwest Cameroon. The Zoo Cologne Nature Conservation project was started in Vietnam and is linked with the new tropical house. Herrmann left the Zoo in 2002 to begin a study on the endangered goliath frog in Cameroon with his wife Patricia, supported by the San Diego Zoo.

Publication: *Zeitschrift des Kölner Zoo* was started in 1958. There are many articles on herpetology.

Personal Reflections: Hans-Werner is precisely the sort of visionary who is desperately needed in the zoo profession. His scientific achievements are broad in scope and reflect a strong commitment to conservation. At this stage of his career, it took courage to leave the Zoo and begin a sustained initiative in Cameroon with his family. I admire his willingness to take on this monumental task, and he has since returned to Arizona.

Historical Overview

Dieckmann, R., T. Pagel, and J. Wolters. 2000. Der REGENWALD - Ein neuartiges Tropenhaus im Kölner Zoo (The RAIN FOREST—a new type of tropical environment in the Cologne Zoo). Z. Köln. Zoo 43:55-73. [Article included color photographs of many exhibits and a list of the plants and animals contained therein; mammals, birds, amphibians, reptiles, fishes and a few invertebrates are featured.]

Herrmann, H.-W., and P. A. Herrmann. 2002. Herpetological conservation at the Cologne Zoo. Herpetol. Rev. 33:168-169.

Jes, H. 1981. 10 Jahre KÖLNER AQUARIUM AM ZOO (10 years COLOGNE AQUARIUM AT THE ZOO). Z. Köln. Zoo 24:3-5 [discussion of the first 10 years of the facility, including visitation, notable reproduction and successful exhibits].

Jes, H. 1997. 25 Jahre Aquarium (25 years Aquarium). Z. Köln. Zoo 40:35-38. [Description covered the 25-year- history of the insectarium and terrarium in the Aquarium.]

Leder, K., and R. Kämper. 1989. Vom Wässer - zum Landleben - Ein Unterrichtsgang durch das Kölner Aquarium (From water - to living on land - an educational tour through the Cologne Aquarium). Z. Köln. Zoo 32:97-104.

Nogge, G. 1986. Jahresbericht 1985 der Aktiengesellschaft Zoologischer Garten Köln (Annual Report of the Cologne Zoo, Inc.). Z. Köln. Zoo 29:3-25.

Morphology, Systematics & Taxonomy

Herrmann, H.-W., and U. Joger. 1995. Molecular data on the phylogeny of African vipers--preliminary results. Llorente et al. (eds.). Scientia Herpetologica:92-97

Herrmann, H.-W., and U. Joger. 1997. Evolution of viperine snakes, p. 43-61. In R. S. Thorp, W. Wüster, and A. Malhotra (eds.), Venomous Snakes. Ecology, Evolution and Snakebite. Symp. Zool. Soc. London No. 70.

Herrmann, H.-W., U. Joger, and G. Nilson. 1992. Molecular phylogeny and systematics of viperine snakes I: General phylogeny of European vipers (*Vipera* sensu stricto), p. 219-224. In Z. Korsós, and I. Kiss (eds.), Proc. 6th Ord.Gen. Meet. S. E. H., Budapest.

Herrmann, H.-W., U. Joger, P. Lenk, and M. Wink. 1999. Morphological and molecular phylogenies of viperines: Conflicting evidence? Kaupia Darmstädter Beiträge zur Naturgeschichte 8:21-30.

Herrmann, H.-W., U. Joger, G. Nilson, and C. Sibley. 1987. First steps toward a biochemical reconstruction of the phylogeny of the genus *Vipera*, p. 195-200. In J. J. van Gelder, H. Strijbosch, and P. J. M. Bergers (eds.), Proc. 4th Ord. Gen. Meet. S. E. H., Nijmegen.

Joger, U., and H.-W. Herrmann. 1993. Ein immunologigische rekonstruierter Stammbaum der paläarktischen Vipern (Viperinae) (An immunologically reconstructed family tree of the Paleo-Arctic vipers [Viperinae]). Verhandlungen der Deutschen Zoologischen Gesellschaft (Discussions of German Zoological Society), 86: 52.

Joger, U., H.-W. Herrmann, and G. Nilson. 1992. Molecular phylogeny and systematics of viperine snakes II: A revision of the *Vipera ursinii* complex, p. 239-244. In Z. Korsós, and I. Kiss (eds.), Proc. 6th Ord. Gen. Meet. S. E. H., Budapest.

Joger, U., H.-W. Herrmann, S. Kalyabina, P. Lenk, and M. Wink. 1999. Molecular phylogeny and systematics of true vipers (Reptilia: Viperinae) - a multi-dimensional approach. Zoology 102. Supplement II (DZG 92.1).

Lenk, P., H.-W. Herrmann, U. Joger, and M. Wink. 1999. Phylogeny and taxonomic subdivision of *Bitis* (Reptilia: Viperidae) based on molecular evidence. Kaupia Darmstädter Beiträge zur Naturgeschichte 8:31-38.

Nilson, G., B. Tuniyev, C. Andrén, N. Orlov, U. Joger, and H.-W. Herrmann. 1999. Taxonomic position of the *Vipera xanthina* complex. Kaupia Darmstädter Beiträge zur Naturgeschichte 8:99-102.

Ziegler, T., H.-W. Herrmann, P. David, N. L. Orlov, and O. S. G. Pauwels. 2000. *Triceratolepidophis sieversorum*, a new genus and species of pitviper (Reptilia: Serpentes: Viperidae: Crotalinae) from Vietnam. Russian J. Herpetol. 7: 199-214.

Behavior

Jes, H. 1984. Beobachtungen an einigen ovovivipar gebärenden Riesenschlangen (Boidae) (Observations made on several ovoviparous birth-giving giant snakes [Boidae]). Salamandra 20:268-269. [Female garden tree boa ate four undeveloped eggs after giving birth and a female Madagascar tree boa ate a Siamese twin after birth.]

Husbandry

Jes, H. 1987. Lizards in the Terrarium: Buying, Feeding, Care, Sicknesses, with a Special Chapter on Setting Up Rain-forest, Desert, and Water Terrariums. Barron's, New York [translation of *Echsen als Terrarientiere*].

Reproduction

Jes, H. 1971/1972. Jahresbericht Aquarium 1971 (Aquarium Annual Report for 1971). Z. Köln. Zoo 14:137-145. [Reproduction in emerald tree boa; there were many photographs of exhibits.]

Jes, H. 1981. Haltung und Zucht von Panzerechsen (Care and breeding of armored lizards). Aquar.-u. Terra.-Z. 34:173-178.

Conservation

Gruschwitz, M., S. Lenz, H. Jes, and G. Nogge. 1992. Die Nachzucht der Würfelnatter (*Natrix tessellata* Laurenti, 1768) im Aquarium des Kölner Zoos - Ein Beitrag zum Artenschutz (Breeding the diced snake (*Natrix tessellata* Laurenti, 1768) at the Aquarium at the Cologne Zoo). Z. Köln. Zoo 35:117-125. [Plan was developed by the staff to improve the habitat and initiate a captive breeding program for this critically endangered species.]

Herrmann, H.-W., and T. Pagel. 2000. Phong Nha - Ke Bang: Ein Naturschutzprojekt des Kölner Zoo in Zentralvietnam (Phong Nha - Ke Bang: A nature conservation project of the Cologne Zoo in Central Vietnam). Z. Köln. Zoo 43:79-88.

General

Herrmann, H.-W. 1993. Bemerkungen zum Beutespektrum von *Vipera monticola* (Notes concerning the range of prey of *Vipera monticola*). Salamandra 29:90-91.

Herrmann, H.-W., W. Böhme, and P. Herrmann. 2000. The ALSCO Cameroon expedition 1998: The sampling of a mountain rainforest, In G. Rheinwald (ed.), Isolated Vertebrate Communities in the Tropics. Proc. 4th Int. Symp., Bonn, Bonn zool. Monogr. 46:95-103.

Herrmann, H.-W., P. Herrmann, and W. Böhme. 1999. Die ALSCO- Kamerun-Expedition I: Amphibien und Reptilien vom Mt. Nlonako (The ALSCO-Cameroons Expedition I: Amphibians and reptiles from Mt. Nlonako). Z. Köln. Zoo 42:181-197. [The article contains a number of black-and-white and color photographs of the expedition, including some disturbing images of poached animals.]

Hirsch, U. 1980. Riesenschildkröten (*Testudo* [*Geochelone*] *elephantopus*) auf Galapagos (Giant tortoises (*Testudo* [*Geochelone*] *elephantopus*) on Galapagos). Z. Köln. Zoo 23:111-117 [discussion of previous exploitation of tortoises and needed conservation measures].

Mebs, D., K. Holada, F. Kornalik, J. Simak, H. Vankova, D. Müller, H. Schoenemann, H. Lange, and H.-W. Herrmann. 1998. Severe coagulopathy after a bite by a green bush viper (*Atheris squamigera*): Case report and biochemical analysis of the venom. Toxicon 36:1333-1340.

Schürer, U. 1982. ZOODOM, der Nationalzoo der Dominikanischen Republik und einige Bemerkungen über die Fauna des Landes (ZOODOM, the National Zoo of the Dominican Republic and some comments about the fauna of that country). Z. Köln. Zoo 25:59-63 [description of the animal collection, conservation programs and buildings in the National Zoological Garden in Santo Domingo, Dominican Republic].

Ziegler, T., and H.-W. Herrmann. 2000. Preliminary list of the herpetofauna of the Phong Nha - Ke Bang area in Quang Binh province, Vietnam. Zoogeographica 76(2):49-62.

Zoologischer Garten Leipzig (1878)

History and Mission: During World War II, the Aquarium, constructed early in the 20th century, suffered bomb damage on three occasions but was rebuilt after the conflict. Starting in 1984, there were several reconstruction projects to repair deterioration of the infrastructure and add new exhibitry. In the Terrarium, naturalistic enclosures with living foliage contain a variety of amphibians, crocodilians, turtles, venomous and giant snakes, and an array of large tortoises.

Facility and Collection: The Aquarium opened on 15 May 1910 and the collection mostly consisted of indigenous and marine fishes, displayed in 30 concrete aquariums. Roughly three years later, an enormous greenhouse terrarium was added which housed reptiles, amphibians, and insects in 25 individual exhibits and a crocodilian enclosure with living plants provided a tropical setting. In 1919, the Aquarium was expanded and 37 aquariums were added to display tropical fishes, including the ornamental angel fish, a cichlid from South America, which was seen for the first time in Europe. Remarkably, one of the oldest and largest captive Japanese giant salamanders lived 70 years at the Zoo.

According to *International Zoo Yearbook* (1998:Volume 28), composition of the collection was 65 taxa of reptiles with 185 specimens and 13 taxa of amphibians numbering 50.

Staff and Scientific Achievements: Karl Max Schneider, highly esteemed amongst international zoo experts and author of numerous herpetological papers, was director of the zoo from 1934 to 1955. His assistant, Heinrich Dathe, also wrote herpetological papers. The first curator was Peter Müller from 1963 to 1973. Wolf-Eberhard Engelmann is currently curator and is interested in snakes, especially venomous taxa.

Publication: *Panthera* started in 1970.

Historical Overview

Engelmann, W.-E. 1994. From the history of Leipzig Zoo-Aquarium. Congrés E.U.A.C., Leipzig, Mem. Inst. Océano. P. Richard :75-79.

Engelmann, W.-E. 2000. Ein neues, altes Kriechtierhaus in Leipzig (A new, old reptile house in Leipzig). Zoomagazin/Nord-Ost 2:55-56.

Engelmann, W.-E. 2004. Zur Geschichte des Aquarien- und Terrarienhauses im Zoologischen Garten Leipzig (On the history of the Aquarium- and Terrariumhouses in the Leipzig Zoological Garden). Sekretär 4:15-27.

Henning, A. S. 1998. Das Terrarium im Zoologischen Garten Leipzig (The Terrarium in the Leipzig Zoo). Reptilia (D) 3(3 Nr 11):69-71.

Kniesche, G. 1915. Ein Kriegs-Terrarium (A War Terrarium). Bl. Aquarien-Terrarienk. 26:193-194.

Morphology, Systematics & Taxonomy

Engelmann, W.-E. 1982. Der Einsatz serologischimmunologischer Methoden in der Lacerten-Taxonomie (Employment of serologic-immunological methods in lizard taxonomy). Vertebrata Hungarica 21:111-115.

Engelmann, W.-E., and H. Schäffner. 1981. Serologisch-immunologische Untersuchungen zu einigen taxonomischen Problemen innerhalb der Sammelgattung *Lacerta* (Sauria, Lacertidae) (Serological and immunological investigations of some taxonomic problems in the genus *Lacerta* [Sauria, Lacertidae). Zool. Jb. Syst. 108:139-161.

Hacker, G., and G.-H. Schumacher. 1955. Die Muskeln und Faszien des Mundbodens bei *Testudo graeca* (The muscles and fascia of the floor of the mouth in *Testudo graeca*). Anat. Anz. 101:234-305.

Kabisch, K., and W.-E. Engelmann. 1975. Vergleichende elektrophoretische Untersuchung der a-Naphthylacetat-Esterase im Serum mitteleuropäischer Discoglossiden und Bufoniden (Amphibia, Anura) (Comparative electrophoretic examination of the a-Naphthylacetate-Estarase in the serum of Central European Discoglossids and Bufonids [Amphibia, Anura]). Biol. Zbl. 94:211-214.

Schnabel, R. 1955. Über den Schlüpfvorgang und die Resorption des Eizahns von *Natrix natrix* (Concerning the hatching process and the re-absorption of the egg tooth of *Natrix natrix*). Z. für mikroskopisch-anatomische Forschung (Periodical for Microscopic-Anatomical Research) 61:487-511.

Schnabel, R. 1956. Beitrag über die frühen Entwicklungsstadien des Eizahns und das Abortivgebiss des Zwischenkiefers bei *Natrix natrix* (Monograph concerning early developmental stages of the egg tooth and the abortive teeth of the interim jaw in *Natrix natrix*). Z. für mikroskopisch-anatomische Forschung (Periodical for Microscopic-Anatomical Research) 62:40-50.

Schnabel, R., and K. Herschel. 1955. Über die Entwicklung des Eizahns von *Natrix natrix* (Concerning the development of the egg tooth of *Natrix natrix*). Z. für mikroskopisch- anatomische Forschung (Periodical for Microscopic-Anatomical Research) 61:246-280.

Schneider, K. M. 1941. Über Fettlager im Schwanz der Krusten- (*Heloderma* WIEGM.) und Stutz-Echse (*Trachydosaurus* GREY) (About fat deposits in the tail of the venomous (*Heloderma* WIEGM.) and Shingleback lizard [*Trachydosaurus* GREY]). Zool. Gart. (N.F.), Leipzig 13:236-247.

Schneider, K. M. 1942. Fettlager im Krokodilschwanz (Fat deposits in the crocodile tail). Zool. Gart. (N.F.), Leipzig 14:263-267.

Schuhmacher, G. H. [1955]. Beiträge zur Kiefermuskulatur der Schildkröten. (Contributions concerning the jaw muscles of turtles) II. Mitteilung. - Ebenda 4, Mathematisch-naturwissenschaftliche Reihe Nr. 5, 1954/1955 [1955] :501-518.

Behavior

Engelmann, W.-E., W. Völkl, and H.-J. Biella. 1990. Zum Paarungsverhalten der Schlingnatter, (*Coronella austriaca*) LAUR. (Reptilia, Serpentes, Colubridae). (On the mating behavior of the smooth snake, [*Coronella austriaca*] LAUR. [Reptilia, Serpentes, Colubridae]). Zool. Abh. Staatl. Mus. Tierk. Dresden 45:137-139.

Schneider, K. M. 1942. Zur Tauchdauer des Hechtalligators (About the diving duration of the pike-snouted alligator). Zool. Gart. (N.F.), Leipzig 14:156-157.

Schneider K. M. 1943. Von den Entleerungen einiger Panzerechsen (About the eliminations of some armored lizards). Zool. Anz. 142:95-101.

Husbandry

Elze, K., and W.-E. Engelmann. 1981. Zur planmässigen Peressigsäure-Desinfektion in mit Reptilien besetzten Terrarien (Concerning systematic disinfection of reptile-containing terraria with paracetic acid). Verhandlungsbericht XXIII, Internat. Symp. Erkrankungen Zootiere Halle/Saale 1981 (Conference Report XXIII, International Symposium about Illnesses of Zoo Animals at Halle an der Saale 1981):109-113 [use of paracetic acid solution for disinfection at Zoo].

Reproduction

Anonymous. 1893. Brütende Riesenschlange (Brooding giant snake). Zool. Gart., Frankfurt a. M. 34:319-320.

Engelmann, W.-E. 1988. Beobachtungen zum Fortpflanzungsgeschehen bei Schlangen im Terrarium des Zoologischen Gartens Leipzig (Observations regarding the reproduction of snakes in the Terrarium at the Leipzig Zoological Garden). Zool. Gart. (N.F.), Jena 58:175-181.

Engelmann, W.-E. 2001. Keeping and breeding of dwarf crocodiles (*Osteolaemus tetraspis*) in the Leipzig Zoo Aquarium. Bull. Inst. Oceanogr. Monaco 20:331-336 [male assisted in caring for young].

Engelmann, W.-E., and H. G. Horn. 2003. Erstmalige Nachzucht von Karl-Schmidt's Waran, *Varanus jobiensis* im Zoo Leipzig Zoo, (First breeding of

Karl Schmidt's monitor, *Varanus jobiensis* at the Leipzig Zoo). Zool-Gart (N.F.), Jena 73:353-358.

Schneider, K. M. 1949. Verzögerte Befruchtung bei einer Goldnatter? (Delayed fertilization in a gold viper?) D. Aquar. Terrar.-Z. 2:205-207.

Medical Management

Elze, K., A. Bergmann, K. Eulenberger, H. J. Selbitz and W.-E. Engelmann. 1977. Auswertung des Krankheitsgeschehens bei Reptilien im Zoologischen Garten Leipzig 1957-1976 (Evaluation of the incidence of illness in reptiles in the Leipzig Zoological Garden 1957 1976). Verhandlungs-bericht XIX. Internat. Symp. Erkrankungen Zootiere Poznan 1977 (Conference Report XIXth International Symposium about Illness in Zoo Animals, Poznan, 1977):45-55.

Grimpe. G. 1929. Merkwürdige Todesursache einer Elefantenschildkröte (Strange cause of death of an elephant tortoise). Zool. Gart. (N.F.), Frankfurt a. M. 1:225-226.

Mertens, R. 1940. Der Knochenpanzer einer kyphotischen Weichschildkröte (The bony armor of a kyphotic trionychine turtle). Senckenbergiana 22:236-243.

Mertens, R. 1942. Bemerkungen zur Sektion einer Elefantenschildkröte (Notes on the dissection of an elephant tortoise). Zool. Gart. (N.F.), Leipzig 14:276-277.

Meyn, A. 1942. Actinomyces-Infektion bei einer Schlange (Actinomyces infection in a snake). Zool. Gart. (N.F.), Leipzig 14:251-253.

Schneider. K. M. 1932. Zum Tode des Leipziger Riesensalamanders *Megalobatrachus maximus* SCHLEGEL (Regarding the death of the Leipzig giant salamander *Megalobatrachus maximus* SCHLEGEL). Zool. Gart., Leipzig 5:142-145.

Schneider, K. M. 1943. Ein Hecht-Alligator mit künstlichen Kiefern (An alligator with artificial jaws). Zahnärztl. Mitteilungen (Dentistry Notes) 34: 120-121.

Selbitz, H.-J., and W.-E. Engelmann. 1984. Die häufigsten bakteriellen Infektionen bei Reptilien (The most frequent bacterial infections in reptiles). Mh. Vet.-Med. 39:383-386.

General

Bischoff, W., and W.-E. Engelmann. 1976. Herpetologische Ergebnisse einiger Sammelreisen im Kaukasus und in Transkaukasien (Herpetological results of some collection travels in the Caucasus and in Transcaucasia). Zool Jb. Syst. 103:361-376 [results of 10 expeditions].

Bischoff, W., and W.-E. Engelmann. 1978. Zur aktuellen Entwicklung der Smaragdeidechsen-Population von Lieberose, Kr. Beeskow (Reptilia, Lacertidae) (Concerning the current development of the emerald lizard population of Lieberose, Kr. Beeskow [Reptilia, Lacertidae]). Faun. Abh. Mus. Tierk. Dresden 7:93-94.

Blahak, S., H. Brücher, W. E. Engelmann, P.-H. Grünwaldt, I. Pauler, W. Rades, M. Riebe, and R. Wicker. 1997. Gutachten über Mindestanforderungen an die Haltung von Reptilien vom 10. Januar 1997. Erstellt im Auftrag des Bundesministeriums für Ernährung, Landwirtschaft und Forsten, Referat Tierschutz. (Expert Opinion Concerning Minimum Requirements for the Keeping of Reptiles, dated January 10, 1997. Prepared at the Request of the Federal Ministry for Nutrition, Agriculture and Forestry, Department for the Protection of Animals):1-78.

Engelmann, W.-E. 1974. Vergleichende Enzymuntersuchungen der Seren mitteleuropäischer Raniden (Amphibia, Anura) (Comparative enzyme examinations of the serum of Central European frogs [Amphibia, Anura]). Experientia 30:870-872.

Engelmann, W.-E. 1985. Herpetologische Reiseeindrücke aus dem Süden Vietnams (Herpetological journey impressions from the South of Vietnam). Aquar. Terrar. (No. 5):170-175 [photograph of an enormous banded krait].

Engelmann, W.-E. 1993. *Coronella austriaca* (Laurenti, 1768) - Schlingnatter, Glatt- oder Haselnatter (*Coronella austriaca* (Laurenti, 1768) - smooth snake). In: W. Böhme (ed.). Handbuch der Reptilien und Amphibien Europas, 3/1, Schlangen (Serpentes) I. (Handbook of the Reptiles and Amphibians of Europe, 3/1, Snakes [Serpentes] I.). Aula-Verlag, Wiesbaden.

Engelmann, W.-E., and F. J. Obst. 1976. Partielle Pigmentlosigkeit bei *Bufo viridis viridis* (Amphibia, Anura, Bufonidae) (Partial absence of pigmentation in *Bufo viridis viridis* [Amphibia, Anura, Bufonidae]). Zool. Abh. Staatl. Mus. Tierk. Dresden 34:39-41.

Engelmann, W.-E., and F. J. Obst. 1981. Mit gespaltener Zunge: Aus der Biologie und Kulturgeschichte der Schlangen (With Forked Tongue. From the Biology and Cultural History of Snakes). Verlag Edition, Leipzig.

Engelmann, W.-E., and F. J. Obst. 1982. Snakes: Biology, Behavior and Relationship to Man. Exeter Books, New York. [At time of publication, Fritz Jürgen Obst was Keeper of the Herpetological Collection at the Dresden State Museum of Zoology and published a number of books on chelonians.]

Engelmann, W.-E., J. Fritzsche, R. Günther, and F. J. Obst. 1985. Lurche und Kriechtiere Europas. Beobachtungen und Bestimmungen (Amphibians and Reptiles of Europe. Observations and determinations). Neumann Verlag, Radebeul.

Mertens, R. 1926. Herpetologische Mitteilungen. XIV. Zur Kenntnis der Herpetofauna von Angola (Herpetological Notes. XIV. Concerning knowledge of the herpetofauna of Angola). Senckenbergiana :151-154.

Obst, F. J., K. Richter, and U. Jacob. 1988. The Completely Illustrated Atlas of Reptiles and Amphibians for the Terrarium. T.F.H. Publications, Neptune City NJ. [W.-E. Engelmann prepared the accounts for anatomy and physiology, ethology and genetics.].

Schneider, K. M. 1926. Vom Leipziger Riesensalamander (About the Leipzig giant salamander). Leipzig. Eine Monatsschrift 3:62-64.

Simroth, H. 1888-89 [1890]. Das Vorkommen der gemeinen Teichschildkröte, *Emys europaea,* bei Leipzig (Incidence of the common pond turtle, *Emys europaea*, near Leipzig). Sitzungsberichte der Naturforschenden Gesellschaft zu Leipzig 15/16:61-64.

Carl Hagenbeck's Tierpark at Stellingen (1907)

History and Mission: In Hamburg, Carl Hagenbeck developed the first exhibits without bars for zoos and served as a design consultant for zoos throughout the world. A shrewd numbering system for exhibits takes the visitor back-and-forth but in reality, only a small area is traversed.

Facility and Collection: Hagenbeck's first reptile building used the same material for walls, ceilings and floors in an attempt to immerse the visitor in the setting. The new facility, Das Troparium, opened during March, 1960 and houses various amphibians and reptiles.

According to *International Zoo Yearbook* (1998:Volume 28), composition of the collection was 37 taxa of reptiles with 146 specimens and 9 taxa of amphibians numbering 28.

Staff and Scientific Achievements: Harald Jes was employed at the Zoo from 1960-1969 and his interests were crocodiles and boid snakes. Uwe Richter, interested in green tree pythons, followed Jes from 1969 to the present. The staff comprises Richter and two colleagues who maintain the collection.

Personal Reflections: The most striking features of this Zoo are the extraordinary displays for mammals and birds which, in my opinion, have never been duplicated in any zoo. I was intrigued by the attention paid to sight lines and landscaping which gave the illusion of a much larger park.

Historical Overview

Hagenbeck, C. 1909. Beasts and Men, Being Carl Hagenbeck's Experiences for Half a Century Among Wild Animals. Longmans, Green, London, New York, Bombay, Calcutta. [Abridged translation by Hugh S. R. Elliot and A. G. Thacker, A.R.C.S. (Lond.), with an introduction by P. Chalmers Mitchell includes accounts of animal training, menageries, zoos and collection and preservation of zoological specimens.]

Hoage, R. J., and W. A. Deiss (eds.). 1996. New Worlds, New Animals. From Menagerie to Zoological Park in the Nineteenth Century. The Johns Hopkins University Press, Baltimore and London [expanded account of Carl Hagenbeck and his Tierpark].

Kreger, M. D. 2001. Hagenbeck, Carl, Jr. 1844-1913, p. 535-537. In C. E. Bell (ed.), Encyclopedia of the World's Zoos. Fitzroy Dearborn Publishers, Chicago, London.

Richter, U. 1995. Zu Besuch in Carl Hagenbecks Tierpark - Troparium (A visit to Carl Hagenbeck's Zoo - Troparium). D. Aquar. Terrar.-Z. 48:54-55.

Wiese, E. 1995. Das Hagenbeck-Buch (The Hagenbeck Book). Historika Photoverlag, Hamburg [discussion with many historical photographs of Tierpark].

General

Anon. 1913. Schlangenkämpfe in Carl Hagenbecks Tierpark in Stellingen (Snake battles in Carl Hagenbeck's Zoo in Stellingen). Zool. Gart., Frankfurt a. M. 54:7-8.

Wilhelma, Zoologischer und Botanischer Garten (1949)

History and Mission: Wilhelma is located in Stuttgart and is known for its extensive collection of tropical plants and botanical holdings. Damage was extensive after World War II; the first building constructed afterward was the combination aquarium and terrarium. During 1963-1967, a large new aquarium was built.

Facility and Collection: The facility has large planted enclosures for crocodilians, many aquariums housing aquatic organisms including breeding marine clown fish, and a number of smaller glass-fronted herpetological exhibits. In 1969, one of the finest collections of Indo-Australian reptiles ever assembled was on display, including many varanids and pythons. The Mertens' monitor was bred for the first time in captivity and the ridge-tailed monitor was bred for the first time in Europe.

According to *International Zoo Yearbook* (1998:Volume 28), composition of the collection was nearly 80 taxa of reptiles with 387 specimens and 27 taxa of amphibians numbering 185.

Staff and Scientific Achievements: Jürgen Lange was head of the department from 1970 to 1977 and was interested in general herpetoculture. Richard Podloucky followed Lange and specialized in turtles and tortoises. Dieter Jauch followed as curator and eventually became director; his interest was captive management. Uwe Fritz, who has published extensively on turtles, was at the Zoo until departing to the Dresden Museum in 1997. Isabelle Koch is the current curator and is interested in captive management.

Publication: *Wilhelma-Magazin* started in 1993.

Personal Reflections: In 1969, Curator Wilbert Neugebauer showed me an amelanotic saltwater crocodile in his collection. The coloration was yellowish-white with blackish spots; its eyes were pigmented. The crocodile was over one meter in length and eventually grew to approximately four meters after 16 years. He was also very proud of the young Mertens' monitors on display which were hatched at the Zoo. He made sure that I saw the breeding groups of clown fish behind the scenes and the extensive filtration system.

Historical Overview

Jauch, D. 2000. Wilhelma; der zoologisch-botanische Garten in Stuttgart (Wilhelma, the Zoological-Botanical Garden in Stuttgart). Stehn, Stuttgart-Bad Cannstatt.

Neugebauer, W. 1993. Die Wilhelma, ein Paradies in der Stadt (The Wilhelma, a Paradise in the City). Theiss, Stuttgart.

Morphology, Systematics & Taxonomy

Fritz, U. 1988. Zur Variabilität der Carapaxzeichnung von *Trachemys decorata* (Barbour & Carr, 1940) und *Trachemys stejnegeri vicina* (Barbour & Carr, 1940) (On the variability of carapace markings of *Trachemys decorata* [Barbour & Carr, 1940] and *Trachemys stejnegeri vicina* [Barbour & Carr, 1940]). Salamandra 24:175-178.

Fritz, U. 1990. Balzverhalten und Systematik in der Subtribus Nectemydina. 1. Die Gattung *Trachemys*, besonders *Trachemys scripta callirostris* (Gray, 1855) (Mating behavior and taxonomy in the sub-tribe Nectemydina. 1. The species *Trachemys*, especially *Trachemys scripta callirostris* [Gray, 1855]). Salamandra 26:221-245. [Account described courtship, breeding, systematics, taxonomy, and sexual dimorphism.]

Fritz, U. 1991. Balzverhalten und Systematik in der Subtribus Nectemydina. 2. Vergleich oberhalb des Artniveaus und Anmerkungen zur Evolution (Mating behavior and taxonomy in the subtribe Nectemydina. 2. Comparison above the species level and notes concerning evolution). Salamandra 27:129-142. [Account described courtship behavior, systematics, zoogeography, and evolution of turtle genera *Chrysemys, Deirochelys, Graptemys, Malaclemys, Pseudemys* and *Trachemys.*]

Fritz, U. 1992. Zur innerartlichen Variabilität von *Emys orbicularis* (Linnaeus, 1758) 2. Variabilität in Osteuropa und Redefinition von *Emys orbicularis orbicularis* (Linnaeus, 1758) und *E. o. hellenica* (Valenciennes, 1832) (Reptilia, Testudines, Emydidae) (Concerning the species-specific variability of *Emys orbicularis* [Linnaeus, 1758]. 2. Variability in Eastern Europe and redefinition of *Emys orbicularis orbicularis* [Linnaeus, 1758] and *E. o. hellenica* [Valenciennes, 1832] [Reptilia, Testudines, Emydidae]). Zool. Abh. (Dresden) 47(1):37-77.

Fritz, U. 1994. Gibt es in Nordafrika zwei verschiedene Formen der Europäischen Sumpfschildkröte (*Emys orbicularis*)? (Are there two different forms of the European fresh-water turtle [*Emys orbicularis*] in North Africa?). Salamandra 30:76-80.

Fritz, U., and I. Pauler. 1992. *Phrynops chacoensis* spec. nov. (Reptilia, Chelidae), eine neue Krötenkopfschildkröte (*Phrynops chacoensis* spec. nov. [Reptilia, Chelidae], a new toad head turtle).

Mitt. Zool. Mus. Berlin 68(2):299-307.

Rummler, H.-J., and U. Fritz. 1991. Geographische Variabilität der Amboina-Scharnierschildkröte *Cuora amboinensis* (Daudin, 1802), mit Beschreibung einer neuen Unterart, *C. a. kamaroma* subsp. nov. (Geographical variability of the Amboina hinged turtle Cuora *amboinensis* [Daudin, 1802], with description of a new subspecies, *C. a. kamaroma* subsp. nov.). Salamandra 27:17-45.

Husbandry

Fritz, U. 1993. Perlite - das ideale Standard-Brutsubstrat? (Perlite - the ideal standard brood substratum?). Salamandra 28:279-281.

General

Fritz, U. 1991. Beitrag zur Kenntnis der Hispaniola-Schmuckschildkröte, *Trachemys decorata* (Barbour & Carr 1940) (Contributions to our knowledge of the Hispaniola Pseudemid turtles, *Trachemys decorata* [Barbour & Carr 1940]). Sauria 13(3):11-14.

Fritz, U. 1992. *Podarcis p. pityusensis* (Bosca, 1883) eingeschleppt in Cala Ratjada (NO-Mallorca) (Squamata: Sauria: Lacertidae) (*Podarcis p. pityusensis* [Bosca, 1883] introduced in Cala Ratjada [N.E. Mallorca] [Squamata: Sauria: Lacertidae]). Herpetozoa 5:131-133.

Fritz, U. 1993. Der seltenste Frosch Europas kehrt aus der Wilhelma und dem Naturkundemuseum in die Heimat zurück! (The rarest frog of Europe returns home from the Wilhelma and the Natural History Museum!). D. Aquar. Terrar.-Z. 46:622.

Fritz, U., and I. Pauler. 1992. Erstnachweis von *Acanthochelys pallidipectoris* (Freiberg, 1945) für Paraguay (Testudines: Pleurodira: Chelidae) (First evidence of *Acanathochelys pallidipectoris* [Freiberg, 1945] for Paraguay [Testudines: Pleurodira: Chelidae]). Herpetozoa 5:135-137.

Fritz, U., S. L. Kuzmin, O. W. Kolobajewa, and W. F. Orlowa. 1995. Zur Variabilität der Sumpfschildkröte (*Emys orbicularis*) im Gebiet zwischen der Manytsch-Niederung und Don (Russland). (Concerning the variability of freshwater turtles [*Emys orbicularis*] in the area between the Manych Lowlands and the Don[(Russia]). Salamandra 31:231-236.

Fritz, U., O. Picariello, R. Gunther, and F. Mutschmann. 1993. Zur Herpetofauna Süditaliens. Teil 1. Flussmündungen und Feuchtgebiete in Kalabrien, Lucanien und Sudapülien (Concerning

the herpetofauna of Southern Italy. Part 1. River deltas and wetlands in Calabria, Lucania and South Apulia). Herpetofauna, Weinstadt 15(84):6-14.

Picariello, O., G. Scillitani, U. Fritz, R. Gunther, and F. Mutschmann. 1993. Zur Herpetofauna Süditaliens. Teil 2. Die Amphibien und Reptilien des Picentini-Gebirges (Apennin, Kampanien). 1. Allgemeines und Amphibien (Concerning the herpetofauna of Southern Italy. Part 2: The amphibians and reptiles of the Picenini Range [Apennines, Campagnia]. 1. General and amphibians). Herpetofauna, Weinstadt 15(85):19-26.

Tierpark Berlin-Friedrichsfelde (1955)

History and Mission: The zoological garden Tierpark Berlin was founded in 1955 in the eastern sector of a divided Berlin, Germany. On 2 July 1955 when the Tierpark opened, the herpetological section at Tierpark Berlin was started. Heinrich Dathe was the first Director of the Tierpark and published over one thousand papers on zoo biology, including a number dealing with herpetology. After the fall of the Berlin Wall, the name was changed to Tierpark Berlin-Friedrichsfelde.

Facility and Collection: In 1956, the snake farm or vivarium opened, housing a large number of venomous snakes. The crocodile and hummingbird house opened on 23 August 1987, with crocodiles and turtles in two sections. An improved Snake Farm was remodeled in 1995.

According to *International Zoo Yearbook* (1998:Volume 28), composition of the collection was 121 taxa of

Fig. 113. Exterior of Terrarium at Tierpark Berlin on 13 April 1995. *Photograph by Klaus Rudloff, provided by Falk Dathe and Gisela Petzold.*

Fig. 114. Exhibits in Terrarium at Tierpark Berlin on 4 January 1996. *Photograph by Klaus Rudloff, provided by Falk Dathe and Gisela Petzold.*

reptiles with 637 specimens and 6 taxa of amphibians numbering over 100.

Staff and Scientific Achievements: Werner Krause started working at the Zoo shortly after the building opened and was known as the "Snake Farmer." The late Hans-Günter Petzold, former curator and later also deputy director, published a number of herpetological papers. Curator Falk Dathe followed Petzold. His interests center on venomous snakes. His father, Heinrich Dathe (7 Nov 1910-6 Jan 1991), wrote over a thousand papers about biology, herpetology and the Tierpark [see Milu, 7(6):3-35, 1993 for list and Urania 56 (11):14-16, 1980]. Klaus Dedekind, born 21 May 1941, was an employee of the Tierpark since 1962; he is Headkeeper of the Terrarium.

Publication: The journal *Milu* was started in 1960 and includes many herpetological papers.

Historical Overview

Conant, R. 1997. A Field Guide to the Life and Times of Roger Conant. Selva, Tyler TX [biography of Heinrich Dathe, p. 487].

Dathe, F. 1981. Schlangenfarm in neuem Glanze (The Snake Farm in its new, improved form). Aquar. Terrar. 28:292-293.

Dathe, F. 1990. Das Haus für Krokodile, Kolibris und Schildkröten im Tierpark Berlin (The building housing crocodiles, hummingbirds and turtles in the Zoo). Aquar. Terrar. 37:362-364. [Account included color photographs and floor plan.]

Dathe, F. 1995. Die Amphibien- und Reptilienhaltung in 40 Jahren Tierpark (Care of amphibians and reptiles during 40 years in the Tierpark). Milu 8:307-325 [historical photographs of Pia and Werner Krause, Heinrich Dathe, Klaus Dedekind, H.- G. Petzold, Robert Mertens, Z. Vogel, Heinz Wermuth, Heinz Heck, and Werner Schröder].

Dathe, F. 1996. Das Konzept der neuen Schlangenfarm (The concept of the new snake farm). Milu 8:717-732. [Account of new facility, constructed in 1995, included many photographs of exhibits and graphic presentations at Tierpark.]

Dathe, H. 1978. Dr. rer. nat. HANS-GÜNTER PETZOLD, Biologe und Kulturpolitiker (Dr. rer. nat. HANS-GUENTER PETZOLD, biologist and cultural politician). Aquar. Terrar. 25:364-365.

Dathe, H. 1983. DR. HANS-GÜNTER PETZOLD zum Gedenken (In memory of DR. HANS-GUENTER PETZOLD). Aquar. Terrar. 30:40-41.

Engelmann, W.-E., and F. J. Obst. 1982. Snakes: Biology, Behavior and Relationship to Man. Exeter Books, New York [story of Werner Krause (1905-1970), "snake farmer" at the Tierpark (p. 184)].

Klemmer, K. 1983. Hans-Günter Petzold. 28. September 1931-19. November 1982 (Hans-Günter Petzold, September 28, 1931-November 19, 1982). Salamandra 19:105-107 [obituary of Petzold].

Petzold, G. 1993. Liste der Publikationen von Hans-Günter Petzold (List of publications of Hans-Günter Petzold). Milu 7:393-408. [Gisela Petzold compiled this bibliography.]

Petzold, H.-G. Im Tierpark Berlin 1955 vorhandene Tierarten. Milu 1960, 1:2-14. Later: Im Tierpark Berlin 19. erstmalig gehaltene Tierformen (Animals kept for the first time in the Berlin Tierpark). Milu 1961-1983, continued by Klaus Rudloff. [Petzold published a number of annual accounts in this series which pictured a variety of mammals, birds, amphibians, reptiles, fishes and invertebrates.]

Petzold, H.-G. 1971. WERNER KRAUSE. Aquar. Terrar. 18:57. [Werner Krause was born on 14 December 1905. He died on 28 October 1970.]

Petzold, H.-G. 1965. *Cuora galbinifrons* und andere südostasiatische Schildkröten im Tierpark Berlin (*Cuora galbinifrons* and other Southeast Asian turtles in the Berlin Zoological Garden). Aquar.-u. Terra.-Z. 18:87-91, 119-121. [English translation in Chelonian Documentation Center Newsletter, 1984, 3(1-4):4-7.]

Books by Hans-Günter Petzold

Petzold, H.-G. 1967. Der Guppy (The Guppy). Die Neue Brehm-Bücherei, vol. 372. A. Ziemsen, Wittenberg [4th edition published in 1990].

Petzold, H.-G. 1971. Blindschleiche und Scheltopusik: Die Familie *Anguidae* (Slowworm and Scheltopusik: The Family *Anguidae*). D. Neue Brehm-Büch. Bd. 448. Ziemsen. Lutherstadt Wittenberg. [Reprint 1995. Petzold described systematics, morphology, natural history, reproduction, behavior, and representatives of the family. The book series Die Neue Brehm-Bücherei covered aspects of various amphibian and reptile taxa by other authors.]

Petzold, H.-G. 1972. Rätsel um Delphine (The Mysteries of Dolphins). A. Ziemsen. Wittenberg [9th edition published 1985].

Petzold, H.-G. 1984. Die Anakondas, Gattung *Eunectes* (The Anacondas, Genus *Eunectes*). D. Neue Brehm-Büch. Bd. 554. Ziemsen. Lutherstadt Wittenberg. [Reprint 1955 in which Petzold described systematics, morphology, natural history, reproduction, behavior, and representatives of the genus.]

Petzold, H.-G. 1984. Aufgaben und Probleme bei der Erforschung der Lebensäusserungen der Niederen Amnioten (Reptilien) (Tasks and Problems Connected with Research into the Life Expressions of the Lower Amniotic Animals [Reptiles]. [This scientific paper was first published in Milu, the Berlin Tierpark periodical (Milu, Bd. **5**, Heft 4/5: 485-786 [1982]) and reprinted by BINA in western Berlin in 1984. This book (Nr. 38) is in the series Berliner Tierpark-Buch and covered reptile development and reproduction.]

Morphology, Systematics & Taxonomy

Dathe, F. 1976. Beobachtung der Überwinterung einer Steppenschildkröte (*Agrionemys horsfieldi*) im Freiland (Observations concerning the hibernation of a tortoise of the steppes [*Agrionemys horsfieldi*] in nature). Aquar. Terrar. 23:384.

Dathe, F. 1990. Bemerkungen zur Anatomie der Bauchdrüsenotter *Maticora bivirgata flaviceps* (Cantor, 1839) (Notes concerning the anatomy of the belly gland viper [*Maticora bivirgata flaviceps*] [Cantor, 1839]). Salamandra 26:316-318.

Petzold, H.-G. 1968. Zur Kenntnis der kubanischen Antillen-Schmuckschildkröte (*Pseudemys terrapen rugosa*) (Information regarding the Cuban-Antilles Pseudemid turtle [*Pseudemys terrapen rugosa*]). Salamandra 4:73-90. [description of systematics, morphometrics and captive management].

Behavior

Dathe, H. 1971. Zum Schwimmen der Blindschleiche, *Anguis fragilis* (Concerning the swimming of slow-worms, *Anguis fragilis*). Aquar. Terrar. 18:65.

Petzold, H.-G. 1964. Ungewöhnliche Fressleistung eines Dunklen Pythons (Exceptional feeding performance of a dark python). Zool. Gart. (N.F.), Leipzig 28:200-202.

Petzold, H.-G. 1976. Bemerkungen zur Fütterungstechnik bei nahrungsspezialisierten Giftnattern und einige Angaben zur Haltungsdauer von Elapiden im Tierpark Berlin (Comments on the feeding techniques with elapid snakes specialized on certain foods and some data on the longevity of elapids in the Berlin Zoo). Zool. Gart. (N.F.), Jena 46:9-23.

Petzold, H.-G. 1983. Über einen bemerkenswerten Fall von Keratophagie bei Schlangen (Concerning a remarkable case of keratophagy in snakes). Zool. Gart. (N.F.), Jena 58:41-48 [puff adder devoured its own shed skin].

Petzold, H.-G., and G. Budich. 1972. Beutestreit bei Levanteottern, Eine Fotodokumentation aus dem Tierpark Berlin (Conflict over prey among vipers from the Levant, a photo documentation from the Berlin Tierpark). Aquar. Terrar. 19:8-11 [snake cannibalism].

Petzold, H.-G., and W.-E. Engelmann. 1975. Über zwei weitere Fälle von Kannibalismus bei Schlangen (Concerning two additional instances of cannibalism among snakes). Zool. Gart. (N.F.), Jena 45:513-515.

Reproduction

Dathe, F. 1998. Die Haltung von Schlammschildkröten, *Kinosternidae*, im Tierpark Berlin-Friedrichsfelde von 1956 bis 1998 (The care of musk turtles, *Kinosternidae,* in the Berlin Tierpark Zoo from 1956 to 1998). Milu 9:526-532.

Husbandry

Dathe, F. 1999. Die Krokodilhaltung im Tierpark Berlin-Friedrichsfelde von 1955 bis 1999 (The care of crocodiles in the Berlin Tierpark from 1955 to 1999). Milu 9:646-657 [photographs of many crocodilian taxa].

Dathe, F. 2000. Die Haltung von Stumpfkrokodilien, *Osteolaemus tetraspis* Cope, 1861, im Tierpark Berlin-Friedrichsfelde von 1956 bis 2000 (The care of dwarf crocodiles, *Osteolaemus tetraspis* Cope, 1861, in the Berlin Tierpark from 1956 to 2000). Milu 10:63-71.

Petzold, H.-G. 1958. Erfahrungen bei der Haltung des Bungars (*Bungarus fasciatus* Schn.) (Experiences with the maintenance of kraits [*Bungarus fasciatus* Schn.]). Aquar. Terrar. 5:257-260.

Petzold, H.-G. 1974. Erfolg und Misserfolg mit javanischen Flugdrachen, *Draco volans* (Success and failure with Javanese flying dragon, *Draco volans*). Aquar. Terrar. 21:158-163.

Medical Management

Dathe, F. 1988. Untersuchungen zum Auftreten von Kokzidienoozysten in den Fäzes lebender Schlangen und Echsen im Reptilienbestand des Tierparks Berlin (Examination of the appearance of coccidial cysts in the feces of living snakes and lizards in the reptile holdings of the Berlin Zoo). Berlin, Humboldt- Universität, Agrarwiss. Fak. Diss.

Dathe, H. 1961. Über die Wirkung von Wofatox auf Lurche und Kriechtiere (Concerning the effect of Wofatox on amphibians and reptiles). Aquar. Terrar. 8:379-380.

Ippen, R. 1987. Pathology survey of reptiles and amphibians. Proc. Amer. Assoc. Zoo Vet. :479-480.

Ippen, R., and G. P. Wildner. 1984. Comparative pathological investigations of thyroid tumors of animals in zoos and in the wild, p. 280-295. In O. A. Ryder, and M. L. Byrd (eds.), One Medicine. A Tribute to Kurt Benirschke, Director, Center for Reproduction of Endangered Species, Zoological Society of San Diego, and Professor of Pathology and Reproductive Medicine, University of California, San Diego/From his Students and Colleagues. Springer-Verlag, Berlin, Heidelberg, etc.

Ippen, R., H.-D. Schröder, and K. Elze. 1985. Handbuch der Zootierkrankheiten. Band 1: Reptilien. (Handbook of Zoo Animal Diseases. Volume 1: Reptiles). Akademie-Verlag, Berlin.

Priemer, J. 1980. Funde von Bandwurmlarven der Gattung *Mesocestoides* (Tetrathyridien) im Tierpark Berlin (Discoveries of tapeworm larvae of the species *Mesocestoides* [Tetrathyridien] in the Berlin Zoo). Milu 5:252-260.

Schröder, H.-D., and R. Ippen. 1970. Ein Beitrag zu den Arizonainfektionen der Reptilien (A contribution to the Arizona infections of reptiles). Zool. Gart. (N.F.), Leipzig 39:248-254.

Schröder, H.-D., and Reckies, H. 1975. Zum Nachweis von Hemmstoffen (Antibiotika) in den Organen von Zootiere (Evidence of antibiotics in the organs of zoo animals), p. 341-345. In R. Ippen, and H.-D. Schröder (eds.), Erkrankungen der Zootiere (Diseases of Zoo Animals) Wroclaw 1972. Akademie-Verlag, Berlin.

Conservation

Dathe, H. 1977. Ein Wort zum Reptilienschutz (A word about the protection of reptiles). Aquar. Terrar. 24:24.

General

Dathe, F. 1984-1990. AT-Terrarientierlexikon (AT Dictionary of Terrarium Animals). Aquar. Terrar. 31-37. [Between the years 1984 and 1990, Falk Dathe published 77 herpetofaunal accounts in this journal series.]

Dathe, F. 1984. Index to the papers on chelonians that have appeared in Der zoologische Garten (N.F.) Volume 1-52, from 1928-1982. Chelonian Documentation Center Newsletter 3(1-4): 20-21. [Journal, featuring mostly European zoos and staff, includes papers on ethology, captive maintenance and management. The "New Series" began in 1928. The list of prominent authors who have contributed to this journal over the years is impressive.]

Dathe, F. 1994. Bemerkenswertes Alter einer Matamata, *Chelus fimbriatus* (Schneider, 1783) (Remarkable age of a matamata, *Chelys fimbriatus*) (Schneider, 1783). Milu 8:103-105.

Dathe, F., and O. Czupalla. 1998. Bemerkenswertes Alter eines Afrikanischen Dreiklauers, *Trionyx triunguis* (Forskål, 1775) (Remarkable age of an African three-clawed turtle, *Trionyx triunguis* [Forskål, 1775]). Milu 9:402-404.

Dathe, F., L. Dedekind, and K. Dedekind. 1993. Tierbeobachtungen in Kasachstan (Animal observations in Kazakhstan). Milu 7:439-450.

Dathe, H. 1944. [Nachruf auf DR. WILLY WOLTERSTORFF] (Eulogy for DR. WILLY WOLTERSTORFF). Zool. Gart. (N.F.), Leipzig 15:277.

Dathe, H. 1959. Erstaunliche Lebenszähigkeit einer Riesenschlange (Remarkable tenacity of a giant snake in clinging to life). Bijdr. Dierk. (Amsterdam) 29:71-72.

Dathe, H. 1960. Schwanz-Regeneration beim Brillenkaiman (Tail regeneration in caimans). Natur u. Volk 90:289-292.

Dathe, H. 1961. Tauchende Smaragdeidechse (The diving emerald lizard). Aquar. Terrar. 8:25.

Dathe, H. 1964. Kampf einer Glattnatter mit einer Zauneidechse (Battle between a smooth snake and a sand lizard). Aquar. Terrar. 11:215-216.

Dathe, H. 1966. ROBERT MERTENS 70 Jahre (ROB-

ERT MERTENS at 70.) Zool. Gart. (N.F.), Leipzig 32:257-260.

Dathe, H. 1968. Herpetologische Skizzen aus Kuba (1). Der Ochsenfrosch, *Rana catesbeiana* Shaw (Herpetological sketches from Cuba (1). The American bull frog, *Rana catesbeiana* Shaw). Aquar. Terrar. 15:48-49.

Dathe, H. 1971. Rückkehr des Nilkrokodils, *Crocodylus niloticus*, in altes Verbreitungsgebiet (Return of the Nile crocodile, *Crocodylus niloticus*, to its former habitat). Salamandra 7:156. [With the construction of the Aswan Dam, Nile crocodiles have recovered part of their former range along the Nile River.]

Dathe, H. 1973. Herpetologische Skizzen aus Südostasien. 1. "Snake charming" in Rangun (Herpetological sketches from Southeast Asia. 1. "Snake charming" in Rangoon). Aquar. Terrar. 20:298-301.

Dathe, H. 1974. Herpetologische Skizzen aus Kuba (2). Die Krokodilfarm (Herpetological sketches from Cuba [2]. The crocodile farm). Zool. Gart. (N.F.), Jena 44:295-309.

Dathe, H. 1990. Eidechsen-Plastiken der Antike? (Ancient lizard sculptures?). Aquar. Terrar. 37:239.

Lau, D. 1977. Bemerkenswertes Alter einer Rotbauchunke, *Bombina bombina* (L.) (Remarkable age of a fire-bellied toad, *Bombina bombina* [L])). Zool. Gart. (N.F.), Jena 47:445-446. [At time of publication, Dieter Lau was associated with Tierpark.].

Obst, F. J., K. Richter, and U. Jacob. 1988. The Completely Illustrated Atlas of Reptiles and Amphibians for the Terrarium. T.F.H. Publications, Neptune City NJ. [Wolf- Eberhard Engelmann from the Leipzig Zoo prepared the accounts for anatomy and physiology, ethology and genetics. Petzold reviewed the book which was originally published in German as *Lexikon der Terraristik* four years carlier.].

Petzold, H.-G. 1959. Gewöllbildung bei Krokodilen (Formation of pellets in crocodiles). Zool. Anz. 163:76-82.

Petzold, H.-G. 1963. Über einige Schildkröten aus Nord-Vietnam im Tierpark Berlin (Regarding some turtles from North Vietnam in the Berlin Tierpark). Senckenb. Biol. 44(1):1-20.

Petzold, H.-G. 1964. Eine Marokko-Schleiche (*Ophisaurus koellikeri*) im Tierpark Berlin (A Moroccan glass lizard [*Ophisaurus koellikeri*] in the Berlin Tierpark). Zool. Gart. (N.F.), Leipzig 29:184-188.

Petzold, H.-G. 1965. Über die Widerstandsfähigkeit von Geckoneneiern und einige andere Beobachtungen an *Hemidactylus frenatus* Dum. & Bibr. 1836 (Concerning the resistance capability of gecko eggs and several other observations made on *Hemidactylus frenatus* Dum. & Bibr. 1836). Zool. Gart. (N.F.), Leipzig 31:262-265.

Petzold, H.-G. 1967. Ein Nashorn-Leguan, *Cyclura cornuta*, mit Gabelschwanz im Tierpark Berlin (A rhinoceros iguana, *Cyclura cornuta*, with forked tail in the Berlin Zoo) Aquar. Terrar. 14:306-309.

Petzold, H.-G. 1967. Notizen zur Gewöllbildung bei einem Bindenwaran (*Varanus salvator*) und einige allgemeine Bemerkungen über Reptiliengewölle (Notes on pellet formation in a two-banded monitor [*Varanus salvator*] and several general comments about reptile pellets). Zool. Gart. (N.F.), Leipzig 34:134-138. [Account described pellet-formation in crocodilians, monitor lizards and giant snakes.]

Petzold, H.-G. 1969. *Cricosaura typica* Gundlach & Peters, eine herpetologische Kostbarkeit aus Kuba (*Cricosaura typica* Gundlach & Peters, a herpetological treasure from Cuba). Aquar.-u. Terra.-Z. 22:82-85.

Petzold, H.-G. 1972. Eine total-melanotische Zauneidechse (*Lacerta agilis*) aus dem Raum Berlin (A totally melanotic sand lizard [*Lacerta agilis*] from the region of Berlin). Salamandra 8:123-127.

Petzold, H.-G. 1973. Forschende Kamera: Das Geheimnis der Chamäleonzunge (The investigative camera: The secret of the chameleon tongue). Urania 49:14-17 [photographs of chameleons feeding].

Petzold, H.-G. 1975. Eine albinotische Vierstreifennatter, *Elaphe quatuorlineata sauromates*, aus Bulgarien (An albino four-lined rat snake, *Elaphe quatuorlineata sauromates* from Bulgaria). Salamandra 11:113-118.

Petzold, H.-G. 1975-1983. AT-Terrarientierlexikon (AT-Terrarium Dictionary). Aquar. Terrar. 22-30. [Between the years 1975 and 1983, Petzold published 92 herpetofaunal accounts in this journal series. Falk Dathe continued the series after Petzold's death.]

Petzold, H.-G. 1978. Nigrinos von *Lacerta vivipara* aus der Umgebung Berlins (Reptilia: *Sauria: Lacertidae*) (Nigrinos of *Lacerta vivipara* from the Berlin region [Reptilia: *Sauria: Lacertidae*]). Salamandra 14:98-100.

Petzold, H.-G. 1982. Scientific contribution to the zookeeper training textbook „Wild-tiere in Menschenhand (Grundlagen) (Wild animals in human hands", vol. 1 (fundamentals)). Deutscher Landwirtschaftsverlag, Berlin, 5th edition published in 1982.

Petzold, H.-G., H.-A. Pederzani, and H. Szidat. 1970. Einige Beobachtungen zur Biologie des kubanischen Rollschwanzleguans, *Leiocephalus carinatus* (Gray) (Some observations regarding the biology of the Cuban flat-headed iguana, *Leiocephalus carinatus* [Gray]). Zool. Gart. (N.F.) 39:304-322. [Account described systematics, habitat, captive breeding and behavior.]

Der zoologische Garten (The Zoological Garden), an international journal which started publication in 1859, has been an important vehicle for disseminating information on captive reptiles and amphibians. Around 1913, the publication ceased to exist but was rejuvenated and carries the initials (N.F. = Neue Folge [New Edition]). A number of prominent herpetologists have written papers for this journal which are listed below, including Oskar Boettger, Johann von Fischer, Günther Freytag, Paul Kammerer, Paul Krefft, Robert Mertens, Heinz Wermuth, Franz Werner and Willy Wolterstorff. For information on its history, see Grummt (2001).

Adamy, F.-W. 1966. Ein Beitrag zur Therapie der Salmonellose bei Schlangen (An essay on the therapy of salmonellosis in serpents). Zool. Gart., Leipzig (N.F.) 32:67.

Andres, A. 1909. Reptilien und Batrachier des Zoologischen Gartens in Gizeh bei Cairo (Reptiles and Batrachians of the Zoo in Gizeh near Cairo). Zool. Gart., Frankfurt a. M. 50:43-46.

Anonymous. 1913. Der Zoologische Garten in Gizeh (The Zoological Garden in Gizeh). Zool. Gart., Frankfurt a. M. 54:187-189.

Bally, P. R. O. 1952. Einige Beobachtungen an der ostafrikanischen Pantherschildkröte, *Testudo pardalis* (Some observations concerning the East African leopard tortoise, *Testudo pardalis*). Zool. Gart. (N.F.), Leipzig 19:236-238.

Bauhof, J. 1891. Die Paarungsweise der griechischen Landschildkröte (The mating behavior of the Greek true tortoise). Zool. Gart., Frankfurt a. M. 32:274-278.

Beckmann, L. 1878. Beobachtungen an Reptilien und Amphibien in der Gefangenschaft (Observations made of reptiles and amphibians in captivity). Zool. Gart., Frankfurt a. M.19:82-89.

Bedriaga, J. von. 1877. Bemerkung über die Umwandlung des Axolotl in ein *Amblystoma* (Comments on the metamorphoses of the axolotl into an *Amblystoma*). Zool. Gart., Frankfurt a. M. 18:132-133.

Bedriaga, J. von.1878. Beobachtungen an Reptilien und Amphibien in der Gefangenschaft (Observations made of reptiles and amphibians in captivity). Zool. Gart., Frankfurt a. M. 19 :82-89.

Berg, J. 1896. Ein neues Tropen-Terrarium (A new tropical terrarium). Zool. Gart., Frankfurt a. M. 37:140-143.

Berg, J. 1901. Indische Dryophiden im Terrarium (Indian Dryophides in the Terrarium). Zool. Gart., Frankfurt a. M. 42:204-215.

Berg, J. 1902. Der Yucatan-Dornschwanz (*Cachryx defensor* Cope). (The Yucatan spiny-tailed lizard [*Cachryx defensor* Cope]). Zool. Gart., Frankfurt a. M. 43:86-92.

Biella, H.-J., P. Eckardt, and W. Ruosch. 1989. Zur Fortpflanzungsbiologie von *Bitis arietans* und *Bitis gabonica*, nebst Bemerkungen zum Nahrungsbedürfnis, Wachstum und zur Haltung beider Arten (Concerning the reproductive biology of *Bitis arietans* and *Bitis gabonica* along with observations regarding requirements for nourishment, growth, and care of both species). Zool. Gart. (N.F.), Jena 59:37-48.

Blum, J. 1890. Das Vorkommen der Aspis-Viper, *Vipera aspis* L., im südlichen Schwarzwalde (Occurrence of the aspis-viper, *Vipera aspis* L. in the southern Black Forest). Zool. Gart., Frankfurt a. M. 31:265-266.

Boettger, O. 1876. Über die äusseren Kiemenöffnungen bei jungen Exemplaren des japanischen Riesenmolches (Concerning the outer gill openings in young specimens of the Japanese giant newt). Zool. Gart., Frankfurt a. M. 17:432-435.

Boettger, O. 1877. Über Rassenunterschiede beim Laubfrosch (Racial differences in the tree toad). Zool. Gart., Frankfurt a. M. 18:27-34.

Boettger, O. 1879. Bemerkungen über das Leben der ungleichzehigen Landschildkröte Asiens, *Testudo Horsfieldii*, in der Gefangenschaft (Notes about the life in captivity of the Asiatic true tortoise with an unequal number of toes *Testudo Horsfieldii*). Zool. Gart., Frankfurt a. M. 20:289-293.

Boettger, O. 1879. Über das Gefangenenleben der gehörnten Krötenechse, *Phrynosoma cornutum*, aus Mexiko (Life in captivity of the horned lizard, *Phrynosoma cornutum* from Mexico). Zool. Gart., Frankfurt a. M. 20:331-336.

Boettger, O. 1885. Zur Naturgeschichte des Grüneders (*Lacerta viridis* L.) (Contribution to the natural history of the green lizard [*Lacerta viridis* L.]). Zool. Gart., Frankfurt a. M. 26:140-147.

Boettger, O. 1893. Ein neuer Beutelfrosch (A new marsupial frog). Zool. Gart., Frankfurt a. M. 34:129-132.

Boettger, O. 1894. F. W. Urich und R. R. Moles Beobachtungen an einer gefangenen Klapperschlange (F. W. Urich and R. R. Moles, observations of a captive rattlesnake). Zool. Gart., Frankfurt a. M. 35:215-217.

Boettger, O. 1906. Kleinere Mitteilungen. Bemerkenswerte Frösche aus Südkamerun (Lesser notes. Remarkable frogs from Southern Cameroon). Zool. Gart., Frankfurt a. M. 47:154-156.

Boettger, O. 1906. Kleinere Mitteilungen. Vom Chinesischen Alligator (*Alligator sinensis* Fauv.) (Lesser notes. About the Chinese alligator [*Alligator sinensis* Fauv.]). Zool. Gart., Frankfurt a. M. 47:182.

Boettger, O. 1906. Kleinere Mitteilungen. Die Frösche von Neu-Guinea (Lesser notes. The frogs of New Guinea). Zool. Gart., Frankfurt a. M. 47:315-316.

Boettger, O. 1907. Kleinere Mitteilungen. Dauer der Eireife bei der Ringelnatter (*Tropidonotus natrix* L.) (Lesser notes. Length of egg maturation in the ringed snake [*Tropidonotus natrix* L.]). Zool. Gart., Frankfurt a. M. 48:60.

Boettger, O. 1907. Kleinere Mitteilungen. Brutpflege eines brasilianischen Laubfrosches (Lesser notes. Brood care of a Brazilian leaf frog). Zool. Gart., Frankfurt a. M. 48:253.

Boettger, O. 1909. Kleinere Mitteilungen. Die Aeskulapschlange (*Coluber longissimus* Laur.) im Böhmerwald und die Zornnatter (*Zamenis gemonensis* Laur.) im Böhmerwald, Wienerwald, in den Kleinen Karpathen, Süd-Steiermark und Kärnten (Lesser notes. The Aesculapius snake [*Coluber longissima* Laur.] in the Bohemian Forest, and the racer [*Zamenis gemonensis* Laur.] in the Bohemian Forest, the Vienna Woods, the Small Carpathian Mountains, southern Styria and Carinthia). Zool. Gart., Frankfurt a. M. 50:341-342.

Boettger, O. 1910. Kleinere Mitteilungen. Zur Entwicklungsgeschichte des Alpensalamanders (*Salamandra atra* Laur.) (Lesser notes. About the developmental history of the Alpine salamander [*Salamandra atra* Laur.]). Zool. Gart., Frankfurt a. M. 51:23.

Bolau, H. 1888. Der neue Reptilienbau im Zoologischen Garten zu Hamburg (The new reptile house in the Hamburg Zoological Garden). Zool. Gart., Frankfurt a. M. 29:201-208 [floor plan and list of specimens].

Brahme Jr., A. H., and G. E. Freytag. 1963. Ein Halbalbino von *Chioglossa lusitanica* (A semi-albino of *Chioglossa lusitanica*). Zool. Gart. (N.F.), Leipzig 27:130-131.

Bruch, C. 1864. Die Riesen- und Zwergformen bei den Batrachiern (Giant and dwarf forms of the batrachians). Zool. Gart., Frankfurt a. M. 5:349-359.

Burtscher, J. 1931. Über die Mundfäule der Schlangen (Regarding mouth rot among snakes). Zool. Gart. (N.F.), Leipzig 4:235-244. [Mouth rot is a disease of captives and is not found in wild snakes.]

Busono, M. S. 1974. Facts about the *Varanus komodoensis* at the Gembira Loka Zoo at Yogyakarta. Zool. Gart. (N.F.), Jena 44:62-63. [Gembira Loka Zoo in Indonesia has successfully bred this species for many years, producing well over 100 neonates. The offspring are maintained in large groups in movable enclosures and are placed in sunlight daily.]

Castellanos, R. 1979. Zur Hämatologie des Kubanischen Rautenkrokodils, *Crocodylus rhombifer* [Cuvier] (On the hematology of the Cuban crocodile, *Crocodylus rhombifer* (Cuvier)). Zool. Gart. (N.F.), Jena 49:69-74.

Christoph, H.-J. 1952. Chirurgische Behandlung einer Wunde bei einer Hieroglyphenschlange (*Python sebae* Gm.) (Surgical treatment of an injury of a hieroglyphic snake (*Python sebae* Gm.)). Zool. Gart. (N.F.), Leipzig 19:195-197 [African rock python].

Cyrén, O. 1909. Herpetologisches von einer Balkanreise (Herpetological data from a voyage in the Balkans). Zool. Gart., Frankfurt a. M. 50:265-271, 295-300.

Deckert, K. 1963. Die Paarung des Beutelfrosches (*Gastrotheca marsupiata*) (The mating of the pouched frog [*Gastrotheca marsupiata*]). Zool. Gart. (N.F.), Leipzig 28:12-17.

Effeldt, R. 1873. Die Reptiliensammlung der Herren Effeldt und Wagenführ (The reptile collection of Messrs. Effeldt and Wagenführ). Zool. Gart., Frankfurt a. M. 14:66-70.

Effeldt, R. 1874. Meine Beobachtungen über die Wasser-Mokassin-Schlange, *Cenchris piscivorus*, in Gefangenschaft (My observations of the water moccasin snake, *Cenchris piscivorus* in captivity). Zool. Gart., Frankfurt a. M. 15:1-5.

Effeldt, R. 1875. Würfelnatter und Gelbe Natter (Diced snake and yellow snake). Zool. Gart., Frankfurt a. M. 16:134-136.

Eiffe, O. E. 1885. Einige Beobachtungen an Schlangen

in der Gefangenschaft (Some observations of snakes in captivity). Zool. Gart., Frankfurt a. M. 26:43-51.

Eismann, G. 1899. Nilwaran (*Varanus niloticus* L.) in der Gefangenschaft (Nile monitor [*Varanus niloticus* L.] in captivity). Zool. Gart., Frankfurt a. M. 40:145-147.

Enderlein, R. 1963. Freilandhaltung und Zucht des Grünen Leguans, *Iguana iguana,* in Schweden (Open-air care and breeding of the true iguana, *Iguana iguana*, in Sweden). Zool. Gart. (N.F.), Leipzig 28:1-7.

Fischer-Sigwart, H. 1893. Die Europäische Sumpfschildkröte *Emys europaea* (The European fresh-water turtle, *Emys europaea*). Zool. Gart., Frankfurt a. M. 34:162-174.

Franz, V. 1912. Ringelnatter und Frosch (Ringed snake and frog). Zool. Gart., Frankfurt a. M. 53:119-122.

Freye, H.-A. 1957. Wirbelsäulenverkrümmung beim Krallenfrosch (*Xenopus laevis* Daudin) (Curvature of the spine in the clawed frog [*Xenopus laevis* Daudin]). Zool. Gart. (N.F.), Leipzig 23:45-49.

Freyse, K., and G. Müller. 1962. Die Parthenogenese bei *Lacerta saxicola armeniaca* Mehely, 1909 (The parthenogenesis in *Lacerta saxicola armeniaca* Mehely, 1909). Zool. Gart. (N.F.), Leipzig 26:348.

Freytag, G. E. 1963. Bewegungsphysiologische Beobachtungen an einem anomalen Kammolch (*Triturus cristatus dobrogigus*) (Motion-physiological observations in an abnormal newt [*Triturus cristatus dobrogigus*]). Zool. Gart. (N.F.), Leipzig 27:127-130.

Freytag, G. E. 1978. Zur Problematik der Freisetzung und Einbürgerung von Schwanzlurchen (Problems encountered in setting free and acclimatizing newts). Zool. Gart. (N.F.), Jena 48:288-292.

Freytag, G. E., and W. Mudrack. 1988. Ein analytischer Beitrag zu Wolterstorffs Methode der Wassermolchhaltung (An analytical contribution to Wolterstorff's method of caring for water salamanders). Zool. Gart. (N.F.), Jena 58:275-280 [discussion of amphibian facility and lighting for newts].

Friedel, E. 1869. Entdeckung lebender Krokodile in Palästina (Discovery of living crocodiles in Palestine). Zool. Gart., Frankfurt a. M. 10:129-135, 10:161-166.

Fröhlich, C. 1881. Der Gecko, *Platydactylus mauritanicus,* in Gefangenschaft (The gecko, *Platydactylus mauritanicus,* in captivity). Zool. Gart., Frankfurt a. M. 22:24-25.

Galstuan, B. 1973. Eiablagen des Komodowarans (*Varanus komodoensis*) im Zoologischen und Botanischen Garten Jakarta (Egg deposits of the Komodo dragon (*Varanus komodoensis*) in the Zoological and Botanical Garden of Jakarta. Zool. Gart. (N.F.), Leipzig 43:136-139.

Geisenheyner, L. 1898. Zum Kapitel „Hausratte und Würfelnatter." (Concerning the chapter „Domestic rat and diced snake"). Zool. Gart., Frankfurt a. M. 39:1-4.

Geisenheyner, L. 1907. Zwei Schlangengeschichten (Two snake stories). Zool. Gart., Frankfurt a. M. 48:24-25.

Gewalt, W. 1977. Einige Bemerkungen über Fang, Transport und Haltung des Goliathfrosches (*Conraua goliath*) Boulenger (Some remarks on capture, transport and care of the goliath frog [*Conraua goliath*] Boulenger. Zool. Gart. (N.F.), Jena 47:161-192. [contribution from Duisburg Zoological Garden, Germany].

Glaser, L. 1871. Beobachtungen betreffend Wassermolche im Stubenaquarium und im Freien (Observations concerning water newts in the home aquarium and in open air enclosures). Zool. Gart., Frankfurt a. M. 12:257-263.

Grijs, P. de. 1898. Beobachtungen an Reptilien in der Gefangenschaft (Observations of reptiles in captivity). Zool. Gart., Frankfurt a. M. 39:265-282.

Grijs, P. de. 1901. Beobachtungen an Reptilien in der Gefangenschaft (Observations of reptiles in captivity). Zool. Gart., Frankfurt a. M. 42:33-46, 42:65-76, 42:97-109.

Grote, H. 1911. Kurze biologische Notizen über einige Säuger und Reptilien Ostafrikas (Brief biological notes about some mammals and reptiles in East Africa). Zool. Gart., Frankfurt a. M. 52:346-349.

Haacke, W. 1883. Fliegenfallen als Zimmerterrarienfüsse (Fly traps as room terrarium feet). Zool. Gart., Frankfurt a. M. 24:357-358. [Wilhelm Haacke was Director of the South Australian Museum in Adelaide.]

Hanau, A. 1896. Einige Beobachtungen an gefangenen Reptilien und Batrachiern (Some observations concerning captive reptiles and amphibians). Zool. Gart., Frankfurt a. M. 37:306-313.

Hanau, A. 1898. Beobachtungen an gefangenen Reptilien und Batrachiern II. (Observations concerning captured reptiles and batrachians II.). Zool. Gart., Frankfurt a. M. 39:41-53.

Heller, K. M. 1888. Amphibiologische Notizen (Amphibian-biological notes). Zool. Gart., Frankfurt a. M. 29:177-181.

Hennicke, C. R. 1895. Ein Beitrag zur Fortpflanzungsgeschichte von *Salamandra maculosa* (A contribution to the reproductive history of *Salamandra maculosa*). Zool. Gart., Frankfurt a. M. 36:203-205.

Heyden, C. H. G. von. 1863. Über das Vorkommen von *Calopeltis flavescens* Scop. bei Schlangenbad und von *Tropidonotus tesselatus*, Laur. bei Ems (Concerning the occurrence of *Calopeltis flavescens* Scop., near Schlangenbad and of *Tropidonotus tesselatus* Laur., near Ems). Zool. Gart., Frankfurt a. M. 4:13-14.

Hiller, A. 1984. Tierärztliche Erfahrungen bei der Zucht der Europäischen Sumpfschildkröte (*Emys orbicularis*) (Veterinary experience gathered in the breeding of the swamp turtle [*Emys orbicularis*]). Zool. Gart. (N.F.), Jena 54:128-130.

Horn, H.-G., M. Gaulke, and W. Böhme. 1994. New data on ritualized combat in monitor lizards (*Sauria: Varanidae*), with remarks on their function and phylogenetic implications. Zool. Gart. (N.F.), Jena, Stuttgart, New York 64:265-280.

Hornung, V. 1896. Fortpflanzung der Blindschleiche (*Anguis fragilis*) in der Gefangenschaft (Reproduction of anguine lizards [*Anguis fragilis*] in captivity). Zool. Gart., Frankfurt a. M. 37:137-140.

Hornung, V. 1900. Fütterung der Mauereidechse (*Lacerta muralis*) in der Gefangenschaft (Feeding the wall lizard (*Lacerta muralis*) in captivity). Zool. Gart., Frankfurt a. M. 41:93-94.

Hornung, V. 1902. Ein Fundort der blaugefleckten Blindschleiche (*Anguis fragilis* L.) (Discovery of a habitat of blue-spotted slow worms [*Anguis fragilis* L.]). Zool. Gart., Frankfurt a. M. 43:323-325.

Kammerer, P. 1901. Zur Biologie der Giftschlangen, mit besonderer Rücksicht auf ihr Gefangenleben (Concerning the biology of the poisonous snakes, with special reference to their life in captivity). Zool. Gart., Frankfurt a. M. 42:139-147, 42:169-180.

Kammerer, P. 1904. Kleinere Mitteilungen. Eine Blindschleiche als Schlangenfresser (Lesser notes. A slow-worm as eater of snakes). Zool. Gart., Frankfurt a. M. 45:352-353.

Kammerer, P. 1904. Kleinere Mitteilungen. Verletzter Laubfrosch (*Hyla arborea* L.) (Lesser notes. Injured leaf frog [*Hyla arborea* L.]) . Zool. Gart., Frankfurt a. M. 45:353-354.

Kanngiesser, E. 1911. Zur Physiologie der gefangenen Schlangen (Concerning the physiology of captive snakes). Zool. Gart., Frankfurt a. M. 52:248-251.

Kanngiesser, E. 1912. Riesenschlangen in der Gefangenschaft (Giant snakes in captivity). Zool. Gart., Frankfurt a. M. 53:1-4.

Kirsche, W. 1969. Frühzeitigung bei *Testudo hermanni hermanni* Gmelin (Early gestation of *Testudo hermanni hermanni* Gmelin). Zool. Gart. (N.F.), Leipzig 37:1-11 [description of behavior and development of the embryo].

Kirsche, W. 1976. Beitrag zur Biologie der Sternschildkröte (*Testudo elegans* Schoepff) (A contribution to the biology of the star tortoise (*Testudo elegans* Schoepff)). Zool. Gart. (N.F.), Jena 46:66-81.

Klingelhöffer, W. 1906. Süsswasserschildkröten (Fresh-water turtles). Zool. Gart., Frankfurt a. M. 47:15-25.

Knauer, F. K. 1879. Das Lebendgebären des Feuersalamanders, *Salamandra maculata* Schr., und die äussere Entwicklung der Jungen von der Geburt bis zum Abschluss ihrer Verwandlung (Live birthing of the fire salamander, *Salamandra maculata* Schr., and the external development of the young from birth to the conclusion of their metamorphosis). Zool. Gart., Frankfurt a. M. 20:97-103.

Knoblauch, A. 1903. Unsere einheimischen Schwanzlurche in der Gefangenschaft und ihre Entwicklung (Our domestic salamanders in captivity and their development). Zool. Gart., Frankfurt a. M. 44:386-404.

Knoblauch, A. 1904. Die Art der Fortpflanzung des Alpen- und des Feuersalamanders und das Anpassungsvermögen der beiden Salamanderarten an äussere Lebensbedingungen (Reproduction methods of the alpine and fire salamanders and the adaptability of the two salamander species to external living conditions). Zool. Gart., Frankfurt a. M. 45:329-336, 45:361-368.

Koch, W. 1931. Über Haltung und Zucht der *Emys europaea* (Concerning care and breeding of the *Emys europaea*). Zool. Gart. (N.F.), Leipzig 4:153-157.

Kolar, K. 1957. Jugendentwicklung von (*Uromastyx acanthinurus* Bell (Juvenile development of *Uromastyx acanthinurus* Bell). Zool. Gart. (N.F.), Leipzig 23:18-27.

Kornalik, F. 1964. Amphigonia retardata bei einer Wassermokassinschlange (Amphigonia retardata in a cottonmouth snake). Zool. Gart. (N.F.), Leipzig 29:272.

Kornalik, F., F. Kornalik Jr., and V. Seichrt. 1967. Die anatomische Grundlage des Schlangenbissmechanismus (The anatomical basis of the snake bite mechanism). Zool. Gart. (N.F.), Leipzig 33:309-319.

Krefft, P. 1904. Herpetologische Reiseerlebnisse in

Hinterindien (Herpetological travel experiences in Indochina). Zool. Gart., Frankfurt a. M. 45:169-178, 45:201-207, 45:241-248.

Krefft, P. 1905. Batrachier- und Reptilienleben in Japan (The life of amphibians and reptiles in Japan). Zool. Gart., Frankfurt a. M. 46:144, 46:161-173.

Krefft, P. 1912. Gehörnte Chamäleons (Horned chameleons) Zool. Gart., Frankfurt a. M. 53:272-277.

Lachmann, H. 1890. Die gesprenkelte Kettennatter oder Sprenkelnatter (*Coronella Sayi*, Deck.) (The speckled common king snake [*Coronella Sayi*, Deck.]. Zool. Gart., Frankfurt a. M. 31:74-83.

Lachmann, H. 1896. Zur Pflege der deutschen Reptilien und Amphibien im Freien (About the care of German reptiles and amphibians in nature). Zool. Gart., Frankfurt a. M. 37:264-272.

Landois, H. 1883. Ein ebenso sinnreicher wie zweckmässiger Behälter für Laubfrösche (A container for tree toads as ingenious as it is practical). Zool. Gart., Frankfurt a. M. 24:103-105.

Leydig, F. 1892. Springfrosch, *Rana agilis*; Ellritze, *Phoxinus laevis* (Leaping frog, *Rana agilis*; Ellritze, *Phoximus laevis*). Zool. Gart., Frankfurt a. M. 33:321-326.

Loewis, O. v. 1889. Mitteilungen über die Kreuzotter (Information about the common viper). Zool. Gart., Frankfurt a. M. 30:129-135.

Marno, E. 1874. Das Nilkrokodil (The Nile crocodile). Zool. Gart., Frankfurt a. M. 15:131-134.

Mebs, D. 1965. Zur Pathologie und Therapie einer bei Wasserschildkröten häufig auftretenden Augenerkrankung (Concerning the pathology and therapy of an illness of the eyes frequently found in water turtles). Zool. Gart. (N.F.), Leipzig 31:304-309.

Méhely, L. v. 1896. Auf welchem Wege ist die Mauereidechse (*Lacerta muralis* Laur.) in Ungarn eingewandert? (By what route did the common lizard [*Lacerta muralis* Laur.] migrate to Hungary?). Zool. Gart., Frankfurt a. M.37:109-114.

Meier, J., P. Hössle, and A. Sandoz-Ogata. 1995. Unfallverhütung bei der Haltung und Pflege von Giftschlangen (Accident prevention in keeping and care of poisonous snakes). Zool. Gart. (N.F.), Jena 65:293-313 [discussion of venomous snakebite procedures and first aid protocols].

Merk-Buchberg, M. 1912. Zur Ernährungsweise der Ringelnatter, *Tropidinotus natrix* (Concerning the nutrition of the ringed snake, *Tropidinotus natrix*). Zool. Gart., Frankfurt a. M. 53:210-212.

Mertens, R. 1922. Farbabänderung beim Alpensalamander (Color change in Alpine salamander). Zool. Gart., Frankfurt a. M. 63:30.

Mertens, R. 1922. Zur Verbreitung der Schwanzlurche in Afrika (About the distribution of salamanders in Africa). Zool. Gart., Frankfurt a. M. 63:30.

Mertens, R. 1922. Bemerkenswerte Tiere im Leipziger Zool. Garten (Remarkable animals in the Leipzig Zoo). Zool. Gart., Frankfurt a. M. 63:173.

Mertens, R. 1922. Eine melanotische Zauneidechse (*Lacerta agilis*) (A melanotic sand lizard [*Lacerta agilis*]). Zool. Gart., Frankfurt a. M. 63:174.

Mertens, R. 1922. Schmuckhornfrösche im Frankfurter Zoologischen Garten (Decorative horned frogs in the Frankfurt Zoo). Zool. Gart., Frankfurt a. M. 63:208-.

Mertens, R. 1923. Ein Bergmolch (*Triturus alpestris* Laurenti) mit geflecktem Bauch (An alpine newt [*Triturus alpestris* Laurenti] with spotted belly). Palasia, Dresden 1:51.

Mertens, R. 1924. Ein Beitrag zur Kenntnis der melanotischen Inseleidechsen des Mittelmeeres (A contribution to knowledge of the melanotic island lizards of the Mediterranean). Palasia, Dresden 2:40-52.

Mertens, R. 1924. Ueber die Reptilien- und Amphibien-Sammlung des Zoologischen Gartens in London (About the reptiles and amphibian collection of the London Zoo). Palasia, Dresden 2:67-72.

Mertens, R. 1925. Ueber einige *Lacerta*-Formen aus Süditalien und Sizilien (About certain *Lacerta* species from Southern Italy and Sicily). Palasia, Dresden 3:75-80.

Mertens, R. 1925. Kleine Mitteilungen. Zwei Mißbildungen bei Schwanzlurchen (Small notes. Two deformations in salamanders). Palasia, Dresden 3:82-83.

Mertens, R. 1930. Zur Biologie der Iguanidengattung *Polychrus* G. Cuvier (Concerning the biology of the iguanid genus *Polychrus* G. Cuvier). Zool. Gart. (N.F.), Leipzig 3:269-279.

Mertens, R. 1952. Eine Fluchtreaktion bei fressenden Schildkröten und ihre Bedeutung (A flight reaction in feeding tortoises and its significance). Zool. Gart. (N.F.), Leipzig 19:257-258.

Mertens, R. 1957. Zur Naturgeschichte des venezolanischen Riesen-Beutelfrosches, *Gastrotheca ovifera* (Concerning the natural history of the Venezuelan giant marsupial frog, *Gastrotheca ovifera*). Zool. Gart. (N.F.), Leipzig 23:110-133.

Mertens, R. 1964. Neukaledonische Riesengeckos (*Rhacodactylus*) (New Caledonian giant geckos [*Rhacodactylus*]). Zool. Gart. (N.F.), Leipzig 29:49-57.

Mettenheimer, C. 1873. Über das Vorkommen des Krokodils in Palästina (Concerning the occurrence of crocodiles in Palestine). Zool. Gart., Frankfurt a. M. 13:237-238.

Meyer, R. 1866. Fang einer vor drei Jahren aus der Gefangenschaft entwischten Alligatorschildkröte, (*Chelydra serpentina* Lacep.) (Capture of an alligator turtle, [*Chelydra serpentina* Lacep.], which had escaped from captivity three years earlier). Zool. Gart., Frankfurt a. M. 7:414-416.

Meyer, R. 1873. Über die Giftschlangen Indiens (Concerning the poisonous snakes of India). Zool. Gart, Frankfurt a. M. 14:306-308.

Mohr, E. 1930. Vom >>Wegnarren<< - dem Alpensalamander (*Salamandra atra* Laur.) (About the "Road Fool" - The European black salamander [*Salamandra atra* Laur.]). Zool. Gart. (N.F.), Leipzig 2:202-204.

Müller, C. 1896. Über Vergiftungen durch Schlangen (About poisoning by snakes). Zool. Gart., Frankfurt a. M. 37:161-170.

Nehring, A. 1880. Einige Notizen über das Vorkommen von *Lacerta viridis, Alytes obstetricans, Pelobates fuscus rec.* und *foss., Coluber flavescens* (Some notes about the occurrence of *Lacerta viridis, Alytes obstetricans, Pelobates fuscus rec.* and *foss. Coluber flavescens*). Zool. Gart., Frankfurt a. M. 21:298-303.

Noll, F. C. 1869. Die Würfelnatter, *Tropidonotus tesselatus,* eine deutsche Schlange (The diced snake, *Tropidonotus tesselatus,* a German snake). Zool. Gart., Frankfurt a. M. 10:299-304.

Nybelin, O. 1951. Das Aquarium in Göteborg (The Aquarium in Göteborg). Zool. Gart. (N.F.), Leipzig 18:99-102.

Otto, H. 1909. Schlangen am Niederrhein (Snakes along the Lower Rhine). Zool. Gart., Frankfurt a. M. 50:47-54.

Poglayen-Neuwall, I. 1983. Geglückte Zucht der Panther-Schildkröte (*Geochelone pardalis babcocki*) (Successful breeding of the leopard tortoise [*Geochelone pardalis babcocki*]). Zool. Gart. (N.F.), Jena 53:217-225. [At time of publication, author employed at Gene Reid Zoological Park, Tucson, Arizona.]

Preisser, B. 1987. Geschichte des Aquariums des Zoologischen Garten Dresden in Verbindung mit der Entwicklung der öffentlichen Aquaristik in der Stadt Dresden (The history of the Dresden Zoological Garden Aquarium, viewed in conjunction with developing public interest in aquaristics in the City of Dresden). Zool. Gart. (N.F.), Jena 57:257-268.

Reichenbach, L. 1866. Nachricht über einen hochgelben Triton (Note concerning a high yellow triton). Zool. Gart., Frankfurt a. M. 7:61-68.

Rein, J. J., and A. v. Roretz. 1876. Beitrag zur Kenntnis des Riesensalamanders, *Cryptobranchus japonicus* (Hoev.) (Contribution to knowledge of the giant salamander, *Cryptobranchus japonicus* (Hoev.)). Zool. Gart., Frankfurt a. M. 17:33-37.

Rotter, J. 1962. Biologische Beobachtungen an der Nördlichen Johannisechse, *Ablepharus kitaibelii fitzingeri* Mertens 1952 (*Sauria / Scincidae*) (Biological observations on a northern scincoid lizard, *Ablepharus kitaibelii fitzingeri* Mertens 1952 (*Sauria / Scincidae*)). Zool. Gart. (N.F.), Frankfurt a. M. 26:312-318.

Sassenburg, L. 1972. Albinotische Strumpfbandnattern (*Thamnophis s. sirtalis*) (Albinistic garter snakes *Thamnophis s. sirtalis*]). Zool. Gart. (N.F.), Leipzig 42:331-334.

Schiche, O. E. 1917. Kasuistische Beiträge zur Pathologie der Reptilien und Amphibien (Casuistic contributions to the pathology of reptiles and amphibians). Zool. Gart., Frankfurt a. M. 58:73-83.

Schifter, H. 1965. Erfahrungen mit einem Pantherchamäleon (*Chamaeleo pardalis* Cuvier, 1829) (Experiences with a panther chameleon (*Chamaeleo pardalis* Cuvier, 1829)). Zool. Gart. (N.F.), Leipzig 30:179-181.

Schifter, H. 1967. Langjährige Haltung eines Stachelschwanzwarans, *Varanus acanthurus brachyurus* (Long-time care of a spiny-tailed monitor, *Varanus acanthurus brachyurus).* Zool. Gart. (N.F.), Leipzig 33:262-264.

Schmidt, M. 1864. Fütterung der Klapperschlange (Feeding the rattlesnake). Zool. Gart., Frankfurt a. M. 5:258-259.

Schmidt, M. 1900. Die Dahl'sche Natter (*Zamenis dahli* Fitz.) in der Gefangenschaft (The Dahl viper (*Zamenis dahli* Fitz.) in captivity). Zool. Gart., Frankfurt a. M. 41:217-218.

Schmidt, P. 1912. Eine grüne Varietät der *Calotes versicolor* (Daudin)? (A green variety of *Calotes versicolor* [Daudin]?). Zool. Gart., Frankfurt a. M. 52:350-352.

Schmidt-Hoensdorf, F. 1937. Das neue Aquarium des Zoologischen Gartens Halle (The new Aquarium of the Zoo in Halle). Zool. Gart. (N.F.), Leipzig 9:107-111.

Schnee, P. 1900. Kleine Mitteilungen über das Freileben einiger australischer Reptilien (Small notes about life in the wild of some Australian reptiles). Zool. Gart., Frankfurt a. M. 41:17-19.

Schnee, P. 1900. Kleinere Mitteilungen. Aufrechter Gang der Kragenechse (Small notes: upright walk of King's lizard). Zool. Gart., Frankfurt a. M. 41:61.

Schnee, P. 1900. Kann eine Sumpfschildkröte überhaupt ausserhalb des Wassers fressen? (Is a fresh-water turtle able to feed outside the water?). Zool. Gart., Frankfurt a. M. 41:215-217.

Schnee, P. 1900. Mimikry bei Schlangen? (Mimicry in snakes?). Zool. Gart., Frankfurt a. M. 41:219-222.

Schnee, P. 1900. Kleinere Mitteilungen. Der Rüssel der Peitschenschlangen (Small notes: The trunk of the eastern coachwhip snake). Zool. Gart., Frankfurt a. M. 41:228-229.

Schnee, P. 1900. Mimikry bei südamerikanischen Schildkröten? (Mimicry in South American turtles?). Zool. Gart., Frankfurt a. M. 41:315-317.

Schnee, P. 1900. Große Geschwulst im Magen einer Schildkröte (Large tumor in the stomach of a turtle). Zool. Gart., Frankfurt a. M. 41:358-359.

Schnee, P. 1900. Kleinere Mitteilungen. *Ancistrodon blomhoffi* Boie (Small notes. *Ancistrodon blomhoffi* Boie). Zool. Gart., Frankfurt a. M. 41:395-396.

Schnee, P. 1902. Beobachtungen aus meinem Terrarium (Observations from my terrarium). Zool. Gart., Frankfurt a. M. 43:348-350.

Schnee, P. 1902. Die Kriechtiere der Marshallinseln (The reptiles of the Marshall Islands). Zool. Gart., Frankfurt a. M. 43:354-362.

Schnee, P. 1903. Riesenschildkröten auf einer polynesischen Insel (Giant turtles on a Polynesian island). Zool. Gart., Frankfurt a. M. 44:129-131.

Schnee, P. 1903. Kleinere Mitteilungen. Angebliche Abrichtung von Brillenschlangen (Small notes. Alleged training of the common cobra). Zool. Gart., Frankfurt a. M. 44:234-235.

Schnee, P. 1903. Das Wachstum der Suppenschildkröte (The growth rate of the green turtle). Zool. Gart., Frankfurt a. M. 46:221- 222.

Schreiber, E. 1878. Über den Rippenmolch, *Pleurodeles Waltlii* (Concerning the ribbed newt, *Pleurodeles Waltlii*). Zool. Gart., Frankfurt a. M.19:321-328.

Schreiber, E. 1912. Über Chamaeleon vulgaris Daud. (Concerning Chamaeleon vulgaris Daud). Zool. Gart., Frankfurt a. M. 53:149-152.

Simons, R. 1877. Tonäusserung des Scheltopusik, *Pseudopus Pallasii* (Sounds made by the Scheltopusik, *Pseudopus Pallasii*). Zool. Gart., Frankfurt a. M.18:230-233.

Stemmler, O., and A. Zingg. 1969. Das Betäuben von Schlangen (The narcotization of snakes). Zool. Gart. (N.F.), Leipzig 37:76-80.

Stemmler-Morath, C. 1956. Beitrag zur Gefangenschafts- und Fortpflanzungsbiologie von *Python molurus* L. (Contributions concerning the biology of captivity and reproduction of *Python molurus* L.). Zool. Gart. (N.F.), Leipzig 21:347-364.

Stricker, W. 1889. Sprachwissenschaft und Naturwissenschaft, XX. Frosch (Linguistics and natural science). Zool. Gart., Frankfurt a. M. 30:267-270.

Sumichrast, F. 1865. Über die Sitten einiger Reptilien in Mexiko (Concerning the habits of some reptiles in Mexico). Zool. Gart., Frankfurt a. M. 6:196-198, 6:237-238.

Vosseler, J. 1908. Die Puffotter Usambaras (*Bitis gabonica* [D.B.]) (The puff adder of Usambara [*Bitis gabonica* (D.B.)]. Zool. Gart., Frankfurt a. M. 49:167-172. [photograph of a large specimen restrained by a capture pole].

Weinland, D. F. 1861. Unsere Schnappschildkröte (Our snapping turtle). Zool. Gart., Frankfurt a. M. 2:69-75.

Weinland, D. F. 1863. Unser Riesensalamander, (*Salamandra maxima*, Schlegel) (Our giant salamander, [*Salamandra maxima* Schlegel]). Zool. Gart., Frankfurt a. M. 4:137-143.

Weismann, A. 1876. Axolotl und *Amblystoma* (Axolotl and *Amblystoma*). Zool. Gart., Frankfurt a. M. 17:1-8.

Wermuth, H. 1962. *Testudo hypselonota* Bourret, eine revalidierte Schildkröten-Art aus Indochina (*Testudo hypselonota* Bourret, a revalidated species of tortoise from Indochina). Zool. Gart. (N.F.), Leipzig 26:228-238.

Werner, F. 1890. Die Nahrung der giftlosen europäischen Schlangen (The nourishment of non-poisonous European snakes). Zool. Gart., Frankfurt a. M. 31:134-143.

Werner, F. 1890. Lebensweise einiger nordafrikanischer Reptilien in Gefangenschaft (Habits of some North-African reptiles in captivity). Zool. Gart., Frankfurt a. M. 31:335-341.

Werner, F. 1891. Biologische Beobachtungen an Reptilien von Istrien und Dalmatien (Biological observations of reptiles from Istria and Dalmatia). Zool. Gart., Frankfurt a. M. 32:225-232.

Werner, F. 1892. Das Vivarium in Wien (The Vivarium in Vienna). Zool. Gart., Frankfurt a. M. 33:22-26.

Werner, F. 1892. Über die Lebensweise des Wüsten-Warans und der Hufeisennatter in Gefangenschaft (About the habits of the desert monitor and the horseshoe snake in captivity). Zool. Gart., Frankfurt a. M. 33:304-306.

Werner, F. 1893. Die Krankheiten der Reptilien und

Amphibien (The illnesses of reptiles and amphibians). Zool. Gart., Frankfurt a. M. 34:65-71.

Werner, F. 1893. Beobachtungen an *Sphenodon* (*Hatteria*) *punctatus* (Observations of *Sphenodon* [*Hatteria*] *punctatus*). Zool. Gart., Frankfurt a. M. 34:335-339.

Werner, F. 1894. Beiträge zur Reptilien-Psychologie (Contributions to the psychology of reptiles). Zool. Gart., Frankfurt a. M. 35:174-179.

Werner, F. 1896. Über die Sandschlange (*Eryx jaculus* L.) (About the javelin sand boa [*Eryx jaculus* L.]). Zool. Gart., Frankfurt a. M. 37:85-88.

Werner, F. 1897. Neues aus dem Wiener Vivarium (News from the Vienna Vivarium). Zool. Gart., Frankfurt a. M. 38:257-263.

Werner, F. 1899. Allerlei aus dem Kriechtierleben im Käfig II. (Miscellany from reptile life in cage II.). Zool. Gart., Frankfurt a. M. 40:12-24.

Werner, F. 1899. Des Wiener Vivariums Ende (The end of the Vienna Vivarium). Zool. Gart., Frankfurt a. M. 40:33-38.

Werner, F. 1899. Auf der Reptilienjagd in Ägypten (Hunting reptiles in Egypt). Zool. Gart., Frankfurt a. M. 40:277-288.

Werner, F. 1900. Riesenschlangen in Gefangenschaft (Giant snakes in captivity). Zool. Gart., Frankfurt a. M. 41:233-243, 41:274-287.

Werner, F. 1901. Noch einmal das Vivarium in Wien (Once more the Vienna Vivarium). Zool. Gart., Frankfurt a. M. 42:1-5.

Werner, F. 1902. Riesenschlangen in Gefangenschaft (Giant snakes in captivity). Zool. Gart., Frankfurt a. M. 43:328-329.

Werner, F. 1903. Mensch und Kriechtier in den Mittelmeerländern (Man and reptile in the Mediterranean countries). Zool. Gart., Frankfurt a. M. 44:1-6.

Werner, F. 1917. Die Kleintierwelt der südlichen Balkanländer (The world of small animals in the Southern Balkan countries). Zool. Gart., Frankfurt a. M. 58:129-137.

Werner, F. 1920. Vom Feuersalamander und seinem Farbenkleid (About the fire salamander and its colorful dress). Zool. Gart., Frankfurt a. M. 61:34-.

Werner, H. 1892. Bemerkungen über den Scheltopusik und die Treppennatter (Comments on the Scheltopusik and the staircase snake). Zool. Gart., Frankfurt a. M. 33:38-41.

Werner, R. M. 1952. Hilfsfütterung von Schlangen (Supplementary feeding of serpents). Zool. Gart. (N.F.), Leipzig 19:210-220.

Wolterstorff, W. 1896. Über die Neotenie der Batrachier (Concerning the neotony of batrachians). Zool. Gart., Frankfurt a. M. 37:327-337.

Zander, A. 1895. Einige transkaspische Reptilien (Some Trans-Cape reptiles). Zool. Gart., Frankfurt a. M. 36:232-238.

Zeller, F. 1957. Über die Haltung von Hundskopfschlingern (*Boa canina*) (Concerning the care of dog-headed snakes [*Boa canina*]). Zool. Gart. (N.F.), Leipzig 23:133-136.

Zimmermann, R. 1909. Die Sumpfschildkröte, *Emys orbicularis* (L.), im Königreich Sachsen und ihr Vorkommen westlich von der Elbe überhaupt (The fresh-water turtle, *Emys orbicularis* [L.], in the kingdom of Saxony and its occurrence west of the Elbe River in general). Zool. Gart., Frankfurt a. M. 50:55-59.

Zimmermann, R. 1914. Von der Glatten Natter (About the smooth snake). Zool. Gart., Frankfurt a. M. 55:121-131.

Zwart, P., B. W. A. Langerwerf, H. Claessen, J. J. C. Leunisse, J. Mennes, C. van Riel, L. Lambrechts, and M. J. L. Kik. 1992. Health aspects in breeding and rearing insectivorous lizards from moderate climatic zones. Zool. Gart. (N.F.), Jena 62:46-52.

Brazilian Horned Frog (*Ceratophrys aurita*)

Chapter 8
Zoological Gardens in the United States and Canada

Common Snapping Turtle (*Chelydra serpentina*)

NORTHEASTERN UNITED STATES

Philadelphia Zoological Garden (1874)

History: America's first zoo has the distinction of having the first reptile collection in a zoological garden in the United States, established in 1874. The animals were kept in the library of the building named "Solitude," the former country home of John Penn, William Penn's grandson. The Superintendent of the Zoo, Arthur E. Brown (1850-1910), was a respected herpetologist. When a new bird house was constructed in 1882, the original aviary was modified by Brown to house reptiles. Two wings were added six years later and a rear wing was opened in 1923. This building was demolished in 1969.

Facility and Collection: Roger Conant published many articles on the reptile building and collection which are listed below. The facility was renovated in 1998.

According to *International Zoo Yearbook* (1998:Volume 28), composition of the collection was 103 taxa of reptiles with 423 specimens and 47 taxa of amphibians numbering 269.

Fig. 115. "Solitude" was the first reptile building in the United States at Philadelphia Zoo where reptiles were displayed in 1874. *Courtesy of Zoological Society of Philadelphia, provided by Brint Spencer.*

Staff and Scientific Achievements: Roger Conant served as president of AZA in 1946-7 and received the R. Marlin Perkins Award for professional excellence

"

Fig. 116. Snake exhibits in old reptile display at Philadelphia Zoo during the 1930s. *Undated photograph by Franklin Williamson, courtesy of Zoological Society of Philadelphia, provided by Brint Spencer.*

Fig. 117. Exhibits in old reptile display at Philadelphia Zoo during the 1930s. *Undated photograph by Franklin Williamson, courtesy of Zoological Society of Philadelphia, provided by Brint Spencer.*

Fig. 118. Entrance to current reptile building, opened in 1972, at Philadelphia Zoo with *Haddonosaurus* statue. *Zoological Society of Philadelphia, provided by Brint Spencer.*

in 1989. J. Kevin Bowler followed Conant as curator and his interests centered on applied technologies to create suitable environments for captive animals. When Kevin left the zoo, John Groves assumed his curatorial position; his focus was directed toward husbandry issues and reproduction of captive amphibians and reptiles. Kevin Wright followed Groves and published a large number of papers on medical management. Recently, curator Brint Spencer has left the zoo and the position is currently unfilled.

Publications: The zoo magazine *Fauna* was started in 1939, followed by *America's First Zoo.* There are a number of articles by prominent herpetologists.

Historical Overview

Adler, K. 1994. The remarkable career of Roger Conant, p. 17- 23. In J. B. Murphy, K. Adler, and J. T. Collins (eds.), Captive Management and Conservation of Amphibians and Reptiles. Society for the Study of Amphibians and Reptiles. Contributions to Herpetology, volume 11, Ithaca NY.

Anon. 1976. Herpetology's loss (an obituary of: Isabelle dePeyster Conant). Herpetol. Rev. 7(4):180.

Brown, A. E. 1878. Guide to the Garden of the Zoological Society of Philadelphia (Fairmount Park) According to the Present Arrangement. Allen, Lane and Scott, Printers, Philadelphia PA.

Card, W., and J. B. Murphy. 2000. Lineages and histories of zoo herpetologists in the United States. Herpetol. Circ. 27:1-45 [portraits of Arthur Brown and Roger Conant].

Conant, R. 1938. Official Illustrated Guide Book to the Philadelphia Zoological Garden. Zool. Soc. Philadelphia :1-108 [nine editions, with last in 1953].

Conant, R. 1957. Arthur Erwin Brown: "Custodian of the Garden" and naturalist of note. America's First Zoo 9(4):3 pp.

Conant, R. 1975. Philadelphia's new Reptile House. Zool. Gart. (N.F.), Jena 54:54-62. [Collection had been located in the historic Penn House and several other sites until this new facility was opened in 1972. Photographs of this modern building featured the entrance, public area, as well as crocodilian, giant tortoise, large lizard and gaboon viper exhibits. The collection numbered 176 reptile species (500 specimens) and 28 amphibian species (98 specimens). A detailed description of the environmental systems and display techniques was included.]

Conant, R. 1997. A Field Guide to the Life and Times of Roger Conant. Selva, Tyler TX.

DeLeon, C. 1999. America's First Zoostory. 125 years at the Philadelphia Zoo. The Donning Co., Virginia Beach VA. [Original reptile house pictured on p. 72. Roger Conant is featured in two photographs: a young man of 26 feeding a banana to a Galapagos tortoise ca. 1936 (p. 32) and doing a radio show in the presence of a live cockatoo (p. 105).]

Kleiman, C. 2001. Roger Conant 1909-. American herpetologist and zoo director, p. 286-288. In C. E. Bell (ed.), Encyclopedia of the World's Zoos. Fitzroy Dearborn Publishers, Chicago, London.

Morphology, Systematics & Taxonomy

Brown, A. E. 1889. Description of a new species of *Eutaenia*. Proc. Acad. Nat. Sci. Philadelphia 41:421-422. [Described as *E. nigrolateris*, now known as Marcy's checkered garter snake *Thamnophis m. marcianus*, and captured in the vicinity of Tucson, Arizona.]

Brown, A. E. 1890. On a new genus of Colubridae from Florida. Proc. Acad. Nat. Sci. Philadelphia 42:199-200. [description of the short-tailed snake *Stilosoma extenuatum*].

Brown, A. E. 1893. Notes on some snakes from tropical America lately living in the collection of the Zoological Society of Philadelphia. Proc. Acad. Nat. Sci. Philadelphia 45:429-435.

Brown, A. E. 1901. A new species of *Coluber* from western Texas. Proc. Acad. Nat. Sci. Philadelphia 53:1-4 [description of Trans-Pecos rat snake *Bogertophis subocularis*].

Brown, A. E. 1901. A review of the genera and species of American snakes, north of Mexico. Proc. Acad. Nat. Sci. Philadelphia 53:10-110.

Brown, A. E. 1902. A collection of reptiles and batrachians from Borneo and the Loo Choo Islands. Proc. Acad. Nat. Sci. Philadelphia 54:175-186.

Brown, A. E. 1902. A new species of *Ophiobolus* from western Texas. Proc. Acad. Nat. Sci. Philadelphia 54:612-613 [description of gray-banded kingsnake *Lampropeltis alterna*].

Brown, A. E. 1902. A list of reptiles and batrachians in the Harrison Hiller collection from Sumatra. Proc. Acad. Nat. Sci. Philadelphia 54:693-695.

Brown, A. E. 1903. Generic types of Nearctic Reptilia and Amphibia. Proc. Acad. Nat. Sci. Philadelphia 55:112-127.

Brown, A. E. 1903. The variation of *Eutaenia* in the Pacific subregion. Proc. Acad. Nat. Sci. Philadelphia 55:286-297.

Brown, A. E. 1903. Note on *Crotalus scutulatus* Kenn. Proc. Acad. Nat. Sci. Philadelphia 55:625.

Brown, A. E. 1905. The identity of *Eutaenia atrata* Kenn. Proc. Acad. Nat. Sci. Philadelphia 57:692-693.

Conant, R. 1937. *Alsophis* from new islands with a description of a new subspecies. Proc. New England Zool. Club 16:81-83 [description of the colubrid snake *Alsophis vudii picticeps* from Bimini Islands, Bahamas].

Conant, R. 1940. A new subspecies of the fox snake, *Elaphe vulpina* Baird and Girard. Herpetologica 2:1-14 [description of eastern fox snake *Elaphe vulpina gloydi*].

Conant, R. 1943. Studies on North American water snakes-I: *Natrix kirtlandii* (Kennicott). Amer. Midl. Natur. 29:313-341.

Conant, R. 1946. Studies on North American water snakes-II: The subspecies of *Natrix valida*. Amer. Midl. Natur. 35:250-275.

Conant, R. 1946. Intergradation among ring-necked snakes from southern New Jersey and the Del-Mar-Va Peninsula. Bull. Chicago Acad. Sci. 7:473-482.

Conant, R. 1949. Two new races of *Natrix erythrogaster*. Copeia 1949:1-15 [description of yellowbelly watersnake *Natrix (Nerodia) erythrogaster flavigaster* and copperbelly watersnake *Natrix (Nerodia) erythrogaster neglecta*].

Conant, R. 1950. On the taxonomic status of *Thamnophis butleri* (Cope). Bull. Chicago Acad. Sci. 9:71-77.

Conant, R. 1953. Three new water snakes of the genus *Natrix* from Mexico. Nat. Hist. Misc. No. 126:1-9 [description of Nazas watersnake *Natrix (=Nerodia) erythrogaster bogerti*, Tabasco watersnake *Natrix (=Nerodia) rhombifera werleri*, and Colima garter snake *Natrix valida isabelleae*, now placed in genus *Thamnophis*].

Conant, R. 1955. Notes on *Natrix erythrogaster* from the eastern and western extremes of its range. Nat. Hist. Misc. No. 147:1-3.

Conant, R. 1955. Notes on three Texas reptiles, including an addition to the fauna of the state. Amer. Mus. Novitates No. 1726:1-6.

Conant, R. (was Committee Chairman). 1956. Common names for North American amphibians and reptiles. Copeia 1956:172-185. [In 1957, there was a special reprint by ASIH. 26 pp.]

Conant, R. 1956. A review of two rare pine snakes from the Gulf Coastal Plain. Amer. Mus. Novitates No. 1781:1-31.

Conant, R. 1958. Notes on the herpetology of the Delmarva Peninsula. Copeia 1958:50-52.

Conant, R. 1961. A new water snake from Mexico, with notes on anal plates and apical pits in *Natrix* and *Thamnophis*. Amer. Mus. Novitates No. 2060:1-22 [description of Tepic garter snake *Natrix valida thamnophisoides*, now known as *Thamnophis valida thamnophisoides*].

Conant, R. 1963. Another new water snake of the genus *Natrix* from the Mexican Plateau. Proc. Biol. Soc. Washington 76:169- 172. [description of Aguanaval watersnake *Natrix* (=*Nerodia*) *erythrogaster alta*.].

Conant, R. 1963. Evidence for the specific status of the water snake *Natrix fasciata*. Amer. Mus. Novitates No. 2122: 1-38.

Conant, R. 1963. Semiaquatic snakes of the genus *Thamnophis* from the isolated drainage system of the Rio Nazas and adjacent areas in Mexico. Copeia 1963:473-499.

Conant, R. 1965. Miscellaneous notes and comments on toads, lizards, and snakes from Mexico. Amer. Mus. Novitates No. 2205:1-38 [notes and comments on one toad, two lizards, and thirteen snakes].

Conant, R. 1969. A review of the water snakes of the genus *Natrix* in Mexico. Bull. Amer. Mus. Nat. Hist. 142(1):1-140. [results of field work over a 10-year span.].

Conant, R. 1977. The Florida water snake (Reptilia, Serpentes, Colubridae) established at Brownsville, Texas, with comments on other herpetological introductions in the area. J. Herpetol. 11:217-220.

Conant, R. 1983. Commentary on a frog and lizard newly recorded from Central Durango, Mexico, p. 399-405. In A. G. C. Rhodin, and K. Miyata (eds.), Advances in Herpetology and Evolutionary Biology; Essays in Honor of Ernest E. Williams. Museum of Comparative Zoology, Harvard University, Cambridge MA [spotted chirping frog *Syrrhophus guttilatus* (now considered a subgenus of *Eleutherodactylus*); southwestern earless lizard *Callisaurus* (*Cophosaurus*) *texanus scitulus*].

Conant, R. 1984. A new subspecies of the pit viper, *Agkistrodon bilineatus* (Reptilia: Viperidae) from Central America. Proc. Biol. Soc. Washington 97:135-141 [named in honor of Howard K. Gloyd, Conant's friend and colleague for nearly 50 years].

Conant, R. 1985 (published 1986). Phylogeny and zoogeography of the genus *Agkistrodon* in North America, p. 89-92. In Z. Roèek (ed.), Studies in Herpetology; Proc. European Herpetological Meeting (Societas Europaea Herpetologica). Charles University, Prague, Czech Republic.

Conant, R. 1992. The type locality of *Agkistrodon halys caraganus*. Asiatic Herpetol. Res. 4:57.

Conant, R. 2003. Observations on garter snakes of the *Thamnophis eques* complex in the Lake's of Mexico's Transvolcanic Belt, with descriptions on new taxa. Amer. Mus. Novitates No.3406: 1-64. [Provides description of *Thamnophis eques cuitzeoensis*, *T. e. insperatus*, *T. e. obscurus*, *T. e. diluvialis*, *T. e. scotti*, *T. e. carmenensis* and *T. e. patzcuaroensis* collected between 1959 through 1965 when Conant was employed at the zoo. Many gravid females located on these trips were brought to the zoo and maintained alive until they gave birth.]

Conant, R., and J. F. Berry. 1978. Turtles of the family Kinosternidae in the southwestern United States and adjacent Mexico: Identification and distribution. Amer. Mus. Novitates No. 642:1-18.

Conant, R., and W. M. Clay. 1937. A new subspecies of water snake from islands in Lake Erie. Occ. Pap. Mus. Zool. Univ. Michigan No. 346:1-10. [description of Lake Erie watersnake *Natrix* (*Nerodia*) *sipedon insularum*].

Conant, R., and W. M. Clay. 1963. A reassessment of the taxonomic status of the Lake Erie water snake. Herpetologica 19:179-184.

Conant, R., and C. J. Goin. 1948. A new subspecies of soft-shelled turtle from the Central United States, with comments on the application of the name *Amyda*. Occ. Pap. Mus. Zool. Univ. Michigan No. 510:1-23 [description of western spiny softshell turtle *Trionyx spiniferus hartwegi*, now known as *Apalone spinifera hartwegi*].

Conant, R., and J. D. Lazell Jr. 1973. The Carolina Salt Marsh snake: a distinct form of *Natrix sipedon*. Breviora No. 400:1-13 [description of Carolina watersnake *Natrix* (*Nerodia*) *sipedon williamengelsi*].

Dunn, E. R., and R. Conant. 1936. Notes on anacondas, with descriptions of two new species. Proc. Acad. Nat. Sci. Philadelphia 88:503-506. [Provides description of *Eunectes deschauenseei* and *E. barbouri* from large specimens living at the zoo. *Eunectes deschauenseei* was named after Rodolphe Meyer de Schauensee, who donated a specimen to the zoo in 1924.]

Gloyd, H. K., and R. Conant. 1934. The broad-banded copperhead: A new subspecies of *Agkistrodon mokasen*. Occ. Pap. Mus. Zool. Univ. Michigan No. 283:1-6 [description of broad-banded copperhead *Agkistrodon contortrix laticinctus*].

Gloyd, H. K., and R. Conant. 1934. The taxonomic status, range and natural history of Schott's racer.

Occ. Pap. Mus. Zool. Univ. Michigan No. 287:1-18 [description of reproduction].

Gloyd, H. K., and R. Conant. 1938. The subspecies of the copperhead, *Agkistrodon mokasen* Beauvois. Bull. Chicago Acad. Sci. 5:163-166.

Gloyd, H. K., and R. Conant. 1943. A synopsis of the American forms of *Agkistrodon* (copperheads and moccasins). Bull. Chicago Acad. Sci. 7(2):147-170 [description of Trans-Pecos copperhead *Agkistrodon contortrix pictigaster*].

Gloyd, H. K., and R. Conant. 1990. Snakes of the *Agkistrodon* Complex. A Monographic Review, Contributions to Herpetology, volume 6. Society for the Study of Amphibians and Reptiles, Athens, GA. [There were nine additional papers by various authors. This project took over 60 years to complete.]

Smith, H. M., L. C. Stuart, and R. Conant. 1971. Request for revision of the 1964 Code to permit valid emendation of certain -ii endings of patronyms. Bull. Zool. Nomencl. 27(Pts. 5/6):250-252.

Strimple, P. D., G. Puorto, W. F. Holmstrom, R. W. Henderson, and R. Conant. 1997. On the status of the anaconda *Eunectes barbouri* Dunn and Conant. J. Herpetol. 31:607-609 [*Eunectes barbouri* placed in synonymy with *E. murinus*].

Tinkle, D. W., and R. Conant. 1961. The rediscovery of the water snake, *Natrix harteri*, in western Texas, with a description of a new subspecies. Southwest. Natur. 6:33-44 [description of Concho watersnake *Natrix (Nerodia) harteri paucimaculata*].

Husbandry

Bowler, J. K. 1980. Modern management and exhibit techniques for reptiles, p. 19-22. In J. B. Murphy, and J. T. Collins (eds.), Reproductive Biology and Diseases of Captive Reptiles. Society for the Study of Amphibians and Reptiles. Contributions to Herpetology, volume 1, Lawrence KS.

Conant, R. 1945. The care of baby turtles. Fauna, Zool. Soc. Philadelphia 7(2):44.

Conant, R. 1971. Reptile and amphibian management practices at Philadelphia Zoo. Inter. Zoo Yearb. 11:224-230. [Account covered caging, food, parasite control, disease and handling.]

Wright, K. M., and T. Minott. 1999. Individual identification of captive Mexican caecilians. Herpetol. Rev. 30(1):32-33.

Wright, K. M., B. Toddes, and S. Donoghue. 1997. To form the more perfect stool: Feeding trials on the Galapagos tortoise (*Geochelone nigra*) and Aldabra tortoise (*Geochelone gigantea*) population at the Philadelphia Zoological Garden. Proc. Amer. Assoc. Zoo Vet.:11-15.

Medical Management

The Penrose Research Laboratory founded in 1901 and directed for years by Herbert Fox, Herbert Ratcliffe and Robert Snyder, was revolutionary for most zoos did not have a research department on site, especially one where investigators focused on animal health, stress and nutrition.

Willette-Frahm, M., K. M. Wright, and B. C. Thode. 1995. Select protozoal diseases in amphibians and reptiles: A Report for the Infectious Diseases Committee, American Association of Zoo Veterinarians. Bull. Assoc. Rept. Amphib. Vet. 5(1):19-29 [comprehensive review from veterinarians from Gladys Porter Zoo and Philadelphia Zoo].

Wright, K. 1991. Husbandry: An essential component of diagnosing disease in reptiles and amphibians. Vivarium 3(3):23-27. [Kevin Wright has published a number of excellent articles on virtually every aspect of captive maintenance in the magazines *Vivarium* and *Reptile and Amphibian Magazine*].

Wright, K. 1994. Amputation of the tail of a two-toed amphiuma, *Amphiuma means*. Bull. Assoc. Rept. Amphib. Vet. 4(1):5.

Wright, K. 1997. Two products useful for tube-feeding herbivorous reptiles. Bull. Assoc. Rept. Amphib. Vet. 7(3):5-6.

Wright, K. 1997. Ivermectin for treatment of pentastomids in the Standing's day gecko, *Phelsuma standingi*. Bull. Assoc. Rept. Amphib. Vet. 7(1):5.

Reproduction

Bowler, J. K. 1975. Galapagos tortoise hatches at Philadelphia Zoo. Herpetol. Rev. 6(4):114.

Conant, R. 1980. The reproductive biology of reptiles: An historical perspective, p. 3-18. In J. B. Murphy, and J. T. Collins (eds.), Reproductive Biology and Diseases of Captive Reptiles. Society for the Study of Amphibians and Reptiles. Contributions to Herpetology, volume 1, Lawrence KS. [In June 1978 at Arizona State University, an SSAR sponsored symposium on captive breeding, management and diseases was convened; this volume was dedicated to Roger Conant and included 15 authors representing zoos.]

Conant, R., and A. Downs Jr. 1940. Miscellaneous notes on the eggs and young of reptiles. Zoologica (New York) 25:33-48 [data on 38 clutches or broods of young at Zoo].

Conservation

Conant, R. 1992. Comments on the survival status of members of the *Agkistrodon* complex. Contributions in Herpetology, Special Publication of the Greater Cincinnati Herpetological Society:29-33.

General

Bowler, J. K. 1977. Longevity of reptiles and amphibians in North American collections. Soc. Study Amphib. Rept. Herp. Circ. No. 6:1-32. [Author is now retired Curator of Reptiles at Audubon Park and Zoological Garden, New Orleans, Louisiana.]

Brown, A. E. 1903. Texas reptiles and their faunal relations. Proc. Acad. Nat. Sci. Philadelphia 55:543-558.

Brown, A. E. 1904. Post-glacial Nearctic centres of dispersal for reptiles. Proc. Acad. Nat. Sci. Philadelphia 56:464-474.

Brown, A. E. 1905. The utility principle in relation to specific characters. Proc. Acad. Nat. Sci. Philadelphia 57:206-209. [Brown discussed positions of Charles Darwin and Alfred Russell Wallace by using specific characters from snakes, i.e., presence or absence of ventral hypapophyses on the posterior trunk vertebrae, variation in characteristics of teeth, arrangement or absence of certain head plates, and differences of body scutellation to enlarge on concept of utility.]

Brown, A. E. 1906. Theories of evolution since Darwin. Proc. Acad. Nat. Sci. Philadelphia 58:1-16 [speech delivered before the Academy of Natural Sciences of Philadelphia on 10 January 1906].

Conant, R. 1932. The educational duty of the zoological park. Zoological Parks and Aquariums I:15-17.

Conant, R. 1932. The zoo museum. Zoological Parks and Aquariums I:91-93.

Conant, R. 1934. The red-bellied water snake, *Natrix sipedon erythrogaster* (Forster), in Ohio. Ohio J. Sci. 34:21-30.

Conant, R. 1934. Two rattlesnakes killed by a cottonmouth. Science 80(No. 2078):382.

Conant, R. 1935. Philadelphia versus Toledo in the field of herpetology. Toledo Field Nat. Assn. Ann. Bull., Spring, 1935:21-23.

Conant, R. 1936. *Virginia valeriae valeriae* in Pennsylvania. Herpetologica 1:17.

Conant, R. 1937. Notes on some European zoos. Part I. Germany. Parks and Recreation 21(1):14-23.

Conant, R. 1937. Notes on some European zoos. Part II. Basle, Switzerland and France. Parks and Recreation 21(2):60-66.

Conant, R. 1937. Notes on some European zoos. Part III. Belgium and the Netherlands. Parks and Recreation 21(3):95-102.

Conant, R. 1937. Notes on some European zoos. Part IV. England. Parks and Recreation 21(4):143-151.

Conant, R. 1938. On the seasonal occurrence of reptiles in Lucas County, Ohio. Herpetologica 1:137-144.

Conant, R. 1939. What Snake is That? A Field Guide to the Snakes of the United States East of the Rocky Mountains, by Roger Conant . . . and William Bridges . . . with 108 Drawings by Edmond Malnate. D. Appleton-Century, New York. 163 pp.

Conant, R. 1940. *Rana virgatipes* in Delaware. Herpetologica 1:176-177.

Conant, R. 1942. Amphibians and reptiles from Dutch Mountain (Pennsylvania) and vicinity. Amer. Midl. Natur. 27:154-170.

Conant, R. 1943. The milk snakes of the Atlantic Coastal Plain. Proc. New England Zool. Club 22:3-24.

Conant, R. 1945. More reptiles in cork shipments. Copeia 1945:233.

Conant, R. 1947. The carpenter frog in Maryland. Maryland J. Nat. Hist. 17(4):72-73.

Conant, R. 1947. Reptiles and Amphibians of the Northeastern States. Zool. Soc. Philadelphia :1-40 [second edition published in 1952, third edition published in 1957].

Conant, R. 1948. Regeneration of clipped subcaudal scales in a pilot black snake. Nat. Hist. Misc. Chicago Acad. Sci. No.13:1-2.

Conant, R. 1951. The Reptiles of Ohio. University of Notre Dame Press, Notre Dame IN [first published in Amer. Midl. Natur. 20(1) in 1938; 2nd edition with revisionary addenda. 284 pp.].

Conant, R. 1951. The red-bellied terrapin, *Pseudemys rubriventris* (LeConte), in Pennsylvania. Ann. Carnegie Mus. 32(Art. 4):281-290.

Conant, R. 1951. A southward range extension for the keeled green snake, *Opheodrys aestivus*. Nat. Hist. Misc. No. 85:84-85.

Conant, R. 1955. Saurian shell crusher. Nature 48(2):85-86 [description and photographs of caiman lizard feeding on a snail].

Conant, R. 1955. Snakes. Canadian Nature 17(5):158-166.

Conant, R. 1956. OBITUARY: E. R. Dunn, Herpetologist. Science 123(No. 3205):975 [see Adler (1989:92) for biography of Emmett Reid Dunn].

Conant, R. 1957. The eastern mud salamander, *Pseudotriton montanus montanus*: A new state record for New Jersey. Copeia 1957:152-153.

Conant, R. 1958. A Field Guide to the Reptiles and Amphibians of the United States and Canada East of the 100th Meridian, Houghton-Mifflin, Boston.

Conant, R. 1958. Philadelphia Zoo's king cobra exceeds record. Bull. Philadelphia Herpetol. Soc. 6(5):2.

Conant, R. 1958. Reptile Study. Reptile Study Merit Badge Pamphlet, Boy Scouts of America (Revised Edition):1-64.

Conant, R. 1959 (published 1960). *Lacerta* colony still extant in Philadelphia. Copeia 1959:335-336 [identified as *Lacerta sicula campestris*].

Conant, R. 1960. The queen snake *Natrix septemvittata,* in the interior highlands of Arkansas and Missouri, with comments upon similar disjunct distributions. Proc. Acad. Nat. Sci. Philadelphia 112(2):25-40.

Conant, R. 1961. The softshell turtle, *Trionyx spinifer*, introduced and established in New Jersey. Copeia 1961:355-356.

Conant, R. 1965. Snakes of record size from Kelleys Island in Lake Erie. J. Ohio Herpetol. Soc. 5(2):54-55.

Conant, R. 1966. OBITUARY: Henry Weed Fowler, 1878-1965. Copeia 1966:228-269. [Fowler was a famous naturalist and ichthyologist.]

Conant, R. 1966. A second record for *Ungaliophis continentalis* from Mexico. Herpetologica 22:157-160 [Isthmian dwarf boa].

Conant, R. 1968. Zoological exploration in Mexico - The route of Lieut. D. N. Couch in 1853. Amer. Mus. Novitates No. 2350:1-14.

Conant, R. 1975. The Pine Barrens, p. 16-27. In M. Cridland (ed.), Medford: Pioneering Township. Medford Historical Society, Philadelphia.

Conant, R. 1977 (published 1978). Semiaquatic reptiles and amphibians of the Chihuahuan Desert and their relationships to drainage patterns of the region, p. 455-491. In R. H. Wauer, and D. H. Riskind (eds.), Trans. Symposium on the Biological Resources of the Chihuahuan Desert Region, United States and Mexico, Sul Ross State Univ., Alpine Texas, 1975, U.S. Natl. Park Ser., Trans. & Proc. Series No. 3.

Conant, R. 1978. In Memoriam: Howard Kay Gloyd, 1902-1978. Herpetol. Rev. 9(4):127-129 [see also Adler (1989:112) for biography].

Conant, R. 1979. Distributional patterns of North American snakes: Some examples of the effects of Pleistocene glaciation and subsequent climatic changes. Bull. Maryland Herpetol. Soc. 14(4):241-259.

Conant, R. 1979. A zoogeographical review of the amphibians and reptiles of southern New Jersey, with emphasis on the Pine Barrens, p. 467-488. In R. T. T. Forman (ed.), Pine Barrens: Ecosystem and Landscape. Academic Press, New York.

Conant, R. 1980. OBITUARY: Arthur Loveridge, 1891-1980. Herpetol. Rev. 11(4):87-88 [see also Adler (1989:111) for biography].

Conant, R. 1984. *Agkistrodon* in Europe. Salamandra 18:191-194.

Conant, R. 1984. OBITUARY: William Marion Clay, 1906-1983. Copeia 1984:563.

Conant, R. 1991. Attempting the impossible. Bull. Chicago Herpetol. Soc. 26(7):149-152. [Account of Conant's efforts to complete the first field guide covering reptiles and amphibians of eastern North America truly gives a sense of the monumental task undertaken with his late wife, Isabelle Hunt Conant, who was the illustrator for the book. Imagine how difficult it must have been to acquire living specimens for each taxon so that she could photograph and accurately record color, pattern, and ontogenetic variation. Two additional installments in this CHS volume (nos. 8, 10) reflect the whole story.]

Conant, R. 1992. In Memoriam: Charles Mitchill Bogert, June 4, 1908-April 10, 1992. Herpetol. Rev. 23(4):102-105.

Conant, R. 1993. The oldest snake. Bull. Chicago Herpetol. Soc. 28(4):77-78.

Conant, R. 1994. Closing the gap-I. Bull. Chicago Herpetol. Soc. 29(3):49-53. [When Robert C. Stebbins finished his field guide which covered western amphibians and reptiles in the Peterson Field Guide Series, there was a gap between his coverage and the first edition of Conant's field guide in that the western half of Texas and parts of all states north of it were not included. Conant was asked by the publisher, Houghton Mifflin, to fill in this region in his second edition since the coverage in his first book only extended westward to the 100th meridian. This article (and Part II) described his collecting adventures in Big Bend National Park, located in the Trans-Pecos region of Texas.]

Conant, R. 1994. Closing the gap-II. Bull. Chicago Herpetol. Soc. 29(10):221-228.

Conant, R. 1997. The Great Barrancas. Bull. Chicago Herpetol. Soc. 32(9):189-196.

Conant, R. 1998. Pause at Paducah. Bull. Chicago Herpetol. Soc. 33(5):97-100.

Conant, R. 1999. OBITUARY: Joseph Randle Bailey, 1913-1998. Herpetol. Rev. 30(2):70-71. [Joseph Bailey, a long-time supporter of SSAR, willed his herpetological library to the Society.]

Conant, R., and R. M. Bailey. 1936. Some herpetological records from Monmouth and Ocean Counties, New Jersey. Occ. Pap. Mus. Zool. Univ. Michigan No. 328:1-10.

Conant, R., and R. G. Hudson. 1949. Longevity records for reptiles and amphibians in the Philadelphia Zoological Garden. Herpetologica 5:1-8.

Conant, R., E. S. Thomas, and R. L. Rausch. 1945. The plains garter snake, *Thamnophis radix*, in Ohio. Copeia 1945:61-68.

Groves, F., and J. D. Groves. 1978. Fatal toad poisoning in snakes. Herpetol. Rev. 9(1):19-20.

Groves, J. D. 1982. Egg-eating behavior of brooding five-lined skinks, *Eumeces fasciatus*. Copeia 1982:969-971. [Author is now employed at North Carolina Zoological Park, Asheboro.]

Groves, J. D., and W. Altimari. 1977. Keratophagy in the slender vine snake, *Uromacer oxyrhynchus*. Herpetol. Rev. 8(4):124.

Groves, J. D., and R. J. Assetto. 1976. *Lampropeltis triangulum elapsoides*. Herpetol. Rev. 7(3):114.

Groves, J. D., and E. M. Groves. 1972. An unusual accident involving an eastern king snake, *Lampropeltis getulus getulus*. Herpetol. Rev. 4(1):14.

Pough, F. H., and J. D. Groves. 1983. Specializations of the body form and food habits of snakes. Amer. Zool. 23:443-453.

Fauna and *America's First Zoo*

Conant, R. 1940. Frogs and toads found near Philadelphia. Fauna, Zool. Soc. Philadelphia 2(1):14-17.

Conant, R. 1940. For snakes to Georgia. Fauna, Zool. Soc. Philadelphia 2(2):37-38.

Conant, R. 1945. In pursuit of the turtle. Fauna, Zool. Soc. Philadelphia 7(2):34-35.

Conant, R. 1945. Turtles of the northeastern states. Fauna, Zool. Soc. Philadelphia 7(2):36-39.

Conant, R. 1946. The snakes of the northeastern states. Fauna, Zool. Soc. Philadelphia 8(2):48-52.

Conant, R. 1946. Spare that snake. Fauna, Zool. Soc. Philadelphia 8(2):47, 53-55.

Conant, R. 1947. Lizards of the northeastern states. Fauna, Zool. Soc. Philadelphia 9(2):42-44. [This issue has articles by Olive B. Goin on otters and Rozella Smith on collecting herps in Mexico.]

Conant, R. 1951. Unwitting stowaways. America's First Zoo 3(4):6-7.

Conant, R. 1952. The giant snakes. America's First Zoo 4(3):4 pp.

Conant, R. 1953. Snakes-How do we handle them? It's all in knowing how! America's First Zoo 5(4):3-6.

Conant, R. 1954. The incredible chameleon. America's First Zoo 6(3):7.

Conant, R. 1955. Lizards are finicky feeders. America's First Zoo 7(3):4 pp.

Conant, R. 1956. Slow-motion giants. America's First Zoo 8(2):3

Conant, R. 1956. How long do snakes live? America's First Zoo 8(4): 3 pp.

Conant, R. 1958. Stocking the Reptile House. America's First Zoo 10(4):4 pp.

Conant, R. 1960. Water snakes in the desert. America's First Zoo 12(4):31-35.

Conant, R. 1963. Head keeper dies suddenly. America's First Zoo 15(2):11 [Patrick Menichini].

Conant, R. 1965. Old-timers among the reptiles. America's First Zoo 17(4):32-34.

Conant, R. 1967. Vacation yields rare rattlesnake. America's First Zoo 19(4):26.

Conant, R. 1972. Paradise for serpents. America's First Zoo 24(1):3-7 [description of the collection in the new Reptile Building in Philadelphia].

Conant, R. 1973. Our Zoo of yesteryear. America's First Zoo 25(2):9-17. [Announcement that J. Kevin Bowler was new Acting Curator of Reptiles.]

Baltimore Zoo (1876)

History: Outside the main zoo boundaries in Druid Hill Park, a small building formerly used as a pumping station for a reservoir, was modified into a reptile building. This facility opened on 12 August 1948 and Frank Buck of "Bring-Em-Back-Alive" fame was the guest of honor. Although expectations were high to build a new building over 50 years ago, the original building is still in existence and until recently, housed a modest collection of amphibians and reptiles. In the fall of 2004, the department was disbanded and most of the entire collection dispersed to other zoos.

Staff and Scientific Achievements: In 1948, Frank Groves, Head Keeper of Birds and Reptiles, was in charge. He was elevated to Curator in 1973 and remained at the Zoo in this capacity until he retired 16 years later. His replacement, Anthony Wisnieski, expanded the collection and improved the exhibits. When he left, his assistant Vicki Poole was elevated to the

Fig. 119. Photograph of late herpetological curator Frank Groves at Baltimore Zoo in 1976. *Courtesy of John D. Groves.*

curatorial position until the department was discontinued.

***Collection*:** According to *International Zoo Yearbook* (1998:Volume 28), composition of the collection was nearly 90 taxa of reptiles with 365 specimens and 24 taxa of amphibians numbering 234.

Historical Overview

Card, W., and J. B. Murphy. 2000. Lineages and histories of zoo herpetologists in the United States. Herpetol. Circ. No. 27:1-44 [portrait of Frank Groves].

Groves, F. 1960. Reptile House: Baltimore Zoo, Druid Hill Park, Maryland. Bull. Philadelphia Herpetol. Soc. 8(5):15-17.

Murphy, J. B. 2005. The sad tale of two zoos. Herpetol. Rev. 36:6-7.

Reproduction

Groves, F. 1969. Some reptile breeding records from Baltimore Zoo. Inter. Zoo Yearb. 9:17-20.

Groves, F. 1979. Procedures for breeding snakes at the Baltimore Zoo, p. 29-30. In R. A. Hahn (ser. ed.), 1st & 2nd International Herpetological Symposium on Captive Propagation and Husbandry. International Herpetological Symposium, Thurmont MD.

Medical Management

Cranfield, M. R., and T. K. Graczyk. 1995. An update on ophidian cryptosporidiosis. Proc. Amer. Assoc. Zoo Vet.:225-230.

Denver, M. C., M. R. Cranfield, T. K. Graczyk, P. Blank, A. Wisnieski, and V. Poole. 1999. A review of reptilian amoebiasis and current research on the diagnosis and treatment of amoebiasis at the Baltimore Zoo. Proc. Amer. Assoc. Zoo Vet.:11-15.

Graczyk, T. K., M. R. Cranfield, and E. F. Bostwick. 1999. Therapeutic efficacy of hyperimmune bovine colostrum treatment against *Cryptosporidium* infections in reptiles. Proc. Amer. Assoc. Zoo Vet.:6-10.

General

Groves, F. 1959. Notes on *Trimeresurus f. flavoviridis.* Herpetologica 15:182.

Groves, F. 1961. A feeding record of the palm viper, *Bothrops schlegelii.* Herpetologica 17:277.

Groves, F. 1976. Envenomation in snakes and its treatment with antivenin. Herpetol. Rev. 7(3):108.

Groves, F., and J. D. Groves. 1978. Fatal toad poisoning in snakes. Herpetol. Rev. 9(1):19-20.

Livingston, B. 1974. Zoo Animals, People, Places. Arbor House, New York. [Provides description of "La Casa Grande" in Baltimore which was a private herpetological enterprise, jointly owned by Anthony Wisnieski in his earlier days, now Curator of Herpetology at the Zoo p. 85-91.]

Wisnieski, A., and V. A. Poole. 2001. Bog turtle conservation, research, and education programs at the Baltimore Zoo. Turtle Tortoise Newsletter 4:2-5.

National Zoological Park (1889) (Washington, DC)

***History and Mission*:** In 1829, James Smithson died and gave his fortune to the citizens of the United States for the purpose of establishing an institution devoted to scientific discovery and the diffusion of knowledge for the betterment of man. Since he was British and had never visited the United States, the gift was unexpected but it provided the impetus for the creation of the Smithsonian Institution which bears his name. Today, the Smithsonian Institution has the world's largest museum complex, four research centers, an astrophysics laboratory, library, several publications, educational department, traveling exhibitions and a plethora of research scientists. The National Zoological Park is one of these units.

***Facility and Collection*:** The development and construction of the Reptile House, built in 1931, exemplifies the Italian Byzanto-Romanesque ecclesiastical style. The building which still stands is unique as the facade abounds with stylized reptiles and amphibians: turtles supporting columns, crocodiles as decorative quoins, toady creatures peering down from the top of the archway, and an array of turtles, toads and lizards scattered on top and around the building. It is a spectacular edifice which probably could not be duplicated today. A strong emphasis on interactive education culminated with the opening of the Reptile Discovery Center several years ago. This unique exhibit consists of a series of interesting modules and displays addressing the biology of amphibians and reptiles in a user-friendly way. Docents interact with zoo visitors by answering questions and leading tours. During the planning stage, an important element was the inclusion of polling data derived from visitors to quantitatively assess the effectiveness of this exhibit.

According to *International Zoo Yearbook* (1998:Volume 28), composition of the collection was 63 taxa of reptiles with nearly 350 specimens and 15 taxa of amphibians numbering nearly 140. The collection is diverse, stable, and includes Cuban crocodiles, gharials, Cayman Island iguanas, Aldabran and radiated tortoises, large constrictors and a variety of amphibians. Komodo dragons reproduced for the first time in any zoo in the Western Hemisphere. Reproduction over

Fig. 121. Carved rock tortoise supporting column at entrance to reptile building at Smithsonian Institution's National Zoological Park. *Photograph by Dennis Desmond.*

Fig. 120. Entrance to reptile building at Smithsonian Institution's National Zoological Park, opened in 1931. *Photograph by Dennis Desmond.*

Fig. 122. Toady creature on top of column at entrance to reptile building at Smithsonian Institution's National Zoological Park. *Photograph by Dennis Desmond.*

Fig. 123. Mosaic above doorway of visitor entrance to reptile building at Smithsonian Institution's National Zoological Park. *Photograph by Dennis Desmond.*

multiple generations has been followed in several taxa: emerald tree boas, green tree pythons, Brazilian rainbow boas, African beaked snakes, Chinese water dragons, and Madagascan giant day geckos.

Staff and Scientific Achievements: Before curatorial positions were established, Mario (Jack) Deprato was the Head Keeper. The first curator was Jaren Horsley, who later became General Curator. Recently retired, he has a strong background in limnology and invertebrate biology. His replacement was Dale Marcellini, also retired, whose research interests were primarily directed toward public education and visitor behavior, as well as gekkonid and iguanid ethology. Louis (Trooper) Walsh recently retired from the Zoo as Biologist/Museum Specialist and is known for his work with Komodo dragons and boid snakes. Michael Davenport is the current supervisor whose overriding concern is for the protection and captive breeding of crocodilians. After retiring from the Dallas Zoo, I am now a research associate at this Zoo.

Long-term investigation of essential dietary ingredients for captive and wild reptiles by Mary Allen and Olav Oftedal includes evaluating the diets of desert tortoises. Important work is being done by pathologists looking at emergent diseases. Recently, the pathogenic chytrid fungus attacking captive and wild amphibians was isolated and described by pathologist Donald Nichols.

Personal Reflections: Christen Wemmer, retired Director of the Zoo's Conservation and Research Center in Front Royal, Virginia, and his associates were instrumental in creating Zoo Biology Training Courses held in many developing countries, Since I have been an instructor for many of these, I was struck by the power and influence of the name Smithsonian Institution worldwide and the universal respect accorded the Zoo as a result of these programs.

Historical Overview

Card, W., and J. B. Murphy. 2000. Lineages and histories of zoo herpetologists in the United States. Herpetol. Circ. 27:1-45 [portraits of Jaren Horsley and Dale Marcellini].

Conant, R. 1957. A tribute to William M. Mann. Parks and Recreation 40(3):30. [William M. Mann was Director at the Zoo.]

Ewing, H. P. 1990. An Architectural History of the National Zoological Park. [Heather P. Ewing was a summer intern in the Office of Architectural History and Historic Preservation, Smithsonian Institution. As part of her senior thesis, she consulted the Archives to compile a history of the Zoo, including the reptile building.]

Mann, L. Q. 1934. From Jungle to Zoo. Adventures of a Naturalist's Wife. Dodd, Mead, New York. [Lucile Mann was the wife of the late Director William M. Mann from the Zoo. There are a number of references to snakes, lizards and Surinam toads as she recalled their adventures in the field.]

Mann, W. M. 1930. Wild Animals In and Out of the Zoo, by William M. Mann. Smithsonian Scientific Series, vol. 6, New York [history of the Zoo, its animals and their behaviors].

Thomson, P. 1988. Keepers and Creatures at the National Zoo. Thomas Y. Crowell, New York. [Chapter 18 was a popularized portrait of the operations of a major zoo herpetological facility which described the husbandry and breeding programs for dart poison frogs, green tree pythons, and Chinese water dragons.]

Morphology, Systematics & Taxonomy

Frazier, J. G. 1983. Analisis estadistico de la tortuga golfina *Lepidochelys olivacea* (Eschscholtz) de Oaxaca, Mexico. (Statistical analysis of the Gulf turtle (*Lepidochelys olivacea* (Eschscholtz) from Oaxaca, Mexico). Ciencia Pesquera. Inst. Nal. Pesca. Sria. Pesca. Mexico (Fisheries Science, Natl. Institute of Fisheries; Fish Series. Mexico) 4:49-75. [John (Jack) Frazier has studied chelonians since the 1960s. After receiving his D. Phil. (Oxon.) under Niko Tinbergen, he was associated with the Division of Reptiles and Amphibians, US National Museum of Natural History. Later, he was involved with the Zoo for many years in the Department of Zoological Research and as a Smithsonian research associate. His major focus centers on the conservation of marine turtles and he has published widely in this area. In addition, his earlier studies have included behavioral and ecological aspects of the Aldabran tortoise.]

Longcore, J. E., A. P. Pessier, and D. K. Nichols. 1999. *Batrachochytrium dendrobatidis* gen. et. sp. nov., a chytrid pathogenic to amphibians. Mycologia 91:219-227.

Oftedal, O. T. 1974. A revision of the genus *Anadia* (Sauria, Teiidae). Arquivos de Zoologia 25(4):203-265. [Olav Oftedal is a Nutritionist at the Zoo who specializes in the feeding ecology of the endangered desert tortoise.]

Behavior

Eisenberg, J. F., and J. Frazier. 1983. A leatherback turtle *Dermochelys coriacea* feeding in the wild. J. Herpetol. 17:81-82.

Frazier, J., M. D. Menegheh, and F. Achaval. 1985. A clarification on the feeding habits of *Dermochelys coriacea*. J. Herpetol. 19:159-160.

Kleiman, D. G. 1992. Behavior research in zoos: past, present, and future. Zoo Biol. 11:301-312. [Devra Kleiman, now retired, was employed as the Senior Scientist in The Department of Zoological Research at the Zoo. She is the senior editor of *Wild Mammals in Captivity*, an update of Crandall's earlier book of the same name. She is coordinator of the golden lion tamarin reintroduction program in Brazil.]

Marcellini, D. L., and L. R. Schettino. 1987. Notes on the natural history of the unusual Cuban lizard, *Anolis lucius*. Herpetol. Rev. 18(3):52-53.

Robinson, M. H. 1989. Homage to Niko Tinbergen and Konrad Lorenz: Is classical ethology relevant to zoos? Zoo Biol. 8:209-221. [Michael Robinson retired as Director of Zoo.].

Wikramanayake, E. D. 1989. Thermoregulatory influences on the ecology of two sympatric varanids in Sri Lanka. Biotropica 21:74-79. [Eric Wikramanayake was research associate at Zoo.]

Wikramanayake, E. D., W. Ridwan, and D. Marcellini. 1999. The thermal ecology of free ranging Komodo dragons, *Varanus komodoensis,* on Komodo Island, Indonesia. Advances in Monitor Research II-Mertensiella 11:157-166.

Husbandry

Walker, E. P. 1941. Care of captive animals. Ann. Rep. Smithsonian Inst.:305-366. [Ernest P. Walker, Assistant Director of the Zoo, was best known for his three volume set entitled *Mammals of the World* which was written in the very office at the Zoo now occupied by me. This paper included sections on general care, handling, foods, enclosures, and medical care for mammals, birds, reptiles, amphibians, fishes and invertebrates.]

Weldon, P. J., B. Demeter, T. Walsh, and J. S. E. Kleister. 1994. Chemoreception in the feeding behavior of reptiles: considerations for maintenance and management, p. 61-70. In J. B. Murphy, K. Adler, and J. T. Collins (eds.), Captive Management and Conservation of Amphibians and Reptiles. Society for the Study of Amphibians and Reptiles. Contributions to Herpetology, volume 11, Ithaca NY. [Paul Weldon was a research associate at the Zoo from January 1991 to January 1993. Over the years, he has worked closely with zoo workers on research projects and has collected glandular materials from a variety of reptiles in zoos.]

Reproduction

Hildebrandt, T. B., F. Göritz, C. Pitra, L. H. Spelman, R. C. Cambre, T. Walsh, R. Rosscoe, and N. C. Pratt. 1996. Sonomorphological sex determination in subadult Komodo dragons. Proc. Amer. Assoc. Zoo Vet. :251-254 [study at Zoo].

Marcellini, D. L., and S. W. Davis. 1982. Effects of handling on reptile egg hatching. Herpetol. Rev. 13(2):43-44.

Wildt, D. E. 1989. Reproductive research in conservation biology: Priorities and aims for support. J. Zoo Wildl. Med. 20:391-395. [David Wildt is employed at Zoo as reproductive physiologist.]

Medical Management

Brooks, D. R., and J. Frazier. 1980. New host and locality for *Kathlania leptura* (Rudolphi) (Nematoda: Oxyurata: Kathlanidae). Proc. Helminth. Soc. Washington 47:267-268 [olive ridley sea turtle].

Brownstein, D. G. 1978. Reptilian Mycobacteriosis, p. 265-268. In R. J. Montali (ed.), Mycobacterial Infections of Zoo Animals. Smithsonian Institution Press, Washington DC. [Richard Montali was chief pathologist at the Zoo which supported this symposium.]

Cambre, R. C., D. E. Green, E. E. Smith, R. J. Montali, and M. Bush. 1980. Salmonellosis and arizonosis in the reptile collection at the National Zoological Park. J. Amer. Vet. Med. Assoc. 177:800-803.

Gray, C. W., J. Davis, and W. G. McCarten. 1966. Treatment of Pseudomonas infections in the snake and lizard collection at Washington Zoo. Inter. Zoo Yearb. 6:278.

Nichols, D. K. 2003. Tracking down the killer chytrid of amphibians. Herpetol. Rev. 34:101-104.

Nutrition

Allen, M. E. 1983. Geckos, tree frogs, crickets and calcium. Proc. Amer. Assoc. Zoo Vet. 1983:189-192.

Allen, M. E. 1984. A review of the composition of live prey in zoo animal diets. Proc. Amer. Assoc. Zoo Vet. 1984.:41.

Allen, M. E. 1988. The effect of three light treatments on growth in the green iguana *Iguana iguana*. Proc. Amer. Assoc. Zoo Vet. 1988:64.

Allen, M. E. 1989. Dietary induction and prevention of

osteodystrophy in an insectivorous reptile, *Eublepharis macularius:* characterization by radiography and histopathology. Third International Colloquium on the Pathology of Reptiles and Amphibians :83.

Allen, M. E., and O. T. Oftedal. 1982. Calcium and phosphorus levels in live prey. Proc. Amer. Assoc. Zoo Vet. 1982:120-128.

Allen, M. E., and O. T. Oftedal. 1989. Dietary manipulation of the calcium content of feed crickets. J. Zoo Wildl. Med. 20:26-33.

Allen, M. E., and O. T. Oftedal. 1994. The nutrition of carnivorous reptiles, p. 71-82. In J. B. Murphy, K. Adler, and J. T. Collins (eds.), Captive Management and Conservation of Amphibians and Reptiles. Society for the Study of Amphibians and Reptiles. Contributions to Herpetology, volume 11, Ithaca NY.

Allen, M. E., S. D. Crissey, and B. J. Demeter. 1986. The effect of diet on growth and bone development in the leopard gecko. Proc. Amer. Assoc. Zoo Vet. 1986:44-45.

Allen, M. E., O. T. Oftedal, and D. E. Ullrey. 1993. Effect of dietary calcium concentration on mineral composition of fox geckos (*Hemidactylus garnoti*) and Cuban tree frogs (*Osteopilus septentrionalis*). J. Zoo Wildl. Med. 24:118-128.

Allen, M. E., O. T. Oftedal, and D. I. Werner. 1990. Management of the green iguana (*Iguana iguana*) in Central America. Proc. Amer. Assoc. Zoo Vet. 1990:19-22.

Allen, M. E., D. E. Ullrey, and M. S. Edwards. 1999. The development of raw meat-based carnivore diets. Proc. Amer. Assoc. Zoo Vet. 1999:317-319.

Allen, M. E., O. T. Oftedal, D. J. Baer, and D. Werner. 1989. Nutritional studies with the green iguana, p. 73-81. In T. P. Meehan, and M. E. Allen (eds.), Proceedings, 8th Dr. Scholl Conference on the Nutrition of Captive Wild Animals. Lincoln Park Zoological Society, Chicago.

Allen, M. E., M. Bush, O. T. Oftedal, R. Rosscoe, T. Walsh, and M. D. Holick, 1994. Update on vitamin D and ultraviolet light in basking lizards. Proc. Amer. Assoc. Zoo Vet. 1994:314-316.

Barnard, J. B., O. T. Oftedal, P. S. Barbosa, C. E. Mathias, M. E. Allen, S. B. Citino, D. E. Ullrey, and R. J. Montali. 1991. The response of vitamin-d-deficient green iguanas *Iguana iguana* to artificial ultraviolet light. Proc. Amer. Assoc. Zoo Vet. 1991:147-155.

Oftedal, O. T. 1993. Nutritional research on the desert tortoise *Gopherus agassizii* in southern Nevada. Dept. Zoological Research, National Zoological Park, Smithsonian Institution, Washington DC.

Oftedal, O. T., and M. E. Allen. 1996. Nutrition as a major facet of reptile conservation. Zoo Biol. 15:491-497.

Oftedal, O. T., S. Hillard, and D. J. Morafka. 2002. Selective spring foraging by juvenile desert tortoises (*Gopherus agassizii*) in the Mojave Desert: Evidence of a selective nutritional strategy. Chelonian Conserv. Biol. 4:341-352.

Richman, L. K., R. J. Montali, M. E. Allen, and O. T. Oftedal. 1995. Paradoxical pathologic changes in vitamin D deficient green iguanas (*Iguana iguana*). Proc. Amer. Assoc. Zoo Vet. 1995:231-232.

Conservation

Beck, B. B. 1991. Managing zoo environments for reintroduction. Proc. Amer. Assoc. Zool. Parks Aquariums 1991:260-264. [Ben Beck, now retired, was employed at Zoo.]

Frazier, J. G. 1979. Marine turtle management in Seychelles: a case-study. Environmental Conserv. 6(3):225-230.

Frazier, J. G. 1980. Exploitation of marine turtles in the Indian Ocean. Human Ecol. 8(4).

Frazier, J. G. 1980. "Sea-turtle faces extinction": Crying "Wolf" or saving sea turtles? Environmental Conserv. 7(3):239-240.

Frazier, J. G. 1982. Tortugas marinas en Chile (Marine turtles of Chile). Bol. Mus. Nac. Hist. Nat. Chile (Bulletin of the National Natural History Museum, Chile) 39:63-73.

Frazier, J. G. 1982. Status of sea turtles and subsistence hunting in Indian Ocean, p. 385-396. In K. Bjorndal (ed.), Biology and Conservation of Sea Turtles: Proceedings of the World Conference on Sea Turtle Conservation. Smithsonian Institution Press, Washington DC.

Frazier, J. G. 1984. Contemporary problems in sea turtle biology and conservation. Central Marine Fisheries Research Institute Spec. Publ. No. 18:77-91.

Frazier, J. G. 1997. Sustainable development: modern elixir or sack dress? Environmental Conserv. 24(2):182-193. [Author explores the meaning of the term "sustainable development" and pleads for a specific definition of its meaning. This paper deserves to be read carefully.]

Frazier, J. G. 2003. Why do we do this? Marine Turtle Newsletter 100:9-15 [perceptive account of sea turtle biologists and their commitment to research and conservation].

Heyer, W. R., and J. B. Murphy. 2005. Declining Amphibian Population Task Force, pp. 17–21. In M. Lannoo (ed.), Amphibian Declines. The Conservation Status of United States Species. Univ. California Press, Berkeley and Los Angeles.

Kleiman, D. G. 1985. Criteria for the evaluation of zoo research projects. Zoo Biol. 4:93-98.

Kleiman, D. G., R. P. Reading, and F. Felleman. 2000. Improving the value of conservation projects. Conserv. Biol. 14:356-365.

Kleiman, D. G., M. R. Stanley Price, and B. B. Beck. 1984. Criteria for reintroductions, p. 287-303. Creative Conservation: Interactive Management of Wild and Captive Animals. Chapman and Hall, London.

Ralls, K. 1997. On becoming a conservation biologist: Autobiography and advice, p. 356-372. In J. R. Clemmons, and R. Buchholz (eds.), Behavioral Approaches to Conservation in the Wild. Cambridge University Press, Cambridge UK [author employed at Zoo].

Robinson, Michael and others. 1992. Conserving Biodiversity. A Research Agenda for Development Agencies. Report of a Panel of the Board on Science and Technology for International Development. US National Research Council. National Academy Press, Washington DC.

Seal, U. S. 1985. The realities of preserving species in captivity, p. 71-96. In R. J. Hoage (ed.), Animal Extinctions. Smithsonian Institution Press, Washington DC. [Robert Hoage has edited or authored a number of books on zoo history, conservation, and animals and human cultures. He was employed at Zoo as Director of Public Affairs until retirement.]

Captive Management

Ballou, J. D. 1984. Strategies for maintaining genetic diversity in captive populations through reproductive technology. Zoo Biol. 3:311-323. [Jonathan Ballou is Population Manager at Zoo.].

Ballou, J. D. 2001. Genetic and demographic considerations in endangered species captive breeding and reintroduction programs, p. 262-275. In Wildlife Populations. Elsevier Science Publications, Barking UK.

Ballou, J. D., and T. J. Foose. 1996. Demographic and genetic management of captive populations, p. 263-283. In D. G. Kleiman, M. E. Allen, K. V. Thompson, S. Lumpkin, and H. Harris (eds.), Wild Mammals in Captivity. University of Chicago Press, Chicago.

Ballou, J. D., and K. Ralls. 1982. Inbreeding and juvenile mortality in small populations of ungulates: a detailed analysis. Biol. Conserv. 24:239-272.

Ballou, J. D., M. Gilpin, and T. J. Foose (eds.). c. 1995. Population Management for Survival and Recovery. Columbia Univ. Press, New York.

Foose, T. J., and J. Ballou. 1988. Population management: Theory and practice. Inter. Zoo Yearb. 27:26-41. [Provides excellent overview of aspects to be considered by zoo workers relative to animal management.]

Haig, S., J. Ballou, and S. Derrickson. 1990. Management options for preserving genetic diversity: reintroduction of Guam rails to the wild. Conserv. Biol. 4:290-300.

Montgomery, M. E., J. D. Ballou, R. K. Nurthen, P. R. England, D. A. Briscoe, and R. Frankham. 1997. Minimizing kinship in captive breeding programs. Zoo Biol. 16:377-389.

Ralls, K., K. Brugger, and J. Ballou. 1979. Inbreeding and juvenile mortality in small populations of ungulates. Science 206:1101-1103. [This seminal paper alerted zoo workers about the dangers of inbreeding captive populations.]

Ryder, O. A., and R. C. Fleischer. 1996. Genetic research and its application in zoos, p. 255-262. In D. G. Kleiman, M. E. Allen, K. V. Thompson, S. Lumpkin, and H. Harris (eds.), Wild Mammals in Captivity. University of Chicago Press, Chicago.

Education

Doering, Z. 1994. From reptile houses to reptile discovery centers. Institutional Studies Smithsonian Institution Report 94-4.

Doolittle, R. L., and T. I. Grand. 1995. Benefits of the zoological park to the teaching of comparative vertebrate anatomy. Zoo Biol. 14:453-462. [Ted Grand is research associate at Zoo.]

Marcellini, D. L., and T. A. Jenssen. 1988. Visitor behavior in the National Zoo's Reptile House. Zoo Biol. 7:329-338. [Seminal study showed that zoo visitors spend an astonishingly brief time viewing exhibits.]

Marcellini, D. L., and J. B. Murphy. 1998. Education in a zoological park or aquarium: An ontogeny of learning opportunities. Herpetologica 54 (Suppl.):S12-S16.

Robinson, M. H. 1995. Zoo and aquarium messages, meanings and contexts, p. 1-24. In C. M. Wemmer (ed.), The Ark Evolving. Zoos and Aquariums in Transition. Smithsonian Institution, Conservation and Research Center, Front Royal VA.

Rowan, A., and R. Hoage. 1995. Public attitudes toward wildlife: the awakening awareness, p. 32-60. In C. M. Wemmer (ed.), The Ark Evolving. Zoos and Aquariums in Transition. Smithsonian Institution, Conservation and Research Center, Front Royal VA.

Wemmer, C., C. Pickett, and J. A. Teare. 1990. Training zoo biology in tropical countries: A report on the method and progress. Zoo Biol. 9:461-470. [Zoo herpetologists are participants in the various training courses which are hosted by zoos in developing countries throughout the world. The program was developed by staff at the Zoo in 1987.]

White, J., and S. Barry. 1984. Families, Frogs, and Fun. Developing a Family Learning Lab in a Zoo. HERPlab: A Case Study. Smithsonian Institution, Washington DC.

White, J., and D. L. Marcellini. 1986. HERPlab: a family learning centre at National Zoological Park. Inter. Zoo Yearb. 24/25:340-343.

General

Avery, J., A. Shafagati, A. Turner, J. W. Wheeler, and P. J. Weldon. 1993. Beta-springene in the paracloacal gland secretions of smooth-fronted caimans (*Paleosuchus trigonatus*). Biochem. Syst. Evol. 21:533-534.

Banta, M. R., T. Joanen, and P. J. Weldon. 1993. Foraging responses by the American alligator to meat scents, p. 413-417. Chemical Signals in Vertebrates VI. Plenum Press, New York.

Block, J. A. 1982. Rules and regulations: Another perspective, p. 76-87. In D. L. Marcellini (ed.), 6th International Herpetological Symposium on Captive Propagation and Husbandry. International Herpetological Symposium, Thurmont MD. [Judith Block was Registrar at the Zoo until retirement. She is a founding member of the Zoo Registrars Association, a group of over 80 members throughout the world, who have met annually since 1985 to discuss zoo philosophy, improved animal record keeping, collection policy and planning, permit procedures, information retrieval and animal management issues.]

Brooks, D. R. 1979. Testing hypotheses of evolutionary relationships among parasites: The digeneans of crocodilians. Amer. Zool. 19:1225-1238. ["The digeneans of crocodilians apparently have co-evolved as a faunal unit with their host group, at least since the end of the Mesozoic."]

Dunn, B. S., P. J. Weldon, R. Howard, and C. A. McDaniel. 1993. Lipids in the paracloacal gland secretions of the Chinese alligator (*Alligator sinensis*). Lipids 28:75-78.

Eisenberg, J. F. 1975. Design and administration of zoological research programs, p. 12-18. Research in Zoos and Aquariums. National Academy of Sciences, Washington DC. [John Eisenberg was employed at Zoo.]

Frazier, J. G. 1971. Observations on sea turtles at Aldabra Atoll. Phil. Trans. R. Soc. London B. 260:373-410.

Frazier, J. G. 1981. Recaptures of marine turtles tagged in East Africa: evidence for a non-migratory green turtle population? Afr. J. Ecol. 19:369-372.

Frazier, J. G. 1982. Tortugas marinas en Chile (Marine turtles of Chile). Bol. Mus. Nac. Hist. Nat. Chile (Bulletin of the National Natural History Museum, Chile) 39:63-73.

Frazier, J. G. 1984. Las tortugas marinas en el oceano Atlantico Sur Occidental (The marine turtles of the South-West Atlantic Ocean). Asociacion Herpetologica Argentina Serie Divulgacion (Herpetological Association Argentina Publishing Series) No. 2:3-21.

Frazier, J. G. 1984. Marine turtles in the Seychelles and adjacent territories, p. 417-468. In D. R. Stoddart (ed.), Biogeography and Ecology of the Seychelles Islands. Dr W Junk Publishers, The Hague.

Frazier, J. G. 1985. Marine Turtles in the Comoro Archipelago. Proc. Koninkl. Nederl. Akademie van Wetenschappen Amsterdam [review by C. K. Dodd Jr. in Herpetologica 1987, vol. 43:129-131].

Frazier, J. G., and S. Salas. 1984. The status of marine turtles in the Egyptian Red Sea. Biol. Conserv. 30:41-67.

Hutchins, M., J. B. Murphy, and N. Schlager (eds.). 2003. Grzimek's Animal Life Encyclopedia, 2nd Edition, Volume 7, Reptiles. Gale Group, Farmington Hills MI.

Jenssen, T. A., D. L. Marcellini, and E. P. Smith. 1988. Seasonal micro-distribution of sympatric *Anolis* lizards in Haiti. J. Herpetol. 22:266-274.

Marcellini, D. L. 1979. Activity patterns and densities of Venezuelan caiman (*Caiman crocodilus*) and pond turtles (*Podocnemis vogli*), p. 263-271. In J. Eisenberg (ed.), Vertebrate Ecology in the Neotropics. Smithsonian Institution Press, Washington DC.

Marcellini, D. L. 1980. Records no longer a luxury, p. 3-14. In R. A. Hahn (ser. ed.), 3rd International Herpetological Symposium on Captive Propagation and Husbandry. International Herpetological Symposium, Thurmont MD.

Marcellini, D. L. 1994. Collection-management, captive-breeding, and conservation programs in zoo herpetological collections, p. 397-400. In J. B. Murphy, K. Adler, and J. T. Collins (eds.), Captive Management and Conservation of Amphibians and Reptiles. Society for the Study of Amphibians and Reptiles. Contributions to Herpetology, volume 11, Ithaca NY.

Marcellini, D., T. Jenssen, and C. Paque. 1985. Distribution of the lizard *Anolis cooki*, with comments on possible future extinctions. Herpetol. Rev. 16:99-102.

Murphy, J. B., and W. Card. 1998. A glimpse into the life of a zoo herpetologist. Herpetol. Rev. 29(2):85-90.

Weldon, P. J. 1996. Thin-layer chromatography of the skin secretions of vertebrates, p. 105-130. In B. Fried, and J. Sherma (eds.), Practical Thin-Layer Chromatography: A Multidisciplinary Approach. CRC Press, Boca Raton FL.

Weldon, P. J., and M. W. J. Ferguson. 1993. Chemoreception in crocodilians: anatomy, natural history, and empirical observations. Brain Behav. Evol. 41:239-245.

Weldon, P. J., and T. L. Leto. 1995. A comparative analysis of proteins in the scent gland secretions of snakes. J. Herpetol. 29:406-408.

Weldon, P. J., W. G. Brinkmeier, and H. Fortunato. 1992. Gular pumping responses by immature American alligators (*Alligator mississippiensis*) to meat scents. Chemical Senses 17:79-83.

Weldon, P. J., B. J. Demeter, and R. Rosscoe. 1993. A survey of shed skin-eating (dermatophagy) in amphibians and reptiles. J. Herpetol. 27:219-228.

Weldon, P. J., R. Ortiz, and T. R. Sharp. 1992. The chemical ecology of crotaline snakes, p. 309-319. Biology of the Pitvipers. Selva Press, Tyler TX.

Wemmer, C., and S. Thompson. 1995. A short history of scientific research in zoological gardens, p. 70-94. In C. M. Wemmer (ed.), The Ark Evolving. Zoos and Aquariums in Transition. Smithsonian Institution, Conservation and Research Center, Front Royal VA.

Wemmer, C., M. Rodden, and C. Pickett. 1997. Publication trends in Zoo Biology: A brief analysis of the first 15 years. Zoo Biol. 16:3-8. [Research articles on reptiles comprised only 7% of total papers published.]

Wikramanayake, E. D., and G. L. Dryden. 1988. The reproductive ecology of *Varanus indicus* on Guam. Herpetologica 44:338-344.

Wood, T. C., J. M. Parker, and P. J. Weldon. 1995. Volatile components in scent gland secretions of garter snakes. J. Chem. Ecol. 21:213-219.

Wildlife Conseervation Society/Bronx Zoo (1899)

History and Mission: The New York Zoological Society, now the Wildlife Conservation Society headquartered at the Bronx Zoo, has one of the richest traditions of any zoological garden in the United States. The conservation program, global in scope, is without peer. The Society supports a number of field researchers engaged in studies focused on conservation and habitat protection, spearheaded by the recently retired Director, William G. Conway. When one visits this zoo, some of the finest naturalistic exhibits ever developed in any institution, may be enjoyed. The Society oversees the Aquarium for Wildlife Conservation, Central Park Wildlife Center and several other zoo operations in the vicinity.

Facility and Collection: The Reptile Building was opened on 8 November 1899 and some of the original collection was donated by R. L. Ditmars. He also traveled to Florida to bring other reptiles to the Zoo. In 1952, the building was renovated. The ground floor consisted of the public area and exhibits and a second level (mezzanine) was constructed above as a working area. The collection has always been diverse. In the 1920s, there was a major effort to protect Galápagos

Fig. 124. Reptile house at New York Zoological Society's Bronx Zoo in 1909. *Photograph courtesy of Wildlife Conservation Society, headquartered at Bronx Zoo.*

Fig. 125. Interior of Reptile house at New York Zoological Society's Bronx Zoo in 1909. *Photograph courtesy of Wildlife Conservation Society, headquartered at Bronx Zoo.*

Fig. 126. Interior of Reptile house at New York Zoological Society's Bronx Zoo in 1909. *Photograph courtesy of Wildlife Conservation Society, headquartered at Bronx Zoo.*

tortoises and secure specimens for distribution to other zoos. Since there are large crocodilian enclosures, significant research in the reproductive biology and conservation of these reptiles has been achieved. Of particular importance was the ongoing propagation program at the Rockefeller Wildlife Refuge in Louisiana with the endangered Chinese alligator. The facility closed and the crocodilians were moved to other facilities. Many tentacled snakes have been produced and sent to many zoos. Efforts are underway to repro-

duce Batagurine turtles and other aquatic chelonians consistently. At the St. Catherine Island facility, now closed, radiated tortoises regularly bred. The Zoo, in concert with the Detroit Zoological Institute, has been involved in saving the endangered Kihansi spray toad (*Nectophrynoides asperginus*) from the Kihansi Gorge in Tanzania, at risk due to a hydroelectric project which diverted the water supply.

According to *International Zoo Yearbook* (1998:Volume 28), composition of the collection was 115 taxa of reptiles with 627 specimens and 22 taxa of amphibians numbering 158. St. Catherine's Wildlife Survival Center houses five species with over 100 individuals. At the Central Park Wildlife Center, the numbers are: Reptiles, 33 taxa, over 600 individuals; Amphibians, 17 taxa, 264 individuals.

Staff and Scientific Achievements: The professional staff has been first-rate. William Beebe, although most often associated with the Society's Tropical Research Station in South America, published a number of papers in herpetology. C. M. Breder Jr., associated with the New York Aquarium, published extensively on herps as well as fishes. Rarely is a zoo blessed with many prominent curators but this zoo enjoys a worldwide reputation in herpetology because of the writings and diverse backgrounds of five men, in concert with their associates: Raymond L. Ditmars, James A. Oliver, Herndon G. Dowling, F. Wayne King, and John L. Behler. Ditmars (1876-1942) was the first curator. Now retired, Herndon Dowling specialized in systematics and taxonomy, but also published on a variety of other herpetological topics. Early studies included the black swamp snake, nomenclature of the common gartersnake and the ribbonsnake, counting scale characters in snakes, many taxonomic treatments of the ratsnakes, and Pleistocene snakes of the Ozarkian Plateau. His successor, F. Wayne King, published papers on systematics, ecology and natural history, and became a major supporter of crocodilian conservation initiatives. His early papers included studies on the ecology of the Mediterranean gecko, new populations of West Indian reptiles and amphibians in southeastern Florida, occurrence of rafts for dispersal of land animals into the West Indies, Palaeolithic reptile and amphibian remains, vertebra duplication, and several papers on geckos of the genera *Sphaerodactylus* and *Cyrtodactylus*. The late John Behler was involved with chelonian conservation and field projects, and was one of the editors of the superb journal *Chelonian Conservation and Biology*. Showing great staying power but now retired as Curator of the Central Park

Fig. 127. Late curator of herpetology John L. Behler at Delaware Water Gap National Forest in November 2003 during multi-year study of wood turtles. See Herpetol. Rev. 37(2):137, 2006. *Courtesy of Bill Holmstrom.*

Wildlife Center, Peter Brazaitis, currently a Research Associate with the WCS's Science Resource Center, worked as a keeper and as Superintendent of Herpetology for the last four curators over many years but also is known as a deeply committed crocodilian conservationist and forensic specialist in herpetology. John Thorbjarnarson has undertaken a number of conservation initiatives with crocodilians as well. George Amato has investigated crocodilian genetics. Behler's assistant, William Holmstrom, has been involved in a long-term field project studying the ecology and natural history of anacondas. The current curator is Jennifer Pramuk.

Publications: The Zoo has supported two important publications. The first, established in 1907, was the scientific journal *Zoologica* which ceased publication in 1973. The newsletter *Animal Kingdom* includes a number of popular articles dealing with herpetological topics. Over a decade ago, the name was changed to *Wildlife Conservation. Curator's Report* is a newsletter outlining current initiatives, mostly in the conservation arena. In the annual reports, R. L. Ditmars published many observations on the amphibian and reptile collection.

Personal Reflections: When I toured the Zoo in 1969, Curator King reached into an aquarium and pulled out the first living Borneo earless monitor seen by me and placed it gently in my hands. He said that this lizard refused food for many months until live goldfish were placed in a water dish.

Historical Overview

Beetem, D. D., and J. F. Iaderosa. 1994. St. Catherine's Island Wildlife Survival Center: A twenty-one year commitment in conservation. AAZPA Reg. Conf. Proc. 1994:69-75.

Brazaitis, P. 2003. You Belong in a Zoo. Tales from a Lifetime Spent with Cobras, Crocs, and Other Creatures. Villard Books, New York.

Burden, W. D. 1956. Look to the Wilderness. Little, Brown, Boston, Toronto [description of collecting trip to collect Komodo dragons for Zoo in 1926].

Card, W., and J. B. Murphy. 2000. Lineages and histories of zoo herpetologists in the United States. Herpetol. Circ. 27:1-45 [portraits of Brayton Eddy, Raymond Ditmars, James Oliver, Herndon Dowling, F. Wayne King and Peter Brazaitis].

Cherfas, J. 1984. Zoo 2000. A Look Behind the Bars. British Broadcasting Corporation, London. [portrait of breeding facility at St. Catherine's Island which contains breeding groups of Madagascan radiated and angulated tortoises.].

Dowling, H. G. 1960. The Reptile House of the Bronx Zoo. Bull. Philadelphia Herpetol. Soc. 8(1):19-22.

Keeling, C. 2002. Skyscrapers and Sealions. Clam Publications, Guilford UK [history of Central Park Menagerie which evolved to become the Central Park Zoological Garden (or Moses)].

Oliver, J. A. 1954. The most beautiful reptile house in the world. Anim. Kingdom 54(4):98-109 [photographs.].

Morphology, Taxonomy & Systematics

Beebe, W. 1947. Snake skins and color. Copeia 1947:205-206 [preparation of skins].

Brazaitis, P. 1971. News and notes. *Crocodylus intermedius* Graves, a review of the recent literature. Zoologica (New York) 56:71-75.

Brazaitis, P. 1973. The identification of *Crocodylus siamensis* Schneider. Zoologica (New York) 58:43.

Brazaitis, P. 1981. Maxillary regeneration in a marsh crocodile, *Crocodylus palustris.* J. Herpetol. 15:360-362.

Dowling, H. G. 1959. The spotted racer, *Coluber anthricus* Cope, in Arkansas. Southwest. Natur. 4:40.

Dowling, H. G. 1959. Apical papillae on the hemipenes of two colubrid snakes. Amer. Mus. Novitates (No. 1948):1-17.

Dowling, H. G. 1959. Classification of the Serpentes: A critical review. Copeia 1959:38-52.

Dowling, H. G. 1960. A taxonomic study of the rat snakes, genus *Elaphe* Fitzinger. VII. The Triaspis section. Zoologica (New York) 45: 53-80.

Dowling, H. G. 1967. Hemipenes and other characters in colubrid classification. In Symposium on colubrid snake systematics held at Miami Beach in June 1966. Herpetologica 23:138-142.

Dowling, H. G. 1968. The karyotype of the Chinese alligator (*Alligator sinensis*). Mamm. Chromosomes Newsl. 9:81-82.

Dowling, H. G., and J. M. Savage. 1960. A guide to the snake hemipenis: A survey of basic structure and systematic characteristics. Zoologica (New York) 45:17-28.

King, F. W. 1978. A new Bornean lizard of the genus *Harpesaurus*. Sarawak Mus. J. 26(No. 47 (new series):205-209 [description of *Harpesaurus thescelorhinos*].

King, F. W., and P. Brazaitis. 1971. Species identification of commercial crocodilian skins. Zoologica (New York) 56:15-70.

King, W., and F. G. Thompson. 1968. A review of the American lizards of the genus *Xenosaurus* Peters. Bull. Florida State Mus. 12:93-123 [description of new subspecies *Xenosaurus grandis agrenon*; new species *X. platyceps*].

Stejneger, L. 1906. A new lizard of the genus *Phrynosoma,* from Mexico. Proc. U. S. National Mus. 29(1437):565-567 [description of *P. ditmarsi*].

Strimple, P. D., G. Puorto, W. F. Holmstrom, R. W. Henderson, and R. Conant. 1997. On the status of the anaconda *Eunectes barbouri* Dunn and Conant. J. Herpetol. 31:607-609. [*Eunectes barbouri* is placed in synonymy with *E. murinus*.]

Thorbjarnarson, J. B., and P. E. McIntosh. 1987. Notes on a large *Malanosuchus niger* skull from Bolivia. Herpetol. Rev. 18(3):49-50.

Reproduction

Behler, J. L. 1980. The problem of success in reptile propagation. AAZPA Annual Proc. 1980:135-141.

Behler, J. L., P. Brazaitis, K. Gerety, and B. Foster. 1987. Propagation of crocodilians at the Bronx Zoo. p. 63-73. In 11th International Herpetological Symposium on Captive Propagation and Husbandry. Chicago: International Herpetological Symposium.

Conway, W. July 1949. Experiment in hatching reptile eggs. St. Louis Zool. Soc. Bull. :6. [Account is the first herpetological paper by author who wanted his life's work to be herpetology.]

Ditmars, R. L. 1913. Two litters of the Fer-de-Lance. Bull. New York Zool. Soc. 16:957. [Female collected in Trinidad gave birth to 57 young.]

Thorbjarnarson, J., X. Wang, and L. He. 2001. Reproductive ecology of the Chinese alligator (*Alligator sinensis*) and implications for conservation. J. Herpetol. 35:553-558 [data on wild population].

Behavior

Chiszar, D., C. W. Radcliffe, T. Boyer, and J. L. Behler. 1987. Cover-seeking behavior in red spitting cobras (*Naja mossambica pallida*): Effects of tactile cues and darkness. Zoo Biol. 6:161-167. [Clear snake hiding boxes are recommended.]

Ditmars, R. L. 1913. Ferocity of the king cobra. Bull. New York Zool. Soc. 16:957. [Keepers used a small shovel or broom to control the snake when it attacked. A day or two prior to feeding, the snake reared up at the door of the enclosure, observing the keepers through a small opening covered with wire.]

Dowling, H. G. 1959. Egg-eating adaptations in the Chinese ratsnake, *Elaphe carinata* Guenther. Copeia 1959:68-69.

Books by Raymond L. Ditmars

Baskett, J. N. 1902. The Story of the Amphibians and the Reptiles, by James Newton Baskett and Raymond L. Ditmars. D. Appleton and Co., New York. [In Appleton's home reading books, Division 1: Natural History. Baskett wrote chapters 1 and 2 which were general accounts of reptiles and amphibians. Ditmars wrote chapter 3 entitled "A collector's experience with reptiles," which included portraits of his experiences with captive reptiles at the Zoo and various collecting adventures.]

Ditmars, R. L. 1910. Reptiles of the World; The Crocodilians, Lizards, Snakes, Turtles and Tortoises of the Eastern and Western Hemispheres, by Raymond L. Ditmars. With a Frontispiece in Color, and Nearly 200 Illustrations, From Photographs Taken by the Author. Sturgis and Walton, New York.

Ditmars, R. L. 1920. The Reptile Book; A Comprehensive, Popularised Work on the Structure and Habits of the Turtles, Tortoises, Crocodilians, Lizards and Snakes Which Inhabit the United States and Northern Mexico. Doubleday, Page, New York.

[first edition, 1907].

Ditmars, R. L. 1928. Reptiles of the World: Tortoises and Turtles, Crocodilians, Lizards, Snakes, Turtles and Tortoises of the Eastern and Western Hemispheres, by Raymond L. Ditmars . . . With a Frontispiece and nearly 200 Illustrations, from Photographs Taken by the Author. Macmillan, New York.

Ditmars, R. L. 1931. Snakes of the World, by Raymond L. Ditmars . . . with Illustrations From Life. Macmillan, New York. [Reprinted 1946. This book contained numerous personal anecdotes and observations with 84 black & white plates. This book included sections on the serpent world, scope of the clan, distribution and general classification.]

Ditmars, R. L. 1931. Strange Animals I Have Known [by] Raymond L. Ditmars, Curator, New York Zoological Park. Brewer, Warren & Putnam, New York. [Ditmars described experiences with an array of mammals and reptiles including a description of fellow scientists involved in systematics " . . . watching like hawks" for the slightest slip by a colleague in order to toss his work on to the list of synonyms. Ditmars suffered the same fate. His description of *Crotalus pulvis* from near Managua, Nicaragua in 1905 languishes in synonymy as it was an albino *C. durissus* (Amaral, 1929) or more likely *C. unicolor* (Klauber, 1956).]

Ditmars, R. L. 1932. Thrills of a Naturalist's Quest. Macmillan, New York. [Reprint edition done by Halcyon House, Garden City NY in 1947. Description of Ditmars transporting a king cobra on a train sleeping car to go to National Zoological Park for opening of new reptile building in 1931. The snake was requested by Director William M. Mann but train officials were outraged that Ditmars supposedly placed fellow passengers "at risk." Also, there was an account of an aggressive bushmaster being unpacked from a shipping crate. He described the snake as advancing toward him which necessitated using a broom to fend off the serpent. In our experience at the Dallas Zoo, all of our adult bushmasters including breeders, were tractable and placid. Many of the photographs used by Ditmars were from the collection at the Zoo and included a variety of reptiles and mammals including Cuban Solenodon.]

Ditmars, R. L. 1933. Reptiles of the World; The Crocodilians, Lizards, Snakes, Turtles and Tortoises of the Eastern and Western Hemispheres. Macmillan, New York [new, revised edition with photographs by Ditmars and from Zoo files].

Ditmars, R. L. 1933. The Forest of Adventure. Macmillan, New York.

Ditmars, R. L. 1935. Serpents of the Northeastern States; A Guide to the Venomous and Non-Venomous Species of the North Atlantic and New England Areas. New York Zoological Society, New York. [This was first published in the Bulletin of New York Zoological Society of May-June 1929 as *Serpents of the eastern states*.]

Ditmars, R. L. 1936. The Book of Living Reptiles, by Raymond L. Ditmars, Illustrated by Helene Carter; Where the Crocodilians, Lizards, Snakes, Turtles and Tortoises Are Found. J. B. Lippincott, Philadelphia, London.

Ditmars, R. L. 1936. The Reptiles of North America;, A Review of the Crocodilians, Lizards, Snakes, Turtles and Tortoises Inhabiting the United States and Northern Mexico. Doubleday, Garden City NY. [Formerly published as *The Reptile Book*,this is the new and revised edition. In the Introduction, Ditmars said that this work follows a scientific framework but has been simplified to assist the beginner and student of reptile life.]

Ditmars, R. L. 1936. Reptiles of the World: The Crocodilians, Lizards, Snakes, Turtles and Tortoises of the Eastern and Western Hemispheres / by Raymond L. Ditmars. Macmillan, New York.

Ditmars, R. L. 1937. The Making of a Scientist. Macmillan, New York.

Ditmars, R. L. 1938. The Fight to Live. J. B. Lippincott, Philadelphia, New York.

Ditmars, R. L. 1939. A Field Book of North American Snakes. Doubleday, Doran, New York.

Ditmars, R. L. 1939. La lutte pour la vie dans monde animal . . . Avec 46 photgraphies hors texte. Payot, Paris, France [in the series Epistemonike vivliotheke].

Ditmars, R. L. 1939. The Reptile Book: A Comprehensive, Popularised Work on the Structure and Habits of the Turtles, Tortoises, Crocodilians, Lizards and Snakes Which Inhabit the United States and Northern Mexico. Doubleday, Doran, New York.

Ditmars, R. L. 1951. The Reptiles of North America. Doubleday, Garden City NY.

Ditmars, R. L., and L. S. Crandall. 1939. Guide to the New York Zoological Park. New York Zoological Society, New York.

Ditmars, R. L., and L. S. Crandall. 1951. Guide to the New York Zoological Park / by Raymond L. Ditmars and Lee S. Crandall. New York Zoological Society, New York [6th edition].

Lamson, G. H. 1935. The Reptiles of Connecticut, by George Herbert Lamson. Geological and Natural

History Survey, Hartford CT. [Ditmars was commissioned to write the section on turtles and "complete the paper for publication." The series was Bulletin (State Geological and Natural History Survey of Connecticut) No. 54.].

Medical Management

Brazaitis, P., and M. E. Watanabe. 1982. The Doppler, a new tool for reptile and amphibian hematological studies. J. Herpetol. 16:1-6.

Camin, J. H., G. K. Clarke, L. H. Goodson, and H. R. Shuyer. 1964. Control of the snake mite, *Ophionyssus natricis* (Gervais) in captive reptile collections. Zoologica (New York) 49:65-79.

Dolensek, E. P., and B. Burn. 1978. The Penguin Book of Pets. A Practical Guide to Animal-Keeping. Penguin Books, New York [chapter on care of amphibians and reptiles].

Dolensek, E. P., and R. A. Cook. 1987. Clinical challenge. J. Zoo Anim. Med. 18:168-170. [Emil Dolensek was Chief Veterinarian at the Zoo.]

Dolensek, E. P., and R. L. Napolitano. 1977. Amoebic gastroenteritis in reptiles. Proc. Amer. Assoc. Zoo Vet. 1977:66-67.

Goss, L. J. 1940. Report of the Hospital and Laboratory of the New York Zoological Park, 1939. Mortality statistics of the Society's collection. Zoologica (New York) 25:269-279 [data from 135 reptiles collected during necropsies].

Herman, C. M. 1939. Parasites obtained from animals in the collection of the New York Zoological Park during 1938. Zoologica (New York) 24:481-485 [parasites from 20 reptiles collected during 103 necropsies].

Jakowska, S., R. F. Nigrelli, and A. H. Sparrow. 1958. Radiobiology of the newt, *Diemictylus viridescens*. Hematological and histological effects of whole-body X irradiation. Zoologica (New York) 43:155-160 [contribution, in part, by Zoo].

Napolitano, R.L., E. P. Dolensek, and J. L. Behler. 1979. Reptilian amoebiasis. Inter. Zoo Yearb. 19:126-131.

Nigrelli, R. F. 1940. Mortality statistics for specimens in the New York Aquarium, 1939. Zoologica (New York) 25:525-552. [General enteritis caused the greatest number of deaths in reptiles.]

Nigrelli, R. F., and H. Vogel. 1963. Spontaneous tuberculosis in fishes and other cold-blooded vertebrates with special reference to *Mycobacterium fortuitum* Cruz from fish and human lesions. Zoologica (New York) 48:131-143. [Study included references for 11 amphibians and 24 reptiles by authors from New York Aquarium.]

Raphael, B. L., P. P. Calle, M. S. Stetter, B. Mangold, and R. Cook. 1995. Normal variations in selected plasma biochemicals in reptiles. Proc. Amer. Assoc. Zoo Vet. 1995:233-235.

Schairer, M. L., E. S. Dierenfeld, and M. P. Fitzpatrick. 1998. Nutrient composition of whole green frogs, *Rana clamitans* and southern toads, *Bufo terrestris*. Bull. Assoc. Rept. Amphib. Vet. 8(3):17-20.

Schroeder, C. R. 1939. Report of the Hospital and Laboratory of the New York Zoological Park, 1938. Mortality statistics of the Society's collection. Zoologica (New York) 24:265-288 [data from reptiles and one amphibian collected during necropsies].

Stetter, M. D., B. Raphael, F. Indiviglio, and R. A. Cook. 1996. Isoflurane anesthesia in amphibians: Comparison of five application methods. Proc. Amer. Assoc. Zoo Vet. 1996:255-257.

Stunkard, H. W., and C. P. Gandal. 1961. Nematodes and cestodes from the Australian monitor, *Varanus gouldii*. Zoologica (New York) 46:161-166. [Horace Stunkard was a Research Associate at the American Museum of Natural History and Charles Gandal was Zoo veterinarian.]

Stunkard, H. W., and C. P. Gandal. 1966. A digenetic trematode, *Parahaplometroides basiliscae* Thatcher, 1963, from the mouth of the crested lizard, *Basiliscus basiliscus*. Zoologica (New York) 51:91-95 [data from Zoo].

Conservation

Amato, G. D., J. Gatesy, and P. Brazaitis. 1998. PCR assays of variable nucleotide sites for identification of conservation units: Examples from caiman, p. 177-189. In R. DeSalle, and B. Schierwater (eds.), Molecular Evolution: Approaches and Applications, 2nd edition. Birkhauser, Basel, Switzerland.

Behler, J. 1978. Feasibility of the establishment of a captive-breeding population of the American crocodile. National Park Service Report T-509, 94 pp.

Brazaitis, P. 1984. The U.S. trade in crocodilian hides and products, a current perspective, p. 103-107. Proc. 6th Working Meet., IUCN/SSC Crocodile Specialist Group, Victoria Falls, Zimbabwe and St. Lucia Estuary, Repub. South Africa, Sept. 19-30, 1982.

Brazaitis, P. 1984. Problems in the identification of commercial crocodilian hides and products, and the effect upon law enforcement, p. 110-116. Proc. 6th Working Meet., IUCN/SSC Crocodile Specialist Group, Victoria Falls, Zimbabwe and St. Lucia Estuary, Repub. South Africa, Sept. 19-30, 1982.

Brazaitis, P. 1986. An assessment of the current crocodilian hide and product market in the United States, p. 370-383. Proc. 7th Working Meet., IUCN/SSC Crocodile Specialist Group, Caracas, Venezuela Oct. 21-28, 1984.

Brazaitis, P. 1986. Reptile leather trade: The forensic science examiner's role in litigation and wildlife law enforcement. J. Forensic Sci. 31:621-629.

Brazaitis, P. 1986. Biochemical techniques: New tools for the forensic identification of crocodilian hides and products, p. 384-388. Proc. 7th Working Meet., IUCN/SSC Crocodile Specialist Group, Caracas, Venezuela Oct. 21-28, 1984.

Brazaitis, P. 1987. Identification of crocodilian skins and products, p. 373-386. In G. J. W. Webb, S. C. Manolis, and P. J. Whitehead (eds.), Wildlife Management. Crocodiles and Alligators. Surrey, Beatty & Sons, Chipping North, NSW, Australia.

Brazaitis, P. 1989. The caiman of the Patanal, past, present and future, p. 119-124. Crocodiles: Proc. 8th Working Group Meeting of the IUCN/SSC Crocodile Specialist Group, Quito, Ecuador. IUCN Publ. N.S., Gland, Switzerland.

Brazaitis, P. 1989. The trade in crocodilians, p. 196-201. Crocodiles and Alligators. Golden Press, Silverwater, NSW, Australia.

Brazaitis, P. 1990. Trade in crocodilian hides and products in the USA. Traffic USA 10(2):4-5.

Brazaitis, P. 1996. The status of *Caiman crocodilus crocodilus* and *Melanosuchus niger* populations in the Amazonian regions of Brazil. Amphibia-Reptilia 17:377-385.

Brazaitis, P., and T. Joanen. 1984. Report on the status of the captive breeding program for the Chinese alligator *Alligator sinensis* in the United States, p. 117-121. Proc. 6th Working Meet., IUCN/SSC Crocodile Specialist Group, Victoria Falls, Zimbabwe and St. Lucia Estuary, Repub. South Africa, Sept. 19-30, 1982.

Brazaitis, P., and M. Wise. 1994. Conservation of commercially important reptiles today: An analysis based on crocodilians, p. 209-221. In J. B. Murphy, K. Adler, and J. T. Collins (eds.), Captive Management and Conservation of Amphibians and Reptiles. Society for the Study of Amphibians and Reptiles. Contributions to Herpetology, volume 11, Ithaca NY.

Brazaitis, P., G. H. Rêbelo, and C. Yamashita. 1992. Report of the WWF/TRAFFIC USA Survey of Brazilian Amazonian Crocodilians Survey Period July 1988-January 1992, p. 207-220. WWF/TRAFFIC USA, Washington DC (Special Report).

Brazaitis, P., M. Watanabe, and G. Amato. 1998 The caiman trade. Sci. Amer. 278(3):53-58.

Brazaitis, P., C. Yamashita, C., and G. Rebêlo. 1990. A Summary Report of the CITES central South American caiman study: Phase I: Brazil, p. 100-115. Crocodiles: Proc. 9th Working Group Meeting of the Crocodile Specialist Group, Lae, Papua-New Guinea. The World Conservation Union Publ. N.S., Gland, Switzerland.

Brazaitis, P., G. Amato, G. H. Rêbelo, C. Yamashita, and J. Gatesy. 1993. Report to CITES on the Biochemical Systematics Study of Yacare Caiman, *Caiman yacare* of Central South America, Office of CITES Secretariat, Lausanne, Switzerland (Special Report).

Brazaitis, P., G. H. Rebêlo, C. Yamashita, E. A. Odeira, and M. E. Watanabe. 1996. Threats to Brazilian crocodilian populations. Oryx 30:275-284.

Collete, B. B., and F. W. King. 1973. Endangered wildlife. Copeia 1973:390-392.

Conway, W. 1985. The Species Survival Plan and the Conference on Reproductive Strategies for Endangered Wildlife. Zoo Biol. 4:219-223. [AZA-approved Plans are Chinese alligator, Cuban crocodile, Dumeril's ground boa, rock iguana, Virgin Island boa, Wyoming toad, Puerto Rican crested toad, radiated tortoise and Aruba Island rattlesnake.]

Conway, W. G. 1986. The practical difficulties and financial implications of endangered species breeding programmes. Inter. Zoo Yearb. 24/25:210-219.

Conway, W. 1989. The prospect for sustaining species and their evolution, p. 199-209. In D. Western, and M. Pearl (eds.), Conservation for the Twenty-First Century. Oxford University Press, New York.

Conway, W. 1995. The conservation park: A new zoo synthesis for a changed world, p. 259-276. In C. M. Wemmer (ed.), The Ark Evolving. Zoos and Aquariums in Transition. Smithsonian Institution, Conservation and Research Center, Front Royal VA.

Hutchins, M., and W. G. Conway. 1995. Beyond Noah's ark: the evolving role of modern zoological parks and aquariums in field conservation. Inter. Zoo Yearb. 34:117-130 [excellent overview].

King, F. W. 1973. Summary of the status of crocodilian species in South America undertaken by Professor F. Medem, p. 33-36. Crocodiles: Proc. 2nd Working Meeting of the IUCN/SSC Crocodile Specialist Group, New York. IUCN Publ. N.S., Morges, Switzerland. [Federico Medem, now deceased, published a number of papers on South American chelonians and crocodilians, including the 2 volume set entitled *Los crocodylia de Sur América*.]

King, F. W. 1974. International trade and endangered

species. Inter. Zoo Yearb. 14:2-13.

King, F. W. 1974. Trade in live crocodilians. Inter. Zoo Yearb. 14:52-56.

King, F. W. 1975. Slaughter in the wild. Can man's profit ever balance nature's loss? Anim. Kingdom (April/May):27-32.

Soulé, M., M. Gilpin, W. Conway, and T. Foose. 1986. The Millennium Ark: How long a voyage, how many staterooms, how many passengers? Zoo Biol. 5:101-113. [Authors recommended maintenance of 90% of the genetic variation in the source (wild) population over a period of 200 years.]

Stuart, B. L., and J. B. Thorbjarnarson. 2003. Biological prioritization of Asian countries for turtle conservation. Chelonian Conserv. Biol. 4:642-647.

Swingland, I. R., and M. W. Klemens (eds.). 1989. The Conservation Biology of Tortoises. Occasional Papers of IUCN Species Survival Commission (SSC) No. 5. IUCN-The World Conservation Union, Gland, Switzerland. [Michael Klemens is employed by WCS.]

Thorbjarnarson, J. B. 1986. The present status and distribution of *Crocodylus acutus* on the Caribbean island of Hispaniola, p. 195-202. Crocodiles: Proc. 7th Working Meeting of the IUCN/SSC Crocodile Specialist Group, Caracas, Venezuela. IUCN Publ. N.S., Gland, Switzerland.

Thorbjarnarson, J. B. 1988. The status and ecology of the American crocodile in Haiti. Bull. Florida State Mus. Biol. Sci. 33:1-86.

Thorbjarnarson, J. B. 1989. Ecology of the American crocodile, *Crocodylus acutus*, p. 228-258. Crocodiles: Their Ecology, Management and Conservation. A Special Publication of the Crocodile Specialist Group. IUCN-The World Conservation Union Publ. N.S., Gland, Switzerland.

Thorbjarnarson, J. B. 1990. An analysis of the spectacled caiman *Caiman crocodilus* harvest program in Venezuela, Neotropical Wildlife Use and Conservation. University of Chicago Press, Chicago.

Thorbjarnarson, J. B. 1993. Efforts to conserve the Orinoco crocodile in the Capanaparo River, Venezuela, p. 320-322. Zoocria de los Crocodylia. Memorias de la I Reunion del CSG, Grupo de Especialistas en Crocodrilos de la IUCN. IUCN-The World Conservation Union, Gland, Switzerland.

Thorbjarnarson, J. B. 1999. Crocodile tears and skins: international trade, economic constraints, and limits in the sustainable use of crocodilians. Conserv. Biol. 13:465-470.

Thorbjarnarson, J. B., and R. Franz. 1987. *Crocodylus intermedius* (Graves). Orinoco crocodile. Cat. Amer. Amphib. Rept.:406.1-406.2.

Thorbjarnarson, J. B. , and G. Hernández. 1990. Recent investigations into the status of Orinoco crocodiles in Venezuela, p. 308-328. Crocodiles: Proc. 9th Working Meeting of the IUCN/SSC Crocodile Specialist Group, Lae, Papua-New Guinea. The World Conservation Union Publ. N.S., Gland, Switzerland.

Thorbjarnarson, J. B., and G. Hernández. 1992. Recent investigations of the status and distribution of the Orinoco crocodile *Crocodylus intermedius* in Venezuela. Biol. Conserv. 62:179-188.

Thorbjarnarson, J. B., and G. Hernández. 1993. Reproductive ecology of the Orinoco crocodile (*Crocodylus intermedius*) in Venezuela. I. Nesting ecology and egg and clutch relationships. J. Herpetol. 27:363-370.

Thorbjarnarson, J. B., and G. Hernández. 1993. Reproductive ecology of the Orinoco crocodile (*Crocodylus intermedius*) in Venezuela. II. Reproductive and social behavior. J. Herpetol. 27:371-379.

Thorbjarnarson, J. B., and X. Wang. 1999. The conservation status of the Chinese alligator. Oryx 33:152-159.

Timmins, R. J., and K. Khounboline. 1999. Occurrence and trade of the golden turtle, *Cuora trifasciata*, in Laos. Chelonian Conserv. Biol. 3:441-447 [contribution of Wildlife Conservation Society Laos Program].

Yamashita, C., P. Brazaitis, and G. Rêbelo. 1992. The crocodilians of Brazil and the identification of the species, p. 207-220. Proc. 3rd Workshop on Conservation and Management of the Broad-nosed caiman *Caiman latirostris*. Sao Paulo, Brazil 26-28 Oct. 1992.

Husbandry

Barker, D. 1997. Preliminary observations on nutrient composition differences between adult and pinhead crickets, *Acheta domestica*. Bull. Assoc. Rept. Amphib. Vet. Vet. 7(1):11-13.

Brazaitis, P., and J. L. Behler. 1973. A durable housing facility for reptiles. Laboratory Anim. Sci. 23:866-868.

Dowling, H., and S. Spencook. 1960. The care of pet turtles. New York Zool. Soc. :16 pp. [Stephen Spencook was Head Keeper.]

Greenhall, A. M. 1936. The care of the bushmaster and of certain lizards in the New York Zoological Park. Herpetologica 1:66-67.

King, F. W. 1971. Housing, sanitation, and nutrition of reptiles. J. Amer. Vet. Med. Assoc. 159:1612-1615.

General

Babich, H, J. LoBue, and H. G. Dowling. 1968. Principles of Biology I. HISS Publ., New York. [Dowling published five laboratory manuals, with additional revisions, between 1978-1988.]

Beebe, W. 1918. Jungle Peace. Henry Holt and Co., New York. [Book was dedicated to Colonel Theodore Roosevelt and his wife. Roosevelt was a supporter of Beebe's plan to develop a tropical research laboratory in British Guiana. After the facility was functional, Roosevelt and his wife spent time with Beebe observing the region. On p. 35, he described his unsuccessful attempts to collect lizards with a net in St. Thomas until a young girl taught him how to use a noose. His encounter with a bushmaster in South America was a bit more stressful until he overpowered the snake, eventually sent alive to the Zoo.]

Beebe, W. 1919. The higher vertebrates of British Guiana with special reference to the fauna of the Bartica District. Zoologica (New York) 2:205-238. [No. 7 was a list of Amphibia, Reptilia, and Mammalia; No. 8 was a list of birds, and No. 9 covered the lizards of the genus *Ameiva*.]

Beebe, W. 1921. Contributions of the Tropical Research Station of the New York Zoological Society. Zoologica (New York) 3:15-32. [List of publications from staff. Henry Fairfield Osborne, President of the Society, listed the goals of the Station: study living organisms in their natural environment at a single locality and use staff and collaborators to engage in these studies for many years.]

Beebe, W. 1924. Galapagos: World's End. G. P. Putnam' Sons, New York [9 color plates, mostly of reptiles].

Beebe, W. 1927. The Log of the Sun. A Chronicle of Nature's Year. Garden City Publishing Co., Inc., Garden City NY. [Beebe published 52 lovely short essays, grouped by seasons of the year, and several described the natural history of amphibians and reptiles. The essay "Dwellers in the dust" mostly covered the habits of garter and ribbon snakes (p. 75). "Polliwog problems" portrayed the development of the toad (p. 110). "Turtle traits" (p. 128) and "The names of animals, frogs, and fish" (p. 252) need no explanation.]

Beebe, W. 1942. Book of Bays. Harcourt, Brace, New York [chapter titled "The bay of sea snakes" (p. 193)].

Beebe, W. 1944. Field notes on the lizards of British Guiana and Venezuela. Part I. Gekkonidae. Zoologica (New York) 29(14).

Beebe, W. 1945. Field notes on the lizards of British Guiana and Venezuela. Part 3. Amphisbaenidae and Scincidae. Zoologica (New York) 30(2).

Beebe, W. 1946. Field notes on the snakes of British Guiana and Venezuela. Amphisbaenidae and Scincidae. Zoologica (New York) 31(4).

Beebe, W. 1949. High Jungle. Duell, Sloan and Pearce, New York.

Beebe, W. 1952. Introduction to the ecology of the Arima Valley, Trinidad, B.W.I. Zoologica (New York) 37:157-183.

Beebe, W. 1953. Unseen Life in New York as a Naturalist Sees It. Duell, Sloan and Pearce, New York.

Beebe, W., G. I. Hartley, and P. G. Howes. 1917. Tropical Wild Life in British Guiana. New York Zoological Society, New York. [Introduction by Colonel Theodore Roosevelt. Part I by Beebe was a generalized treatment of British Guiana, including the last chapter on "alligators." The next two parts were written by Hartley and Howes.]

Behler, J. L. 1970. The bog turtle (*Clemmys muhlenbergi*) in Monroe County, Pennsylvania. Bull. Maryland Herpetol. Soc.; September :52-53.

Behler, J. L. and D. A. Behler. 1998. Alligators & Crocodiles. Voyageur Press, Stillwater MN.

Behler, J. L. and D. A. Behler. 2005. Frogs! A Chorus of Colors. Sterling Publishing, New York.

Behler, J. L., and F. W. King. 1979. The Audubon Society Field Guide to North American Reptiles and Amphibians. Alfred A. Knopf, New York.

Behler, J. L., B. Winn, and R. H. Hayes Jr. 1997. Snake fauna of St. Catherines Island, Georgia. Herpetol. Rev. 28(3): 152. [St. Catherines Island was the satellite breeding site for the Wildlife Conservation Society.]

Brazaitis, P. 1969. Occurrence and ingestion of gastrolith in two crocodilians. Herpetologica 25:63-64.

Brazaitis, P., and M. E. Watanabe. 1992. Snakes of the World. Crescent Books, Avenel NJ.

Brazaitis, P., and M. E. Watanabe. 1993. The World of Snakes. Tetra Press, Blacksburg VA.

Brazaitis, P., and M. E. Watanabe (eds.). 1994. The Fight for Survival: Animals In Their Natural Habitats. P. Friedman Group, New York.

Breder, C. M., Jr. 1942. A consideration of evolutionary hypotheses in reference to the origin of life. Zoologica (New York) 27:131-143.

Breder, C. M., Jr. 1946. Amphibians and reptiles of the Rio Chucunaque drainage, Darien, Panama, with notes on their life histories and habits. Bull. Amer. Mus. Nat. His. 86:375-436.

Breder, C. M., Jr., and R. B. Breder. 1923. A list of fishes, amphibians and reptiles collected in Ashe

County North Carolina. Zoologica (New York) 4:3-23.

Breder, C. M., Jr., R. B. Breder, and A. C. Redmond. 1927. Frog tagging: A method of studying anuran life habits. Zoologica (New York) 9:201-229 [cord tied around "waist" with tag attached].

Breder, R. B. 1927. Turtle trailing: A new technique for studying the life habits of certain Testudinata. Zoologica (New York) 9:231-243 [spool of thread attached to carapace].

Bridges, W. 1968. The Bronx Zoo Book of Wild Animals. A Guide to Mammals, Birds, Reptiles and Amphibians of the World. New York Zoological Society and Golden Press, New York. [Curator F. Wayne King advised Bridges on the herpetological section.]

Bridges, W. 1971. Zoo Careers. William Morrow, New York. [Chapter 6, "The Reptile Department phone never stops ringing," is an overview of the operation of the herpetological division at the Bronx Zoo. The employees are identified only by first names so I will hazard a guess as to their identities: Wayne (F. Wayne King), John (John Behler) and Peter (Peter Brazaitis). There are many photographs of employees busy at work, unlike the Dallas Zoo where I spent most of my time smoking cigarettes, drinking coffee and talking on the telephone.]

Ditmars, R. L. 1896. The snakes found within fifty miles of New York City. Proc. Linn. Soc. New York No. 8:9-24.

Ditmars, R. L. 1905. The reptiles of the vicinity of New York City. Amer. Mus. J. No. 19.

Ditmars, R. L. 1905. The batrachians of the vicinity of New York City. Amer. Mus. J. No. 20.

Ditmars, R. L. 1906. Our collection of amphibians. Reprinted from the Tenth Annual Report of the New York Zoological Society, pp 1-17.

Ditmars, R. L. 1912. The feeding habits of serpents. Zoologica (New York) 1:197-238.

Ditmars, R. L. 1934. A review of the box turtles. Zoologica (New York) 17:1-44.

Dowling, H. G. 1957. A Review of the Amphibians and Reptiles of Arkansas. University of Arkansas [Occasional Papers of the University of Arkansas Museum, No. 3., Fayattesville].

Dowling, H. G. 1958. Geographic relations of Ozarkian amphibians and reptiles. Southwest. Natur. 1:174-189.

Dowling, H. G. 1959. The groundsnake, *Sonora episcopa*, in Arkansas. Southwest. Natur. 3:231-232.

Dowling, H. G., and I. Gilboa. 1966. A zoo record system: the method used in the department of herpetology at New York Zoo. Inter. Zoo Yearb. 8:405-408.

Hornaday, W. T. 1894 [c.1891]. Taxidermy and Zoological Collecting; A Complete Handbook for the Amateur Taxidermist, Collector, Osteologist, Museum-Builder, Sportsman, and Traveller. Charles. Scribner's Sons, New York. [William Temple Hornaday was Director of Zoo.]

Hornaday, W. T. 1899. Popular Official Guide to the New York Zoological Park, as Far as Completed. New York Zoological Society, New York.

Hornaday, W. T. 1914. The American Natural History. A Foundation of Useful Knowledge of the Higher Animals of North America. Volume IV - Reptiles, Amphibians and Fishes. Charles Scribner's Sons, New York. [This lavish work contained 225 original drawings by several artists, 151 photographs, 16 color plates, charts and maps. He published books on his field adventures in the American southwest and Mexico, India, Sri Lanka, Borneo and the Malay Peninsula. Some of his other books dealt with taxidermy, extinction and conservation, zoological collecting and preservation of specimens.]

King, W. 1968. Danger - do not move. Inter. Turtle Tortoise Soc. J. 2(2):17-35 [discussion of giant tortoises].

King, W., and T. Krakauer. 1966. The exotic herpetofauna of southeast Florida. Q. J. Florida Acad. Sci. 29:144-154.

Lloyd, M., R. F. Inger, and F. W. King. 1968. On the diversity of reptile and amphibian species in a Bornean rain forest. Amer. Natur. 102(No. 928):497-515.

Minton, S. A., H. G. Dowling, and F. E. Russell. 1968. Poisonous Snakes of the World. U.S. Govt. Printing Office, Washington DC.

Oliver, J. A. 1955. The Natural History of North American Amphibians and Reptiles, D. Van Nostrand Co., Inc., Princeton NJ. [Chapter 12 treated growth, size and longevity; the next chapter covered amphibians and reptiles as pets.]

Oliver, J. A. 1958. Snakes in Fact and Fiction. Macmillan, New York.

Platt, S. G., R. J. Lee, and M. W. Klemens. 2001. Notes on the distribution, life history, and exploitation of turtles in Sulawesi, Indonesia, with emphasis on *Indotestudo forstenii* and *Leucocephalon yuwonoi*. Chelonian Conserv. Biol. 4:154-159.

Platt, S. G., Saw Tun Khaing, Win Ko Ko, and Kalyar. 2001. A tortoise survey of Shwe Settaw Wildlife Sanctuary, Myanmar, with notes on the ecology of

Geochelone platynota and *Indotestudo elongata.* Chelonian Conserv. Biol. 4:172-177.

Tashian, R. E., and C. Ray. 1957. The relation of oxygen consumption to temperature in some tropical, temperate and boreal anuran amphibians. Zoologica (New York) 42:63-68. [At time of publication, authors were associated with Zoo.]

Thorbjarnson, J. B., H. Messel, F. W. King, and J. P. Ross (compiled accounts). 1992. Crocodiles. An Action Plan for Their Conservation. IUCN-The World Conservation Union, Gland, Switzerland.

Vinegar, A., and V. H. Hutchison. 1965. Pulmonary and cutaneous gas exchange in the green frog, *Rana clamitans.* Zoologica (New York) 50:47-53. [Allen Vinegar and Victor Hutchison were Research Fellow and Research Associate respectively in the Department of Herpetology at Zoo.]

Animal Kingdom: A magazine for popular consumption.

Behler Jr., J. L. 1970. Galapagos tortoise *Geochelone elephantopus.* Anim. Kingdom 73(5):33.

Behler, J. L. 1971. Coahuilan box turtle *Terrapene coahuila.* Anim. Kingdom 74(5):33.

Behler, J. L. 1971. Pine Barrens treefrog *Hyla andersoni.* Anim. Kingdom 74(5):33.

Behler, J. L. 1972. Toads. Anim. Kingdom 75(5):33.

Behler, J. L. 1975. To brew a deadly dose. Anim. Kingdom 70(5):16-19 [toxins].

Boylan, T. 1970. Thorny devil. Anim. Kingdom 73(1):24-27.

Brazaitis, P. 1966. Snake hunting in South Carolina. Anim. Kingdom 69:175-179, 189.

Chace, G. E. 1971. The reptiles of Paha Sapa. A herpetologist's report from the Black Hills of South Dakota. Anim. Kingdom 74(4):20-28.

Dowling, H. G. 1959. Rattles, hisses, and rasping scales. Anim. Kingdom 62(5):159-160.

Dowling, H. G. 1960. A bushmaster at the zoo again. Anim. Kingdom 63(3):109-111.

Dowling, H. G. 1960. The curious feeding habits of the Java wart snake. Anim. Kingdom 63(1):13-15.

Dowling, H. G. 1960. Thermistors, air-conditioners, and insecticides. Anim. Kingdom 63(6):206-207.

Dowling, H. G. 1961. Vanishing giants and enduring dwarfs. The tortoises. Anim. Kingdom 64(3):66-75.

Dowling, H. G. 1961. Our air-conditioned bushmaster. Anim. Kingdom 64(5):155-156. [Lower environmental temperatures contributed to greater success in maintaining this difficult species.]

Dowling, H. G. 1961. How old are they and how big do they grow? Anim. Kingdom 64(6):171-175.

Dowling, H. G. 1962. Long-lived bushmaster dies short of the record. Anim. Kingdom 65(1):31.

Dowling, H. G. 1962. Some like it hot--some like it cold. Anim. Kingdom 65(2):41-47.

Dowling, H. G. 1962. Their day begins at sundown: Reptiles after dark. Anim. Kingdom 64(4):119-123.

Dowling, H. G. 1962. World's end revisited: The Zoological Society's fifth expedition to the Galapagos Islands. Anim. Kingdom 65(5):130-142.

Dowling, H. G. 1962. Sea dragons of the Galapagos: The marine iguanas. Anim. Kingdom 65(6):169-174.

Dowling, H. G. 1963. Madagascar: Distant home of "American" reptiles. Anim. Kingdom 66(1):18-22.

Dowling, H. G. 1963. A giant step toward keeping giant chameleons. Anim. Kingdom 66:178-181.

Dowling, H. G. 1964. An exercise in zoogeography. Anim. Kingdom 67(3):88-91.

Dowling, H. G. 1964. Goats and hawks--a new theory of predation on the land iguana. Anim. Kingdom 67(2):51-56.

Dowling, H. G. 1964. Our first Calabar python. Anim. Kingdom 67(1):14-15.

Dowling, H. G. 1964. Zoologist's tour of Martinique. Anim. Kingdom 67(6):182-186.

Dowling, H. G. 1965. Our crocodilians like it hot. Anim. Kingdom 68(3):74-78.

Dowling, H. G. 1965. Boa constrictors: From tropical menace to popular pet. Anim. Kingdom 68(6):183-185.

Dowling, H. G. 1965. The puzzle of the *Bothrops*: Or a tangle of serpents. Anim. Kingdom 68(1):18-21.

Dowling, H. G. 1966. Poisonous snakes of Vietnam. Anim. Kingdom 69(2):34-43.

Dunton, S. 1963. A morning in the reptile house. Anim. Kingdom 66(3):83-85.

Eibl-Eibesfeldt, I. 1964. The art of keeping marine iguanas. Anim. Kingdom 67(3):75-77. [This taxon is exceedingly difficult to maintain in captivity and was most successfully done at Brookfield Zoo (Chicago Zoological Park) in 1960s.]

Goss, L. J. 1939. Snake parasites in man. Anim. Kingdom 42(6):178-179.

King, W. 1968. Ora-giants of Komodo. Anim. Kingdom 71(4):2-9.

King, W. 1969. Texas blind salamander *Typhlomolge rathbuni* Stejneger. Anim. Kingdom 72(2):33.

King, W. 1969. Adaptations for nocturnal survival. Anim. Kingdom 72(3):29-32.

King, W. 1969. American alligator *Alligator mississippiensis* (Daudin). Anim. Kingdom 72(6):33.

King, W. 1976. The year of the dragon. Anim. Kingdom 79(1):1.

Oliver, J. A. 1951. Caiman lizard--a reptile rarity. Anim. Kingdom 54(5):151-153.

Oliver, J. A. 1953. The timeless tuatara. Anim. Kingdom 51(1):2-8, 31.

Oliver, J. A. 1956. Hortense has no tail to tell. Anim. Kingdom 59(3):77-79.

Oliver, J. A. 1957. Feeding king cobras is easy (they say)--you can give them milk. Anim. Kingdom 60(4):126-127.

Oliver, J. A. 1957. Now we know how to feed baby king cobras. Anim. Kingdom 60(5):155-156.

Oliver, J. A. 1957. Is temperature the secret? How to feed a bushmaster. Anim. Kingdom 60(6):175-177.

Oliver, J. A. 1958. The taipan, Australia's deadliest snake. Anim. Kingdom 61(1):23-26.

Oliver, J. A. 1959. The matamata: Surely all turtles can swim . . . ? Anim. Kingdom 62(6):167-170.

Watanabe, M. E. 1986. A change of fortune for the Chinese alligator. Anim. Kingdom 89(5):34-39.

Zoologica: A number of prominent herpetologists who were not employed at the Zoo published papers in this journal. Thomas Barbour generated several lists of Antillean reptiles and amphibians. Charles Bogert investigated body temperatures of the tuatara under natural conditions. Fred Cagle described the status of the ringed map turtle. Howard Campbell observed the acoustic behavior of the American alligator and crocodile, black and spectacled caimans. Carl Gans presented the functional morphology of the African egg-eating snake. R. E. Gordon, J. A. MacMahon, and D. B. Wake demonstrated relative abundance, microhabitat and behavior of some southern Appalachian salamanders. Edmond Malnate studied the yellow-lipped snake. Jay Savage covered the iguanid lizard genera *Urosaurus* and *Uta* and described a new genus for the lizard family Xantusiidae [*Klauberina*]. He also elaborated on the family position of Neotropical frogs, genus *Pseudis*. Hobart Smith published several papers on Mexican snakes. The eight titles below have relevance to zoo biology.

Barton, A. J., and W. B. Allen Jr. 1961. Observations on the feeding, shedding and growth rates of captive snakes (Boidae). Zoologica (New York) 46:83-87 [study at Highland Park Zoological Gardens in Pittsburgh, Pennsylvania].

Charipper, H. A., and J. J. Martorano. 1948. The morphology of the pituitary gland of the South African clawed toad, *Xenopus laevis* Daudin. Zoologica (New York) 33:157-162 [many specimens provided by Zoo].

Cochran, D. M. 1956. A new species of frog from Kartabo, British Guiana. Zoologica (New York) 41:11-12. [*Eleutherodactylus beebei* named in honor of William Beebe. The locality data were in error so this taxon is now *Eleutherodactylus inoptatus* from Hispaniola.]

Fox, W., C. Gordon, and M. H. Fox. 1961. Morphological effects of low temperatures during the embryonic development of the garter snake, *Thamnophis elegans*. Zoologica (New York) 46:57-71. [This is a seminal study of environmental factors altering scutellation patterns which is of importance to zoo biologists.]

Noble, G. K. 1923. 14. New batrachians. 15. New lizards. From the Tropical Research Station. British Guiana. Zoologica (New York) 3:289-305. [Collaborator G. Kingsley Noble was from the American Museum of Natural History in New York.]

Noble, G. K., and M. K. Brady. 1933. Observations on the life history of the marbled salamander, *Ambystoma opacum* Gravenhorst. Zoologica (New York) 11:89-132 [study from the American Museum of Natural History and Smithsonian National Zoological Park].

Noble, G. K., and R. C. Noble. 1923. The Anderson tree frog (*Hyla andersonii* Baird). Observations on its habits and life history. Zoologica (New York) 11:413-455.

Quaranta, J. V. 1952. An experimental study of the color vision of the giant tortoise. Zoologica (New York) 37:295-312. [at time of publication, author was Research Associate at Zoo.].

Staten Island Zoo (1936)

Facility and Collection: According to *International Zoo Yearbook* (1998: Volume 28), composition of the collection was 70 taxa of reptiles with 167 specimens and 9 taxa of amphibians numbering 25.

Historical Overview

Adler, K. 1989. Herpetologists of the past, p. 5-141. In K. Adler (ed.), Contributions to the History of Herpetology. Society for the Study of Amphibians and Reptiles. Contributions to Herpetology, volume 5, Oxford OH [biographical sketch of Carl Kauffeld].

Card, W., and J. B. Murphy. 2000. Lineages and histories of zoo herpetologists in the United States. Herpetol. Circ. 27:1-45 [biography of Kauffeld].

Fig. 128. Reptile wing at Staten Island Zoo pictured in "News Bulletin of the Staten Island Zoological Society" in March 1937. *Staten Island Zoological Society, provided by Ken Kawata.*

Fig. 129. Mural backgrounds depicting natural habitats of snakes kept in different cages at Staten Island Zoo pictured in 1937. *Staten Island Zoological Society, provided by Ken Kawata.*

Fig. 130. Undated postcard of cobra from Staten Island Zoo. *Provided by Brint Spencer.*

Conant, R. 1975. OBITUARY: Carl Frederick Kauffeld, April 17, 1911-July 10, 1974. Herpetol. Rev. 6(1):27-28.

Conant, R. 1997. A Field Guide to the Life and Times of Roger Conant. Selva, Tyler TX [biography of Kauffeld, p. 466].

Kawata, K. 2003. New York's Biggest Little Zoo. A History of the Staten Island Zoo. Kendall/Hunt Publishing Co., Dubuque IA [wonderful account of Kauffeld's career].

1984. Carl F. Kauffeld memorial issue. HERP 18(1) [collection of Kauffeld articles from various publications].

Morphology, Systematics & Taxonomy

Gloyd, H. K., and C. F. Kauffeld. 1940. A new rattlesnake from Mexico. Bull. Chicago Acad. Sci. 6(2):11-14 [description of *Crotalus durissus totonacus* from northeastern Mexico].

Kauffeld, C. F., and H. K. Gloyd. 1939. Notes on the Aruba rattlesnake, *Crotalus unicolor*. Herpetologica 1(6):156-160. [Account covers feeding, reproduction and defensive behavior. Comparison with the Neotropical rattlesnake *Crotalus d. durissus* led authors to conclude that *C. unicolor* is a distinct species.]

Husbandry

Kauffeld, C. F. 1943. Growth and feeding of newborn Price's and green rock rattlesnakes. Amer. Midl. Natur. 29:607-614. [specimens of the former gave birth to six neonates and the latter to four young which were maintained in captivity. Kauffeld mentioned that food lizards were frozen in ice cube trays and thawed later, a technique commonly used today for storing prey items which are often difficult to locate when needed.]

Kauffeld, C. F. 1953. Methods of feeding captive snakes. Herpetologica 9:129-131.

Kauffeld, C. F. 1953. Newer treatment of mouthrot in snakes. Herpetologica 9:132.

Kauffeld, C. F. 1953. Removal of abnormal snake eggs by sectioning. Herpetologica 9:161-163.

Kauffeld, C. F. 1954. Mites and ticks in captive snakes with remarks on cage sanitation. Herpetologica 10:103-107.

Kauffeld, C. F. 1954. Manipulation of odor as an aid in feeding captive snakes, with special reference to king cobras. Herpetologica 10:108-110.

Kauffeld, C. 1969. The effect of altitude, ultraviolet light, and humidity on captive reptiles. Inter. Zoo Yearb. 9:8-9. [Altitudinal changes have little effect, snakes do not require ultraviolet light, high humidity should be avoided.]

General

Curran, C. H., and C. Kauffeld. 1937. Snakes and Their Ways. Harper & Brothers, New York, London. [C. H. Curran was an entomologist at the American Museum of Natural History. This is a general treatment on snake biology, based on questions received at the Museum from the public.]

Kauffeld, C. F. 1943. Field notes on some Arizona reptiles and amphibians. Amer. Midl. Natur. 29:342-359.

Kauffeld, C. F. 1960. The search for subocularis. Bull. Philadelphia Herpetol. Soc. 8(2):13-19 [in two parts, second installment in 8(3):9-15].

Kauffeld, C. F. 1961. Massasauga land. Bull. Philadelphia Herpetol. Soc. 9(3):7-13. [Account described collecting experiences at Cicero Swamp, New York. This issue also contained a letter by Kauffeld decrying irresponsible collecting practices.]

Kauffeld, C. F. 1963. More at home with cobras. Bull. Philadelphia Herpetol. Soc. 11(1-2):49-52.

Kauffeld, C. F. 1965. The rattlesnake collection at Staten Island Zoo. Inter. Zoo Yearb. 5:168-170 [collection reached 34 forms in 1963].

Rubio, M.. 1998. Rattlesnake: Portrait of a Predator. Smithsonian Institution Press, Washington, London. [Manny Rubio dedicated this book to Kauffeld who was a major influence during his early years. The quality of the photographs is outstanding and the text is easily read.].

Zappalorti, R. T. 1976. The Amateur Zoologist's Guide to Turtles and Crocodilians. Stackpole Books, Harrisburg PA. [At time of publication, Robert Zappalorti was Associate Curator of Herpetology and Education at the Zoo.]

Animaland: Each issue cost a princely sum of 5 cents in 1938.

Clements, H. 1966. Chelonian giants of the sea. Animaland 33(1). [Clements replaced David Zucconi who became Curator of Herpetology at Milwaukee County Zoo.]

Gans, C. 1956. Frogs and paradoxes. Animaland 23(4). [Carl Gans was Research Associate, Carnegie Museum, Pittsburgh, Pennsylvania and described a six month trip to South America where he observed the Paradox frog *Pseudis paradoxa*. This anuran is unusual since the adult is smaller than the tadpole and has an opposable thumb, most unique for a lower vertebrate.]

Kauffeld, C. F. 1938. Pythons survive shipwreck. Animaland 5(4).

Kauffeld, C. F. 1939. Have you a pet alligator? Animaland 6(2).

Kauffeld, C. F. 1939. Home care of American chameleons. Animaland 6(4).

Kauffeld, C. F. 1949. Reptile Wing News and Notes: Births and deaths in the Reptile Wing. Animaland 16(3).

Kauffeld, C. F. 1950. Horned lizards. Reptilian cacti. Animaland 17(3). [Account on the biology of horned lizards with a description of five species at the Zoo which seldom lived more than a few months. In another article in this issue describing his career, Kauffeld was responsible for acquiring over 600 kinds of reptiles during a 14-year span.]

Kauffeld, C. F. 1951. Reptile Department in 1950. Animaland 18(1).

Kauffeld, C. F. 1951. Longevity of snakes in the Staten Island Zoo. Animaland 18(6). [64 species and subspecies, totaling 122 specimens, exceeded two years. 20 specimens of 14 forms had attained or exceeded 5 years and 8 had reached or exceeded 10 years.]

Kauffeld, C. F. 1951. Gila monster. Animaland 21(5). [Kauffeld described its biology and mentioned that this species was given protection in the State of Arizona, a measure which should be extended to other endangered reptiles, especially snakes.]

Kauffeld, C. F. 1951. Reptile Department in 1950. Animaland 18(1).

Kauffeld, C. F. 1952. To South Carolina for snakes. Animaland 19(3). [Account of snake hunting experiences with the observation that a very large cottonmouth moccasin, when captured, disgorged an adult canebrake rattlesnake.]

Kauffeld, C. F. 1952. The Reptile Wing in 1951. Animaland 19(1).

Kauffeld, C. F. 1953. The Reptile Wing in 1952. Animaland 20(1).

Kauffeld, C. F. 1953. Snakes have personality. Animaland 20(5).

Kauffeld, C. F. 1954. The Reptile Department in 1953. Animaland 21(1). [Acquisitions included Palestine viper, European horned viper, European asp viper, mottled and green rock rattlesnakes, Jararaca, Asiatic green tree viper, matamata and five White's Cay iguana. Kauffeld described a new 14-foot king cobra, normally ophiophagus, which was weaned to a horsemeat and rat diet by applying snake scent to the prey. The black cobra died of a malignant bone tumor in the spinal column. One problem still plaguing reptile keepers today is an array of parasitic flukes, linguatalids and filarial blood worms; Kauffeld struggled with various remedies to treat his charges but with limited success.]

Kauffeld, C. F. 1955. The care of pet alligators. Animaland 22(3).

Kauffeld, C. F. 1955. Off with their skins. Animaland 22(5). [This account covers mechanics of shedding and environmental factors such as temperature. Kauffeld mentioned that wounds or skin abrasions resulted in an increase in shedding frequency. In another section, he stated that a group of fence lizards *Sceloporus undulatus hyacinthinus*, originally from the pine barrens of southern New Jersey (60-70 miles from Staten Island), were released on zoo grounds as a potential source of food lizards; the lizards reproduced and extended their range from the Zoo at least a mile (probably more).]

Kauffeld, C. F. 1956. Spring snake hunts. Animaland 23(2). [Kauffeld and staff regularly mounted annual trips to South Carolina to hunt snakes for the Zoo and for sister institutions. From this article, it is clear that this region was a favorite haunt.]

Kauffeld, C. F. 1957. Snake hunts in the Southwest. Animaland 24(6). [Account of snake hunting experiences in southeastern Arizona, a favorite rattlesnake collecting spot for Kauffeld. All species from the United States were in the collection at the Zoo as of 1 October 1957. Also, the new oil burning heaters were installed so the emergency charcoal heaters no longer needed to be used.]

Kauffeld, C. F. 1958. The keeper and the kept. Animaland 25(4). [tBreder, R. B. 1927. Turtle trailing: A new technique for studying the life habits of certain Testudinata. Zoologica (New York) 9:231-243 [spool of thread attached to carapace].Title of this article was later used for his seminal book on the care and maintenance of reptiles in captivity. He stressed several important factors for success in keeping snakes: need for security, maintaining them singly, and offering pre-killed food. Further, he said that elaborate naturalistic exhibits can be harmful in that cleaning and sanitation are compromised; such aesthetic efforts are of no value to the snake (see Chiszar et al. 1993 for a different view).]

Kauffeld, C. F. 1959. The rattle of the rattlesnake. Animaland 26(1). [Kauffeld described structure of rattle and new exhibit of a mechanical rattle using a doorbell clapper arm.]

Kauffeld, C. F. 1959. Snake eyes. Animaland 26(4).

Kauffeld, C. F. 1960. "What's in a name." Animaland 27(1) [overview of classification and scientific nomenclature].

Kauffeld, C. F. 1961. Longevity of snakes in Staten Island Zoo-II. Animaland 28(5). [Kauffeld listed 55 species and subspecies, representing 84 specimens, which exceeded 5 years and 24 which attained or exceeded 10 years. Although he was justifiably proud of this accomplishment, he worried that the mere mention of success might send specimens into immediate decline and compilation of this list might unleash this "evil jinx." He goes on to say that the extremes of climate, construction projects and various other aspects related to the nature of the subjects mitigated against success but the staff could take pride in these longevities.]

Kauffeld, C. F. 1962. From a snake hunter's diary--I. Animaland 24(3) [account of feeding problems with captive bull snake].

Kauffeld, C. F. 1967. So you own a boa. Animaland 34(5). [title of this article was used as a chapter heading for his book titled *Snakes: The Keeper and the Kept* on the care and maintenance of reptiles in captivity. The rattlesnake collection continued to expand with 38 species and subspecies. A duplicate group of all 32 forms of rattlesnakes from the United States was assembled.]

Kauffeld, C. F. 1969. The notorious bushmaster. Animaland 36(3). [Bushmasters were very difficult to maintain in captivity over extended periods and zoo workers at several zoos, including Staten Island, experimented with bushmasters during the late 1960s and early 1970s with mixed results. In 1967, the Steinhart Aquarium in San Francisco, California, hatched nine eggs from a wild female. This accomplishment was recognized by the American Association of Zoological Parks and Aquariums through the issuance of the Edward H. Bean Award for the most significant captive breeding but caused an outcry from the zoo herpetological community as this was not a captive breeding. Now several institutions, such as the Dallas Zoo, have breeding colonies of bushmasters.]

Kauffeld, C. F. 1971. Life expectancy of snakes. Animaland 38(2 & 3). [Sixty-four individuals exceeded 10 years.]

Kurtz, B. 2001. The Kauffeld connection. The lineages

of herpetologists at the Staten Island Zoo. Animaland 55:1-3, 14-15.

O'Connor, P. 1954. The Veterinary Department. Animaland 21(1). [Patricia O'Connor was the first full-time woman zoo veterinarian and she was honored at a lavish retirement dinner by the staff for her contributions to the Zoo. In this article she described the sanitation, quarantine, and nutritional procedures used at the Zoo.]

O'Connor, P. 1956. Nutrition of zoo animals. Animaland 23(3). [She described the exacting requirements for captive snakes and stressed that snakes often go for substantial periods without feeding.]

National Aquarium in Baltimore (1981)

Facility and Collection: Several features at the Aquarium deserve mention. First is a series of small naturalistic exhibits housing Neotropical reptiles and amphibians. Second is a large walk-through rainforest depicting tropical America. Third is a superb Amazonian display. Visitors are able to view the fish collection in the lower aquatic portion and many arboreal reptiles in the upper regions. The latest exhibit shows Australian amphibians and reptiles in natural settings: an extraordinary chelid turtle collection, many varanids, large agamid lizards, and Fly River turtles.

The staff has successfully reproduced Neotropical hylid and dendrobatid frogs. Many of the species are kept behind the scenes in small aquariums and terrariums. Reproduction has occurred in many of the captive dendrobatid frogs, but captive frogs did not produce defensive alkaloids similar to wild counterparts, a finding which precludes successful reintroduction.

According to *International Zoo Yearbook* (1998:Volume 28), composition of the collection was 40 taxa of reptiles with 140 specimens and 52 taxa of amphibians numbering over 1000.

Staff and Scientific Achievements: General Curator John Cover Jr. has been involved in field projects in Latin America. An important initiative has been the long-term project in Surinam studying and collecting Azure poison dart frogs for captive breeding at the Aquarium. Cover assisted the staff at the Zoo Nacional Simon Bolivar in San Jose, Costa Rica in remodeling the small herpetological exhibit and collecting specimens for display. Former herpetological curator at the Baltimore Zoo, Vicki Poole, now works at the Aquarium.

Personal Reflections: Some of the nicest mixed exhibits utilizing living plants and fiberglass dioramas for smaller amphibians and reptiles may be found in this building.

Historical Overview

Wisnieski, A. P., and J. F. Cover Jr. 1989. Hidden Life: The new amphibian and reptile exhibits at the National Aquarium in Baltimore. AAZPA Annual Conf. Proc. 1989:547-554.

Husbandry

Barnett, S. L. 1992. Husbandry of Neotropical lizards at the National Aquarium in Baltimore. AAZPA Reg. Conf. Proc. 1992:33-40.

Barnett, S. 1993. Husbandry of Neotropical iguanid lizards at the National Aquarium in Baltimore, p. 19-34. In M. Bumgardner (ed.), Captive Propagation and Husbandry of Reptiles and Amphibians. Northern California Herpetological Society.

Medical Management

Stamper, M. A., and B. R. Whitaker. 1994. Medical observations and implications on "healthy" sea turtles prior to release into the wild. Proc. Amer. Assoc. Zoo Vet. 1994:182-185.

Stoskopf, M. K. 1994. Veterinary issues in aquatic exhibit design. Proc. Amer. Assoc. Zoo Vet. 1994:173-178.

Stoskopf, M. K., A. Wisnieski, and L. Pieper. 1985. Iodine toxicity in poison arrow frogs. Proc. Amer. Assoc. Zoo Vet. 1985:86-88.

Whitaker, B. R. 1993. The use of poly vinyl chloride glues and their potential toxicity to amphibians. Proc. Amer. Assoc. Zoo Vet. 1995:16-18.

General

Daly, J. W., C. W. Myers, J. E. Warnick, and E. X. Albuquerque. 1980. Levels of batrachotoxin and lack of sensitivity to its action in poison-dart frogs (*Phyllobates*). Science 208:1383-1385.

Daly, J. W., S. I. Secunda, H. M. Garraffo, T. F. Spande, A. Wisnieski, C. Nishihira, and J. F. Cover Jr. 1992. Variation in alkaloid profiles in Neotropical poison frogs (Dendrobatidae): genetic versus environmental determinants. Toxicon 30:887-898.

SOUTHEASTERN UNITED STATES

Zoo Atlanta (1889)

History: In 1966, I entered the profession as a keeper at this Zoo, which was exciting as a new reptile building had just been completed and was being stocked with specimens. At that time, John (Steve) Dobbs was the Curator and his assistant was R. Howard Hunt. My delight quickly changed to apprehension for there were major problems. The exhibits had been built with a toxic bonding agent that proved deleterious to reptiles. A protracted legal struggle between the architect and zoo officials delayed the opening for several years. As a result, I only stayed briefly in Atlanta before going to the Dallas Zoo. During my stay there, I had developed a nice working relationship with Dobbs and Hunt so my leaving was bittersweet.

Facility and Collection: The building is large with exhibits on the perimeter and a center island with a variety of mixed exhibits. When the building opened, Dobbs and Hunt specialized in crocodilians as there were several large solariums with living plants available for large reptiles. Some exceedingly large Morelet's crocodiles lived in the enclosure and later bred regularly. In fact, they were so fecund that Howard worked with officials in Latin America to try to reintroduce them in suitable areas. The collection has been diverse and well-maintained since the beginning, with a number of amphibians on display. Today, the facility shows signs of aging. The solariums have had problems with water leakage and the only reptile being exhibited there is an alligator snapping turtle. Plans are being developed to renovate the entire building.

According to *International Zoo Yearbook* (1998:Volume 28), composition of the collection was over 122 taxa of reptiles with over 390 specimens and 11 taxa of amphibians numbering 20.

Staff and Scientific Achievements: When Dobbs was elevated to Zoo Director, Hunt became the Curator and Dennis Herman was his assistant. Herman, prior to leaving the Zoo for a museum position in North Carolina, was involved in an extensive monitoring study of the bog turtle with Bern Tryon of the Knoxville Zoo. Susan Barnard, who specializes in parasitology, has published a number of papers and a two-volume book on this topic. Terry Maple, a primatologist, became Director when Dobbs retired. Dwight Lawson, who spent five years doing a herpetofaunal survey and developing a coherent conservation strategy in Cameroon, Africa, has been hired as General Curator. Recently, Hunt has retired but continues his long-term study of alligators in the Okefenokee Swamp. His successor was Gordon Schuett who left shortly thereafter. The current curator is Joseph Mendelson III, whose research interests have focused on taxonomy and sys-

tematics. Recently, he has spearheaded the amphibian recovery program in Panama.

Personal Reflections: Hunt has a rapport with crocodilians that I have rarely seen and he taught me a great deal about their behaviors and habits. One day, he asked if crocodiles have acute vision. When I was uncertain, he threw some mealworms to a large male estuarine crocodile which was resting in its pool. The crocodilian quickly picked the insects off the surface of the water and swallowed them.

Historical Overview

Dobbs, J. S. 1967. Reptile house at Atlanta Zoo. Inter. Zoo Yearb. 7:75-77.

Maple, T. L. 1992. In memoriam. Prof. Dr. Heini Hediger (1908-1992). Zoo Biol. 11:369-372. [Hediger was late Director of Zürich Zoo, Switzerland.]

Morphology, Systematics & Taxonomy

Lawson, D. P. 1999. A new species of arboreal viper (Serpentes: Viperidae: *Atheris*) from Cameroon, Africa. Proc. Biol. Soc. Washington 112:793-803 [description of *Atheris broadleyi*].

Lawson, D. P. 2000. A new caecilian from Cameroon, Africa (Amphibia: Gymnophiona: Scolecomorphidae). Herpetologica 56:77-80 [description of *Crotaphatrema tchabalmbaboensis*].

Lawson, D. P., and P. C. Ustach. 2000. A redescription of *Atheris squamigera* (Serpentes: Viperidae) with comments on the validity of *Atheris anisolepis*. J. Herpetol. 34:386-389.

Narins, P. M., E. R. Lewis, A. P. Purgue, P. J. Bishop, L. R. Minter, and D. P. Lawson. 2001. Functional consequences of a novel middle ear adaptation in the Central African frog *Petropedetes parkeri* (Ranidae). J. Exp. Biol. 204:1223-1232.

Upton, S. J., and S. M. Barnard. 1986. Experimental transmission of *Caryospora simplex* (Apicomplexa: Eimeriidae) to Palestine vipers, *Vipera xanthina palestinae* (Serpentes: Viperidae). J. Protozool. 33:129-130.

Upton, S. J., and S. M. Barnard. 1987. Two new species of coccidia (Apicomplexa: Eimeriidae) from Madagascar gekkonids. J. Protozool. 34:452-454.

Upton, S. J., W. L. Current, and S. M. Barnard. 1984. A new species of *Caryospora* (Apicomplexa: Eimeriidae) from the green lizard, *Anolis carolinensis*. Trans. Amer. Micros. Soc. 102:245-248.

Upton, S. J., W. L. Current, and S. M. Barnard. 1984. A new species of *Caryospora* (Apicomplexa: Eimeriidae) from *Elaphe* spp. (Serpentes: Colubridae) of

Southeastern and Central United States. Trans. Amer. Microsc. Soc. 103:240-244.

Upton, S. J., W. L. Current, and S. M. Barnard. 1986. A review of the genus *Caryospora* Leger, 1904 (Apicomplexa: Eimeriidae). Systematic Parasitol. 8:3-21.

Upton, S. J., W. L. Current, S. M. Barnard, and J. V. Ernst. 1984. *in vitro* excystation of *Caryospora simplex* (Apicomplexa: Eimeriidae). J. Protozool. 31:293-297.

Upton, S. J., J. V. Ernst, S. M. Barnard, and W. L. Current. 1983. Redescription of the oocysts of *Caryospora simplex* (Apicomplexa: Eimeriidae) from an Ottoman viper, *Vipera xanthina xanthina*. Trans. Amer. Microsc. Soc. 102:258-263.

Husbandry

Barnard, S. M. 1996. Reptile Keeper's Handbook, Krieger Publishing Co., Malabar FL.

Dobbs, J. S. 1969. A note on reinforced heterogenous polyester reptile cages at Atlanta Zoo. Inter. Zoo Yearb. 9:70-71.

Conservation

Herman, D. W. 1989. Captive management and conservation of the bog turtle at Zoo Atlanta. AAZPA Reg. Conf. Proc. 1989:597-600.

Lawler, H. E. 1977. The status of *Drymarchon corais couperi* (Holbrook), the eastern indigo snake, in the southeastern United States. Herpetol. Rev. 8(3):76-79 [excellent overview by Howard Lawler, who was employed at Zoo at time of publication].

Lawson, D. P. 2000. Local harvest of hingeback tortoises, *Kinixys erosa* and *K. homeana* in Southwestern Cameroon. Chelonian Conserv. Biol. 3:722-729.

Lawson, D. P., and M. W. Klemens. 2001. Herpetofauna of the African rain forest, p. 291-307. In W. Webcr, L. J. T. White, A. Vedder, and L. Naughton-Treves (eds.), African Rainforest Ecology & Conservation. Yale Univ. Press, New Haven CT.

Medical Management

Barnard, S. M. 1983. A review of some fecal pseudoparasites of reptiles. J. Zoo Anim. Med. 14:79-88.

Barnard, S. M. 1986. An annotated outline of commonly occurring reptilian parasites. Acta Zool. Pathol. Antverpiensia 79:39-72.

Barnard, S. M., and L. Durden. 2000. A Veterinary Guide to the Parasites of Reptiles: Arthropods (Excluding the Mites). Vol. II. Krieger, Malabar FL.

Barnard, S. M., and S. J. Upton. 1994. A Veterinary Guide to the Parasites of Reptiles: Vol. I, Protozoa. Krieger, Malabar, FL.

Upton, S. J., C. T. McAllister, P. S. Freed, and S. M. Barnard. 1989. Cryptosporidiosis in wild and captive reptiles. J. Wildl. Dis. 25:20-30.

Wilson, N., and S. M. Barnard. 1985. Three species of *Aponomma* (Acari: Ixodidae) imported to the United States on reptiles. Florida Entomologist 68:478-480.

Reproduction

Honegger, R. E., and H. Hunt. 1990. Breeding crocodiles in zoological gardens outside the species range, with some data on the general situations in European zoos, p. 200-228. Crocodiles: Proc. 10th Working Meeting of the Crocodile Specialist Group, Gainesville, Florida. IUCN. The World Conservation Union Publ. N.S., Gland, Switzerland.

King, F. W., and J. S. Dobbs. 1975. Crocodilian propagation in American zoos and aquaria. Inter. Zoo Yearb. 15:272-277.

General

Finlay, T. W., and T. L. Maple. 1986. A survey of research in American zoos and aquariums. Zoo Biol. 5:261-268. [Behavioral research and reproductive research are most common topics and nonprimate mammals the most likely subjects.].

Hoff, M. P., and T. L. Maple. 1982. Sex and age differences in the avoidance of reptile exhibits by zoo visitors. Zoo Biol.1:263-269. [Females were more often reluctant to enter exhibit than males. Teenagers more often entered and spent more time in exhibit than visitors of other ages.]

Lawson, D. P. 2004. Partners in saving turtles: The Turtle Survival Alliance. Herpetol. Rev. 35:111.

Schuett, G. W., M. Höggren, M. E. Douglas, and H. W. Greene (eds.). 2002. Biology of the Vipers. Eagle Mountain Publishing, Eagle Mountain CO.

Stoinski, T. S., K. E. Lukas, and T. L. Maple. 1998. A survey of research in North American zoos and aquariums. Zoo Biol. 17:167-180. [Surveys were sent to 173 AZA institutions which showed that research has increased during the last decade.]

Audubon Park and Zoological Garden (1914) (New Orleans)

History: Over the past several decades, there have been major improvements at this Zoo including an accelerated building program to replace outmoded and inadequate exhibits. In 1990, the impressive Aquarium of the Americas was opened. The Zoo sustained some damage from Hurricane Katrina in 2005 and was forced to close for a while but has now reopened.

Facility and Collection: In the late 1980s, J. Kevin Bowler, now retired, was hired to direct the debut of "Reptile Encounter," a significant herpetological facility. Included in the plan was a series of exhibits displaying many different rattlesnakes. Bowler appealed to zoo colleagues to help stock the building which resulted in many interesting species, already acclimated to captivity, being available for display. The Zoo is known for amelanotic American alligators, donated as hatchlings, which have been sent to many zoos for temporary exhibition. Public interest in these striking crocodilians is dramatic and increased visitation inevitably occurs. At the Aquarium of the Americas, biologist and amphibian specialist Ian Hiler has developed a captive breeding program for dendrobatid frogs.

According to *International Zoo Yearbook* (1998:Volume 28), composition of the collection was 118 species of reptiles with 424 specimens and 18 species of amphibians numbering 93.

Staff and Scientific Achievements: Janice Perry was the first Assistant Curator when the building opened and Andrew Snider was her successor. Dino Ferri currently occupies this slot.

Personal Reflections: When I visited the building shortly after it opened, the health of the specimens and overall cleanliness were notable.

Historical Overview
Bowler, J. K. 1989. Reptile Encounter: a new herpetological facility at Audubon Zoo. Inter. Zoo Yearb. 28:216-221. [Account covered planning and design, graphics, service facilities, safety and husbandry.]

Medical Management
Jacobson, E. R., J. M. Gaskin, S. Wells, K. Bowler, and J. Schumacher. 1992. Epizootic of ophidian paramyxovirus in a zoological collection: pathological, microbiological, and serological findings. J. Zoo Wildl. Med. 23:318-327.

General
Bowler, J. K. 1977. Longevity of reptiles and amphibians in North American collections. Soc. Study Amphib. Rept. Herp. Circ. No. 6:1-32.

Snider, A. T. and J. K. Bowler. 1992. Longevity of reptiles and amphibians in North American collections. Second Edition. Soc. Study Amphib. Rept. Herp. Circ. No. 21:1-40.

Ross Allen's Reptile Institute (1931)

History: The Institute, located in Silver Springs, Florida, was a mixture of exhibits and shows designed to appeal to tourists. There were many large outdoor enclosures with a nice collection of native and exotic crocodilians, venomous snake demonstrations and venom extraction, alligator wrestling pavilion, recreated Seminole Indian village, gift shop and retail outlet for purchasing reptiles as pets. The Institute is now closed.

Historical Overview
Conant, R. 1981. OBITUARY: Ensil Ross Allen, 1908-1981. Herpetol. Rev. 12(4):99.

Liner, E., C. J. McCoy, and D. L. Auth. 1993. Biographical sketch and bibliography of Wilfred T. Neill. Smithson. Herpetol. Inf. Serv.:1-24.

Mertens, R. 1951. Zwischen [Between] Atlantik und Pazifik; zoologische Reiseskizzen aus Nordamerika. A. Kernen, Stuttgart, Germany. [In 1949, Robert Mertens from the Senckenberg Museum in Frankfurt am Main, Germany, traveled to the United States to visit museums, zoos, aquariums and spend some time in the field. One of the places he visited was central Florida where he spent time with colleagues in Gainesville and at Ross Allen's Reptile Institute in Silver Springs. As Walter Auffenberg later told me, his visit was legendary. During the time he was in the field, naturally he identified all of the herpetofauna. This was not unusual as the reputation for his prodigious memory was well known; it has been said that he could identify virtually all herps from throughout the world on sight. But his colleagues were surprised to find that he was able to identify the other Florida animals and plants as well. As he called out one technical name after the other, he was asked how he knew so much about the biota of Florida since his direct experience was limited. He told his astonished companions that prior to leaving Germany on his trip, he consulted the literature to memorize the technical names and diagnostic characters. This book is a travelogue of his adventures and includes photographs of the taxa which he encountered.]

Neill, W. T. 1949. Ross Allen's Reptile Institute. Ross

Allen's Reptile Institute, Silver Springs FL.

Neill, W. T. 1950. Ross Allen's Reptile Institute. Ross Allen's Reptile Institute, Silver Springs FL.

Neill, W. T. 1951. Herpetology and nature training school at Silver Springs, Florida. Ross Allen's Reptile Institute, Silver Springs FL.

Who's Who in the South and Southwest. 14th Edition, 1975-1976. Marquis Who's Who. [Ross Allen was listed in four editions between 1975-1981.]

Morphology, Taxonomy & Systematics

Allen, E. R., and W. T. Neill. 1949. A new subspecies of salamander (genus *Plethodon*) from Florida and Georgia. Herpetologica 5:112-114 [description of *Plethodon glutinosus grobmani,* now known as *Plethodon grobmani*].

Neill, W. T. 1949. A new subspecies of rat snake (genus *Elaphe*), and notes on related forms. Herpetologica 5 (second suppl.):1-12 [description of *Elaphe obsoleta rossalleni*].

Neill, W. T. 1949. The status of Baird's chorus frog. Copeia 1949 3:227-228.

Neill, W. T. 1950. A new species of salamander, genus *Desmognathus,* from Georgia. Publ. Res. Div. Ross Allen's Reptile Institute. 1:1-6. [description of *Desmognathus perlapsus,* now relegated to a single monotypic species *Demognathus ochrophaeus*].

Neill, W. T. 1950. Taxonomy, nomenclature, and distribution of southeastern cricket frogs, genus *Acris.* Amer. Midl. Natur. 43:152-156.

Neill, W. T. 1951. The taxonomy of North American soft-shelled turtles, genus *Amyda.* Publ. Res. Div. Ross Allen's Reptile Institute, Silver Springs FL. 1:7-24.

Neill, W. T. 1951. A new subspecies of dusky salamander, genus *Desmognathus,* from south-central Florida. Publ. Res. Div. Ross Allen's Reptile Institute 1:25-38 [description of *Desmognathus fuscus carri,* now known as *Desmognathus apalechicolae*].

Neill, W. T. 1951. A new subspecies of salamander, genus *Pseudobranchus,* from the Gulf Hammock region of Florida. Publ. Res. Div. Ross Allen's Reptile Institute I:39-46 [description of *Pseudobranchus striatus lustricolus*].

Neill, W. T. 1954. A new species of frog, genus *Nyctimystes,* from Papua. Copeia 1954:83-85 [description of *Nyctimystes loveridgei*].

Neill, W. T. 1954. Ranges and taxonomic allocations of amphibians and reptiles in the southeastern United States. Publ. Res. Div. Ross Allen's Reptile Institute 1:75-96.

Neill, W. T. 1956. The possibility of an undescribed crocodile on New Britain. Herpetologica 12:174-176.

Neill, W. T., and E. R. Allen. 1949. A new kingsnake (genus *Lampropeltis*) from Florida. Herpetologica 5 (5 (special)):101-106 [description of *Lampropeltis getulus goini,* now not recognized].

Behavior

Allen, E. R. 1950. Sounds produced by the Suwannee terrapin. Copeia 1950:62.

Neill, W. T. 1952. Burrowing habits of *Hyla gratiosa.* Copeia 1952:196.

Neill, W. T. 1952. Remarks on salamander voices. Copeia 1952:195-196.

Neill, W. T. 1957. Notes on metamorphic and breeding aggregations of the eastern spadefoot, *Scaphiopus holbrooki* (Harlan). Herpetologica 13:185-187.

Neill, W. T. 1960. The caudal lure of various juvenile snakes. Q. J. Florida Acad. Sci. 23:173-200.

Husbandry

Allen, E. R. 1966. Keep Them Alive. Great Outdoors, St. Petersburg FL. [When this pamphlet was published, Allen had kept 525,000 captive reptiles during a 36- year span.]

Allen, E. R., and W. T. Neill. 1950. Keep Them Alive! How to keep snakes, lizards, turtles, alligators, and crocodiles in captivity. Special Publication No. 1. Ross Allen's Reptile Institute, Silver Springs FL [3 editions].

General

Allen, E. R. 1938. Notes on Wright's bullfrog, *Rana heckscheri* (Wright). Copeia 1938:50.

Allen, E. R. 1949. Range of cane-brake rattlesnake in Florida. Copeia 1949:73-74.

Allen, E. R. 1967. Alligator Farming. Ross Allen's Reptile Institute, Silver Springs FL.

Allen, E. R. 1967. Where to buy or sell wild animals. Ross Allen's Reptile Institute Bulletin (7): 1.

Allen, E. R., and E. Maier. 1948. Reprints of snake venom publications by Ross Allen's Reptile Institute. Ross Allen's Reptile Institute, Silver Springs FL.

Allen, E. R., and W. T. Neill. 1949. Increasing abundance of the alligator in the eastern portion of its range. Herpetologica 5:109-112.

Allen, E. R., and W. T. Neill. 1950. The alligator snapping turtle *Macrochelys temminckii* in Florida. Ross Allen's Reptile Institute Spec. Publ. (4):1-15.

Allen, E. R., and W. T. Neill. 1950. The life history of the Everglades rat snake, *Elaphe obsoleta rossalleni*. Herpetologica 6:109-112.

Allen, E. R., and W. T. Neill. 1953. Juveniles of the tortoise, *Gopherus polyphemus*. Copeia 1953:128.

Allen, E. R., and W. T. Neill. 1953. The crocodile in the Everglades National Park. Copeia 1953:54-59.

Allen, E. R., and W. T. Neill. 1953. The treefrog, *Hyla septentrionalis*, in Florida. Copeia 1953:127-128.

Allen, E. R., and W. T. Neill. 1954. Juveniles of Brooks' kingsnake, *Lampropeltis getulus brooksi*. Copeia 1954:59.

Allen, E. R., and W. T. Neill. 1956. Some color abnormalities in crocodilians. Copeia 1956:124.

Allen, E. R., and W. T. Neill. 1957. Another record of the Atlantic leatherback, *Dermochelys c. coriacea*, nesting on the Florida coast. Copeia 1957:143-144.

Allen, E. R., and W. T. Neill. 1957. Some interesting rattlesnakes from southern British Guiana. Herpetologica 13:67-74.

Allen, E. R., and W. T. Neill. 1959. Doubtful locality records in British Honduras. Herpetologica 15:227-233.

Allen, E. R., and D. Swindell. 1948. Cottonmouth moccasin of Florida. Herpetologica 4 (first suppl.):1-16.

Goin, C. J., and J. N. Layne. 1958. Notes on a collection of frogs from Leticia, Colombia. Publ. Res. Div. Ross Allen's Reptile Institute 1 (8):97-114.

Heilman, R. E. 1953. A comparative study of the eggs and tadpoles of *Hyla phaeocrypta* and *Hyla versicolor* in Florida. Publ. Res. Div. Ross Allen's Reptile Institute 1 (6):61-74.

Neill, W. T. 1946. Notes on *Crocodylus novae-guineae*. Copeia 1946:17-20 [distribution, vocalization, parental care, ecology, movement, food, reproduction, size, olfaction, morphology].

Neil, W. T. 1947. Size and habits of the cottonmouth moccasin. Herpetologica 3:203-205.

Neill, W. T. 1949. A checklist of the amphibians and reptiles of Florida. Ross Allen's Reptile Institute, Silver Springs FL.

Neill, W. T. 1949. A checklist of the amphibians and reptiles of Georgia. Ross Allen's Reptile Institute, Silver Springs FL, 4 pp.

Neill, W. T. 1949. Juveniles of *Siren lacertina* and *S. i. intermedia*. Herpetologica 5:19-20.

Neill, W. T. 1950. How to preserve reptiles and amphibians for scientific study. Ross Allen's Reptile Institute Special Publication No. 2.

Neill, W. T. 1950. Reptiles and amphibians in urban areas of Georgia. Herpetologica 6:113-116.

Neill, W. T. 1950. The status of the Florida brown snake, *Storeria victa*. Copeia 1950:155-156.

Neill, W. T. 1951. Amphibians and reptiles of a fifteen-acre tract in Georgia. Amer. Midl. Natur. 45:241-244.

Neill, W T. 1951. A bromeliad herpetofauna in Florida. Ecology 32:140-143.

Neill, W. T. 1951. Notes on the natural history of certain North American snakes. Publ. Res. Div. Ross Allen's Reptile Institute 1 (5):47-60.

Neill, W. T. 1952. New records of *Rana virgatipes* and *Rana grylio* in Georgia and South Carolina. Copeia 1952:194-195.

Neill, W. T. 1952. The reptiles of Florida. Florida Nat. 25:11-16.

Neill, W. T. 1954. Evidence of venom in snakes of the genera *Alsophis* and *Rhadinaea*. Copeia 1954:59-60.

Neill, W. T. 1957. The eggs of the crowned snake, *Tantilla coronata*. Herpetologica 13:77-78.

Neill, W. T., and E. R. Allen. 1950. *Eumeces fasciatus* in Florida. Copeia 1950:156.

Neill, W. T., and E. R. Allen. 1955. Metachrosis in snakes. Q. J. Florida Acad. Sci. 18:207-215.

Neill, W. T., and E. R. Allen. 1959. Additions to the British Honduras herpetofaunal list. Herpetologica 15:235-240.

Neill, W. T., and E. R. Allen. 1959. The rediscovery of *Thamnophis praeocularis* (Bocourt) in British Honduras. Herpetologica 15:223-227.

Neill, W T., and E. R. Allen. 1959. Studies on the amphibians and reptiles of British Honduras. Publ. Res. Div. Ross Allen's Reptile Institute 2 (1):1-76.

Neill, W. T., and F. L. Rose. 1949. Nests and eggs of the southern dusky salamander, *Desmognathus fuscus auriculatus*. . Copeia 1949:234.

Neill, W. T., and F. L. Rose. 1953. Records of the green watersnake, *Natrix cyclopion floridana*, in South Carolina. Copeia 1953:127.

Knoxville Zoological Gardens (1947)

Facility and Collection: The main zoo building comprises a number of exhibits which have housed rarities such as Madagascar tortoises, Fiji Island iguanas, olive pythons and a variety of exotic boas and pythons. A fabulous outdoor exhibit contains bog and spotted turtles living with indigenous plants.

According to *International Zoo Yearbook* (1998:Volume 28), composition of the collection was over 100 taxa of reptiles with over 450 specimens and 4 taxa of amphibians numbering 8.

Staff and Scientific Achievements: There have been three curators at this Zoo since the opening of the reptile building: John Arnett, Howard Lawler and Bern Tryon, who is currently General Curator. Bern Tryon has been involved in a number of field projects to conserve the endangered bog turtle. Jon Whitehead is Tryon's assistant.

Personal Reflections: In 1982, this zoo provided an excellent example of the type of collaborative study which can be accomplished in a zoo setting when academic researchers Neil Greenberg and Thomas Jenssen investigated the captive colony of Fiji banded iguanas.

Historical Overview

Arnett, J. R. 1976. The new reptile complex at Knoxville Zoo. Inter. Zoo Yearb. 16:210-211 [building completed in 1975].

General

Greenberg, N., and T. A. Jenssen. 1982. Displays of captive banded iguanas, *Brachylophus fasciatus*, p. 232-251. In G. M. Burghardt, and A. S. Rand (eds.), Iguanas of the World. Their Behavior, Ecology, and Conservation. Noyes Publications, Park Ridge NJ.

Tryon, B. W. 1984. Snake hibernation: In and out of the zoo. Brit. Herpetol. Soc. Bull. 10:22-29.

Tryon, B. W. 2004. The final point. Turtle Tortoise Newsletter 7: 3-6 [bog turtle field study].

CENTRAL UNITED STATES

Lincoln Park Zoological Gardens (1868)

History: The Lincoln Park Zoo is situated on the shores of Lake Michigan in Chicago. Although there is a limited number of herpetological publications emanating from this Zoo, there has been a long historical tradition in exhibiting unusual reptiles and amphibians.

Facility and Collection: The original building was a modified aquarium before the John G. Shedd Aquarium opened in the early 1930s. The converted exhibit cages were originally aquariums with concrete dioramas and natural lighting. Many of the enclosures were spacious which allowed large reptiles to be kept: many crocodilians, boid snakes and the largest king cobra I have ever seen in captivity. In the basement were several large aquatic exhibits and a section housing smaller "zoo gems." Since the building was old and

the heating and cooling systems outdated, plans were developed in the 1950s to totally upgrade the facility. An elaborate model was constructed to show zoo visitors how the remodeling was to proceed. During that time, funding was precarious and as the years passed without change, the model slowly deteriorated until only a pile of rubble was left. Many years later, the new Regenstein Small Mammal-Reptile House was built.

According to *International Zoo Yearbook* (1998:Volume 28), composition of the collection was 65 taxa of reptiles with nearly 300 specimens and 13 taxa of amphibians numbering nearly 100.

Staff and Scientific Achievements: Former Director R. Marlin Perkins started his television program *"Zooparade"* which later evolved into the long-running *Mutual of Omaha's Wild Kingdom*. Ray Pawley was hired to assist with the show before traveling across town to the Brookfield Zoo. Zoologist Gene Hartz and his assistant Ed Almandarz, keeper Ed Maruska who later was Director of the Cincinnati Zoo, and Ray Pawley were employed at the zoo during the 1950s. When Hartz left the zoo to work for an animal dealer, Alamandarz was elevated to his position and published three chapters on reptile husbandry and manual re-

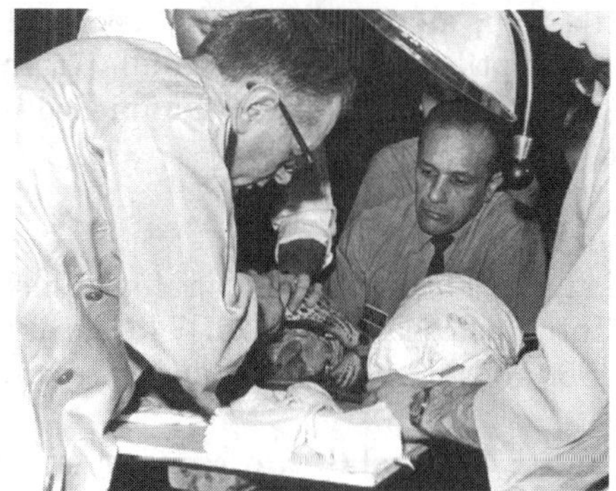

Fig. 131. Photograph of zoo veterinarian Eric Maschgan (left) and late herpetological curator Edward Almandarz (center) working on a leopard tooth extraction in 1970. *Courtesy of Chicago Park District/Lincoln Park Zoo.*

straint in Murray Fowler's book *Zoo and Wild Animal Medicine*. After he died, Clarence Wright assumed his position but left to take another job. The new Regenstein Small Mammal-Reptile House replaced the old facility and was operated by John Gramieri, Curator of Small Mammals and Reptiles, until he left the Zoo. The curatorial position is currently unfilled.

Personal Reflections: During the 1950s occasionally when my parents decided that my collection of reptiles at home had to go, mostly due to untimely escapes, my herpetological treasures always went to this Zoo. With heavy heart, I periodically carried my charges to the reptile building. I met R. Marlin Perkins, Edward Maruska, Gene Hartz, Ray Pawley and Ed Almandarz during the first purge, who remained colleagues for years thereafter. They subsequently spent hours explaining the intricacies of the zoo business, encouraging me to give it a try as a career. Since it was clear that I was ill-equipped to do anything else, their wise counsel prevailed. Marlin Perkins was the only dissenter; he insisted that the zoo business would thrust me into a life of insolvency, chaos and frustration. Predicting two out of three is not bad!

Historical Overview

Card, W., and J. B. Murphy. 2000. Lineages and histories of zoo herpetologists in the United States. Herpetol. Circ. No. 27:1-44 [portraits of Gene Hartz, R. Marlin Perkins and Edward Almandarz].

Gold, D. 1988. Zoo: A Behind-the-Scenes Look at the Animals and the People who Care for Them. Contemporary Books, Chicago. [In chapter 4 titled "Looking for Mr. Wright," herpetological curator Clarence Wright is featured. Chapter 14, "The Reptile Rap," details the daily tasks and experiences of reptile keeper Art Maraldi. Wright is currently Director of the Potawatomi Zoo in South Bend, Indiana.]

Reproduction

Almandarz, E. 1969. Reptile birth rate at Lincoln Park Zoo. Inter. Zoo Yearb. 9:23-24. [749 births between Jan 1960 and end of Dec 1966, 464 were hatchlings.]

Cincinnati Zoo and Botanical Garden (1875)

History and Mission: The Monkey House, built in 1875, is the oldest existing zoo building in the United States and is listed on the National Register of Historic Places. This smallish facility was converted into a reptile building. There is a center crocodilian pool surrounded by glass-fronted modules which have been upgraded into a series of attractive amphibian and reptile exhibits.

Collection: A number of rare species have been maintained over the years, including the Borneo earless monitor. Since the salamander collection is so diverse and comprehensive, an impressive number of

Fig. 132. The old monkey house at Cincinnati Zoological and Botanical Gardens, built in 1875, is oldest existing zoo building in United States and is listed on National Register of Historic Places. It is now current reptile building. The original Reptile House is now the bird house. *Cincinnati Zoological and Botanical Gardens Archives, provided by Winston Card.*

longevities have been compiled and many husbandry and medical protocols have been developed. In the second edition of "Longevity of Reptiles and Amphibians in North American Collections" (SSAR Herpetol. Circ. No. 21), 66 taxa are listed as longevities. Salamanders of 32 species lived over 10 years and of these, 5 passed 20 years. This is a remarkable accomplishment.

According to *International Zoo Yearbook* (1998:Volume 28), composition of the collection was 97 taxa of reptiles with 365 specimens and 66 taxa of amphibians numbering over 560.

Staff and Scientific Achievements: Donald Way was supervisor of the reptile collection in the early 1960s. When he was hired for a similar post at the Dallas Zoo, John Arnett took his position but left to oversee the herpetological collection at the Knoxville Zoo. David Jardine filled Arnett's slot but moved to the mammal department when Arnett returned to the Zoo. Don Gillespie, former veterinarian now at the Nashville Zoo in Tennessee, is well known for his work on the immunological system and microbial organisms in the toxic saliva of the Komodo dragon. Director Edward Maruska recently retired from the Zoo. Winston Card, former supervisor at the Dallas Zoo, began employment as the curator of the department in June 2002 but has since left the profession.

Fig. 133. John Arnett and his charges at Cincinnati Zoological and Botanical Gardens in 1966. *Cincinnati Zoological and Botanical Gardens Archives, provided by Winston Card.*

Personal Reflections: When the annual SSAR meeting was held in Miami, Ohio, in 1976, a symposium on the reproductive biology of amphibians was scheduled which attracted many prominent amphibian biologists from the museum and academic community. Some had heard that Edward Maruska had amassed a salamander collection at the Cincinnati Zoo. Since a tour was scheduled, anticipation was high but the attendees were not prepared for the width and breadth of this caudate assemblage. Salamanders were housed in every nook and cranny throughout the Zoo. As they viewed olms, giant salamanders, rare cave salamanders, and many exotic forms infrequently seen alive, their collective impression was one of awe. Since Ed is a skilled photographer, he had prepared many dioramas with living plants and other natural objects for his backgrounds. Many excited colleagues told me that this was without doubt the most overwhelming living salamander collection ever!

Historical Overview

Ehrlinger, D. 1993. The Cincinnati Zoo and Botanical Garden from Past to Present. The Cincinnati Zoo and Botanical Garden, Cincinnati OH. [Foreword by Roger Conant. A description, a drawing and a photograph of the building are included.]

Morphology, Systematics & Taxonomy

Brandon, R. A., E. J. Maruska, and W. T. Rumph. 1981. A new species of neotenic *Ambystoma* (Amphibia, Caudata) endemic to Laguna Alchichica, Puebla, Mexico. Bull. S. California Acad. Sci. 80:112-125. [Provides a description of *Ambystoma taylori*. Captive specimens were maintained at the Zoo.]

Reproduction

Gillespie, D., B. L. Dresser, and E. J. Maruska. 1988/1989. Sexing goliath frogs *Gigantorana goliath* by laparoscopy. Proc. Amer. Assoc. Zoo Vet.:62.

Kramer, L., B. L. Dresser, and E. J. Maruska. 1983. Sexing aquatic salamanders by laparoscopy. Proc. Amer. Assoc. Zoo Vet.:192-194.

Behavior

Cooper Jr., W. E. 1989. Prey odor discrimination in the varanoid lizards *Heloderma suspectum* and *Varanus exanthematicus*. Ecology 81:250-258 [observations at Zoo].

Cooper, W. E., Jr. 2000. Food chemical discriminations by the omnivorous lizards *Tiliqua scincoides* and *Tiliqua rugosa*. Herpetologica 56:480-488. [observations at Cincinnati, Dallas, Ft. Worth, Indianapolis, St. Louis Zoos, and National Zoological Park.].

Cooper, W. E., Jr. C. S. DePerno, and J. Arnett. 1994. Prolonged post-bite elevation in tongue-flicking rate with rapid onset in the gila monster, *Heloderma suspectum*: Relation to diet and foraging and implications for evolution of chemosensory searching. J. Chem. Ecol. 20:2587-2601.

Saint Louis Zoological Park (1890)

History: In 1927, a modern reptile building was constructed at the St Louis Zoo and still remains one of the premier facilities in the United States. Part of its charm lies in its beauty, with warm and inviting public areas. Open and spacious exhibits festooned with living tropical plants add to its appeal. The original hand-painted murals depicting habitats have been preserved. Although there have been several renovations over the years which could have destroyed the original architecture and its historical importance, these changes have been done carefully. Within the past few years, many of the smaller enclosures have been enlarged and sophisticated thermal gradients have been installed. As a result, reptiles have become more active. In 2005, the building was renamed the Charles H. Hoessle Herpetarium.

Collection: Since the facility is large, larger varanids, chelonians, crocodilians, and constrictors can be accommodated. The collection has always been representative and well displayed. Over the past years, many true vipers have been added to the collection, with impressive reproductive successes. A colony of tuataras has lived off exhibit for many years.

According to *International Zoo Yearbook* (1998:Volume 28), composition of the collection was 167 taxa of reptiles with 418 specimens and 28 taxa of amphibians numbering 111.

***Staff and Scientific Achievements*:** Two pioneers in zoo herpetology started their careers at this Zoo: R. Marlin Perkins (1905-1986) and his assistant Moody Lentz (1901-1970). Retired director Charles Hoessle started as the head of the reptile department. His curatorial replacement, the late Ronald Goellner, became Deputy General Curator. Jeff Ettling is currently Curator of Herpetology and Aquatics and AZA Snakes Taxon Advisory Group Chair.

Joel Wallach was Pathologist and Director of Research. After he left the Zoo, he was Director of the Memphis and Jacksonville Zoos where he published studies on ulcerative shell disease in turtles and overviews of reptile nutritional diseases and physiology.

***Personal Reflections*:** This historic building, with its beauty and grace, has been one of my favorites. When I walked into the tuatara enclosure with Goellner and Hoessle years ago, it was the first time I was able to have a close encounter with these magnificent reptiles.

Historical Overview

Card, W., and J. B. Murphy. 2000. Lineages and histories of zoo herpetologists in the United States. Herpetol. Circ. 27:1-45 [portraits of Moody Lentz and R. Marlin Perkins].

Conant, R. 1980. The reproductive biology of reptiles: An historical perspective, p. 3-18. In J. B. Murphy, and J. T. Collins (eds.), Reproductive Biology and Diseases of Captive Reptiles. Society for the Study of Amphibians and Reptiles. Contributions to Herpetology, volume 1, Lawrence KS [photographs of visitor exhibition halls].

Conant, R. 1997. A Field Guide to the Life and Times of Roger Conant. Selva, Tyler TX [memories of R. Marlin Perkins].

Behavior

London, G. D., K. L. Bauman, and C. S. Asa. 1998. Time-lapse infrared videography for animal behavior observations. Zoo Biol. 17:535-543.

Husbandry

Dempsey, J. L., and J. B. Bernard. 1999. Everything you never wanted to know, but should, about feedstuffs or the importance of chemical analysis in feed quality control. Proc. Amer. Assoc. Zoo Vet. 1999:307-311.

Goellner, R. 1976. Experiences in reptilian and amphibian nutrition. Proc. Amer. Assoc. Zoo Vet. 1976:64-69.

Goellner, R. 1979. Hatching techniques for reptile eggs at the St. Louis Zoo, p. 27-28. In R. A. Hahn (ser. ed.), 1st & 2nd International Herpetological Symposium on Captive Propagation and Husbandry. International Herpetological Symposium, Thurmont MD.

Hoessle, C. 1969. Simple incubators for reptile eggs at St. Louis Zoological Park. Inter. Zoo Yearb. 9:13-14.

Lentz, M. J. R., and C. H. Hoessle. 1971. Mite control at St. Louis Zoo. Inter. Zoo Yearb. 11:235. [Account detailed the use of Shell No-Pest Strips.]

Patton, R. S., and R. Goellner. 1979. Research results of development of a prepared reptile diet, p. 21-26. In R. A. Hahn (ser. ed.), 1st & 2nd International Herpetological Symposium on Captive Propagation and Husbandry. International Herpetological Symposium, Thurmont MD.

Medical Management

Wallach, J. D. 1983. Diseases of Exotic Animals: Medical and Surgical Management. J. D. Wallach, and W. J. Boever (eds.). Saunders, Philadelphia [editors associated with Zoo].

Wallach, J. D., and C. Hoessle. 1967. Visceral gout in captive reptiles. J. Amer. Vet. Med. Assoc. 151:897-899.

Reproduction

Ettling, J. 1991. An overview of amphibian and reptile reproduction at the St. Louis Zoological Park, p. 96-105. In A. W. Zulich (ed.), 14th International Herpetological Symposium on Captive Propagation and Husbandry. International Herpetological Symposium.

Conservation

Ettling, J. 2005. Saint Louis Zoo Wildcare Institute: Center for Conservation of Near East Mountain Vipers. Herpetol. Rev. 36:231-232.

Milwaukee County Zoological Gardens (1892)

***History*:** The new combination reptile building and aquarium opened May 1968 and featured a large aquatic display in the center of the lobby which housed Wisconsin native fishes. Around the perimeter were a number of elaborate exhibits such as the Amazon forest which had reptiles and fishes, the largest captive yellow-spotted side-necked turtle I have ever seen, and a crocodilian enclosure.

Collection: According to *International Zoo Yearbook* (1998:Volume 28), composition of the collection was 44 taxa of reptiles with over 150 specimens and 9 taxa of amphibians numbering over 20.

Staff and Scientific Achievements: The first curator, David Zucconi, began his career working for Carl Kauffeld and later was installed as Director of Tulsa and El Paso Zoos. His successor Sherman Ketchum, worked for William Austin at the Detroit Zoo. When he retired, Richard Sajdak assumed the curatorial job until he retired in 1996. Craig Berg is the current curator.

Personal Reflections: The naturalistic display of native fishes of Wisconsin is one of my favorite zoo exhibits.

Historical Overview

Bruhin, T. A. 1874. Die Amphibien des County-Milwaukee (The amphibians of the Milwaukee County Zoo). Zool. Gart., Frankfurt a. M. 15:312-314.

Card, W., and J. B. Murphy. 2000. Lineages and histories of zoo herpetologists in the United States. Herpetol. Circ. 27:1-45 [portrait of Richard Sajdak].

Miller, D. 2004. Sherm Ketcham (1926–2004). Herpetol. Rev. 35:217.

Nickerson, M. A., A. Williams, and S. Ketchum. 1980. Herpetology in Milwaukee. Herpetol. Rev. 11(2):27-28 [discussion of Zoo].

Speidel, G. 1969. Aquarium and reptile building at Milwaukee Zoo. Inter. Zoo Yearb. 9:71-72.

Zucconi, D. 1967. Impressive new aquarium and reptile building for Milwaukee County Zoological Park. Herpetol. Rev. 1(1):9.

Zucconi, D. G. 1968. Reptile & amphibian exhibits in the Milwaukee Zoo's new Reptile/Aquarium Building. Bull. Philadelphia Herpetol. Soc. 16:21-30. [Contribution was the last on reptile buildings in the United States. Photographs of the building, public areas, rear section, a very young Zucconi and a few alligators were included.]

Medical Management

Beehler, B. A., and A. M. Sauro. 1983. Aerobic bacterial isolates and antibiotic sensitivities in a captive reptile population. Proc. Amer. Assoc. Zoo Vet. 1983:198-201.

Conservation

King, R., C. Berg, and B. Hay. 2004. A repatriation study of the eastern massasauga (*Sistrurus catenatus catenatus*) in Wisconsin. Herpetologica 60:429-437.

Sajdak, R. A. 2001. Patriarchs of the bluffs. Fauna May/June:8-21 [description of long-term monitoring study on timber rattlesnakes in Wisconsin].

General

Henderson, R. W., and R. A. Sajdak. 1996. Diets of West Indian racers (Colubridae: *Alsophis*): composition and biogeographical implications, p. 327-338. In R. Powell, and R. W. Henderson (eds.), Contributions to West Indian Herpetology: A Tribute to Albert Schwartz. Society for the Study of Amphibians and Reptiles, Contributions to Herpetology, volume 12, Ithaca NY.

Henderson, R. W., M. H. Binder, and R. A. Sajdak. 1982. Ecological relationships of the tree snakes *Uromacer catesbyi* and *U. oxyrhynchus* (Colubridae) on Isla Saona, República Dominicana. Amphibia-Reptilia 2:153-163.

Nickerson M. A., R. A. Sajdak, R. W. Henderson, and S. Ketchum. 1978. Notes on the movements of some Neotropical snakes (Reptilia, Serpentes). J. Herpetol. 12:419-422.

Sajdak, R. A. 1983. Herpetological research in zoos: A literature survey, 1977-1981. Zoo Biol. 2:149-152.

Sajdak, R. A. 2003. The shape of things, p. 185-189. In R. W. Henderson and R. Powell (eds.), Islands and the Sea: Essays on Herpetological Exploration in the West Indies. Society for the Study of Amphibians and Reptiles, Contributions to Herpetology, volume 20, Ithaca NY [description of field work on vine snakes (*Uromacer*) in Hispaniola in 1980].

Sajdak, R. A., and R. W. Henderson. 1991. Status of West Indian racers in the Lesser Antilles. Oryx 25:33-38.

Sajdak, R. A., M. A. Nickerson, R. W. Henderson, and M. W. Moffett. 1980. Notes on the movements of *Basiliscus plumifrons* (Sauria: Iguanidae) in Costa Rica. Milwaukee Public Mus. Contrib. Biol. Geol. No. 36.

Toledo Zoological Gardens (1900)

History: In 1929, the United States suffered through the Great Depression. In an attempt to reverse the downward financial spiral and return to a spirit of optimism, President Franklin Delano Roosevelt started the Works Progress Administration, a program to put citizens back to work on civic projects. The Toledo Zoo benefited from his initiative and a new reptile building was conceived.

Facility and Collection: Zoo administrators were fortunate to have Roger Conant on board as curator of reptiles for he was intimately involved in the design and execution of this new reptile building. This edifice, which stands today, is a beautiful combination of stone and wood. Elaborate displays were built which complemented the varied collection. Although this building is basically unchanged, there have been improvements. The center exhibit has been replaced with computer-enhanced interactive modules. An extension to the building includes increased animal holding enclosures, laboratory, and office space.

Staff and Scientific Achievements: One of the most gratifying aspects of any zoo conservation program is the successful merging of captive and field components. The Zoo has two, both stellar models. The first, directed by Curator of Conservation Peter Tolson, is broad-based for it focuses on the reproductive and conservation biology of the insular boid snake genus *Epicrates,* including reintroduction of the Mona Island boa. The second, by Curator of Reptiles R. Andrew Odum, was critical in highlighting the perilous status of the endangered Aruba Island rattlesnake. Elements include captive management, with founder snakes distributed to other zoos for breeding purposes. Since this taxon is particularly susceptible to the pathogenic paramyxovirus, studies continue to develop a vaccine. Now departed Director William Dennler, formerly the reptile curator, has been instrumental in supporting these enterprises.

Personal Reflections: There has been a long history of herpetological research at this zoo. When I visited some years ago, the collection of *Epicrates* assembled by Pete Tolson was the most complete that I had ever seen in a zoo. Andy Odum, the most technically proficient zoo person I know, demonstrated his newly designed tuatara enclosure and interactive computer programs for visitors. He teaches a course on data collection and management for AZA.

Historical Overview

Conant, R. 1934. Guide to the Reptile House in the Toledo Zoological Park. 1-25. [When the Reptile Building opened in 1934, Curator Conant wrote this account of the biology and natural history of amphibians and reptiles which is more than just a guide book. There is an introduction by Alexander G. Ruthven asking why a zoo should have a reptile house: for educational purposes! There are many photographs by Conant and others and a final page explaining snake bite first aid.]

Conant, R. 1934. Toledo's Museum of Natural History. Toledo Field Nat. Assn. Ann. Bull. :28-29.

Conant, R. 1982. Herpetology in Ohio-Fifty Years Ago. Toledo Herpetological Society, Toledo OH. [Special publication of the Toledo Herpetological Society covered history, collectors and collecting in Ohio. 64 pp.]

Conant, R. 1997. A Field Guide to the Life and Times of Roger Conant. Selva, Tyler TX [memories of Conant's early years in Toledo].

Morphology, Taxonomy & Systematics

Conant, R. 1938. The reptiles of Ohio. Amer. Midl. Natur. 20(1):1-200. [This study, updated in 1951, became the standard for publications dealing with state herpetofauna.]

Tolson, P. J. 1987. Phylogenetics of the boid snake genus *Epicrates* and Caribbean vicariance theory. Occ. Pap. Mus. Zool. Univ. Michigan 715:1-68.

Tolson, P. J., and R. W. Henderson. 1993. The Taxonomy and Natural History of West Indian Boas. R & A Publ., Ltd., Portishead UK.

Conservation

Odum, R. A., and M. J. Goode. 1994. The Species Survival Plan for *Crotalus durissus unicolor*: a multifaceted approach to conservation of an insular rattlesnake, p. 363-368. In J. B. Murphy, K. Adler, and J. T. Collins (eds.), Captive Management and Conservation of Amphibians and Reptiles. Society for the Study of Amphibians and Reptiles. Contributions to Herpetology, volume 11, Ithaca NY.

Quick, J. S., H. K. Reinert, E. R. de Cuba, and R. A. Odum. 2005. Recent occurrence and dietary habits of *Boa constrictor* on Aruba, Dutch West Indies. J. Herpetol. 39:304-307.

Tolson, P. J. 1988. Critical habitat, predator pressures, and the management of *Epicrates monensis* (Serpentes: Boidae) on the Puerto Rico Bank: a multivariate analysis, p. 228-238. Management of Amphibians, reptiles, and Small Mammals in North America. Gen. Tech. Rept. U.S. Dept. Agriculture, Washington DC.

Tolson, P. J. 1996. Conservation of *Epicrates monensis*

on the satellite islands of Puerto Rico, p. 407-416. In R. Powell, and R. W. Henderson (eds.), Contributions to West Indian Herpetology: A Tribute to Albert Schwartz. Society for the Study of Amphibians and Reptiles, Contributions to Herpetology, volume 12, Ithaca NY. [Twenty-eight captive-born snakes from seven zoos were released on Cayo Ratones.]

Husbandry

Conant, R. 1929. Notes on a water moccasin in captivity *Agkistrodon piscivorus* (female). Bull. Antivenin Inst. Amer. 3(3):61-64. [This journal was published from 1927 until 1932 and mostly covered taxonomy and toxicology; several papers on behavior and husbandry were included.]

Conant, R., and R. M. Perkins. 1931. This mite question. Bull. Antivenin Inst. Amer. 5(2):36-39.

Odum, A. 1984. Water quality, an often overlooked parameter for the amphibian enclosure, p. 33-58. In R. A. Hahn (ser. ed.), 8th International Herpetological Symposium on Captive Propagation and Husbandry. International Herpetological Symposium, Thurmont MD.

Medical Management

Lloyd, M. L. 1990. Reptilian dystocias review - causes, prevention, management and comments on the synthetic hormone vasotocin. Proc. Amer. Assoc. Zoo Vet. 1990:290-294.

Lloyd, M. L. 1999. Crocodilian anesthesia, p. 205-216. In M. E. Fowler, and R. E. Miller (eds.), Zoo and Wild Animal Medicine: Current Therapy. W. B. Saunders Co., Philadelphia, London, etc.

Lloyd, M. L., and J. Flanagan. 1991. Recent developments in ophidian paramyxovirus research and recommendations on control. Proc. Amer. Assoc. Zoo Vet. 1991:151-155. [Toledo and Houston Zoos.].

Lloyd, M. L., T. Reichard, and R. A. Odum. 1994. Gallamine reversal in Cuban crocodiles (*Crocodylus rhombifer*) using neostigmine alone versus neostigmine with hyaluronidase. Proc. Amer. Assoc. Zoo Vet.:117-120 [Roger Williams Park Zoo, Providence, RI, and Toledo Zoo].

General

Anon. 1978. A report on the restoration of the preserved amphibian and reptile collection of the Toledo Zoological Society. Herpetol. Rev. 9(2): 48-49.

Conant, R. 1930. Field notes on a collecting trip. Bull. Antivenin Inst. America 4(3):60-64.

Conant, R. 1933. Report of reptiles and amphibians collected during 1933. Toledo Field Nat. Assn. Ann. Bull., Spring 1933:7-8.

Conant, R. 1934. Herpetological activities - 1934. Toledo Field Nat. Assn. Ann. Bull., Spring 1934 :p. 23.

Columbus Zoological Gardens (1927) (Ohio)

Collection: According to *International Zoo Yearbook* (1998:Volume 28), composition of the collection was 135 taxa of reptiles with 832 specimens and 26 taxa of amphibians numbering 155.

Staff and Scientific Achievements: When curator Lou Pistoia died, his assistant J. Michael Goode (1945-1998) became curator and developed one of the finest zoo collections of freshwater turtles. Reproduction quickly outstripped available holding enclosures so many eggs and young were sent to outside researchers for their studies. Presently, the curatorial position is unfilled.

Fig. 134. Photograph of late herpetological curator J. Michael Goode at Columbus Zoological Gardens in 1993. *Courtesy of Dan Bagley, Columbus Zoo.*

Historical Overview

Anon. 1976. Obituary of: Louis J. Pistoia (1917-1976). Herpetol. Rev. 7(3):120.

Card, W., and J. B. Murphy. 2000. Lineages and histories of zoo herpetologists in the United States. Herpetol. Circ. 27:1-45. [biographies of Louis Pistoia and J. Michael Goode.].

Goode, M. 1980. The reptile and amphibian collection at the Columbus Zoo, p. 15-22. In R. A. Hahn (ser. ed.), 3rd International Herpetological Symposium on Captive Propagation and Husbandry. Interna-

tional Herpetological Symposium, Thurmont MD.
Pistoia, L. 1968. The Columbus Zoo's new Reptilia-Amphibia Hall: A new approach to design. Herpetol. Rev. 3:4-6.

General

Ewert, M. A., R. E. Hatcher, and J. M. Goode. 2004. Sex determination and ontogeny in *Malacochersus tornieri*, the pancake tortoise. J. Herpetol. 38:291-295.

Detroit Zoological Institute (1928)

History: The Holden Museum of Living Reptiles, opened in 1960, was a beautiful combination of elaborate exhibits, large lobby receptacles with living plants, exposed wooden ceiling beams, and planted crocodilian and aquatic turtle enclosures. When I visited this facility shortly after its opening, several features remain vivid in my memory. The first was a group of "subterranean" modules which had sliding metal tops that could be moved to the rear by a zoo visitor in order to see the snake within. All day long, there was a constant roar of metal upon metal as the tops clanged against the façade. The second was like a rotating "Lazy-Susan" used in restaurants to display food. When I asked Curator Bill Austin if the herps displayed suffered dizziness, he said that the inhabitants fared better than many in conventional enclosures. The third was the use of double-doors so the keeper would be better protected while servicing venomous snakes. Today, the Zoo is now called the Detroit Zoological Institute and the first international amphibian breeding facility has been opened.

Fig. 135. Holden Museum of Living Reptiles at Detroit Zoological Institute constructed in 1960. *Photograph by Bill Flanagan.*

Fig. 136. National Amphibian Conservation Center at Detroit Zoological Institute. *Photograph by Bill Flanagan.*

Facility and Collection: The early collection was varied: goliath frog, tuatara, bushmaster, many Australian elapids, pythons, varanids and skinks, and two Bibron's softshell turtles. The new National Amphibian Conservation Center is surely the most ambitious project for amphibian conservation and research ever attempted by zoo workers and it is magnificent. There are several world class programmatic elements related to this facility: superb exhibits within a building located in a natural Michigan wetland setting, strong educational component, conservation orientation, breeding endangered species, the future establishment of a chair and field support for prominent amphibian biologists, and community outreach. It is refreshing to see that an innovative zoo director, Ron Kagan, and his talented staff have chosen amphibians to highlight in a zoo; it rarely happens today. The Institute received the AZA Edward H. Bean Award in 1998 for reproduction of the emperor newt. In 2002, the amphibian building received an award as the best new zoo exhibit by AZA.

According to *International Zoo Yearbook* (1998:Volume 28), composition of the collection was 72 taxa of reptiles with nearly 230 specimens and 35 taxa of amphibians numbering 242.

Staff and Scientific Achievements: William Austin, now deceased, was the Curator of Reptiles when the Holden Museum of Living Reptiles was opened and James Langhammer was Curator of Education. Later, they switched positions until Langhammer went to head the Belle Isle Aquarium. The next curator, Gordon Schuett, stayed for a short time, then left to enroll in a doctoral program to study snake behavior. Steve

Fig. 137. Photograph of late herpetological curator William A. Austin in front of king cobra exhibit at Detroit Zoo around 1961–1962. *Photography Department, Detroit Zoo.*

Connors filled the vacant position but later moved to the Birmingham Zoo. Andrew Snider moved from the herpetology department at the Audubon Park Zoo but has recently moved to another zoo. Kevin Zippel filled the newly created curatorial job to oversee the new amphibian facility but has now left the zoo to work as Amphibian Ark Coordinator. The position is currently vacant.

Personal Reflections: I began visiting the zoo in the late 1950s to see Austin and Langhammer. When I asked Bill what herpetological curators discuss when they meet, he said the most important topic was the best brand of coffee, followed by the most palatable beer, the best cut of steak, the most exciting night-clubs, the most attractive movie actresses, and finally the best glass cleaner to polish exhibit glass. When I asked the more conservative Langhammer the same question, he said "retirement!"

Historical Overview

Austin, W. A. 1961. The Holden Museum of Living Reptiles, Detroit Zoological Park, Detroit, Michigan (Dedication Day - August 17, 1960). Bull. Philadelphia Herpetol. Soc. 9(1):13-16. [Contribution was the fifth on reptile buildings in the United States. Photographs of the building facade, public and rear area, and several exhibits were included.]

Austin, W. A. 1963. The Holden Museum of Living Reptiles (Detroit). Inter. Zoo Yearb. 5:247-251. [Account described exhibits, temperature control, filtration, pest control. reserve facilities and public reaction.]

Austin, W. A. 1974. The First Fifty years . . . an Informal History of the Detroit Zoological Park and the Detroit Zoological Society. Detroit Zoological So-

ciety, Detroit MI. [Description of Holden Museum of Living Reptiles on p. 61 which attracted an unbelievable 30,000 visitors on 30 Aug 1960, shortly after opening day.]

Card, W., and J. B. Murphy. 2000. Lineages and histories of zoo herpetologists in the United States. Herpetol. Circ. 27:1-45 [biographies of William Austin and James Langhammer].

Morphology, Systematics & Taxonomy

Langhammer, J. K. 1983. A new subspecies of boa constrictor: *Boa constrictor melanogaster*, from Ecuador. Tropical Fish Hobbyist 32(4):70-79 [not recognized].

Zippel, K. C., H. B. Lillywhite, and C. R. J. Mladinich. 1998. Contribution of the vertebral artery to cerebral circulation in the rat snake *Elaphe obsoleta*. J. Morphol. 238:39-51.

Zippel, K. C., H. B. Lillywhite, and C. R. J. Mladinich. 2001. New vascular system in reptiles: Anatomy and postural hemodynamics of the vertebral venous plexus in snakes. J. Morphol. 250:173-184.

Zippel, K. C, H. B. Lillywhite, and C. R. J. Mladinich. 2003. Anatomy of the crocodilian spinal vein. J. Morphol. 258:327-335.

Reproduction

Connors, J. S. 1986. A captive breeding of the Great Basin gopher snake, *Pituophis malanoleucus deserticola*. Herpetol. Rev. 17(1):12-13. [At time of publication, author was at the Cheyenne Mountain Zoological Park, Colorado.]

Pitman, C. R. S. 1974. A Guide to the Snakes of Uganda. Wheldon & Wesley, Codicote UK [information on reproduction of ball pythons from the captive colony at Zoo].

Husbandry

Langhammer, J. 1962. Care in captivity series: Care of the ratsnake genus *Elaphe* (I). Bull. Philadelphia Herpetol. Soc. 10(4):33-34.

Conservation

Zippel, K. C. 2002. Conserving the Panamanian Golden Frog: Proyecto Rana Dorada. Herpetol. Rev. 33:11-12.

Zippel, K. C. 2002. Emperor Newt (*Tylototriton shanjing*) Studbook. Detroit Zoological Institute, Royal Oak MI. 162 pp.

General

Snider, A. T., and J. K. Bowler. 1992. Longevity of reptiles and amphibians in North American collections. Second Edition. Soc. Study Amphib. Rept. Herp. Circ. No. 21:1-40.

Snider, A. T., and K. Zippel. 2000. Amphibian conservation at the Detroit Zoological Institute. Froglog No. 40:2. [Zoo underwrites cost of this publication.]

Zippel, K., and A. Snider. 2001. The Detroit Zoo makes a bold statement for amphibians. AZA Communique March 2001:5-6, 51.

Chicago Zoological Park (1934)

History and Mission: Besides being a productive center for herpetology for many years, the Chicago Zoological Park, also known as the Brookfield Zoo, is recognized globally as a major force in international conservation, due in large part to of the efforts of retired Director George Rabb.

Facility and Collection: The building is traditional with exhibits around the outer wall. These cages are very tall and somewhat difficult to service. The rear areas are small and rather cramped. In the center, an array of enclosures house aquatic turtles, large lizards, giant constrictors, and smaller crocodilians. On both sides of the building, two large open displays with skylights and living plants accommodate crocodilians and giant tortoises. There are a number of small amphibians and reptiles on display. In the past, the collection has featured Arafuran wart snakes, a large alligator snapping turtle, many Indo-Australian chelid turtles, Weber's water dragon, Chinese cobras, and an impressive assemblage of Russian reptiles. There were several mixed exhibits such as king cobras and tokay geckos, mangrove snakes and blood pythons, and rattlesnakes and tortoises.

According to *International Zoo Yearbook* (1998:Volume 28), composition of the collection was 102 taxa of reptiles with 172 specimens and 17 taxa of amphibians numbering 72.

Staff and Scientific Achievements: Grace Olive Wiley was hired as curator at the Zoo in the 1930s. Her successor was Robert Snedigar and I met him often during the early 1950s. He had an annual standing order for 20 snakes of each species from an animal dealer: copperheads, cottonmouths, eastern and western diamondback rattlesnakes and an assortment of ratsnakes and kingsnakes. Attrition was rather high as snakes were not maintained singly or in small groups but rather as "snake dens" in large exhibits. Emil Rokosky and Bert Tschambers, employees in the

Fig. 138. Curator Robert Snedigar, probably in the late-1950s, lecturing group of zoo visitors with reptile house in background at Chicago Zoological Park (Brookfield Zoo). *Provided by George Rabb, Chicago Zoological Society.*

department at this time, published several papers on reptiles in the collection, which included green tree pythons rarely seen in the United States.

When Snedigar retired in 1963, Ray Pawley came across town from the Lincoln Park Zoo. Improved husbandry protocols were installed and thematic exhibits were designed. Some of the most memorable were garter snakes in a trash dump to show environmental degradation, Russian ratsnakes in a human dwelling with a hammer and sickle hanging from the wall, boa constrictor on a log hanging from the roof, and king cobra living in an Indian pagoda. A number of important husbandry techniques were developed at the Zoo by Ray, who retired in 1998.

After Ray retired, John Cadle assumed the curatorial mantle. His research has centered on snake evolutionary history. Papers in print cover molecular systematics of xenodontine colubrid snakes from South and Central America xenodontines; overview of xenodontine phylogeny and the history of New World snakes; phylogenetic patterns, lineage components, biogeography, and ecological structure of Neotropical colubrid snake fauna; biogeography, phylogenetic relationships among advanced snakes; albumin immunological evidence and relationships of sea snakes; distribution patterns of some amphibians, reptiles, and mammals of the eastern Andean slope of southern

Peru; and phylogenetic relationships and molecular evolution in uropeltid snakes and New World coral snakes. Cadle is no longer at the zoo, and the position remains unfilled.

Since 1976, George Rabb had been Director. He began his professional career as research coordinator at the Zoo in 1956, was former Chairman of the Species Survival Commission (SSC) of IUCN-The World Conservation Union beginning in 1989. In this role, he was crucial in forming and supporting the Declining Amphibian Populations Task Force under the aegis of IUCN. For his accomplishments, he was awarded the Peter Scott Award in 1996 by SSC. Rabb and his staff have been at the forefront in developing protocols for successful captive management programs. Much of their effort has centered on designing techniques for surveying and maintaining genetic diversity in zoo animal colonies. He was instrumental in establishing the International Species Information System (ISIS), the inventory listing of captive animal populations which is crucial in providing data to insure proper pairings of potential breeders. He is internationally known as a conservationist and herpetologist, serving as president of the American Society of Ichthyologists and Herpetologists. He has worked closely with his colleague, the late Hymen Marx at the Field Museum of Natural History, on many seminal papers on snake evolutionary biology. Rabb published a number of important papers, some with his late wife Mary, on amphibian reproductive behavior.

Personal Reflections: George Rabb and Ray Pawley were important mentors for me during my adolescence and they encouraged me to consider zoo herpetology as a career.

When I was in high school during the early 1950s, I was astounded to see several green tree pythons, obtained from Sir Edward Hallstrom of the Taronga Zoo in Australia. These snakes were on display for years. Two species known to be delicate were successfully kept: Goliath frog and marine iguana. These were the first living examples I had ever seen.

Historical Overview

Card, W., and J. B. Murphy. 2000. Lineages and histories of zoo herpetologists in the United States. Herpetol. Circ. 27:1-45 [biographies of Grace Olive Wiley, Robert Snedigar and Ray Pawley].

International Union of Directors of Zoological Gardens (of the World Zoo Organization). 1993. The World Zoo Conservation Strategy: The Role of Zoos and Aquaria of the World in Global Conservation, p. 1-76. Chicago Zoological Society, Brookfield IL. [Rabb believed that zoos must change from menageries to conservation centers and this summary incorporates his vision.]

Pawley, R. L. 1967. Mixing it up in Brookfield's reptile house. Anim. Kingdom 70:90-95.

Pawley, R. L. 1971. Mixed species exhibits in the reptile building at Brookfield Zoo. Inter. Zoo Yearb. 11:220-224.

Ross, A. F. 1997. Let the Lions Roar. The Evolution of Brookfield Zoo. Chicago Zoological Society, Chicago. [Provides accounts of reptile curator Grace Olive Wiley (p. 32), picture of subsequent curator Robert Snedigar (p. 83), measuring an anaconda (p. 84), picture of newest and recently retired curator Ray Pawley as a young man (p. 126) and a portrait of George Rabb as the new Director, installed in 1976 (p.173).]

Ross, A. 2001. Rabb, George B. 1930 - American director of the Chicago Zoological Park, p. 1035-1037. In Bell, C. E. (ed.), Encyclopedia of the World's Zoos. Fitzroy Dearborn Publishers, Chicago, London.

Morphology, Systematics & Taxonomy

Liem, K. F., H. Marx, and G. B. Rabb. 1971. The viperid snake *Azemiops*: Its comparative cephalic anatomy and phylogenetic position in relation to Viperinae and Crotalinae. Fieldiana Zoology 59:62-126.

Lombard, R. E., H. Marx, and G. B. Rabb. 1986. Morphometrics of the ectopterygoid in advanced snakes (Colubroidea): a concordance of shape and phylogeny. Biol. J. Linn. Soc. 27: 133-164.

Marx, H., and G. B. Rabb. 1965. Relationships and zoogeography of the viperine snakes (Family Viperidae). Fieldiana Zoology 44:161-206.

Marx, H., and G. B. Rabb. 1970. Character analysis: An empirical approach applied to advanced snakes. J. Zool. (London) 161:525-548.

Marx, H., and G. B. Rabb. 1972. Phyletic analysis of fifty characters of advanced snakes. Fieldiana Zoology 63:1-321.

Marx, H., G. B. Rabb, and S. J. Arnold. 1982. *Pythonodipsas* and *Spalerosophis*, colubrid snake genera convergent to the vipers. Copeia 1982:553-561.

Myers, C. W., and J. E. Cadle. 2003. On the snake hemipenis, with notes on *Psomophis* and techniques of eversion: A response to Dowling. Herpetol. Rev. 34:295-302.

Rabb, G. B. 1955. A new salamander of the genus *Parvimolge* from Mexico. Breviora No. 42:1-9 [description of *Parvimolge praecellens*, genus now known as *Pseudoeurycea*].

Rabb, G. B. 1956. A new plethodontid salamander from Nuevo León, Mexico. Fieldiana Zoology 39:11-20 [description of *Chiropterotriton prisca*, now *C. priscus*].

Rabb, G. B. 1957. A new race of the iguanid lizard *Leiocephalus carinatus* from Cayman Brac, B.W.I. Herpetologica 13:109-110 [description of *Leiocephalus carinatus granti*].

Rabb, G. B. 1959. A new frog of the genus *Plectrohyla* from the Sierra de los Tuxtlas, Mexico. Herpetologica 15:45-47 [description of *Plectrohyla pycnochila*].

Rabb, G. B., and H. Marx. 1973. Major ecological and geographical patterns in the evolution of colubroid snakes. Evolution 27:69-83.

Rabb, G. B., and J. E. Mosimann. 1955. The tadpole of *Hyla robertsorum*, with comments on the affinities of the species. Occ. Pap. Mus. Zool. Univ. Michigan (563):1-9.

Reproduction

Pawley, R. 1999. Parthenogenesis in snakes: A case for investigation. Inter. Zoo News 46 (No. 293):205-207 [Arafuran file snake].

Husbandry

Blakely, R. L. 1964. Notes on new diets for birds and reptiles at Chicago Zoological Park. Inter. Zoo Yearb. 6:105-106 [covers Galápagos tortoise, snakes].

Pawley, R. 1969. Observations on a prolonged food refusal period of an adult fer-de-lance *Bothrops atrox asper*. Inter. Zoo Yearb. 9:58-59. [Snake refused food for nearly 18 months.]

Pawley, R. 1998. It's new, it's high-tech, it's a thermometer gun. Inter. Zoo News 45(No.285):212-219.

Snedigar, R. 1939. Our Small Native Animals, Their Habits and Care, Random House, New York [reprinted by Dover Publ., New York in 1964].

Tschambers, B. 1948. Feeding of the mud snake, *Farancia abacura reinwardtii*, in captivity. Herpetologica 4:210.

Medical Management

Rokosky, E. J. 1948. A bot-fly parasitic in box turtles. Nat. Hist. Misc. Chicago Acad. Sci. :32.

Behavior

Rabb, G. B. 1972. An issue about colubrid snakes. Brookfield Bandarlog No. 39:1-14 [series of articles,

including one by Gordon Burghardt, on feeding behavior].

Rabb, G. B., and R. Snedigar. 1960. Notes on feeding behavior of an African egg-eating snake. Copeia 1960:59-60.

Captive Management

Lacy, R. C. 1987. Loss of genetic diversity from managed populations: Interesting effects of drift, mutation, immigration, selection, and population subdivision. Conserv. Biol. 1:143-158.

Lacy, R. C. 1989. Analysis of founder representation in pedigrees: Founder equivalents and founder genome equivalents. Zoo Biol. 8:111-123.

General

Pawley, R., J. Ott-Joslin, and R. W. Torgerson. 1986. The reptiles, their curators and the veterinarian > harmony or cacophony. Proc. Amer. Assoc. Zoo Vet. 1986:123-124.

Pough, F. H., R. M. Andrews, J. E. Cadle, M. L. Crump, A. H. Savitsky, and K. D. Wells. 1998. Herpetology. Prentice Hall, Upper Saddle River NJ.

Rabb, G. B., and E. B. Hayden. 1957. The Van Voast-American Museum of Natural History Bahama Islands Expedition record of the expedition and general features of the Islands. Amer. Mus. Novitates (No.1836):1-53.

Rabb, G. B., and C. D. Saunders. 2005. The future of zoos and aquariums: Conservation and caring. Inter. Zoo Yearb 29:1-26.

Tschambers, B. 1948. Death of a snake from toad venom. Herpetologica 4:214.

Tschambers, B. 1949. *Boa constrictor* eats porcupine. Herpetologica 5:141.

Wiley, G. O. 1930. Notes on the Neotropical rattlesnake *Crotalus terrificus basiliscus* in captivity. Bull. Antivenin Inst. Amer. 3(4):103-105. [Author described taming this rattlesnake species.]

SOUTHWESTERN UNITED STATES

Dallas Zoo (1888) and Aquarium (1936)

History: When I arrived at the Zoo in March 1966, only a few giant tortoises and some local reptiles were scattered throughout the park. Construction was nearly finished on the innovative Pierre A. Fontaine Reptile, Bird and Rainforest building but a sizeable number of new amphibians and reptiles were arriving almost daily before its completion. This was a problem as these animals had to be held in the small zoo hospital, a facility not designed to accommodate them. Temporary cages and aquariums were strewn everywhere, and it was exciting and challenging to handle giant constrictors, venomous snakes, large crocodilians and lizards in cramped quarters but we all survived. Since the collection was recent, it required a great deal of effort to acclimate and exhibit hundreds of our charges for the grand opening in a few months. Although we were all sleep-deprived, the debut went without incident.

Collection: Beginning in the 1970s, Jonathan Campbell and Barry Armstrong took annual collecting trips to Mexico and Central America, in part to secure specimens for our Zoo. As they were very successful, the collection comprised many Neotropical pitvipers and rattlesnakes, unusual colubrid and protocolubrid snakes, as well as a surprising variety of lizards and chelonians. Mostly through their efforts, over 50 varieties of rattlesnakes were maintained. The Aquarium staff has successfully reproduced several endangered Texas cave salamanders.

According to *International Zoo Yearbook* (1998:Volume 28), composition of the Zoo collection was 107 taxa of reptiles with 389 specimens and 16 taxa of amphibians numbering 120. The Aquarium held 6 taxa of reptiles with 11 specimens and 5 taxa of amphibians numbering nearly 200.

Staff and Scientific Achievements: When the building opened, Donald Way was Supervisor but left a short time later. I was elevated to his position and Michael Edwards was Lead Keeper. When he left to pursue a country music career, Ardell Mitchell filled his job. In 1977, my curatorial slot was added so Ardell became Supervisor and David Barker became Lead Keeper. After Barker returned to graduate school, another reorganization occurred: Mitchell became Assistant Curator, and Donal Boyer became Supervisor until he left for the San Diego Zoo. When Mitchell retired as Curator, Winston Card became titular head of the Department but he left to go to the Cincinnati Zoo. Ruston Hartdegen currently fills the curatorial post.

There have been a number of studies generated on reproduction and behavior in rattlesnakes and other Neotropical pitvipers, colubrid snakes, as well as monitor lizards,

Personal Reflections: As I reflect over the nearly 30 years I spent in Dallas, one consistent thought remains. The skills, talents and dedication of my co-workers were awesome. Over the years, the staff was composed of individuals who specialized in art, photography, exhibit design, herpetological history, behavioral research, field biology, herpetoculture, education, and conservation. This energy and synergy were critical in developing broad-based departmental initiatives. Although many were suspect in polite company, their intellectual and personal attributes contributed mightily to a pleasant workplace and I never regretted coming to the Zoo.

Historical Overview

Card, W., and J. B. Murphy. 2000. Lineages and histories of zoo herpetologists in the United States. Herpetol. Circ. No. 27:1-44 [biography and portrait of me after retirement].

Fontaine, P. A. 1967. Bird, reptile and rain forest building at Dallas Zoo. Inter. Zoo Yearb. 7:75. [One of the best features of this building is the spacious and bright keeper service area.]

Mitchell, L. A. 1980. Reptile reproduction at the Dallas Zoo, p. 44-46. In R. A. Hahn (ser. ed.), 3rd International Herpetological Symposium on Captive Propagation and Husbandry. International Herpetological Symposium, Thurmont MD.

Murphy, J. B. 1979. Herpetology at the Dallas Zoo. Herpetol. Rev. 10(4):111-112.

Murphy, J. B., and L. A. Mitchell. 1989. The other side: An experiment in reptile appreciation at the Dallas Zoo. Inter. Zoo Yearb. 28:205-207.

Morphology, Systematics & Taxonomy

Campbell, J. A., and J. B. Murphy. 1977. A new species of *Geophis* (Reptilia, Serpentes, Colubridae) from the Sierra de Coalcoman, Michoacan. J. Herpetol. 11:397-403. [*Geophis pyburni* was named in honor of amphibian biologist William Pyburn.]

Card, W., and A. G. Kluge. 1995. Hemipenal skeleton and varanid lizard systematics. J. Herpetol. 29:275-280.

Behavior

Chiszar, D., C. Castro, J. B. Murphy, and H. M. Smith. 1989. Discrimination between thermally different rodent carcasses by bull snakes (*Pituophis melanoleucus*) and cobras (*Naja pallida* and *Aspidelaps scutatus*). Bull. Chicago Herpetol. Soc. 24(10):181-183.

Garrett, C. M., and D. T. Roberts. 1990. Cannibalism in two species of arboreal pitviper, *Trimeresurus wagleri* and *Bothriechis schlegelii*. Herpetol. Rev. 21(3):54-55.

Tryon, B. W., and R. K. Guese. 1984. Death-feigning

in the gray-banded kingsnake *Lampropeltis alterna*. Herpetol. Rev. 15(4):108-109. [Raymond Guese was employed at Zoo.]

Husbandry

Boyer, T., and D. Boyer. 1992. Aquatic turtle care. Bull. Assoc. Rept. Amphib. Vet. 2(2):13-17.

Edwards, M. S. 1969. Notes on some tropidophid snakes in captivity. Inter. Zoo Yearb. 9:53-54.

Jordan, T. 1969. Notes on keeping arboreal and terrestrial amphibians in captivity. Inter. Zoo Yearb. 9:14-16.

Murphy, J. B. 1969. Notes on iguanids and varanids in a mixed exhibit at Dallas Zoo. Inter. Zoo Yearb. 9:39-41.

Murphy, J. B. 1971. A method for immobilizing venomous snakes at Dallas Zoo. Inter. Zoo Yearb. 11:233 [description of clear acrylic plastic tubing].

Murphy, J. B., and J. A. Campbell. 1987. Captive maintenance, p. 165-183. In R. A. Seigel, J. T. Collins, and S. S. Novak (eds.), Snakes: Ecology and Evolutionary Biology. Macmillan, New York.

Warwick, C., F. L. Frye, and J. B. Murphy (eds.). 1995. Health and Welfare of Captive Reptiles. Chapman & Hall, New York.

Reproduction

Barker, D. G., and T. M. Barker. 1996. The Lesser Sunda python (*Python timoriensis*): taxonomic history, distribution, husbandry, and captive reproduction, p. 103-108. In P. D. Strimple (ed.), Advances in Herpetoculture. Special Publication of the International Herpetological Symposium, Inc., No. 1. International Herpetological Symposium, Inc., Des Moines IA. [In 1975, the first captive reproduction of this species occurred at the Zoo where Dave Barker was employed. He was instrumental in inducing the snakes to reproduce successfully.]

Medical Management

Boyer, D., and T. H. Boyer. 1991. Trichlorfon spray for snake mites (*Ophionyssus natricis*). Bull. Assoc. Rept. Amphib. Vet. 1(1):2-3. [Donal Boyer and I worked together at the Zoo and his brother Tom, a recently graduated DVM, would visit periodically. I remember well the genesis of their idea to create an association of veterinarians who specialized in herpetological medicine. The Boyer brothers, with a few colleagues, decided to launch this Bulletin by soliciting manuscripts from their peers, never an easy proposition when a new venture is involved. In fact to prove this point, they asked me but I could never get my act together. Tom was the first editor and the Boyer team wrote many of the articles for this first issue. Don's wife, Sally, designed the Surinam toad logo. Today, the journal is highly respected and since 1994, an annual meeting has been held in the United States. Quite an accomplishment from such a modest beginning!]

Gamble, K. C., T. P. Alvarado, and C. L. Bennett. 1996. Plasma Itraconazole pharmacokinetics in spiny lizards (*Sceloporus* spp.). Proc. Amer. Assoc. Zoo Vet. 1986:245-246.

Hartdegen, R. W., M. J. Russell, and R. Buice. 1999. An enteric parasite survey of Neotropical herpetofauna. Herpetol. Rev. 30(1):26-28. [Samples were collected at La Selva Research Station, Heredia Province, Costa Rica, by employees at Zoo.]

Murphy, J. B., and J. T. Collins. 1983. A Review of the Diseases and Treatments of Captive Turtles. AMS Publishing, Lawrence, KS.

Murphy, J. B., and J. E. Joy. 1973. Removal of a deficient fang mechanism in a captive puff adder, *Bitis arietans*. Brit. J. Herpetol. 4:314-316.

Raphael, B. L. 1984. Diagnosis and management of renal gout in captive snakes. Proc. Amer. Assoc. Zoo Vet. 1984:53.

Raphael, B. L., and R. M. Robinson. 1984. Diagnosis and treatment of liver disease in a Boelen's python. Proc. Amer. Assoc. Zoo Vet. 1984:54.

Shadduck, J. A., and J. B. Murphy. 1979. Suggestions for the post mortem examination of reptiles. Herpetol. Rev. 10(4):113-115.

Upton, S. J., C. T. McAllister, and C. M. Garrett. 1993. A new species of *Eimeria* (Apicomplexa) from *Cordylus cataphractus* (Sauria: Cordylidae), from South Africa. J. Egyptian Soc. Parasitol. 23:189-193. [I was delighted to have a fecal parasite named in my honor.].

Philosophy

Chiszar, D., J. B. Murphy, and W. Iliff. 1990. For zoos. Bull. Psychon. Soc. 27(3):3-13 [response to Dale Jameison's article "Against Zoos"].

Murphy, J. B., and D. Chiszar. 1989. Herpetological masterplanning for the 1990s. Inter. Zoo Yearb. 28:1-7. [Topics included research, conservation, collection management and composition, exhibit development, and education.]

Murphy, J. B., K. Adler, and J. T. Collins. 1997. Herpetologists vs. herpetoculturists: Cooperation or competition? Reptiles March 1997:72-79.

General

Armstrong, B. L., and J. B. Murphy. 1979. The natural history of Mexican rattlesnakes. Univ. Kansas Mus. Nat. Hist. Spec. Publ.:1-88. [Study was done in part at Zoo which employed a captive and field component.].

Boyer, D. M. 1995. Venomous reptiles: An overview of families, handling, restraint techniques, and emergency protocol. Proc. Assoc. Rept. Amphib. Vet. 1995:27-29, 83-90. [Donal Boyer has been the editor of the AZA Antivenin Index for many years.]

Campbell, J. A. 1982. A new species of *Abronia* (Sauria, Anguidae) from the Sierra Jáurez, Oaxaca, México. Herpetologica 38:355-361 [description of *Abronia mitchelli*, named in honor of Ardell Mitchell, who collected the only known specimen].

Campbell, J. A., and W. W. Lamar 1989. The Venomous Reptiles of Latin America. Comstock Pub. Associates, Ithaca NY. [Jonathan Campbell started his professional career, earlier than he would care to admit, as Assistant Reptile Supervisor at the Fort Worth Zoo. During the 1970s, he spent much time in Mexico and Central America with the indescribable Barry Armstrong collecting a variety of herps, many of which they placed on loan at the Zoo. A variety of studies on reproductive biology and behavior resulted from their generosity. In this book, many of the specimens photographed were at the Zoo, including the astonishing yellow-blotched palm pitviper featured on the dust jacket. Campbell's stellar career has been described earlier (Fauna Magazine, 1997, vol 1(1):72-78). Before he started his ecotourism business, Greentracks, William Lamar was Curator of Herpetology at the Caldwell Zoo, Texas.]

Card, W., and D. T. Roberts. 1996. Incidence of bites from venomous reptiles in North American zoos. Herpetol. Rev. 27(1):15-16.

Greene, H. W. 1983. Field studies of hunting behavior by bushmasters. Amer. Zool. 23:897 [staff from Zoo assisted in study].

Greene, H. W. 1986. Diet and arboreality in the emerald monitor, *Varanus prasinus*, with comments on the study of adaptation. Fieldiana (New Series) 31:1-12 [data in part from Zoo].

Murphy, J. B. 1977. Regional herpetological societies. Herpetol. Rev. 8(3):85-87.

Murphy, J. B., and J. T. Collins. (eds.) 1980. Reproductive Biology and Diseases of Captive Reptiles. Society for the Study of Amphibians and Reptiles Con-

tributions to Herpetology, volume 1, Lawrence KS [15 authors associated with zoos].

Murphy, J. B., K. Adler, and J. T. Collins (eds.). 1994. Captive Management and Conservation of Amphibians and Reptiles. Society for the Study of Amphibians and Reptiles. Contributions to Herpetology, volume 11, Ithaca NY. [In August, 1991 at Penn State University, SSAR and Herpetologists' League sponsored a symposium on captive breeding and management which honored Roger Conant; this subsequent volume following the 1980 book included 21 authors representing zoos. The contributions reflected a significant number of zoo projects which included many non-zoo professionals as collaborators.]

Oklahoma City Zoo (1904)

***Facility and Collection*:** Although there are many attractive exhibits, the reptile building infrastructure is old, requiring constant maintenance. Since much of the building and many exhibits are constructed with wood which has been deteriorating, the staff has nonetheless accomplished many laudable milestones in spite of these limitations. Relief was in sight for plans were in motion to construct a major facility but the zoo administration unfortunately decided to eliminate this new building in the master plan. One exhibit features fauna from the Galápagos Islands, including a large exhibit housing tortoises. This exhibit was constructed in 1979 and renovated fifteen years later. The name was changed to Island Life Exhibit.

There are excellent programs in place. Years ago, Eileen Castle developed an impressive initiative for true chameleons, resulting in many captive breedings. Since these lizards are notoriously difficult to keep in captivity and many zoo workers tried unsuccessfully to do so, her accomplishments were noteworthy. A number of pythons, mastigures and other reptiles have reproduced at the Zoo. Significant husbandry protocols have been established.

***Staff and Scientific Achievements*:** One of the former directors, Philip Ogilivie, wrote his doctoral dissertation on chameleons and reported that some taxa secrete a glandular substance which attracts insects. During his tenure, Jaren Horsley was curator of herpetology. Many zoo veterans spent time here before going to other institutions: J. Michael Goode, Hugh Quinn, Gordon Henley, Michael Blakely, Rusty Grimpe, Eric Rundquist and John Walczak. David Grow occupied the curatorial post for many years and was responsible for many of the initiatives currently in place; he left the zoo. Currently, Brian Aucone, formerly of the Dallas Zoo, is the herpetological curator.

Personal Reflections: When I visited the zoo years ago, the number of chameleons being maintained and bred was impressive. Every nook and cranny was filled with chameleon cages. Limited information was available on their needs in captivity and Grow and his associates were experimenting with a variety of husbandry techniques.

Historical Overview

Curtis, L., and V. Hutchison. 1973. Obituary of Hobart Landreth. HISS NJ 1(3):88. [Landreth held a joint research appointment with the Oklahoma City Zoo and University of Oklahoma before his tragic death. The research facility at the Zoo was named in his honor.]

Reproduction

Wheeler, S. 1989. Captive reproduction of pythons at the Oklahoma City Zoo, p. 113-116. In M. J. Uricheck (ed.), 13th International Herpetological Symposium on Captive Propagation and Husbandry. International Herpetological Symposium.

Medical Management

Deakins, D. E. 1972-1973. Diagnosis and treatment of parasites of Amphibia and reptiles. Proc. Amer. Assoc. Zoo Vet. 1972/1973:37-47.

Captive Management

Grow, D. 1991. Management of Chamaeleonidae at the Oklahoma City Zoological Park, p. 67-75. In R. E. Staub (ed.), Captive Propagation and Husbandry of Reptiles and Amphibians. Northern California Herpetological Society.

Grow, D. T. 1992. Lizard management at the Oklahoma City Zoological Park with special reference to *Uromastyx*, *Chamaeleo* and *Heloderma*, p. 161-172. In M. J. Uricheck (ed.), 15th International Herpetological Symposium on Captive Propagation and Husbandry. International Herpetological Symposium, Palo Alto CA.

Sandefer, C. 1992. Preliminary comments on the Galapagos tortoise (*Geochelone elephantopus*) program at the Oklahoma City Zoological Park, p. 1-17. 16th International Herpetological Symposium on Captive Propagation and Husbandry. International Herpetological Symposium.

General

Rundquist, E. M. 1981. Longevity records at Oklahoma City Zoo. Herpetol. Rev. 12(3):87.

Fort Worth Zoological Park (1909)

History: When the Herpetarium opened in 1960, it incorporated many features missing from older reptile buildings. The enormous public area was subdivided into five smaller rooms arranged zoogeographically. The walls were redwood and brick which created a warm and welcoming space for the visitor. Each section had benches and elevated platforms in front of the exhibits so children could readily see the inhabitants. Some of the most elaborate exhibits ever seen in a zoo facility were featured, unusual since most curators favored simple displays as concern about hygiene was paramount. Many of the large enclosures had rock dioramas, waterfalls, living plants and colored lights; even small exhibits were elaborate. A row of interactive push-button exhibits allowed the zoogoer to choose between local venomous and non-venomous snakes. There was a large outdoor alligator exhibit, beautifully planted, in front of the building. The major problem was the cramped service area. Although this building is now showing its age, several new lizard and crocodilian displays have been added recently in another part of the zoo.

Facility and Collection: The collection was unique in the early days: goliath frogs, bushmasters, many exotic turtles, a number of unusual elapids and viperids, sea snakes, wart snakes, amphisbaenians, and an array of Neotropical hylid frogs. As the staff quickly discovered, because there were many displays to fill, any zoogeographic approach was beset with problems as many taxa, especially from Indo-Pacific regions, were rarely available.

According to *International Zoo Yearbook* (1998:Volume 28), composition of the collection was 166 taxa of reptiles with 654 specimens and 19 taxa of amphibians numbering over 50.

Staff and Scientific Achievements: John Mehrtens, author of the book *Living Snakes of the World in Color* published years later, was hired to oversee the building by Director Lawrence Curtis, who published several herpetological papers, including his master's thesis on the color variation in copperheads from the Dallas-Fort Worth area. Mehrtens, a volatile and demanding supervisor, was responsible for developing many of the displays, husbandry protocols and graph-

Fig. 139. Staff members at the Fort Worth Zoo beginning in 1960: (L to R) Lawrence Curtis and J. P. Jones; J. P. Jones and Jonathan Campbell; Lawrence Curtis and Roger and Isabel Hunt Conant; the herpetarium as it appeared shortly after construction; Rick Hudson; David Blody; Bern Tryon. *Photographs from Ft. Worth Zoo Archives, collage prepared by Clay M. Garrett.*

Fig. 140. Photograph of late herpetological curator John M. Mehrtens in Dallas, Texas in early 1960s. *Courtesy of J. Steve Dobbs.*

ics but his employees had relatively little input in the overall direction of the department. A number of prominent zoo workers worked in this fertile yet intimidating setting: Steve Dobbs and his mother Myra Mehrtens, Tim Jones, J. P. Jones, Tommy Logan, John Banks, and Joseph Laszlo. After Mehrtens left to become Director of the Victoria Zoo in Texas, Jonathan Campbell was hired as Assistant Supervisor under J. P. Jones but quickly decided that the zoo profession was too strange. Later, Bern Tryon arrived and was eventually elevated to Supervisor after Jones left the Zoo but he went to the Houston Zoo shortly thereafter. Hugh Quinn was in the education department. David Blody accepted the curatorial post and hired Rick Hudson as his assistant. Hudson, now in the Zoo's conservation department, has been the prime force in the impressive and successful captive breeding/reintroduction program for the critically endangered Jamaican iguana. Blody left the zoo and the curatorial job has been filled by Diane Barber.

Personal Reflections: Since the Dallas and Ft. Worth Zoos are nearby, this proximity offered the opportunity to develop friendships with many of the herpetologists over the years.

Historical Overview

Card, W., and J. B. Murphy. 2000. Lineages and histories of zoo herpetologists in the United States. Herpetol. Circ. No. 27:1-44 [biography of John Mehrtens].

Curtis, L. 1962. The aquarium and herpetarium at Fort Worth Zoological Park. Inter. Zoo Yearb. 4:46-48. [Aquarium opened in 1964, herpetarium in 1960. First use of word "Herpetarium."]

Garrett, C. M. 2006. Herpetology at the Fort Worth Zoo: A 45-Year History. Herpetol. Rev. 37:264-268.

Hudson, R. 1982. The reptile reproduction program at the Ft. Worth Zoo, p. 329-349. In D. L. Marcellini (ed.), 6th International Herpetological Symposium on Captive Propagation and Husbandry. International Herpetological Symposium, Thurmont MD.

Husbandry

Blody, D. A. 1984. An efficient and inexpensive tool for feeding captive reptiles. Inter. Zoo Yearb. 23:206-207.

Campbell, J. A. 1974. Use of a silicone emulsion base in reptile collections. Inter. Zoo Yearb. 14:137 [seminal paper on use of product to keep geckos off glass where they often defecate].

Mehrtens, J. M. 1962. A successful technique for exhibiting amphisbaenids. Inter. Zoo Yearb. 4:123 [*Amphisbaena alba*].

Mehrtens, J. M. 1962. The use of non-reflecting curved glass in reptile exhibits. Inter. Zoo Yearb. 4:159-160.

Mehrtens, J. M. 1965. Leuchtstofflampen Gro Lux in der Terrarienkunde (Use of fluorescent lamps "Gro-Lux" in terrarium science). Salamandra 1:26-27.

Mehrtens, J. M. 1966. The captive softshell (Trionychoidae). Checklist of the world's softshells and their care. Int. Turtle Tortoise J. 1(1):14-17, 3 1-37, 38, 45.

Mehrtens, J. M. 1967. Chelids of Australasia. Int. Turtle Tortoise J. 1(6):14-21.

Reproduction

Blody, D. A. 1985. Maintenance and reproduction of selected species of Neotropical iguanid lizards, p. 77-82. In R. L. Gray (ed.), Captive Propagation and Husbandry of Reptiles and Amphibians. Northern California Herpetological Society & Bay Area Amphibian and Reptile Society.

Brown, D. A. 1964. Nesting of a captive *Gopherus berlandieri* (Agassiz). Herpetologica 20:209.

Hammack, S. H. 1992. New concepts in colubrid egg incubation: A preliminary report, p. 103-108. In M. J. Uricheck (ed.), 15th International Herpetological Symposium on Captive Propagation and Husbandry. International Herpetological Symposium, Palo Alto CA.

Jones, J. P. 1978. Photoperiod and reptile reproduction. Herpetol. Rev. 9(3):95-100.

Mehrtens, J. M. 1952. Notes on eggs and young of *Pituophis melanoleucus sayi*. Herpetologica 8:69-70.

Tryon, B. W. 1978. Some aspects of breeding and raising aquatic chelonians. Part I. Herpetol. Rev. 9(1):15-19.

Tryon, B. W. 1978. Some aspects of breeding and raising aquatic chelonians: part 2. Herpetol. Rev. 9(2):58-61.

Conservation

Hudson, R. 1983. The Species Survival Plan and its application to reptiles, p. 1-9. In P. J. Tolson (ed.), 7th International Herpetological Symposium on Captive Propagation and Husbandry. International Herpetological Symposium, Thurmont MD [Current Lizard Advisory Group Chair for the AZA is responsible for the development of SSPs.]

Hutchins, M., and R. Wiese. 1991. Beyond genetic and demographic management: the future of the Species Survival Plan and related AAZPA conservation efforts. Zoo Biol. 10: 285-292. [Michael Hutchins was employed at AZA and Robert Wiese at Zoo.]

Behavior

Curtis, L. 1952. Cannibalism in the Texas coral snake. Herpetologica 8:27

Tryon, B. W., and R. K. Guese. 1984. Death-feigning in the gray-banded kingsnake *Lampropeltis alterna.* Herpetol. Rev. 15(4):108-109.

General

Curtis, L. 1968. Zoological Park Fundamentals. American Association of Zoological Parks and Aquariums, Wheeling WV.

Quinn, H., and J. P. Jones. 1974. Squeeze box technique for measuring snakes. Herpetol. Rev. 5(2):35.

Tryon, B. W. 1979. An unusually patterned specimen of the gray-banded kingsnake, *Lampropeltis mexicana alterna* (Brown). Herpetol. Rev 10(1):4-5.

Tryon, B. W. 1984. Snake hibernation: In and out of the zoo. Brit. Herpetol. Soc. Bull. 10:22-29.

San Antonio Zoological Gardens (1914)

History, Facility and Collection: The reptile building is small and suffers from age since it was constructed in 1942 by the Works Progress Administration.. The layout is traditional: small exhibits line the outer walls and larger ones are found in the center island. The inner exhibits are serviced from the back and the smaller ones open from the front. In spite of limitations of size and space, significant accomplish-

ments have been achieved in the past through the efforts of two zoo herpetologists, John Werler and Jozsef (Joe) Laszlo.

According to *International Zoo Yearbook* (1998:Volume 28), composition of the collection was 138 taxa of reptiles with over 430 specimens and 17 taxa of amphibians numbering 107.

Staff and Scientific Achievements: John Werler served as departmental head during the 1950s until he took over the reins at the Houston Zoo. Through his efforts, the Zoo became recognized as a place where herpetology flourished. Joe Laszlo followed Werler and carried on the tradition. After his death, John McLain was curator and Alan Kardon was his assistant. When John left, Alan filled his post. McLain died in 2003 after a long bout with liver cancer.

Historical Overview

Card, W., and J. B. Murphy. 2000. Lineages and histories of zoo herpetologists in the United States. Herpetol. Circ. No. 27:1-44 [portraits of John Werler and Joe Laszlo].

Murphy, J. B. 1988. In memoriam. Joseph Laszlo. Herp. Rev. 19(1):5-6.

Morphology, Systematics & Taxonomy

Burger, W. L. and J. E. Werler. 1954. The subspecies of the ring-necked coffee snake, *Ninia diademata,* and a short biological and taxonomic account of the genus. Univ. Kansas Sci. Bull. 36:643-672.

Shannon, F. A., and J. E. Werler. 1955. Report on a small collection of amphibians from Veracruz, with a description of a new species of *Pseudoeurycea.* Herpetologica 11:81-85 [description of *Pseudoeurycea firscheini*].

Werler, J. E., and H. M. Smith. 1952. Notes on a collection of reptiles and amphibians from Mexico. Texas J. Sci. 4:551-573.

Husbandry

Laszlo, J. 1975. Probing as a practical method of sex recognition in snakes. Inter. Zoo Yearb. 15:178-179.

Laszlo, J. 1975. The effect of light and temperature on reptilian mating and reproduction: Recent developments. AAZPA Reg. Conf. Proc. 1975:33-42.

Laszlo, J. 1977. Practical methods of inducing mating in snakes using extended day lengths and darkness. AAZPA Reg. Conf. Proc. 1977:204-210.

Laszlo., J. 1979. Notes on reproductive patterns of reptiles in relation to captive breeding. Inter. Zoo Yearb. 19:22-27.

Laszlo, J. 1979. Notes on photobiology, hibernation, and reproduction of snakes, p. 16-20. In R. A. Hahn (ser. ed.), 1st & 2nd International Herpetological Symposium on Captive Propagation and Husbandry. International Herpetological Symposium, Thurmont MD.

Laszlo, J. 1979. Notes on thermal requirements of reptiles and amphibians in captivity: The relationship between temperature ranges and vertical climatic (life) zone concept, p. 24-43. In R. A. Hahn (ser. ed.), 4th International Herpetological Symposium on Captive Propagation and Husbandry. Thurmont MD.

Laszlo, J. 1980. Notes on thermal requirements of reptiles and amphibians in captivity, p. 24-46. In R. A. Hahn (ser. ed.), 3rd International Herpetological Symposium on Captive Propagation and Husbandry. International Herpetological Symposium, Thurmont MD.

Laszlo, J. 1984. Further notes on reproductive patterns of amphibians and reptiles in relation to captive breeding. Inter. Zoo Yearb. 23:166-174.

Reproduction

Kardon, A. 1988. Viperidae: husbandry and propagation techniques at the San Antonio Zoo, p. 169-173. In K. H. Peterson (ed.), 10th International Herpetological Symposium on Captive Propagation and Husbandry. Zoological Consortium, Inc., Thurmont MD.

Werler, J. E. 1949. Eggs and young of several Texas and Mexican snakes. Herpetologica 5:59-60.

Werler, J. E. 1949. Reproduction of captive Texas and Mexican snakes. Herpetologica 5:67-70.

Werler, J. E. 1951. Miscellaneous notes on eggs and young Texas and Mexican reptiles. Zoologica (NY) 36:34-48.

Medical Management

Fletcher, K. C. 1979. Clinical use of mafenide acetate in reptiles. Proc. Amer. Assoc. Zoo Vet. 1979:27-32.

Richter, A. G., J. Olsen, K. Fletcher, and M. Bogart. 1977. Techniques for collecting blood from Galapagos tortoises and box turtles. Vet. Med./Small Anim. Clin. 70:1376.

Schmidt, R. E., G. B. Hubbard, and K. C. Fletcher. 1986. Systematic survey of lesions from animals in a zoologic collection. J. Zoo Anim. Med. 17(1):8-11. [Papers in this journal issue covered central nervous, musculo-skeletal, cardiovascular, reproductive, and respiratory systems as well as mammary and endocrine glands, and special sense organs.]

Houston Zoological Gardens (1921)

***History*:** When the late John Werler left the San Antonio Zoo in 1956, he established a strong herpetological presence at his new venue. During his tenure as general curator and later as general manager (director), he assembled a distinguished staff and a magnificent collection.

***Facility and Collection*:** The current reptile building was opened in 1964. In February 2002, the building was renovated with larger enclosures for crocodilians and monitors.

According to *International Zoo Yearbook* (1998:Volume 28), composition of the collection was 150 taxa of reptiles with over 500 specimens and 11 taxa of amphibians numbering 62.

***Staff and Scientific Achievements*:** Tommy Logan was curator during the 1960s-1970s. After Logan left the zoo profession, Hugh Quinn was employed in this capacity. When Quinn left to assume the position as general curator at the Cleveland Metroparks Zoo, his assistant, Karl Peterson, became curator. Stan Mays is the current curator and retired Paul Freed was supervisor. Veterinarians Gary Harwell and Joseph Flanagan published many papers, mostly on reptiles. The late Ben Dial began his professional career here, prior to obtaining his doctoral degree. Louis Porras was employed years ago as well before becoming an animal dealer specializing in amphibians and reptiles.

***Personal Reflections*:** Beginning in the mid-1960s, I spent many days with our colleagues at this Zoo, discussing techniques for keeping a living collection. Joe Laszlo had started a number of projects investigating the thermal biology and behavior of snakes and we regularly exchanged specimens. These were exciting and creative times. Many of the techniques used for keeping snakes now in zoos and the private sector were generated from early discussions in Houston.

I encountered the most uncontrollable snakes in my experience at this Zoo. Curator Tommy Logan asked if I would help clean two recently imported Gould's tree cobras, held in small aquariums off exhibit. These alert, agile, and aggressive cobras were completely in control as they dashed from floor to ceiling and it took nearly an hour to return them to their enclosures. Both of us were shaken by this episode and I suggested that shift cages might be a suitable alternative in the future.

Historical Overview

Card, W., and J. B. Murphy. 2000. Lineages and histories of zoo herpetologists in the United States. Herpetol. Circ. No. 27:1-44 [biography of John Werler].

Morphology, Systematics & Taxonomy

Quinn, H. 1978. Sexual dimorphism in tail pattern of Oklahoma snakes. Texas J. Sci. 31:157-160.

Quinn, H. 1983. Two new subspecies of *Lampropeltis triangulum* from Mexico. Trans. Kansas Acad. Sci. 86:113-135 [description of *L. t. campbelli* and *L. t. dixoni*].

Smith, H. M., and J. E. Werler. 1969. The status of the northern red black-headed snake, *Tantilla diabola* Fouquette and Potter. J. Herpetol. 3:172-173.

Werler, J. E. 1957. A new lizard of the genus *Lepidophyma* from Volcán San Martín Pajapan, Veracruz. Herpetologica 13:223-236 [description of *Lepidophyma pajapanensis*].

Werler, J. E., and F. A. Shannon. 1957. A new lizard of the genus *Lepidophyma* from Veracruz, Mexico. Herpetologica 13:119-122 [description of *Lepidophyma tuxtlae*].

Werler, J. E., and F. A. Shannon. 1961. Two new lizards (genera *Abronia* and *Xenosaurus*) from the Los Tuxtlas Range in Veracruz, Mexico. Trans. Kansas Acad. Sci. 64:123-132 [description of *Abronia reidi* and *Xenosaurus sanmartinensis*].

Reproduction

McLain, J. M. 1982. Notes on boid reproduction at the Houston Zoological Park, p. 248-264. In D. L. Marcellini (ed.), 6th International Herpetological Symposium on Captive Propagation and Husbandry. International Herpetological Symposium, Thurmont MD.

Platz, C. C., G. Mengden, H. Quinn, F. Wood, and J. Wood. 1980. Semen collection, evaluation and freezing in the green sea turtle, Galapagos tortoise, and red-eared pond turtle. Proc. Amer. Assoc. Zoo Vet. 1980:47-54 [Cayman Turtle Farm, Grand Cayman Island and Zoo].

Quinn, H., T. Blaesdel, and C. C. Platz Jr. 1989. Successful artificial insemination in the checkered garter snake *Thamnophis marcianus*. Inter. Zoo Yearb. 28:177-183.

Werler, J. E. 1970. Notes on young and eggs of captive reptiles. Inter. Zoo Yearb. 10:105-116 [reproductive data in *Phyllodactylus muralis, Crotaphytus reticulatus, Sceloporus edwardtaylori, S. f. formosus, Urosaurus o. ornatus, Abronia taeniata graminea, Dracaena guianensis, Enyaliosaurus quinquecarinatus, Toluca lineata varians, Coluber constrictor anthicus, Conophis vittatus viduus*].

Medical Management

Flanagan, J. P. 1986. Management problems of tortoises in captivity. Proc. Amer. Assoc. Zoo Vet. 1986:115.

Flanagan, J. P., and G. M. Harwell. 1983. Pathobiology and management of chronic regurgitation in snakes. Proc. Amer. Assoc. Zoo Vet. 1983:208-209.

Harwell, G., and K. C. Fletcher. 1985. Practice tips in reptile medicine - safety considerations. Proc. Amer. Assoc. Zoo Vet. 1985:105 [Houston Zoo and San Antonio Zoo].

Husbandry

Laszlo, J. 1969. Observations on two new artificial lights for reptile displays. Inter. Zoo Yearb. 9:12-13 [observations on Optima and Vita-Lite fluorescent lights at Houston Zoo].

Logan, T. 1969. Experiments with Gro-Lux light and its effect on reptiles. Inter. Zoo Yearb. 9:9-11. [Study was done at Zoo which demonstrated that Gro-Lux bulbs did not cause deleterious effects on reptiles.]

Logan, T. 1972. A method of heating reptile terraria. Inter. Zoo Yearb. 12:91-93.

General

Freed, P. 2003. Of Golden Toads and Serpents' Roads. Texas A & M Press, College Station.

Peterson, K. H. 1989. Further comments on anecdotal data. Bull. Chicago Herpetol. Soc. 24:190-191.

Peterson, K. H. 1996. The global decline in amphibian species: A perceptual deficit in the zoo and conservation community. Int. Zoo News 43 (7):476-482.

Pitman, C. R. S. 1974. A Guide to the Snakes of Uganda. Wheldon & Wesley, Codicote UK. [Provides information on reproduction of ball and Angolan pythons from the captive colonies at Houston Zoo and Detroit Zoo. A number of photographs in this revised edition were taken at the former zoo (plate AC).]

Quinn, H. 1979. The Rio Grande frog, *Syrrhophus cystignathoides campi*, in Houston, Texas. Proc. Kansas Acad. Sci. 82:209-210.

Quinn, H. 1995. Strategic collection planning: A global perspective. Zoo Biol. 14:51-52.

Quinn, H., and H. Quinn. 1993. Estimated number of

snake species that can be managed by Species Survival Plans in North America. Zoo Biol. 12:243-255. [After sending questionnaires to zoos to assess available space, authors concluded that although carrying capacity for snakes in zoos is limited, as many as nine families of snakes in 16 SSPs administered by the AZA might be maintained.]

Schwaner, T. D., P. R. Baverstock, H. C. Dessauer, and G. A. Mengden. 1985. Immunological evidence for the phylogenetic relationships of Australian elapid snakes, p. 177-184. Biology of Australasian Frogs and Reptiles. Royal Zoological Society, New South Wales. [Prior to his work in Australia on elapid snakes, Gregory Mengden was the research biologist at the Zoo.]

Werler, J. E., and J. R. Dixon. 2000. Texas Snakes. Identification, Distribution, and Natural History. University of Texas Press, Austin TX.

Gladys Porter Zoo (1971)

History and Mission: Although this Zoo in Brownsville, Texas, opened relatively recently, impressive breeding accomplishments and captive management programs are already in place. A notable conservation highlight is the head-starting initiative for the endangered Kemp's Ridley sea turtle with Mexican wildlife biologists in Tamaulipas. Another aspect of this joint effort is transport of eggs to reestablish the population on Padre Island, Texas.

Facility and Collection: The Zoo is known for a number of well-designed captive management enterprises: Galápagos and radiated tortoises, king cobra, Taylor's cantil, Coahuilan box turtle, bushmaster, Blomberg's toad and water monitor. The radiated tortoise venture is so successful that many neonates have been placed in other zoos.

According to *International Zoo Yearbook* (1998:Volume 28), composition of the collection was 89 taxa of reptiles with nearly 400 specimens and 8 taxa of amphibians numbering 72.

Staff and Scientific Achievements: Part of the success must be attributed to the longevity of two persons, Patrick Burchfield and his assistant Colette Adams (formerly Hairston). The former started his career at the Columbus Zoo under Lou Pistoia and lived to tell about it. Pat holds the distinction as the person who worked for the longest time with Lou and was fired and rehired most often. Colette is known for her husbandry skills which have been instrumental in establishing these notable breeding programs.

Personal Reflections: I first met Pat Burchfield when he was a keeper at the Columbus Zoo. He told me so many fascinating and amusing stories about his experiences there with Lou that I still chortle when they come to mind.

Husbandry

Burchfield, P. M. 1977. An experimental artificial diet for captive snakes. Inter. Zoo Yearb. 17:172-173 [prepared carnivore mixture stuffed into sausages].

Burchfield, P. M. 1982. Husbandry of reptiles, p. 265-282. In K. Sausman (ed.), Zoological Park and Aquarium Fundamentals. AAZPA, Wheeling WV.

Medical Management

Willette-Frahm, M., K. M. Wright, and B. C. Thode. 1995. Select protozoal diseases in amphibians and reptiles: A Report for the Infectious Diseases Committee, American Association of Zoo Veterinarians. Bull. Assoc. Rept. Amphib. Vet. 5(1):19-29 [comprehensive review from veterinarians at Gladys Porter Zoo and Philadelphia Zoo].

Behavior

Burchfield, P. M., T. F. Beimler, and C. S. Doucette. 1982. An unusual precoital head-biting behavior in the Texas patch-nosed snake, *Salvadora grahamiae lineata* (Reptilia: Serpentes: Colubridae). Copeia 1982:192-193.

General

Burchfield, P. M. 1993. An unusual dietary inclusion for the cat-eyed snake, *Leptodeira septentrionalis septentrionalis*. Bull. Chicago Herpetol. Soc. 28:266-267.

Burchfield, P. M. 2005. Texans, turtles, and the early Kemp's ridley population restoration project, 1963–67. Chelonian Conserv. Biol. 4:835-837.

WESTERN UNITED STATES

Denver Zoological Park (1896)

History: Taking many years to plan, the Tropical Discovery building was opened in 1993. Development and presentation are impressive. Visitors encounter living plants in a simulated tropical rain forest and many elaborate smaller terrestrial and aquatic exhibits may

Fig. 141. Tropical Discovery at Denver Zoo was opened in 1993. *Photograph by Clayton Freiheit, provided by Charles Radcliffe.*

Fig. 142. Exhibit for small reptiles in Tropical Discovery at Denver Zoo. *Photograph by Rick Haeffner, provided by Charles Radcliffe.*

be experienced. In this computer age, the environmental support technology and machinery allow creation and monitoring of enclosures duplicating unique microhabitats to house delicate species.

Facility and Collection: In addition to amphibians and reptiles, the building houses mammals, birds, fishes and invertebrates. The exhibits are naturalistic and attractive. One of my favorites is an aquarium housing African Rift Lake cichlids.

According to *International Zoo Yearbook* (1998:Volume 28), composition of the collection was 92 taxa of reptiles with 335 specimens and 24 taxa of amphibians numbering 77.

Staff and Scientific Achievements: Beginning in 1976, David Chiszar and Assistant Curator Charles Radcliffe began investigating the predatory behavior of snakes, especially rattlesnakes. Part of their plan was to develop simple and elegant experiments to elucidate feeding patterns which could be applied to maintenance of captive specimens. It soon became apparent that this study was to be far more complex than anticipated; one only needs to peruse the titles below to get a sense of the genesis of this project. It can be said without reservation that their publications contributed in a substantial way to our understanding of the subtlety and specialization of snake predation; their findings resulted in improved techniques for feeding captive snakes. Two years later, Hobart Smith joined the team, and the combined efforts of these three investigators and their associates have produced an unparalleled body of work.

Chiszar has been a member of the Zoo's scientific advisory board for many years and Radcliffe continues as assistant curator in this building. The curator is Rick Haeffner, who specializes in ichthyology and herpetology.

Historical Overview

Etter, C., and D. Etter. 1995. The Denver Zoo: A Centennial History. Roberts Reinhart, Boulder CO.

Behavior

Chiszar, D. 1979. Experiments on predatory behavior of rattlesnakes. AAZPA Annual Proc. 1979:28-43.

Chiszar, D., and C. W. Radcliffe. 1976. Rate of tongue flicking by rattlesnakes during successive stages of feeding on rodent prey. Bull. Psychon. Soc. 7:485-486.

Chiszar, D., and C. W. Radcliffe. 1977. Absence of prey-chemical preferences in newborn rattlesnakes (*Cro-*

talus cerastes, C. enyo and *C. viridis*). Behav. Biol. 21:146-150.

Chiszar, D., C. Radcliffe, and F. Feiler. 1986. Trailing behavior in banded rock rattlesnakes (*Crotalus lepidus klauberi*) and prairie rattlesnakes (*C. viridis viridis*). J. Comp. Psychol. 100:368-371.

Chiszar, D., C. W. Radcliffe, and K. M. Scudder. 1977. Analysis of the behavioral sequence emitted by rattlesnakes during predation. I. Striking and chemosensory searching. Behav. Biol. 21:418-425.

Chiszar, D., C. W. Radcliffe, and K. Scudder. 1980. Use of the vomeronasal system during predatory episodes by bull snakes (*Pituophis melanoleucus*). Bull. Psychon. Soc. 15:35-36.

Chiszar, D., C. W. Radcliffe, and H. M. Smith. 1982. Predatory behavior of "nervous" rattlesnakes, *Crotalus durissus terrificus*. Herpetol. Rev. 13(3):90-93. [Search image remains available to snake for at least 30 min, defensive reaction may outlast search image, precluding feeding.]

Chiszar, D., H. M. Smith, and C. C. Carpenter. 1994. An ethological approach to reproductive success in reptiles, p. 147-173. In J. B. Murphy, K. Adler, and J. T. Collins (eds.), Captive Management and Conservation of Amphibians and Reptiles. Society for the Study of Amphibians and Reptiles. Contributions to Herpetology, volume 11, Ithaca NY.

Chiszar, D., D. Duvall, K. M. Scudder, and C. W. Radcliffe. 1980. Simultaneous discriminations between envenomated and nonenvenomated mice by rattlesnakes (*Crotalus durissus* and *C. viridis*). Behav. Neural Biol. 29:518-521.

Chiszar, D., C. W. Radcliffe, T. Byers, and R. Stoops. 1986. Prey capture in nine species of venomous snakes. Psychol. Rec. 36:433-438.

Chiszar, D., C. Radcliffe, B. O'Connell, and H. M. Smith. 1980. Strike-induced chemosensory searching in rattlesnakes (*Crotalus enyo*) as a function of disturbance prior to presentation of prey. Trans. Kansas Acad. Sci. 83:230-234.

Chiszar, D., C. Radcliffe, B. O'Connell, and H. M. Smith. 1981. Strike-induced chemosensory searching in rattlesnakes (*Crotalus viridis*) as a function of disturbance prior to presentation of rodent prey. Psychol. Rec. 31:57-62.

Chiszar, D., C. W. Radcliffe, B. O'Connell, and H. M. Smith. 1982. Analysis of the behavioral sequence emitted by rattlesnakes during feeding episodes II. Duration of strike-induced chemosensory searching in rattlesnakes (*Crotalus viridis, C. enyo*). Behav. Neural Biol. 34:261-270.

Chiszar, D., C. W. Radcliffe, H. M. Smith, and H. Bashinski 1981. Effect of prolonged food deprivation in response to prey odors by rattlesnakes. Herpetologica 37:237-243.

Chiszar, D., K. M. Scudder, L. Knight, and H. M. Smith. 1978. Exploratory behavior in prairie rattlesnakes (*Crotalus viridis*) and water moccasins (*Agkistrodon piscivorus*). Psychol. Rec. 28:363-368.

Chiszar, D., L. Simonson, C. Radcliffe, and H. M. Smith. 1979. Rate of tongue flicking by water moccasins (*Agkistrodon piscivorus*) during prolonged exposure to various food odors; and, strike-induced chemosensory searching in the cantil (*Agkistrodon bilineatus*). Trans. Kansas Acad. Sci. 82:49-54.

Chiszar, D., H. M. Smith, J. L. Glenn, and R. C. Straight. 1991. Strike-induced chemosensory searching in venomoid pit vipers at Hogle Zoo. Zoo Biol. 10:111-117. [venom injection is irrelevant to induce SICS.].

Chiszar, D., C. Radcliffe, R. Overstreet, T. Poole, and T. Byers. 1985. Duration of strike-induced chemosensory searching in cottonmouths (*Agkistrodon piscivorus*) and a test of the hypothesis that striking prey creates a specific search image. Canadian J. Zool. 63:1057-1061.

Chiszar, D., S. V. Taylor, C. W. Radcliffe, H. M. Smith, and B. O'Connell. 1981. Effects of chemical and visual stimuli upon chemosensory searching by garter snakes and rattlesnakes. J. Herpetol. 15:415-424.

Chiszar, D., C. Andren, G. Nilson, B. O'Connell, J. S. Mestas Jr., H. M. Smith, and C. W. Radcliffe. 1982. Strike-induced chemosensory searching in old world vipers and new world pit vipers. Anim. Learning Behav. 10:121-125.

Chiszar, D., C. W. Radcliffe, R. Boyd, A. Radcliffe, H. Yun, H. M. Smith, T. Boyer, B. Atkins, and F. Feiler. 1986. Trailing behavior in cottonmouths (*Agkistrodon piscivorus*) after detecting prey. J. Herpetol. 20:269-272.

Cruz, E., S. Gibson, K. Kandler, and G. C., D. Sanchez. 1987. Strike-induced chemosensory searching in rattlesnakes: A rodent specialist (*Crotalus viridis*) differs from a lizard specialist (*Crotalus pricei*). Bull. Psychon. Soc. 25:136-138.

Duvall, D., D. Chiszar, J. Trupiano, and C. W. Radcliffe. 1978. Preference for envenomated rodent prey by rattlesnakes. Bull. Pyschon. Soc. 11:7-8.

Estep, K., T. Poole, C. W. Radcliffe, B. O'Connell, and D. Chiszar. 1981. Distance traveled by mice after envenomation by a rattlesnake (*Crotalus viridis*). Bull. Psychon. Soc. 18:108-110.

Golan, L., C. W. Radcliffe, T. Miller, B. O'Connell, and

D. Chiszar. 1982. Trailing behavior in prairie rattle-snakes (*Crotalus viridis*). J. Herpetol. 16:287-293.

Keith, J., R. Lee, and D. Chiszar. 1985. Spatial orientation by cottonmouths (*Agkistrodon piscivorus*) after detecting prey. Bull. Maryland Herpetol. Soc. 21:145-149.

Marmie, W., S. Kuhn, and D. Chiszar. 1990. Behavior of captive-raised rattlesnakes (*Crotalus enyo*) as a function of rearing conditions. Zoo Biol. 9:241-246 [analysis of enclosure size relative to rattlesnake behavior].

Radcliffe, C. W., and J. B. Murphy. 1983. Precopulatory and related behaviours in captive crotalids and other reptiles: Suggestions for future investigations. Inter. Zoo Yearb. 23:163-166.

Radcliffe, C. W., D. Chiszar, and B. O'Connell. 1980. Effects of prey size on poststrike behavior in rattlesnakes (*Crotalus durissus, C. enyo* and *C. viridis*). Bull. Psychon. Soc. 16:449-450.

Radcliffe, C. W., D. Chiszar, and H. M. Smith. 1980. Prey-induced caudal movements in boa constrictor with comments on the evolution of caudal luring. Bull. Maryland Herpetol. Soc. 16: 19-22.

Radcliffe, C. W., K. Estep, T. Boyer, and D. Chiszar. 1986. Stimulus control of predatory behavior in red spitting cobras (*Naja mossambica pallida*) and prairie rattlesnakes (*Crotalus v. viridis*). Anim. Behav. 34:804-814.

Radcliffe, C. W., T. Poole, F. Feiler, N. Warnock, T. Byers, A. Radcliffe, and D. Chiszar. 1983. Immobilization of mice following envenomation by cobras (*Naja mossambica pallida*). Bull. Psychon. Soc. 21:243-246.

Husbandry

Radcliffe, C. W. 1975. A method for force-feeding snakes. Herpetol. Rev. 6(1):18.

Radcliffe, C. W., and D. Chiszar. 1983. Clear plastic hiding boxes as a husbandry device for nervous snakes. Herpetol. Rev. 14(1):18.

Medical Management

Poston, J. 1996. Recent experiences with *Salmonella* sp. at the Denver Zoological Gardens. AAZPA Reg. Conf. Proc. 1996:729-733.

San Diego Zoological Society (1916)

History: The first reptile collection was assembled in 1922 and the first reptile house was the modified International Harvester building, built for the Panama-California Exposition during 1915-16. In 1931, C. B.

Fig. 143. Entrance to reptile building at San Diego Zoo, opened in 1936. *San Diego Zoological Society Archives.*

Perkins was placed in charge of the operation and as many curators are wont to do, lobbied vigorously for a new facility. Five years later, he was successful and the grand opening occurred on 4 July. This large quadrangular building, still standing, is surrounded by an arcade where visitors view exhibits. One drawback is the bright outdoor light which may reflect off of the glass at certain times of the day.

Facility and Collection: Because the weather in the region is marvelous, many reptiles are kept outdoors in large enclosures: crocodilians, giant tortoises, and large varanids and iguanids. The Zoo is known for the horticultural collection and all of the areas around these enclosures are heavily planted with mostly tropical foliage which is a spectacular sight. Smaller herps are kept in a series of small buildings named in honor of the late Laurence Klauber, known for his studies on rattlesnakes. The collection over the years has been impressive: tuataras including the recent acquisition of the highly endangered North Brother Island taxon, Komodo dragons, many exotic Indo-Pacific pythons, African water cobras, a variety of Madagascan chameleons, several two-headed California kingsnakes, yellow-bellied sea snakes and a variety of species from Baja California. Large outdoor gharial and Komodo dragon exhibits have been added.

Fig. 144. Reptile Mesa at San Diego Zoo. *San Diego Zoological Society Archives.*

Fig. 145. One of several Klauber buildings at San Diego Zoo which house smaller amphibians and reptiles. Named in honor of Laurence M. Klauber. *San Diego Zoological Society Archives.*

Fig. 146. Laurence M. Klauber. *Courtesy of Kraig Adler.*

According to *International Zoo Yearbook* (1998:Volume 28), composition of the collection was 174 taxa of reptiles and amphibians numbering 841.

Staff and Scientific Achievements: Klauber was the first curator of reptiles in the 1920s, consulting curator beginning in 1931 and on the Board of Trustees from 1943 to 1968.

Charles Shaw followed C. B. Perkins and instituted a successful reproductive program for Galápagos tortoises and other reptiles.

When Shaw died, Jerry Staedeli was installed in his post. Later, James Bacon became curator until he died in 1986. Jim was concerned about conservation issues, especially those related to the potential loss of island populations. His assistant Earl (Tom) Schultz filled the job until he recently retired. The present curator, Donal Boyer, was formerly the supervisor at the Dallas Zoo where we worked together for over a decade.

Research has been steady over many years, especially emanating from the Center for Research of Endangered Species (CRES) by Allison Alberts, Valentine Lance, J. M. Lemm, John (Andy) Phillips, and retired

Fig. 147. Photograph of late herpetological curator James P. Bacon examining Galápagos tortoise at San Diego Zoo in 1986. *Photograph by Ron Garrison, San Diego Zoo.*

Director Kurt Benirschke. In the early days, Perkins and Shaw compiled lists of longevities and published on a variety of herpetological topics.

Publication: The newsletter *ZOONOOZ* is the Bulletin published monthly by the San Diego Zoological Society and has a number of articles dealing with the biology and habits of reptiles and amphibians. If the name is turned upside-down, it remains the same.

Personal Reflections: When I visited in 1969, there

were many holding cages in the rear section filled with local snakes donated by local residents, a group of rare Mona Island iguanas and Philippine tarsiers on exhibit. The only Storm's water cobra from Lake Tanganyika I have ever seen alive was on display.

Historical Overview

Bacon, J. P., and M. Hallett. 1981. Exhibit systems for reptiles and amphibians at the San Diego Zoo: dioramas and graphics. Inter. Zoo Yearb. 21:14-21.

Card, W., and J. B. Murphy. 2000. Lineages and histories of zoo herpetologists in the United States. Herpetol. Circ. No. 27:1-44 [biographies of C. B. Perkins, Charles Shaw, Laurence Klauber and James Bacon.]

Glaser, H. S. R. 1984. Now-it-can-be-told department: "Who the hell was Perkins?" Herpetol. Rev. 15(1):8.

Honegger, R. E. 1971. Charles Edward Shaw. 31. Juli 1918 - 30. Mai 1971 (Charles Edward Shaw. July 31, 1918 - May 30, 1971). Salamandra 7:47-48.

Jones, M. 2003. A Conversation with Marvin Jones. Mark A. Rosenthal, Milwaukee WI. [Interesting account of career of the late registrar Marvin Jones includes extensive descriptions of his travels throughout the world and his visits to over 300 zoos and aquariums. Jones discussed the evolution of AZA conservation programs and his role in the development of expanded record keeping which led to improved management of small animal populations in zoos.]

Klauber, L. M. 1956. Rattlesnakes: Their Habits, Life Histories, and Influence on Mankind. Univ. California Press, Berkeley.

Klauber, L. M.. 1972. Rattlesnakes: Their Habits, Life Histories, and Influence on Mankind. Univ. California Press, Berkeley CA. [2 Vols. 1533 pp. Definitive work on rattlesnakes, abridged edition (1982) is without sections on identification, classification and zoogeography from the original set.]

Mertens, R. 1951. Zwischen [Between] Atlantik und Pazifik; zoologische Reiseskizzen aus Nordamerika. A. Kernen, Stuttgart, Germany. [In 1949, Robert Mertens from the Senckenberg Museum in Frankfurt am Main, Germany, traveled to the United States to visit museums, zoos, and aquariums. This book is a travelogue of his adventures and includes photographs of the taxa which he encountered as well as several images of the outdoor reptile exhibits at the Zoo.]

Scortecci, G. 1953. Animali. Come sono/Dove vivono/ Come vivono. Edizioni Labor, Milan, Italy. [This five- volume set on zoology and animal behavior included many photographs of amphibians and reptiles from the archives of the Zoo.]

Shaw, C. E. 1959. Letter. Bull. Philadelphia Herpetol. Soc. 7(5). [Letter discussed mites on wild snakes and lizards around San Diego.]

Shaw, C. E. 1959. The reptile collection at the San Diego Zoo. Bull. Philadelphia Herpetol. Soc. 7(6):8-12. [Contribution was the first on reptile buildings in the United States. Photographs of the public areas, rear section, outdoor tortoise enclosure, the longevity record for captive snakes (black-lipped cobra at 29 years, one month) and five hatchling Galápagos tortoises were included.]

Wei Yew. 1991. Noah's Art: Zoo, Aquarium, Aviary and Wildlife Park Graphics. Quon Editions, Edmonton [color photographs of herpetological graphics at Zoo].

Morphology, Systematics & Taxonomy

Herrmann, H.-W., P. A. Herrmann, A. Schmitz, and W. Böhme. 2004. A new frog species of the genus *Cardioglossa* from the Tchabal Mbabo Mtns, Cameroon (Anura: Arthroleptidae). Herpetozoa 17:119-125 [description of *Cardioglossa alsco*].

Herrmann, H.-W., T. Ziegler, B. L. Stuart, and N. L. Orlov. 2002. New findings on the distribution, morphology and natural history of *Triceratolepidophis sieversorum* (Serpentes: Viperidae). Herpetol. Nat. Hist. 9:89-94.

Herrmann, H.-W., T. Ziegler, A. Malhotra, R. S. Thorpe, and C. L. Parkinson. 2004. Redescription and systematics of *Trimeresurus cornutus* (Serpentes: Viperidae) based on morphology and molecular data. Herpetologica 60:211-221.

Plath, M., M. Solbach, and H.-W. Herrmann. 2004. Anuran habitat selection and temporal partitioning in a montane and submontane rainforest in Southwestern Cameroon--first results. Salamandra 40:239-260.

Shaw, C. E. 1945. The chuckwallas, genus *Sauromalus*. Trans. San Diego Soc. Nat. Hist. 10:269-306.

Shaw, C. E. 1946. A new locality for the spiny chuckwalla, *Sauromalus hispidus*. Copeia 1946; 254.

Husbandry

Boyer, D. M., and T. H. Boyer. 1994. Tortoise care. Bull. Assoc. Rept. Amphib. Vet. 4(1):16-28.

Boyer, T., and D. Boyer. 1992. Aquatic turtle care. Bull. Assoc. Rept. Amphib. Vet. 2(2):13-17.

Reproduction

Lance, V. A. 1984. Endocrinology of reproduction in male reptiles. Symp. Zool. Soc. London 52:357-383.

Lance, V. A., and I. P. Callard. 1978. Hormonal control of ovarian steroidogenesis in nonmammalian vertebrates, p. 361-407. In R. E. Jones (ed.), The Vertebrate Ovary. Plenum Press, New York, London.

Lieberman, A. 1980. Nesting of the basilisk lizard (*Basiliscus basiliscus*). J. Herpetol. 14:103-105.

Owens, D. W., J. R. Hendrickson, V. Lance, and I. P. Callard. 1978. A technique for determining sex of immature *Chelonia mydas* using radioimmunoassay. Herpetologica 34:270-273.

Phillips, J. A., and B. L. Lasley. 1987. Modification of reproductive rhythm in lizards via GnRH therapy. Ann. New York Acad. Sci. 519:128-136.

Phillips, J. A., and R. P. Miller. 1998. Reproductive biology of the white-throated savanna monitor, *Varanus albigularis*. J. Herpetol. 32:366-377.

Shaw, C. E. 1952. Notes on the eggs and young of some United States and Mexican lizards, I. Herpetologica 8:71-79.

Shaw, C. E. 1963. Notes on the eggs, incubation and young of some African reptiles. Brit. J. Herpetol. 3:63-70.

Shaw, C. E. 1969. Breeding the rhinoceros iguana *Cyclura cornuta cornuta* at San Diego Zoo. Inter. Zoo Yearb. 9:45-48.

Behavior

Alberts, A. C. 1992. Pheromonal self-recognition in desert iguanas. Copeia 1992:229-232.

Alberts, A. C. 1994. Dominance hierarchies in male lizards: Implications for zoo management programs. Zoo Biol. 13:479-490.

Cooper, W. E., Jr. and A. C. Alberts. 1991. Tongue-flicking and biting in response to chemical food stimuli by an iguanid lizard, (*Dipsosaurus dorsalis*) having sealed vomeronasal ducts: vomerolfaction may mediate these behavioral responses. J. Chem. Ecol. 17:135-146.

Cooper, W. E., Jr. and A. C. Alberts. 1993. Postbite elevation in tongue-flicking rate by an iguanian lizard, *Dipsosaurus dorsalis*. J. Chem. Ecol. 19:2329-2336.

Hunsaker, D., and B. R. Burrage. 1969. The significance of interspecific social dominance in iguanid lizards. Amer. Midl. Natur. 81:500-511 [observations at Zoo].

Kaufman, J. D., G. M. Burghardt, and J. A. Phillips. 1996. Sensory cues and foraging decisions in a large carnivorous lizard, *Varanus albigularis*. Anim. Behav. 52:727-736.

Phillips, J. A. 1995. Does cadence of *Iguana iguana* displays facilitate individual recognition? Behav. Ecol. Sociobiol. 37:337-342.

Phillips, J. A. 1995. Movement patterns and density of *Varanus albigularis*. J. Herpetol. 29:407-416.

Schafer, S. F., and C. O. Krekorian. 1983. Agonistic behavior of the Galapagos tortoise, *Geochelone elephantopus*, with emphasis on the relationship to saddle-backed shell shape. Herpetologica 39:448-456 [observations at Zoo and Charles Darwin Research Station, Galápagos].

Shaw, C. E. 1948. The male combat "dance" of some crotalid snakes. Herpetologica 4:137-145.

Shaw, C. E. 1951. Male combat in American colubrid snakes with remarks on combat in other colubrid and elapid snakes. Herpetologica 7:149-168.

Captive Management

Alberts, A. C., A. M. Perry, J. M. Lemm, and J. A. Phillips. 1997. Effects of incubation temperature and water potential on growth and thermoregulatory behavior of hatchling rock iguanas (*Cyclura nubila*). Copeia 1997:766-776.

Benirschke, K. 1977. Genetic management. Inter. Zoo Yearb. 17:50-60.

Benirschke, K. 1980. The scientific background to the use of studbooks. Inter. Zoo Yearb. 20:486-487.

Benirschke, K. 1985. The genetic management of exotic animals. Symp. Zool. Soc. London No. 54:71-87.

Benirschke, K. 1996. The need for multidisciplinary research units in the zoo, p. 537-544. In D. G. Kleiman, M. E. Allen, K. V. Thompson, S. Lumpkin, and H. Harris (eds.), Wild Mammals in Captivity. University of Chicago Press, Chicago.

Conservation

Alberts, A. C. 1995. Developing recovery strategies for West Indian rock iguanas. Endangered Species Update 14:543-553.

Alberts, A. C. (ed.). 2000. West Indian Iguanas: Status Survey and Conservation Action Plan. IUCN-The World Conservation Union, Gland, Switzerland. [Edited publication included conservation strategy, taxonomic accounts and action plan compiled by specialists.]

Alberts, A. C. 2002. Ten years of conservation research on Cuban rock iguanas. Herpetol. Rev. 33:119-120.

Ryder, O. A., J. H. Shaw, and C. M. Wemmer. 1988. Species, subspecies and *ex situ* conservation. Inter. Zoo Yearb. 27:134-140. [analysis of zoo conservation issues].

Medical Management

Alberts, A. C., M. L. Oliva, M. B. Worley, S. R. Telford Jr., P. J. Morris, and D. L. Janssen. 1998. The need for pre-release health screening in animal translocations: A case study of the Cuban iguana (*Cyclura nubila*). Anim. Conserv. 1:165-172.

Anderson, M. P. 1986. Anatomic hints on wild animals and the clinician. Proc. Amer. Assoc. Zoo Vet. 1986:125-126 [contribution on snake anatomy from Zoo].

Effron, M., L. Griner, and K. Benirschke. 1977. Nature and rate of neoplasia found in captive wild mammals, birds, and reptiles at necropsy. J. Natl. Cancer Inst. 59:185.

Ensley, P. K. 1981. Management of cloacal lesions in Galapagos tortoises. Proc. Amer. Assoc. Zoo Vet. 1981:14-15.

Griner, L. A. 1975. Hematopietic neoplasia in animals at San Diego Zoological Gardens, p. 253-259. In R. Ippen, and H.-D. Schröder (eds.), Erkrankungen der Zootiere. Verhandlungsberichte des XVII. Internationalen Symposiums über die Erkrankungen der Zootiere Tunis 1975. Akademie-Verlag, Berlin.

Ippen, R., and G. P. Wildner. 1984. Comparative pathological investigations of thyroid tumors of animals in zoos and in the wild, p. 280-295. In O. A. Ryder, and M. L. Byrd (eds.), One Medicine. A Tribute to Kurt Benirschke, Director, Center for Reproduction of Endangered Species, Zoological Society of San Diego, and Professor of Pathology and Reproductive Medicine, University of California, San Diego / From his Students and Colleagues. Springer-Verlag, Berlin, Heidelberg, etc. [At the time of publication, the honoree had published over 20 books and 325 scientific papers.]

Kuehn, G. 1972-1973. Bilateral transverse mandibular fractures in a turtle. Proc. Amer. Assoc. Zoo Vet. 1972/1983:243.

Robinson, P. T. 1977. Special techniques in reptile medicine. Proc. Amer. Assoc. Zoo Vet. 1977:68-70.

Sutherland-Smith, M., J. A. Miller, and D. D. Oehler. 1994. Use of long-acting ivermectin suspension in snakes. Proc. Amer. Assoc. Zoo Vet. 1994:98-101.

Tarshis, I. B. 1961. The use of the sorptive dust SG 67 for the control of the snake mite, *Ophionyssus natricis* (Gervais). Bull. Philadelphia Herpetol. Soc. 9(2):11-17. [Study was done in part at Zoo. This ectoparasite caused massive problems in zoo collections for years and a number of treatments had been tried to control outbreaks.]

Tarshis, I. B., and L. R. Penner. 1960. Treatment of ectoparasites in captive wild animals. Inter. Zoo Yearb. 2:107-109 [research supported by Zoo].

ZOONOOZ:

Bacon Jr., J. P. 1975. An ecological enigma - the Madagascar red frog. ZOONOOZ 48(4):14-15.

Shaw, C. E. 1955. Aloha, Tuatara. ZOONOOZ 28(1):5

Shaw, C. E. 1955. Technically reptiles. ZOONOOZ 28(1):7-11.

Shaw, C. E. 1955. Unlizard-like lizards. ZOONOOZ 28(5):3-6..

Shaw, C. E. 1955. The dreaded bushmaster. ZOONOOZ 28(9):3-4.

Shaw, C. E. 1956. Lizards live languid lives. ZOONOOZ 29(7):6-8.

Shaw, C. E. 1956. Dudley-Duplex did it again! ZOONOOZ 29(12):12 [description of two-headed California kingsnake].

Shaw, C. E. 1957. Baby, it's cold outside! ZOONOOZ 30(2):10-11.

Shaw, C. E. 1957. Sea serpents *do* exist. ZOONOOZ 30(4):10-12.

Shaw, C. E. 1957. Monitors do admonish. ZOONOOZ 30(7):3-5.

Shaw, C. E. 1959. Double trouble. ZOONOOZ 32(11):10-11 [description of two-headed California kingsnake].

Shaw, C. E. 1959. Future giants. ZOONOOZ 32(12):3-5.

Shaw, C. E. 1960. Lizards with a hundred-million-year history. ZOONOOZ 33(8):3-5.

Shaw, C. E. 1960. The she-dragons of Guiana. ZOONOOZ 33(12):6-7.

Shaw, C. E. 1961. Snakes of the sea. ZOONOOZ 34(7):3-5.

Shaw, C. E. 1961. Another generation of Galapagos giants. ZOONOOZ 34(12):10-15.

Shaw, C. E. 1962. *Bipes* the elusive two-legged lizard. ZOONOOZ 35(8):10-15.

Shaw, C. E. 1963. Dragons big and bold. ZOONOOZ 36(1):3-7 [photograph of Komodo dragon outdoor exhibit].

Shaw, C. E. 1963. Growth of the Galapagos tortoise. ZOONOOZ 36(2):15.

Shaw, C. E. 1963. Boon from Borneo--three earless lizards. ZOONOOZ 36(5):10-12.

Shaw, C. E. 1964. Beaded lizards--dreaded, but seldom deadly. ZOONOOZ 37(3):10-15.

Shaw, C. E. 1965. Why do snakes live so long? ZOONOOZ 38(1):3-7.

Shaw, C. E. 1965. Tentacled fishing snake. ZOONOOZ 38(8):3-5.

Staedeli, J. H. 1961. Raising a giant snake. ZOONOOZ 34(10):4-7.

Staedeli, J. H. 1971. South American green tree snake. ZOONOOZ 44(9):18.

Staedeli, J. H. 1972. Keeping a frog alive and healthy. ZOONOOZ 45(4):4-11.

Staedeli, J. H. 1972. Concern for the Galapagos tortoises. ZOONOOZ 45(8):4-10.

Staedeli, J. H. 1972. Indian cobra. ZOONOOZ 45(8):11.

Staedeli, J. H. 1972. Whipsnakes. ZOONOOZ 45(9):17.

Staedeli, J. H. 1972. The mysterious rear-fanged snakes. ZOONOOZ 45(10):18-19.

Staedeli, J. H. 1972. Captive snakes will eat voluntarily. ZOONOOZ 45(11):15-19.

Staedeli, J. H. 1972. Keeping chameleons. ZOONOOZ 45(12):18.

Staedeli, J. H. 1973. An American monster. ZOONOOZ 46(1):18.

Staedeli, J. H. 1973. Probing the methods of incubating reptile eggs. ZOONOOZ 46(7):4-6.

General

Alberts, A. C. 1989. Ultraviolet visual sensitivity in desert iguanas: Implications for pheromone detection. Anim. Behav. 38:129-137.

Alberts, A. C. 1990. Chemical properties of femoral gland secretions in the desert iguana, *Dipsosaurus dorsalis*. J. Chem. Ecol. 16:13-25.

Alberts, A. C. 1991. Phylogenetic and adaptive variation in lizard femoral gland secretions. Copeia 1991:69-79.

Alberts, A. C. 1992. Constraints on the design of chemical communication systems in terrestrial vertebrates. Amer. Natur. 139S:62-89.

Alberts, A. C. 1992. Density dependent scent gland activity in desert iguanas. Anim. Behav. 44:774-776.

Alberts, A. C. 1993. Chemical and behavioral studies of femoral gland secretions in iguanid lizards. Brain Behav. Evol. 41:255-260.

Alberts, A. C. 1993. Relationship of space use to population density in an herbivorous lizard. Herpetologica 49:469-479.

Alberts, A. 1994. Off to see the lizard: Lessons from the wild. Vivarium 5(5):26-28.

Alberts, A. C. 2003. Conserving the remarkable reptiles of Guantanamo Bay, p. 67-73. In R. W. Henderson and R. Powell (eds.), Islands and the Sea: Essays on Herpetological Exploration in the West Indies. Society for the Study of Amphibians and Reptiles, Contributions to Herpetology, volume 20, Ithaca NY.

Alberts, A. C., and T. D. Grant. 1997. Use of a non-contact temperature reader for measuring skin surface temperatures and estimating internal body temperatures in lizards. Herpetol. Rev. 28:32-33.

Alberts, A. C., and D. I. Werner. 1993. Chemical recognition of unfamiliar conspecifics by green iguanas: functional significance of different signal components. Anim. Behav. 46:197-199.

Alberts, A. C., L. A. Jackintell, and J. A. Phillips. 1994. Effects of chemical and visual exposure to adults on growth, hormones, and behavior of juvenile green iguanas. Physiol. Behav. 55:987-992.

Alberts, A. C., J. A. Phillips, and D. I. Werner. 1993. Sources of intraspecific variability in the protein composition of lizard femoral gland secretions. Copeia 1993:775-781.

Alberts, A. C., N. C. Pratt, and J. A. Phillips. 1992. Seasonal productivity of lizard femoral glands: Relationship to social dominance and androgen levels. Physiol. Behav. 51:729-733.

Alberts, A. C., R. L. Carter, W. K. Hayes, and E. P. Martins (eds.). 2004. Iguanas: Biology and Conservation. Univ. California Press, Berkeley, Los Angeles, London.

Alberts, A. C., T. R. Sharp, D. I. Werner, and P. J. Weldon. 1992. Seasonal variation of lipids in the femoral gland secretions of male green iguanas (*Iguana iguana*). J. Chem. Ecol. 18:703-712.

Bacon, J. P. 1979. Final report of Study Trip to European Zoological Gardens. July - August 1978. Submitted to: National Museum Act Advisory Council (grant #78/251) and Trustees of the Zoological Society of San Diego. [Survey compiled data on visitors impressions, collection composition, staffing, facilities, record keeping, emergency systems and procedures, diets, breeding programs, and veterinary and pathology services for 18 European zoos with herpetological collections.]

Boyer, D. M. 1995. Venomous reptiles: An overview of families, handling, restraint techniques, and emergency protocol. Proc. Assoc. Rept. Amphib. Vet. 1995:27-29, 83-90. [Donal Boyer has been the editor of the AZA Antivenin Index for many years.]

Boyer, D. M. 1997. Phylogenetic allocation of resources and the impact on zoo herpetology. AAZPA Reg. Conf. Proc. 1997:96-99. [Zoo herpetological collections receive a proportionally smaller amount of financial support when compared to other animal departments: mammals first, birds second and so on.]

Esra, G. N., K. Benirschke, and L. A. Griner. 1975. Blood collecting techniques in lizards. J. Amer. Vet. Med. Assoc. 167:555-556.

Herrmann, P. A., and H.-W. Herrmann. 2003. New records and natural history notes for amphibians and reptiles from southern Morocco. Herpetol. Rev. 34:76-77.

Lance, V. A., and D. C. Rostal. 2002. The annual reproductive cycle of the male and female desert tortoise: physiology and endocrinology. Chelonian Conserv. Biol. 4:302-312.

Lance, V. A., N. Valenzuela, and P. von Hildebrand. 1992. A hormonal method to determine the sex of hatchling giant river turtles, *Podocnemis expansa*: application to endangered species research. Amer. Zool. 32:16A.

Mebs, D., U. Kuch, H.-W. Herrmann, and T. Ziegler. 2003. Biochemical and biological activities of the venom of a new genus and species of pitviper from Vietnam, *Triceratolepidophis sieversorum*. Toxicon 41:139-143.

Perkins, C. B. 1943. Notes on captive-bred snakes. Copeia 1943:108-112.

Perkins, C. B. 1947. A note on longevity of amphibians and reptiles in captivity. Copeia 1947:144.

Perkins, C. B. 1948. Hatching snake eggs. Herpetologica 4:184.

Perkins, C. B. 1948. Longevity of snakes in captivity in the United States (as of January 1, 1948). Copeia 1948:217. [Perkins compiled lists between 1948 and 1955: Copeia 1949:223; Copeia 1950:238; Copeia 1951:182; Copeia 1952:280-281; Copeia 1953:243; Copeia 1954:229-230; Copeia 1955:262.].

Perkins, C. B. 1950. Frequency of shedding in injured snakes. Herpetologica 6:35-36.

Perkins, C. B. 1951. Hybrid rattlesnakes. Herpetologica 7:146.

Perkins, C. B. 1952. Incubation period of snake eggs. Herpetologica 8:79.

Phillips, J. A. 1984. Choice as a biological optimum, p. 191-196. In O. A. Ryder, and M. L. Byrd (eds.), One Medicine. A Tribute to Kurt Benirschke, Director, Center for Reproduction of Endangered Species, Zoological Society of San Diego, and Professor of Pathology and Reproductive Medicine, University of California, San Diego / From his Students and Colleagues. Springer-Verlag, Berlin, Heidelberg, etc.

Phillips, J. A. 1985. A holistic view of reptile physiology, p. 25-38. In R. L. Gray (ed.), Captive Propagation and Husbandry of Reptiles and Amphibians.

Northern California Herpetological Society & Bay Area Amphibian and Reptile Society.

Phillips, J. A. 1986. Ontogeny of metabolic processes in blue-tongued skinks, *Tiliqua scincoides*. Herpetologica 42:405-412.

Phillips, J. A., and A. C. Alberts. 1992. Naive ophiophagus lizards recognize and avoid venomous snakes using chemical cues. J. Chem. Ecol. 18:1775-1783.

Phillips, J. A., and H. J. Harlow. 1981. Elevation of upper voluntary temperatures after shielding the parietal eye of horned lizards (*Phrynosoma douglassi*). Herpetologica 37:199-205.

Phillips, J. A., and G. C. Packard. 1994. Influence on temperature and moisture on eggs and embryos of the white-throated savanna monitor *Varanus albigularis*: implications for conservation. Biol. Conserv. 69:131-136.

Phillips, J. A., A. C. Alberts, and N. Pratt. 1993. Differential resource use, growth, and the ontogeny of social relationships in the green iguana. Physiol. Behav. 53:81-88. [Lizards which used supplemental heat were the most dominant, fastest growing (males) and this was not correlated with size at hatching. Study at Zoo.]

Phillips, J. A., A. W. B. Karesh, R. Millar, and B. L. Lasley. 1985. Stimulating male sexual behavior with repetitive pulses of GnRh in female green iguanas, *Iguana iguana*. J. Exp. Zool. 234:481-484 [study done at Zoo].

Phillips, J. A., F. Frye Jr., A. Bercovitz, P. Calle, R. Millar, J. Rivier, and B. L. Lasley. 1987. Exogenous GnRh overrides the endogenous annual reproductive rhythm in green iguanas, *Iguana iguana*. J. Exper. Zool. 241:227-236.

Pratt, N. C., A. C. Alberts, K. G. Fulton-Medler, and J. A. Phillips. 1992. Behavioral, physiological, and morphological components of dominance and mate attraction in male green iguanas. Zoo Biol. 11:153-163 [study at Zoo].

Pratt, N. C., J. A. Phillips, A. C. Alberts, and K. S. Bolda. 1994. Sexual bimaturism in green iguanas: Functional versus physiological puberty. Anim. Behav. 47:1101-1114.

Rostal, D. C., T. Wibbels, J. S. Grumbles, V. A. Lance, and J. R. Spotila. 2002. Chronology of sex determination in the desert tortoise(*Gopherus agassizii*). Chelonian Conserv. Biol. 4:313-318.

Secor, S. M., and J. A. Phillips. 1997. Specific dynamic action in a large carnivorous lizard, *Varanus albigularis*. Comp. Biochem. Physiol. 117A(4):515-522.

Shaw, C. E. 1948. A note on the food habits of *Heloderma suspectum* Cope. Herpetologica 4:145.

Shaw, C. E. 1950. Lizards in the diet of captive *Uma*. Herpetologica 6:36-37.

Shaw, C. E. 1960. Four baby giants. Anim. Kingdom 63(2):59-60.

Shaw, C. E. 1961. Tier bevorzugt (Animal preferred). Freunde Köln. Zoo 4(1):28-29 [photographs of African egg-eating snake swallowing egg].

Shaw, C. E. 1969. Longevity of snakes in North American collections as of 1 January, 1968. Zool. Gart. (N.F.), Leipzig 37:193-196.

Shaw, C. E., and S. Campbell. 1974. Snakes of the American West. Alfred A. Knopf, New York. [Sheldon Campbell was a Zoo supporter and Shaw's colleague for many years.]

Soulé, M. E., A. C. Alberts, and D. T. Bolger. 1992. The effects of habitat fragmentation on chaparral plants and vertebrates. Oikos 63:39-47.

Staedeli, J. H. 1962. „Unsere Warane" („Our monitors"). Freunde Köln. Zoo 5(1):74-76.

Arizona-Sonora Desert Museum (1952)

History: This facility is located in one of most lovely settings imaginable: natural desert in the Tucson Mountains west of downtown Tucson. The Museum specializes in the fauna and flora of the Sonoran Desert. Great care has been taken to develop naturalistic exhibits which nicely fit into the landscape. It may be surprising that this was the site where early experimentation occurred to create fiberglass cages and plastic rockwork dioramas now commonly seen in zoos.

Facility and Collection: The collection comprises amphibians and reptiles from Arizona and the Sonoran Desert. Some of the best naturalistic zoo exhibits are found at this institution.

According to *International Zoo Yearbook* (1998:Volume 28), composition of the collection was 91 taxa of reptiles with 449 specimens and 27 taxa of amphibians numbering 139.

Staff and Scientific Achievements: In 1958, Mervin Larson, General Curator, created four exhibits with artificial plastic rockwork and living plants for amphibians: canyon stream, Mexican river, permanent pond and temporary rain pool.

William Woodin was Director between 1954-1971: the Arizona Mountain kingsnake is named in his honor. After he retired, Larson was Director for four years until he left to found the Larson Company, specializing in naturalistic zoo exhibits. Merritt Keasey III oversaw the herpetological collection during this time. Later, the herpetological curator was Howard Lawler from 1981 to 1996. Craig Ivanyi, then collections manager, currently occupies this post. Thomas Van Devender, senior research scientist, has studied the environmental history and evolution of the Desert and has focused on the ecology and conservation of the desert tortoise.

Personal Reflections: When I saw San Esteban Island chuckwallas for the first time here, I was stunned by their size and beauty. A large male, basking in the early morning sun, was so colorful that I spotted him many meters from the exhibit, situated at the entrance to the Museum.

Historical Overview

Carr, W. H. 1982. Pebbles in Your Shoes: The Story of How the Arizona-Sonora Desert Museum Began and Grew. Arizona-Sonora Desert Museum, Tucson.

Hancocks, D. 2001. A Different Nature. The Paradoxical World of Zoos and Their Uncertain Future. University California Press, Berkeley, Los Angeles, London. [David Hancocks was former Director of the Woodland Park Zoo in Seattle and Desert Museum and his book included an extensive discussion of their history. His basic message is that zoos, other than a few exceptions, have not been effective examples overall in terms of exhibitry, conservation, education and research.]

Woodin, W. H. 1960. Reptile and amphibian exhibits at the Arizona-Sonora Desert Museum. Bull. Philadelphia Herpetol. Soc. 80(4):13-15.

Woodin, W. H. 1962. "Tunnel in the Desert," an underground exhibit for nocturnal animals. Inter. Zoo Yearb. 4:156-158 [western diamondback rattlesnake den].

Husbandry

Perry-Richardson, J. J., and C. S. Ivanyi. 1995. Captive design for reptiles and amphibians, p. 205-221. In E. F. Gibbons Jr., B. S. Durrant, and J. Demarest (eds.), Conservation of Endangered Species in Captivity. An Interdisciplinary Approach. State University of New York Press, Albany.

Smith, C. 1994. Desert tortoise care. Proceed. Assoc. Rept. Amphib. Vet. 4(1):12-15.

Captive Management

Lawler, H. E. 1992. Advanced protocols for the management and propagation of endangered and threatened reptiles, p. 57-66. In M. J. Uricheck (ed.), 15th International Herpetological Symposium on Captive Propagation and Husbandry. International Herpetological Symposium, Palo Alto CA.

Smith, C. 1994. Desert tortoise care. Bull. Assoc. Rept. Amphib. Vet. 4(1):12-15.

General

Altimari, W. 1998. Venomous snakes: A safety guide for reptile keepers. Herpetol. Circ. 26:1-28.

Grismer, L. L., K. R. Beaman, and H. E. Lawler. 1995. *Sauromalus hispidus*. Cat. Amer. Amphib. Rept.:615.1-615.4.

Ivanyi, C., J. Perry, T. R. Van Devender, and H. Lawler. 2000. Reptiles and amphibians, p. 527-585. In S. J. Phillips, and P. W. Comus (eds.), Natural History of the Sonoran Desert. Arizona-Sonora Desert Museum. Arizona-Sonora Desert Museum Press, Tucson; University of California Press, Berkeley and Los Angeles CA. [This splendid book, compiled by staff at the Museum, covered all aspects of flora, fauna and physical characteristics of the region.]

Keasey, M. S. III., 1969. Some records of reptiles at Arizona-Sonora Desert Museum. Inter. Zoo Yearb. 9:16-17.

Lawler, H. E., K. R. Beaman, and L. L. Grismer. 1995. *Sauromalus varius*. Cat. Amer. Amphib. Rept.:616.1-616.4.

Moon, B. R., C. S. Ivanyi, and J. Johnson. 2004. Identifying individual rattlesnakes using tail pattern variation. Herpetol. Rev. 35:154-156.

Poulin, S., and C. S. Ivanyi. 2003. A technique for manual restraint of helodermatid lizards. Herpetol. Rev. 34:43.

Van Devender, T. R. 2002. The Sonoran Desert Tortoise: Natural History, Biology, and Conservation. University of Arizona Press, Tucson.

Phoenix Zoo (1962)

Facility and Collection: In the beginning, approximately 40 small enclosures generally reflected native desert/grassland/woodland themes and featured invertebrates, amphibians and reptiles. Many specimens were in mixed species enclosures but lighting was poor, temperature control was inadequate and keeper access to the upper tier of displays was dangerous. Some of the cages had deteriorated to the point that the ventilation screen on the Mojave rattlesnake exhibit fell off when the cage was moved; fortunately, the snakes did not try to escape.

In 2000, the facility was remodeled and the environmental systems upgraded. To improve the visitor experience, the theme was changed to a more personal and interactive interpretation, from home into the wild. The visitor first encounters a bedroom inside a house, goes out to the backyard, into a vacant lot, and out into the undisturbed lowland and upland desert. The exhibits are about a foot above ground level, and have 8-12 ft viewing areas. The bedroom is a child's bedroom and features appropriate captive bred pets, such as bearded dragons, and an aquarium with swordtails and guppies. The backyard is set up to show how to protect and encourage native wildlife and has several species of toads (*Bufo alvarius, B. cognatus, B. fowleri*), California kingsnakes, and Mediterranean geckos. The traveler moves to the vacant lot with coachwhips, sidewinders, western diamondback rattlesnakes, and Sonoran gopher snakes. The lowland desert display has tiger rattlesnakes, Mojave rattlesnakes, longnose snakes, desert tortoises, and patchnose snakes. The upland desert had speckled and blacktail rattlesnakes, rosy boas, Sonoran whipsnakes, and Gila monsters. Results have been positive because even though there are fewer exhibits, visitation time has increased because zoogoers can see behind-the-scenes and interact with staff by using an intercom. To determine whether graphics are effective, tests are in place to see if people understand the exhibit without more complex graphics. The second phase will feature graphics centered on responsible pet ownership, nature exploration, and conservation of wild spaces.

There is a naturalistic outdoor Gila monster exhibit. The collection includes a breeding pair of Galápagos tortoises that have intermittently produced hatchlings and two pairs of Aldabra tortoises weighing 650 pounds each. There was a pair of American alligators but the male died (12 ft and 865 pounds). A pond contains eastern painted turtles, Blanding's turtles and redbelly shortneck turtles (*Emydura subglobosa*). Interestingly, the front lake outside the perimeter fence of the zoo has hundreds of released pet turtles. A common boa, some radiated tortoises, dwarf caimans, green iguanas, and a rainbow boa round out the exhibit collection.

Conservation: There are two off-exhibit buildings, one for head-starting the Ramsey Canyon leopard frog (*Rana subaquavocalis*) and the other for breeding the Yavapai leopard frog (*Rana yavapaiensis*). A converted refrigerated semi-trailer was added to expand the native ranid frog facilities for collaborative work with the

United States Fish and Wildlife Service and Arizona Game and Fish Department.

Staff and Scientific Achievements: Beginning in 1999, Kevin Wright was Curator of Ectotherms/Arizona Trail and Tropics Trail and later Director of Living Collections, in charge of all animal programs. He has left the zoo. He is an adjunct professor in the Department of Clinical Sciences, University of Pennsylvania College of Veterinary Medicine, in Philadelphia.

Personal Reflections: In my view, Kevin and Brent Whitaker, Director of Animal Health at the National Aquarium in Baltimore, have published the finest book ever assembled and produced on captive amphibian biology, *Amphibian Medicine and Captive Husbandry,* with 27 chapters by top specialists; this tome is certainly the definitive source for the diagnosis and treatment of diseases. Moreover, there is plenty of information about husbandry including environmental parameters, diet and nutrition.

Medical Management

Wright, K. M. 1999. Reptiles—crisis management protocols, p. 353-356. In S. D. Chan, W. K. Baker, Jr., and D. L. Guerrero (eds.), Resources for Crisis Management in Zoos and Other Animal Care Facilities, American Association of Zoo Keepers, Topeka.

Wright, K. M. 2000. Surgery of amphibians. In Veterinary Clinics of North America: Exotic Animal Practice 3(4).

Wright, K. M. (ed.). 2002. AZA Amphibian Taxon Advisory Group North American Regional Collection Plan.

Wright, K. M. and B. R. Whitaker. 2000. Metabolic bone disease in amphibians. Exotic DVM 1(6):23-26.

Wright, K. M., and M. J. Tyler. 2003. Medical problems of amphibians. In F. R. Frye (ed.), A Colour Handbook of Small Exotic Animal Medicine. Manson Publishing, London.

Whitaker, B. R., K. M. Wright, and S. Barnett. 1999. Basic husbandry and clinical assessment of the amphibian patient, In Veterinary Clinics of North America: Exotic Animal Practice 2(2):265-290.

Institute for Herpetological Research (1972) (Santa Barbara, CA)

History and Mission: Richard Ross, a medical doctor by profession, privately developed the Institute to support research studies on the husbandry, behavior and reproductive biology of captive reptiles, especially boas and pythons. The late Thomas Huff was involved with the Institute for many years. Brett Stearns serves on the Board of Directors.

The Institute is an accredited member of the American Zoo and Aquarium Association and received the significant Edward H. Bean Award in 1975 for the captive propagation of the white-lipped python.

Scientific Achievements: In January 1977, questionnaires were sent to those practitioners with captive colonies of pythons to gather information on successful reproductive events: techniques for inducing courtship and copulation and incubation protocols for egg incubation. The results of the survey were published in *The Python Breeding Manual* the next year. Later, an expanded, more sophisticated version included boas and was laced with color photographs.

Ross traveled to New Guinea in 1980 to survey bacterial organisms in green tree pythons and other boid snakes to compare with captive specimens. His findings were presented in the privately published *The Bacterial Diseases of Reptiles* with Research Fellow Gerald Marzec.

Facility and Collection: Currently, the Institute is located in Santa Barbara, California and the research trajectory now includes the systematics, husbandry and biology of freshwater stingrays.

Personal Reflections: I invited Dick Ross to be participant for the 1978 SSAR Symposium on reproductive biology and diseases of captive reptiles in 1978 at Arizona State University in Tempe. We had never met but had talked on the telephone several times. When we did meet at the social, he enthusiastically explained that his personal life was in tatters but nothing would keep him from attending any meeting dealing with captive reptiles.

Husbandry

Moyle, M., and R. A. Ross. 1992. Problems encountered with the husbandry of the emerald tree boa, *Corallus canina*, p. 119-126. In M. J. Uricheck, (ed.), 15th International Herpetological Symposium on Captive Propagation and Husbandry. International Herpetological Symposium, Palo Alto CA.

Ross, R. A. 1984. The husbandry of python egg incubation. Bull. Chicago Herpetol. Soc. 19: 25-26.

Reproduction

Ross, R. A. 1978. The Python Breeding Manual. Institute for Herpetological Research, Stanford CA.

Ross, R. A. 1980. The breeding of pythons (Subfamily Pythoninae) in captivity, p. 135-139. In J. B. Murphy and J. T. Collins, (eds.), Reproductive Biology and Diseases of Captive Reptiles. Society for the Study of Amphibians and Reptiles Contributions to Herpetology, volume 1, Lawrence KS.

Ross, R.A. 1991. Infertility and fecundity disorders of reptiles, p. 91-95. In A. W. Zulich, (ed.), 14th International Herpetological Symposium on Captive Propagation and Husbandry. International Herpetological Symposium.

Medical Management

Ross, R. A. 1973. Fatal gastrointestinal obstruction secondary to ingestion of San-i-cel. HISS NJ. 1(3): 92. [Product was used as substrate in reptile enclosures.]

Ross, R. A. 1983. Thermotherapy as a technique for treatment of respiratory infections, p. 31-33. In P. J. Tolson (ed.), 7th International Herpetological Symposium on Captive Propagation and Husbandry. International Herpetological Symposium, Thurmont MD.

Ross, R. A., and G. Marzec. 1984. The Bacterial Diseases of Reptiles. Institute for Herpetological Research, Stanford CA.

General

Ross, R. A. 1986. Hybridization of snakes: The current status, p. 199-202. In S. McKeown, F. Caporaso, and K. H. Peterson (eds.), 9th International Herpetological Symposium on Captive Propagation and Husbandry. Zoological Consortium, Inc., Thurmont MD.

Ross, R. A., and G. Marzec. 1990. The Reproductive Husbandry of Pythons and Boas. Institute for Herpetological Research, Stanford CA. [Book is based in part on extensive surveys of captive reproduction at zoos and private facilities undertaken by the authors over many years. The book continues to be the definitive treatment of captive husbandry and reproductive biology of boid snakes.]

CANADA

Metro Toronto Zoo (1974)

History: The Zoo is divided into biogeographical regions with four pavilions representing Australasia, Americas, Africa, and IndoMalaya. Large outside paddocks and exhibit biomes include Africa, IndoMalaya, Eurasia, South America, Canadian Domain, and Arctic. All pavilions are climate controlled and architecturally dramatic, the largest (Africa) over an acre under glass. Reptiles and amphibians are located in four pavilions.

Facility and Collection: According to *International Zoo Yearbook* (1998:Volume 28), composition of the collection was 52 taxa of reptiles with 227 specimens and 19 taxa of amphibians numbering nearly 200 specimens.

Staff and Scientific Achievements: When the Zoo opened in 1974, curators were not in the original management structure but six years later, positions were created. Bob Johnson, who started at the zoo a year prior to opening, was the first Curator of Reptiles, Amphibians and Invertebrates and is currently Curator of Amphibians and Reptiles. Curatorial Keeper Andrew Lentini focuses on husbandry issues to support zoo-based amphibian and reptile conservation programs. Johnson served as Chair of the IUCN/SSC Declining Amphibian Populations Task Force from 1993-94 and Herpetology Coordinator, IUCN/SSC Reintroduction Specialist Group. He is Director of the Ontario Herpetological Society. Johnson's involvement with AZA includes the following: Steering Committee, American Zoo Association Amphibian Advisory Group; Species Coordinator, AZA Puerto Rican Crested Toad Species Survival Plan; and Working Member of the AZA Lizard and Snake Advisory Groups. Bob is Co-coordinator of Toronto Zoo's Eastern Massasauga Rattlesnake conservation program and Director of the Zoo's "Adopt a Pond" wetlands conservation program and Frogwatch Ontario with 5,000 schools.

Conservation: The Zoo was the first to breed the endangered Puerto Rican crested toad without hormonally inducing reproduction. This experience stimulated interest in initiating a recovery plan, in concert with Rick Paine at the Buffalo Zoo, for the Puerto Rican crested toad. The first AZA Amphibian SSP was for this species and now 23 zoos participate. The SSP includes release of captive-produced individuals (two artificial release ponds 4,000^2 and 10,000^2 feet in size have been constructed to receive the captive bred tadpoles) and recovery implementation (habitat protection, wetland construction, and public education programs). The Juan Rivero Zoo in Puerto Rico is collaborating with the Toronto Zoo on projects such as toad exhibition and interpretative graphics on island conservation. Soon, habitat conservation signs will be installed in the Guánica State Forest, the Mayaguez Zoo, at the release ponds constructed by the SSP and at the Toronto Zoo. These signs will highlight the Zoo's role in captive breeding and release, explain the life history of toad breeding, demonstrate the importance of the last natural breeding pond for toad survival, and explain the reasons for road closures during reproductive events. Signs will be installed to encourage Guánica Forest users to visit the Mayaguez Zoo, also in Puerto Rico, to see toads and zoo visitors to visit the Guánica Forest to see the toads' natural habitat. These graphics will focus on the role of zoos in education and the importance of constructing release ponds.

Toronto Zoo has been involved with the conservation of the eastern massasauga rattlesnake since 1989. This taxon had no legal protection at that time (until 1994) when it was listed as threatened. Massasaugas were protected within Canadian National Parks and the Toronto Zoo and Parks personnel began a public awareness collaboration that continues to this day. A Zoo study demonstrating that snakes moving outside of home ranges died over winter resulted in a renewed commitment to demonstrate that people and wildlife could coexist. Zoo workshops and published education outreach resources provide factual information and the Zoo financially supports a number of programs. Zoo staff are members of the National Recovery team and coordinate outreach resources that have been used as a model for snake conservation across North America. The Zoo has published a teacher curriculum guide on the conservation of the taxon, printed a "Snakes of Ontario" poster, produced free videos featuring snake biology to distribute to persons living near rattlesnakes, and techniques for living with rattlesnakes. The veterinary staff provides care, advises the recovery team on health issues, and provides disease screening for wild populations.

Declining amphibian populations stimulated development of the Toronto Zoo Adopt-A-Pond wetland conservation program (see www.torontozoo.ca/adoptapond) which focuses on amphibian conservation (and turtle conservation), and the Zoo is the provincial coordinator for a National Frogwatch data collection initiative.

The Zoo has sponsored several research projects and financial and technical support for *in situ* conservation and research on the axolotl in Lake Xochimilco, Mexico. The salamander is declining due to several factors including pollution, collection and the introduction of exotic fish species. This program is carried out in Mexico in collaboration with El Centro de Investigaciones Biologicas y Acuicolas de Cuemanco (CIBAC), Universidad Autonoma Metropolitana--Xochimilco (UAM), Chapultepec Zoo, the Durrell Institute of Conservation and Ecology and community-based conservation groups on habitat restoration and conservation education.

A study began in 1999 to determine snapping turtle needs by using surveys of habitat use and movement in the Rouge Park through use of radio telemetry. Movement data demonstrated that male territories spread linearly along the Rouge River and the importance of small wetlands and seepage areas as over wintering sites was identified.

Experimental plots will soon be constructed to test the design and function of constructed vernal pools. A gray tree frog pond restoration project will include pond design and vegetation structure to provide habitat on the Zoo site. The wetland project will be the focus of graphics for visitors. Two other amphibian projects are underway. The first is a genetic assessment of historic and extant range of Blanchard's cricket frogs using live and museum specimens to determine potential source animals for release on Pelee Island, Ontario (Recovery Plan Objective). The second project continues research on the use of habitat by American toads in the Rouge tablelands. Gaining a better understanding of habitat use by amphibians will aid in protecting and/or restoring their habitats. To educate the public about toad tracking, the Toronto Zoo's Spring Toad Festival features a "Meet the Toad Tracker" activity and a toad mascot who welcomes visitors to the Zoo wetlands.

Publications: *Rattlesnake Tales Newsletter*. 1989-2002. (Vol 1- Vol. 12); *Amphibian Voice Newsletter*. 1992-2002 (Vol. 1-Vol. 10).

Personal Reflections: Bob Johnson has developed a model zoo program for amphibian conservation which incorporates successful educational, research and reintroduction components. In my view, there is no better one in the world.

Medical Management

Crawshaw, G. J., R. R. Johnson, and D. A. Smith. 1996. The veterinarian and the reptile: An introduction to the conservation of amphibians and reptiles. Bull. Assoc. Rept. Amphib. Vet. 6(3):18-22.

Holz, P., I. K. Barker, P. D. Conlon, G. J. Crawshaw, and J. Burger. 1994. The reptilian renal portal system and its effect on drug kinetics. Proc. Amer. Assoc. Zoo Vet. 1994:95-96.

Conservation

Johnson, R. R. 1991. Conservation of threatened amphibians: the integration of captive breeding and field research, p. 33-38. In R. E. Staub (ed.), Captive Propagation and Husbandry of Reptiles and Amphibians. Northern California Herpetological Society.

Johnson, R. R. 1992. Habitat loss and declining amphibian populations, p. 71-75. In C. A. Bishop and K. E. Pettit (eds.), Declines in Canadian Amphibian Populations: Designing a National Monitoring Strategy. Canadian Wildlife Service, Ottawa.

Johnson, R. R. 1992. Release and translocation strategies for the Puerto Rican crested toad, (*Peltophryne lemur*). Endangered Species Update Special Issue 8:54-57, 1990.

Johnson, R. R.. 1993. Eastern massasauga rattlesnake conservation program at Metro Toronto Zoo, p. 89-93. In B. Johnson and V. Menzies (eds.), International Symposium and Workshop on the Conservation of the Eastern Massasauga Rattlesnake *Sistrurus catenatus catenatus*: Proceedings of a Symposium and Workshop Sponsored by the Metro Toronto Zoo, held at the Zoo from May 8-9, 1992. Metro Toronto Zoo, West Hill, Ontario.

Johnson, R. R. 1994. Model programs for reproduction and management: Ex situ and in situ conservation of toads of the family Bufonidae, p. 243-254. In J. B. Murphy, K. Adler and J. T. Collins (eds.), Captive Management and Conservation of Amphibians and Reptiles. Society for the Study of Amphibians and Reptiles. Contributions to Herpetology, volume 11, Ithaca NY.

Johnson, R. R., and F. L. Paine. 1989. The release of captive bred Puerto Rican crested toads: Management implications and the cactus connection. AAZPA Reg. Conf. Proc. 1989:962-967.

Johnson, R. R., and M. Wright (eds.). 1999. Second International Symposium and Workshop on the Conservation of the Eastern Massasauga Rattlesnake (*Sistrurus catenatus catenatus*): Population and Habitat Management Issues in Urban, Bog, Prairie and Forested Habitats. Toronto Zoo, Toronto.

Paine, F. L., J. D. Miller, G. Crawshaw, B. Johnson, R. Lacy, C. F. Smith III, and P. J. Tolson. 1989. Status of the Puerto Rican crested toad *Peltophryne lemur*. Inter. Zoo Yearb. 28:53-58.

General

Baily, K., H. Gosselin, and R. R. Johnson, et al. 1998. Wetland curriculum resource: a toadally awesome wetland guide for educators. Toronto Zoo.

Gosselin, H., and R. R. Johnson. 1995 The urban outback—Wetlands for wildlife: a guide to wetland restoration and frog-friendly backyards. Metro Toronto Zoo, Adopt-A-Pound wetland conservation program.

Johnson, R. R. 1989. Familiar Amphibians and Reptiles of Ontario. Natural History. Natural History Press.

Johnson, R. R. 1995. Global perspective on amphibians, p. 4-5. In K. Coleman and R. Cholmondeley (eds.), Bullfrog management in Ontario. Workshop Proceedings. Southern Region Science Technology Transfer Unit WP-005.

Johnson, R. R. 1996. Reintroductions: Guidelines, Case Studies, Amphibian Ecology, and Landscapes. The Amphibian Reintroduction Symposium. Royal Botanical Gardens, Hamilton Ontario.

Johnson, R. R. 1999. Amphibians and reptiles, p. 187-201. In B. Roots, D. Chant and C. Heidenreich (eds.), Special Places, The Changing Ecosystems of the Toronto Region. UBC Press, Toronto.

Rapley, W. A., and S. E. Oyarzun. 1980. Feeding program at the Metropolitan Toronto Zoo. Proc. Amer. Assoc. Zoo Vet. 1980:23-29.

Reptile Breeding Foundation (1975) (Picton, Ontario)

History: Specializing in reproductive colonies for insular boids of the genus *Epicrates*, the private Reptile Breeding Foundation was a long-time member of the American Association of Zoological Parks and Aquariums. To recognize this important *Epicrates* program, the AAZPA gave its prestigious Edward H. Bean Award to the Foundation in 1977, an annual prize given to an institution for the most significant accomplishment in captive propagation. When the Director, Thomas Huff died prematurely in 1998, the Foundation began a precipitous downward spiral. Unfortunately, the Foundation no longer exists and the collection has been disbanded.

Publication: *Scales and Tales: the Newsletter of the Reptile Breeding Foundation*

Historical Overview

Huff, T. A. 1979. A new captive propagation centre for the Reptile Breeding Foundation. Inter. Zoo Yearb. 19:267-269.

Huff, T. A. 1989. Geckos of the U.S.S.R. in the collection of the Reptile Breeding Foundation, p. 43-47.

In R. Gowen (ed.), Captive Propagation and Husbandry of Reptiles and Amphibians. Northern California Herpetological Society.

Reproduction

Huff, T. A. 1976. The use of oxytocin to induce labour in a Jamaican boa. Inter. Zoo Yearb. 16:82.

Huff, T. A. 1980. Some parameters for breeding boids in captivity, p. 84-90. In R. A. Hahn (ser. ed.), 3rd International Herpetological Symposium on Captive Propagation and Husbandry. International Herpetological Symposium, Thurmont MD.

Huff, T. A. 1989. An update on plated lizard reproduction at the Reptile Breeding Foundation, p. 65-67. In R. Gowen (ed.), Captive Propagation and Husbandry of Reptiles and Amphibians. Northern California Herpetological Society.

General

Huff, T. A. 1985. An overview of lizard husbandry and propagation with emphasis on the work at the Reptile Breeding Foundation, p. 51-76. In R. L. Gray (ed.), Captive Propagation and Husbandry of Reptiles and Amphibians. Northern California Herpetological Society & Bay Area Amphibian and Reptile Society.

Huff, T. 1992. The husbandry and headaches of maintaining large lizards in captivity: An anecdotal approach, p. 173-179. In M. J. Uricheck (ed.), 15th International Herpetological Symposium on Captive Propagation and Husbandry. International Herpetological Symposium, Palo Alto CA.

Chapter 9

Zoological Gardens of Australia and New Zealand

Nile Crocodile (*Crocodylus niloticus*)

Introduction

Research in herpetology in the context of conservation biology in Australia has expanded dramatically over the past several decades, as evidenced by the remarkable increase of literature devoted to this subject such as Stanger et al. (1998) and Tyler (1996). Zoos and wildlife parks have kept pace with this overall resurgence (see Giles and Kelly, 1992; Hamilton and Phelps, 1992; Hopkins, 1992; Jakob-Hoff, 1992; Mumaw, 1992; Weigel, 1992 for examples). In 1990, the Herp Taxon Advisory Group (Herp TAG) was formalized at the Auckland Zoo. The next year, the Australasian Regional Association of Zoological Parks and Aquaria (ARAZPA) was incorporated, and its purpose is to encourage and coordinate conservation and education resources. Today, a number of *ex situ* and *in situ* conservation projects are being developed or are in place. One example of a multifaceted initiative is the recovery effort by the University of Western Australia, Perth Zoological Gardens and the Western Australian Wildlife Research Centre for the highly endangered western swamp tortoise (*Pseudemydura umbrina*), certainly the most threatened freshwater chelonian in the world (Kuchling and Dejose, 1989; Kuchling et al., 1992; Hall et al., 1992). Over 300 individuals have been provided for release as of January 2002.

Numerous innovative exhibits of amphibians and reptiles, many with a strong conservation message, have been created over the past few decades. De Courcy (2001) offered a brief discussion of reptile exhibits at the Adelaide, Melbourne and Sydney Zoos.

Royal Melbourne Zoological Gardens (1857)
Healesville Sanctuary (1934)

History and Mission: The Zoological Parks and Gardens Board administers the Zoo and Healesville Sanctuary, a park specializing in southeastern Australian wildlife. The Sanctuary is located approximately 40 miles from Melbourne. Reptiles have been a part of the Zoo for many years, beginning in 1878 with the report of the death of a boa constrictor. In May 1882, the reptile building was destroyed by fire. Nearly 80 years later, a storage building was converted into a new reptile building. Impressive *in situ* conservation programs began in 1990 and have involved several

Fig. 148. Reptile building at Melbourne Zoo in Australia. *Photograph provided by Chris Banks, Melbourne Zoo.*

Australian and Philippine initiatives: striped legless lizard, native Australian frogs, Philippine crocodile and amphibians, and South East Asian tortoises and freshwater turtles. In concert with Hong Kong researchers, over 800 captive-bred Romer's tree frogs were released in secure sites.

Facility and Collection: In 1968, over 30 naturalistic exhibits utilizing natural light and living plants were constructed which featured animals within appropriate habitats. Nineteen years later, the inadequate climate control system was replaced and the rear service area modified. Eighty-eight taxa have been bred at the zoo since the building's opening with some of the notable ones being green iguana, banded basilisk, elongated tortoise, freshwater and estuarine crocodiles, thorny devil, Arafuran file snake, D'Albert's python, twist-neck turtle, reticulated Gila monster, Fiji banded iguana, striped legless lizard and rhinoceros viper. In 1993, the staff opened the "World of Frogs" exhibit with naturalistic enclosures and separate climatic systems. Housing up to 12 species, the complex included a wetland habitat and off-exhibit breeding areas. Twelve species of amphibians have been reproduced. The name of the section was changed from "Reptile" to "Herpetofauna" to reflect this change of emphasis for amphibians.

According to *International Zoo Yearbook* (1998:Volume 28), composition of the Zoo collection was 78 species of reptiles with 414 specimens and 15 species of amphibians numbering 240. Healesville Sanctuary had 33 species of reptiles with 128 specimens and 2 species of amphibians numbering 10.

Staff and Scientific Achievements: Roy Dunn was Curator from 1969-1975 and was involved in creating the naturalistic exhibits in the early 1970s. Chris Banks has been instrumental since that time in establishing some of the most impressive conservation programs in a zoo setting. Chris and his staff are major coordinators and managers of many Reptile & Amphibian Taxon Advisory Groups (TAGs) administered through the ARAZPA.

Personal Reflections: When I visited the Zoo many years ago, Chris gave me a tour, and I was very impressed with his operation and his commitment to conservation. The exhibits were attractive and clean with animals in apparent good health. Later, I went to Healesville and saw my first living Inland taipans, said to be the most toxic terrestrial snake.

Historical Overview

Banks, C. B. 2002. A thirty-year history of Melbourne Zoo's Herp Department. Herpetol. Rev. 33:256-259.

De Courcy, C. 1995. Zoo Story. Penguin Books, Ringwood.

Wilkie, A. A. W. 1918. Almost Human: Reminiscences from the Melbourne Zoo. Whitcombe & Tombs, Melbourne. [Told by A. A. W. Wilke, Overseer of the Zoo, and written by Mrs. A. R. Osborn (Annie O'Neill), there were amusing accounts of a fight between a snake and cat, feeding behavior of large snakes, heating of snake enclosures, and venomous snakebite research.]

Husbandry

Banks, C. B. 1985. Observations on feeding and sloughing in a collection of captive snakes, p. 495-501. In G. Grigg, R. Shine, and H. Ehmann (eds.), Biology of Australian Frogs and Reptiles. Royal Society of New South Wales, Sydney.

Banks, C. B. 1989. Management of fully aquatic snakes. Inter. Zoo Yearb. 28:155-163. [Results of questionnaire showed that average period of survival is one year, nine months. Skin lesions in genera *Acrochordus* and *Erpeton* may be caused by excessively bright lighting.]

Captive Management

Banks, C. B. 1993. A regional approach to managing reptiles and amphibians in Australian zoos, p. 59-65. In D. Lunney and D. Ayers (eds.), Herpetology in Australia: A Diverse Discipline. Surrey Beatty & Sons Pty Ltd., Chipping Norton, New South Wales.

Banks, C. B. 1999. TAG Action Plan; directing the management of reptiles and amphibians in

Australasian zoos. Australasian Regional Association of Zoological Parks & Aquaria, Inc., Mosman.

Banks, C. B., and H. E. McCracken Captive management and pathology of sharp-snouted dayfrog 2002, *Taudactylus acutirostris* at Melbourne and Taronga Zoos. In Proceedings of the Frogs in the Community Conference, Brisbane, p.94-102.

Reproduction

Banks, C. B. 1986. Reptile breeding records at the Melbourne Zoo, p. 57-71. In S. McKeown, F. Caporaso and K. H. Peterson (eds.), 9th International Herpetological Symposium on Captive Propagation and Husbandry. Zoological Consortium, Inc., Thurmont MD [examples of records used at Zoo to record reproductive data].

Medical Management

Birkett, J., and H. McCraken. 1992. Captive treatment of a Johnstone's crocodile *Crocodylus johnstoni*, under treatment for bilateral mandibular fractures, p. 87-93. In M. J. Uricheck (ed.), 15th International Herpetological Symposium on Captive Propagation and Husbandry. International Herpetological Symposium, Palo Alto CA.

Holz, P. H. 1999. The reptilian renal portal system - a review. Bull. Assoc. Rept. Amphib. Vet. 9(1):4-9.

McCraken, H. E. 1991. The topographical anatomy of snakes and its clinical applications, a preliminary report. Proc. Amer. Assoc. Zoo Vet. 1991:112-119.

McCraken, H. E., and C. A. Birch. 1994. Periodontal disease in lizards - a review of numerous cases. Proc. Amer. Assoc. Zoo Vet. 1994:108-115.

McCraken, H. E., A. D. Hyatt, and R. F. Slocombe. 1994. Two cases of anemia in reptiles treated with blood transfusions: (1) hemolytic anemia in a diamond python caused by an erythrocytic virus; (2) nutritional anemia in a bearded dragon. Proc. Amer. Assoc. Zoo Vet. 1994:47-51.

Conservation

Banks, C. B. 1992. The Striped Legless Lizard Working Group: An interagency initiative to save *Delma impar*, an endangered Australian reptile. Inter. Zoo Yearb. 31:45-49.

Banks, C. B. 1996. Cooperative management of the striped legless lizard (*Delmar impar*): A vulnerable Australian pygopodid, p. 65-68. In P. D. Strimple (ed.), Advances in Herpetoculture. Special Publication of the International Herpetological Symposium, Inc., No. 1. International Herpetological Symposium, Inc., Des Moines IA.

Banks, C. B. 1996. A conservation program for the threatened Romer's tree frog (*Philautus romeri*), p. 1-5. In P. D. Strimple (ed.). Advances in Herpetoculture. Special Publication of the International Herpetological Symposium, Inc., No. 1. International Herpetological Symposium, Inc., Des Moines IA.

Banks, C. B. 1999. Philippine amphibians assessed. Froglog 33:1.

Banks, C. B., and J. Birkett. 1992. A regional herpetofaunal management plan for Australasia, p. 49-66. In M. J. Uricheck (ed.), 15th International Herpetological Symposium on Captive Propagation and Husbandry. International Herpetological Symposium, Palo Alto CA.

Dudgeon, D., and M. W. N. Lau. 1999. Romer's tree frog reintroduction into a degraded tropical landscape, Hong Kong, P. R. China. Re-Introduction News 17:10-11.

Kutt, A., J. Ross, C. Banks, G. Coulson, and A. Webster. 1995. Conservation of an endangered species: The Striped Legless Lizard Working Group as a successful interagency initiative, p. 451-459. In D. A. Saunders, G. L. Craig and E. M. Mattiske (eds.) Nature Conservation 4: The Role of Networks. Surrey Beatty & Sons, Chipping Norton, Australia.

General

Banks, C. B. 1995. "World of Frogs": An environmental resource, p. 222-226. In P. Cust and K. Langham (eds.) Proceedings of the 1994 ARAZPA/ASZK Conference. Territory Wildlife Park, Palmerston.

Dunn, R. W. 1978. Observations on the moloch or thorny devil, *Moloch horridus*. Inter. Zoo Yearb. 18:151-152.

Harcourt, N. 1989. A northern Australian tropical lagoon exhibit. Inter. Zoo Yearb. 28:207-210.

Adelaide Zoological Gardens (1883)

History and Mission: The Royal Zoological Society of South Australia has exhibited reptiles for over a century. C. E. Rix consulted the annual reports and published its history in *Royal Zoological Society of South Australia. 1878-1978.*

Facility and Collection: The following list of names from records compiled by Rix reflect some of the significant holdings at the Zoo since 1878 with the year a taxon was first obtained in parentheses: Murray tortoise (1883), long-necked tortoise (1884), radiated tortoise (1885), American alligator (1914), estuarine crocodile (1900), mugger crocodile (1886), tuatara (1886), scaly-foot lizard (1898), moloch (1935), bearded dragon (1887), Australian bloodsucker (1893), stumpy-tailed or sleepy lizard (1884), blue-tongued lizard (1887), Indian python (1884), reticulated python (1905), diamond or carpet snake (1884), common boa constrictor (prior to 1893), anaconda (1928), Indian cobra (1901), death adder (1885), Australian tiger snake (1885), Australian black snake (1884), Australian brown snake (1901), peninsula brown snake (1893), Australian copperhead (1893) and Russell's viper (1904).

There were several reptile buildings and snake pits in the early days. A snake pit containing about 30 indigenous snakes was built in 1893 but some of the inhabitants were killed by visitors heaving stones at them. This exhibit was modified two years later when a house was provided. The first reptile house was added to the Zoo in 1884, replaced four years later by a second one. A small building (49 x 29 feet) with a pool (21 x 9 feet) in the center, was constructed in 1900. The facility had a glass roof, glass exhibit cases and a slow combustion boiler for heat. Although it is not clear when this building was no longer used, the Annual Report in 1936 listed only four taxa, two tortoises, one alligator and a lizard, so it was likely during the 1920s or early 1930s when it was converted into a giraffe house. In 1950, an outdoor enclosure was built for snakes. Six years later, three naturalistic enclosures were added. In 1985, an octagonal building with 13 temperature-controlled displays, inner service area, and an enclosed public walkway was opened. At the front, three outdoor displays hold American alligators, Johnstone's crocodile and Galápagos tortoises. Other additions to the collection included the common anaconda, Russell's and gaboon vipers, water moccasin, forest cobra, Gila monster, both species of taipan and common death adders. In 1994/95, an extension was built on to Reptile House to include a much larger Amazonian Boid Exhibit, housing anacondas, boa constrictors and Amazonian tree boas. This section abuts a new exhibit following the Amazonian river theme, with underwater viewing of matamata turtles. A larger Aldabra tortoise paddock with a heated house and undercover viewing area was also part of this extension, as was a larger Gila monster display and a large indoor/outdoor exhibit for Fijian banded iguanas.

There have been impressive breedings: Hosmer's skink, woma (*Aspidites ramsayi*), Aruba Island rattlesnake, blood python, Madagascan tree boa, inland taipan (*Oxyuranus microlepidota*) and rhinoceros viper. After the rediscovery of the highly endangered Adelaide pygmy blue-tongue skink (*Tiliqua adelaidensis*), thought to be extinct or nearly so, the Zoo has acquired several individuals to inaugurate a captive-breeding program. The Pernatty knob-tail gecko (*Nephrurus deleani*) has been bred, an endangered species from a very restricted distribution in northern South Australia.

According to *International Zoo Yearbook* (1998:Volume 28), composition of the collection was 31 taxa of reptiles with 121 specimens and one amphibian species with six specimens.

Staff and Scientific Achievements: Roger Ainsley was the first Superintendent of Reptiles in 1984. Eleven years later, he decided to specialize in exhibit design and development and Terry Morley filled his post. The Zoo hired its first research officer, Greg Johnstone, in 2000.

Taronga Zoo (1916)

Curator Uwe Peters at the Taronga Zoo in Sydney described a number of breeding successes at his institution (1982): Aldabran tortoise, rhinoceros and green iguanas, taipan, several python species and leopard tortoise. Other impressive accomplishments included the second generation breeding of the cantil (Peters, 1979), radiated tortoise (Peters, 1969), and first breeding of the Aldabra tortoise (Peters and Finnie, 1979). His successor Terry Boylan continued the tradition with papers on breeding (Boylan, 1984) and captive management (Boylan, 1985) of a population of rhinoceros iguanas. Another contribution chronicled reproduction of the Fijian crested iguana (Boylan, 1989). Boylan (1982, 1989) covered problems associated with maintaining a healthy zoo reptile collection. Timmis (1969) published observations on Pacific boas. Russ et al. (1995) described assisted breeding at the Taronga and Western Plains Zoos.

According to *International Zoo Yearbook* (1998:Volume 28), composition of the collection was 80 taxa of reptiles with 329 specimens and 11 taxa of amphibians numbering 83.

Australian Reptile Park (1958)

In 1958, the late Eric Worrell established the Australian Reptile Park in Gosford, New South Wales. His facility was a public display and research center, best known for extracting snake and funnel-web spider venoms for the preparation of antivenins at the Commonwealth Serum Laboratories in Melbourne. Many scientific papers and books have been published by Worrell and the staff at the Park. Worrell wrote an interesting book in 1963 entitled *Reptiles of Australia*, as well as a handbook on the dangerous snakes of Australia (c1963), *Song of the Snake* in 1958 and others (1961, 1966).

His successor John Weigel has been an active contributor as well with books on the care of Australian reptiles in captivity (1988), and a book entitled *Australian Reptile Park's Guide to Snakes of South-East Australia* two years later. Several papers on the maintenance of elapid snakes (1992), superb dragon (1989), history of the Park (1988), and problems with wildlife legislative practices in Australia (1992) have been published.

Australia Zoo (1970)

History and Mission: Australia Zoo began as a small family run operation when in 1970 Bob and Lyn Irwin purchased a four-acre block near Beerwah on the Sunshine Coast in Queensland. After three years the property, first known as the Beerwah Reptile Park, opened its gates. The collection grew to include a wider range of Australian species but still focusing on reptiles. In 1980, as the park continued to expand, the name was changed to the Queensland Reptile and Fauna Park, and a decision was made to develop the Crocodile Environmental Park. In 1999, the current name, Australia Zoo, was adopted. Beginning with four acres, the zoo now has 20 developed acres with the potential to expand into the 250 acres currently owned. The Irwins' son, Steve Irwin and his wife Terri hosted the television program "The Crocodile Hunter" which often stresses the importance of wildlife protection and conservation by showing interesting animals doing interesting things. On 4 September 2006, Steve Irwin died when a sting ray barb pierced his chest while he was diving during a filming sequence on the northeastern coast of Australia near Port Douglas.

Facility and Collection: In 1984, the first exotic reptiles were displayed: reticulated and Burmese pythons, corn snakes and boa constrictors. Within the next few years, American alligators, Galápagos tortoise ("Harriet" reportedly collected by Charles Darwin in 1835) and green iguanas were added. Currently, the property houses 14 exotic reptile species, comprising of over 100 individuals.

The first enclosures consisted primarily of outdoor pit exhibits and later incorporated extra heating or cooling devices. In 1999, the "snake house" (18m x 6m)

Fig. 149. "The Crocodile Hunter" Steve Irwin avoids injury from one of his charges. *Photograph provided by Australia Zoo*

was constructed with full temperature control; this structure accommodated over 100 specimens. The second stage of development began at the end of 2000 and included the construction of two more buildings with temperature control. One of these provides additional snake holding facilities while the second is specifically for lizards. The "lizard house" is constructed so as the northern wall has four vertically, rolling doors allowing access to direct sunlight. Beginning in 2001, native snake exhibits were upgraded which incorporate sunlight in each exhibit; some of the enclosures include air-conditioning.

Green and rhinoceros iguanas have reproduced at the Zoo. There is a propagation plan in place for the Fiji Island iguana. A number of varanids have bred: canopy goanna, rusty monitor, ridge-tailed monitor, mangrove monitor, Mertens' water monitor, sand goanna and perentie monitor. Reproduction in green tree pythons from the Australian population has occurred.

Staff and Scientific Achievements: In 1970, the staff was largely composed of Irwin family members: Bob,

Lyn, and Steve. The current Director, Wes Mannion, has been at the institution for over 15 years. Brian Coulter is head of the Crocodile section and has been involved with the Zoo since 1995. An amazing period of development within the Zoo has occurred, with the number of employees increasing from four in 1973 to 117 in 2002.

Historical Overview

Chambers, P. 2006. A Sheltered Life. The Unexpected History of the Giant Tortoise. Oxford University Press, Oxford, New York. [After extensively researching the history of Harriet the Galapagos tortoise, Paul Chambers concluded that the belief that this chelonian, reportedly collected or owned by Charles Darwin, is likely untrue.]

Engle, K. 2000. Australia Zoo, more than a name change. Proceedings of the Australasian Regional Association of Zoological Parks and Aquaria Inc./ Australasian Society of Zoo Keeping.

Thomson, S., S. Irwin, and T. Irwin. 1998. Harriet the Galapagos tortoise: Disclosing one and a half centuries of history. Reptilia March/April (2):48-51.

Reproduction

Irwin, B. 1986. Captive breeding of two species of monitor. Thylacinus 11(2):4-5.

Irwin, S. 1996. An innovative strategy for the detection of egg deposition in captive varanid reptiles. Herpetofauna 26(1):31-32.

Irwin, S., and K. Engle. 1996. Nocturnal nesting by captive varanid lizards. Herpetol. Rev. 27(4):192-194.

Conservation

Irwin, T. 1995. The co-operative conservation of the Fijian crested iguana *Brachylophus vitiensis*. Proceedings of the Australasian Regional Association of Zoological Parks and Aquaria Inc./Australasian Society of Zoo Keeping.

General

Covacevich, J., W. Mannion, and S. Irwin. 1997. The browns at Brookfield. Wildlife Australia. Winter:32-33.

Irwin, S. 1996. Survival of a large *Crocodylus porosus* despite significant lower jaw loss. Mem. Queensland Mus. 39(2):328.

Irwin, S. 1996. Capture, field observations and husbandry of the rare canopy goanna *Varanus keithhornei*. Thylacinus 21(2):19.

Irwin, S., and T. Irwin. 1995. Notes on early development in a hatching-assisted frilled lizard *Chlamydosaurus kingii*. Herpetofauna 25(2):60-61.

Irwin, S., T. Irwin, B. Lyons, and S. Lyons. 1998. Crocs vs people. Wildlife Australia. Autumn:36-39.

Porter, R., S. Irwin, T. Irwin, and K. Rodrigues. 1997. Records of marine snake species from the Hey-Embley and Mission Rivers of North Queensland. Herpetofauna 27(2):2-7.

Ballarat Wildlife Park (1985)

The Park, founded by Greg Parker, is located one and a half hours from Melbourne in Ballarat. The Park is privately owned and operated and houses a collection of approximately 250 snakes, lizards, crocodilians, fish, and other animals. The collection consists of a wide variety of Australian native animals, and there is a reptile house devoted to exotic and Australian taxa such as Mozambique spitting cobras, reticulated pythons, Burmese brown tortoises, alligator snapping turtles, Aldabran tortoises and white-lipped pythons. The Park boasts many breeding successes, some of which are American alligators and the second breeding of estuarine crocodiles in Australia.

History of Zoo Herpetology in New Zealand

Although tuataras have been exhibited in zoos, wildlife parks and museums in New Zealand for many years, reproductive successes were not impressive and information on proper care was limited (Blanchard, 1992). In 1967, nine tuatara eggs were laid at the Auckland Zoo but these may have been the result of a wild breeding; six of them hatched (Wood, 1967). Based on observations in the field, Newman et al. (1979) offered recommendations on captive tuatara maintenance: physical environment, burrows, space, temperature, light, relative humidity, food and medical care. Later, Newman (1982) listed reasons for unsuccessful captive breeding programs in the past but mentioned that the exhibits at the Wellington Zoo and Southland Museum seemed to fill the basic needs for successful tuatara breeding. In 1987, Senior Reptile Keeper Vernon Tintinger from Auckland described breeding, followed by husbandry, egg incubation and rearing. Four years later, Boardman and Sibley outlined the captive management, diseases and veterinary care of tuataras in Auckland. Gillingham and Miller (1991) studied reproductive ethology in wild tuataras and recommended that groups of males be allowed to engage in aggressive interactions in large enclosures to stimulate courtship and mating. In 1994, Goetz and Thomas published two papers: use of annual growth and activity patterns to assess management procedures for captives and a larger study on captive maintenance of tuataras.

In October 1990, participants at a workshop spearheaded by the Tuatara Research Group at Victoria University and the New Zealand Department of Conservation recognized the need for a national recovery plan. This plan, in addition to *in situ* interventions, identified the need for a captive management component for tuataras, especially for those on North Brother Island and 30 offshore islands. Barbara Blanchard from the Wellington Zoological Gardens was appointed Captive Breeding Coordinator; her responsibilities included the monitoring of the captive breeding population, identifying improved husbandry protocols such as captive incubation of wild-collected eggs, and assessing the feasibility of head-starting captives for reintroduction.

Based on this clear need for a broad-based initiative, Cree et al. (1994) outlined an ambitious plan to survey genetic variation and population status, begin island restoration and feral animal eradication, and integrate and improve captive management.

In addition to the work with tuataras being undertaken within zoos in New Zealand, four lizards have been identified as candidates for captive breeding programs: robust skink (*Cyclodina alani*), Whitaker's skink (*C. whitakeri*), grand skink (*Leiolopisma grande*), and Otago skink (*L. otagense*) (Butler, 1992).

To my knowledge, there is only one captive-breeding program envisioned for any of the three endemic anurans (*Leiopelma*), although there have been reports of successful captive breeding episodes (Bell, 1985) and laboratory maintenance (Sharbel and Green, 1989, 1992). The Auckland Zoo plans to build a unit for the Coromandel New Zealand frog (*L. archeyi*) (Holyoake et al., 2001; Froglog, December 2002, No. 54).

Long nose tree snake of the genus *Dryophis*

Chapter 10

Zoological Gardens of Japan, China, and India

Leatherback Seaturtle (*Dermochelys coriacea*)

Introduction to Herpetology in Japan

Ken Kawata, former General Curator at the Staten Island Zoo, is a zoo worker specializing in the history of Japanese zoos. In his writings and our conversations, he has often said that the maintenance and exhibition of amphibians and reptiles in Japan has been surprisingly limited (see Kawata, 2003, 2004). This is a strange situation as there are many aquariums in Japan and it would seem logical that at least some of these facilities would specialize in zoo herpetology. The Kyoto Zoo (opened 1903) housed reptiles in a heated facility completed in 1923; the first was a python placed in this new facility on 9 December. The Tennoji Zoo in Osaka (opened 1915) only had eight reptile and two amphibian species three years after opening. In 1971 a reptile house made a debut in Nogeyama Zoo (opened 1951) in Yokohama; the new building was five times larger than a 1965 facility which exhibited only nine species of reptiles. Late in 1973, another reptile house was opened in Nihondaira Zoo (opened 1969) in Shizuoka. The next year, a two-storied building opened at the Asa Zoo in Hiroshima; nocturnal exhibits, mainly for mammals, occupied the first floor while the second

floor was devoted to reptiles. The Yagiyama Zoo (opened 1965) in Sendai opened a reptile house in 1978. In terms of newer exhibit approaches, Osaka's Tennoji Zoo introduced a large-scale indoor application of the "immersion landscape" concept in March 1995 where animals and visitors were juxtaposed. Although reptiles were the featured theme, a wide variety of taxa was on exhibit to depict the world's diverse habitats including the southern swamp of North America and Japanese wetlands. This exhibit complex was later named IFAR, representing invertebrates, fishes, amphibians and reptiles.

Several reptiles survived the turbulence of World War II. An American alligator was said to have arrived at Suwayama Zoo in Kobe in 1939. Seven years later, the Zoo was closed and the animals were managed by a humane society. The Zoo was temporarily reopened in 1950, and in March of the following year a new facility, Kobe Oji Zoo, opened its gate. This alligator, said to be a male, was listed on Oji Zoo's longevity animal inventory as of March 1980. This specimen was no longer listed in the inventory in March 1982, but had lived well over four decades. Near the cities of Osaka and Kobe lies Takarazuka, known for its amusement centers, one of which is a zoo (opened 1924). In the pre-war era an industrialist built an exotic animal collection, including many tropical fish, 50 American alligators, and two alligator snapping turtles. Over the years, many animals were either sent to other institutions or died, and by the mid-1960s only one original animal was left, an alligator snapping turtle which had arrived in January 1933. After a period of more than 10 years of rapidly expanding tumors over its body, which took up 20% of the body weight (20.5 kg), the turtle died on 27 November 1968 (Koto, 1969). "Mr. Long," a Chinese alligator, arrived at the Ritsurin Park Zoo (opened 1929) in 1933. Although it survived the war, a liver tumor took its toll and the reptile died on 10 February 1994 (Kagawa, 1994).

Certainly, there have been some advances with reptiles and amphibians, especially since those early days. Araki et al. (1982) described reproduction in the reticulated python at the Takarazuka Zoological and Botanical Gardens. A total of 30 Indian pythons hatched between 21-31 May 1971 at the Tokuyama Zoo (opened 1960). The dam hatched at the Ueno Zoo in 1967, the first multi-generational breeding of the taxon in Japan. Hosono (1982) successfully bred the Florida kingsnake, ratsnakes and pythons at Kyoto Municipal Zoo. Yanaga et al. (1984) artificially incubated and reared leopard tortoises at Miyazaki Safari Park.

Sea turtles commonly have been kept in aquariums for many years but reproduction had been virtually non-existent, especially indoors. Oka et al. (1983) kept the leather-back sea turtle at the Shimonoseki Municipal Aquarium (opened 1956). Yoshioka and Samejima (1989) from the Nagasakibana Parking Garden (opened 1966) bred the loggerhead sea turtle on the Nagasakibana coast of Kagoshima Prefecture. The Port of Nagoya Public Aquarium, established in 1990 and opened to the public two years later, had kept three species of sea turtles. Sea turtles had been known to lay eggs in an outdoor captive environment, but never indoors. The staff built an indoor turtle beach, which led to the breeding of the loggerhead sea turtle; this successful breeding of a sea turtle in an indoor facility is believed to be the first in the world (Uchida, 1996). Breeding of sea turtles is no longer uncommon in Japanese aquariums today.

Amphibians bred include Ishikawa's frog at Nagasakibana Parking Garden (opened 1966), Kagoshima (Shiihara and Samejima, 1995). There have been successful long-term efforts to breed the Japanese giant salamander. Doi et al. (1999) attempted to breed them under indoor artificial conditions at Kobe Municipal Suma Aqualife Park (opened 1987), Hyogo.

Field studies undertaken by staff and captive propagation of the salamander have been accomplished at the Asa Zoological Park (opened 1971) in Hiroshima and Himeji City Aquarium (opened 1966) which has resulted in a number of scientific papers listed below. The Japanese list this taxon as a National Special Natural Treasure.

Ueno Zoological Gardens (1882)

History: The Zoo is located near the center of Tokyo. During World War II, the institution suffered from shrinking support and attendance dropped precipitously. Exhibit animals and their requirements were in short supply. The Zoo began its rebirth when Tadamichi Koga was named director in 1937 but since he was in the army as a veterinary officer during the war, it was only after he returned that the institution began to thrive.

Facility and Collection: Early in 1932, a 10-foot long salt-water crocodile arrived at the zoo, but due to the lack of a proper facility, it was temporarily kept in the heated hippo house. Later that year a permanent reptile facility was built with stable temperature control and skylights, thus enabling giant snakes and large crocodilians to be kept properly. This zoo received a pair of Komodo dragons on 30 October 1942 which were captured by the Imperial Navy but they died shortly thereafter from hypothermia due to a fuel shortage. In 1949, the Hogle Zoo in Salt Lake City sent four mud turtles and four box turtles to the Zoo. In October 1964, a large, four-storied aquarium, designed to

Fig. 150. Ueno Zoo Vivarium in Japan. *Akiyoshi Nawa, arranged by Ken Kawata.*

Fig. 151. Giant tortoise display at Ueno Zoo Vivarium in Japan. *Akiyoshi Nawa, arranged by Ken Kawata.*

Fig. 152. Aquatic turtle exhibit at Ueno Zoo Vivarium in Japan. *Akiyoshi Nawa, arranged by Ken Kawata.*

exhibit a wide range of taxa including marine invertebrates, fish, amphibians and reptiles, was opened. Patterned after the "vivarium" concept of European zoos, aquatic amphibians were located on the third floor of this complex, and the entire fourth floor was eventually opened in June 1973 for the other amphibians and reptiles. The 1964 building was demolished in 1992, and a new facility for reptiles and amphibians, named "Vivarium," opened its door in July 1999.

According to *International Zoo Yearbook* (1998:Volume 28), composition of the collection was 71 taxa of reptiles with 261 specimens and 7 taxa of amphibians numbering over 280.

Publication: *Animals and Zoos* in Japanese.

Historical Overview

Kawata, K. 2001. Zoological gardens of Japan, p. 295-331. In V. N. Kisling Jr. (ed.), Zoo and Aquarium History. Ancient Animal Collections to Zoological Gardens. CRC Press, Boca Raton, London, New York, Washington DC.

Kawata, K. 2001. Ueno Zoological Gardens, p. 1273-1276. In C. E. Bell (ed.), Encyclopedia of the World's Zoos. Fitzroy Dearborn Publishers, Chicago, London.

Miyashita, M. 1996. The reptile house enters its third year. Animals and Zoos 48(12 No. 561):424-428. [in Japanese.].

Ueno Zoo. 1982. Ueno Zoo: The 100 Year History. Tokyo Metropolitan Government, Tokyo.

Yamamoto, Y. 1997. New reptile house at Ueno Zoo. Animals and Zoos 49(8. No. 569):260-261. [in Japanese.].

General

Sugiura, H. 1977. The thriving Indian python. Animals and Zoos 29. No. 1. (No. 324):6-8. [in Japanese with English summary.].

**Scientific Papers on
Japanese Giant Salamanders
by Staff at ASA Zoological Park,
Himeji City Aquarium and Other Institutions**

Early in 1871, this species of giant salamander was exhibited by the government in Tokyo before the first zoo in the country was opened. When the Ueno Zoo opened an aquarium section on 20 September 1882, a giant salamander was on display. First breedings occurred in the Netherlands, at least in 1902, 1903 and 1905, in the Amsterdam Aquarium.

In 1973, Jiro Kobara, then the director of the Asa Zoo in Hiroshima, began field research on reproductive biology, ecology and natural history, which led to the first captive breeding six years later in Japan. Recently, the zoo staff has been working on *in situ* conservation projects using civil engineering techniques to reconfigure and secure salamander breeding sites.

Another zoo professional who has been conducting research, is Takeyoshi Tochimoto, director of the Himeji City Aquarium, began field work in 1975. Along with ecological studies, he has also been involved with *in situ* cooperative projects with civil engineers to secure habitats. The Japanese Association of Zoological

Fig. 153. Developing embryos of Japanese giant salamander (*Andrias japonica*) at Asa Zoo. *Jiro Kobara at Asa Zoo, arranged by Ken Kawata.*

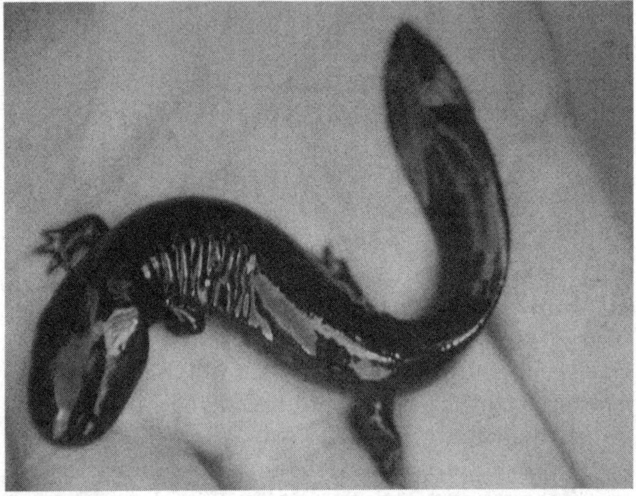

Fig. 154. Fifteen month old Japanese giant salamander (*Andrias japonica*) at Asa Zoo. *Jiro Kobara at Asa Zoo, arranged by Ken Kawata.*

Gardens and Aquariums (JAZGA) includes this taxon in the Species Survival Committee of Japan (SSCJ), a captive species survival management plan (Kawata, 1991).

References

Asa Zoological Park. 1983. Research Report on the Conservation and Propagation of the Giant Salamander. 30 pages. (Zoo) December 1983.

Asa Zoological Park. 1988. The Giant Salamander: Research Records No. 1. 37 pages. (Zoo) 1988.

Ashikaga, K. 1978. The Giant Salamander Habitats in Hiroshima Prefecture. Asa Zoological Park Husbandry Records (7), 5-10. (Society) December 1978.

Doi, T., H. Tsukamoto, and S. Aoyama. 1999. An attempt at breeding the Japanese giant salamander, *Andrias japonicus,* under indoor artificial conditions. J. Japan Assoc. Zoos Aquar. 40(2): 56-64 [description in Japanese with English summary of husbandry and environmental parameters used to reproduce this species at Kobe Municipal Suma Aqualife Park, Hyogo].

Fukumoto, Y. 1990. Sex determination in the Japanese giant salamander, *Andrias japonicus*, by the laparoscope under general anesthesia. J. Japan Assoc. Zoos Aqua. 31(3):85-86 [n Japanese with English summary; contribution from Asa Zoological Park].

Inoue, T. 1971. Survey of the Giant Salamander. Asa Zoological Park Husbandry Records (6), 1-9. (Society) December 1977.

Japanese Association of Zoological Gardens and Aquariums (JAZGA). Research Report on the Conservation and Propagation of Rare Animals: The Giant Salamander. JAZGA, Tokyo, 338 pages. March 1978.

JAZGA Conservation Committee, 1976. Giant Salamander Conservation Committee Report (3). 25 pages. (Zoo) January 1976.

JAZGA. 1978. Giant Salamander Conservation Committee Report. JAZGA, Tokyo, 45 pages. November 1978.

Kobara, J. 1985. The Giant Salamander. Dobutsu-sha, Tokyo.

Kobara, J., K. Ashikaga, T. Inoue, F. Wakabayashi, K. Kuwabara, and N. Suzuki. 1980. The study on the protection of Japanese giant salamander, *Megalobatrachus j. japonicus*, in Hiroshima Prefecture. 5. The egg-laying in Aquarium. J. Japan Assoc. Zoos Aqua. 22:67-71. [In Japanese with English summary; describes breeding group, enclosures, water quality, and food at Asa Aquarium which resulted in successful oviposition.]

Kuwabara, K., T. Inoue, F. Wakabayashi, K. Ashikaga, N. Suzuki, and J. Kobara. 1980. The study on the protection of Japanese giant salamander, *Megalobatrachus japonicus japonicus,* in Hiroshima Prefecture. 4. Observation on the reproductive behaviour in the Stream of Matsuzai-gawa. J. Japan Assoc. Zoos Aqua. 22:55-66. [In Japanese with English summary. Twenty-seven salamanders aggregated around nest site. One male occupied the nest. Resident males demonstrated territorial behavior. Five females laid eggs in sequence in single nest, followed by several males that took part in egg-laying. Males were more active than females and most activity occurred during the day. Asa Zoological Park.]

Kuwabara, K., N. Suzuki, F. Wakabayashi, H. Ashikaga, T. Inoue, and J. Kobara. 1989. Breeding the Japanese giant salamander *Andrias japonicus* at Asa Zoological Park. Inter. Zoo Yearb. 28:22-31.

Suzuki, N. 1974. The Giant Salamander in Hiroshima Prefecture. Asa Zoological Park Husbandry Records (3), 13-21. (Society) December 1974.

Suzuki, N., F. Wakabayashi, K. Ashikaga, and J. Kobara. 1972. The study on the protection of Japanese giant salamander, *Megalobatrachus japonicus japonicus*, in Hiroshima Prefecture. Preliminary Report. J. Japan Assoc. Zoos Aqua. 14:83-86. [In Japanese, this was the historic first paper by the staff of Asa Zoological Park.]

Suzuki, N., F. Wakabayashi, K. Ashikaga, and J. Kobara. 1973. The study on the protection of Japanese giant salamander, *Megalobatrachus j. japonicus*, in Hiroshima Prefecture. 2. A case of abolished habitat. J. Japan Assoc. Zoos Aqua. 15:83-84 [Asa Zoological Park].

Suzuki, N., K. Kuwabara, K. Ashikaga, M. Nakanishi, N. Minamigata, and H. Morimoto. 2000. Breeding of the Japanese giant salamander, *Andrias japonicus*, in a portable artificial nest. J. Japan Assoc. Zoos Aqua. 41(3):83-87. [In Japanese, this contribution is from Asa Zoological Park.].

Tochimoto, T. 1990. Ecological studies on the Japanese giant salamander, *Andrias japonicus*, in the Ichi River in Hyogo Prefecture (1) Marking of animals for recognition. J. Japan Assoc. Zoos Aqua. 31(4):112-116. [Contribution in Japanese with English summary; 356 individuals have been recorded and 176 recaptured. Five salamanders have lived over 10 years.]

Tochimoto, T. 1990. Ecological studies on the Japanese giant salamander, *Andrias japonicus* (Temminck), in the Ichi River in Hyogo Prefecture (II) Growth in the field. J. Japan Assoc. Zoos Aqua. 32(1):14-20 [contribution in Japanese with English summary].

Tochimoto, T. 1990. Ecological studies on the Japanese giant salamander, *Andrias japonicus* (Temminck), in the Ichi River in Hyogo Prefecture (III) On patterns. J. Japan Assoc. Zoos Aqua. 32(4):90-93 [contribution in Japanese].

Tochimoto, T. 1991. Japanese giant salamanders along Ichikawa River in Hyogo Prefecture II, growth in natural habitat. Doubutsuen-Suizokukan Magazine, Association of Japanese Zoos and Aquariums 32(1):14-20.

Tochimoto, T. 1991. Ecological studies on the Japanese giant salamander, *Andrias japonicus*, in the Ichi River in Hyogo Prefecture IV. Movements and permanent residences (Part 1). J. Japan Assoc. Zoos Aqua. 33(3):49-52 [contribution in Japanese with English summary].

Tochimoto, T. 1992. Ecological studies on the Japanese giant salamander, *Andrias japonicus*, in the Ichi River in Hyogo Prefecture IV. Movements and permanent residences (Part 2). J. Japan Assoc. Zoos Aqua. 33(4):85-90 [contribution in Japanese with English summary].

Tochimoto, T. 1993. Ecological studies on the Japanese giant salamander, *Andrias japonicus*, in the Ichi River, Hyogo Prefecture VI. Metamorphosis. J. Japan Assoc. Zoos Aqua. 34(2-3):28-33. [contribution in Japanese with English summary. Author suggested that metamorphosis took between 4-5 years in the wild when salamanders are between 20-22 cm in total length.].

Tochimoto, T. 1994. Ecological studies on the Japanese giant salamander, *Andrias japonicus*, in the Ichi River, Hyogo Prefecture VII. Spawning grounds. J. Japan Assoc. Zoos Aqua. 35(2):33-41. [Contribution in Japanese with English summary. Author observed five active nests and four others which may have been used.]

Tochimoto, T. 1995. Ecological studies on the Japanese giant salamander, *Andrias japonicus*, in the Ichi River, Hyogo Prefecture VIII. Reproductive ecology. Part 2. Fighting. J. Japan Assoc. Zoos Aqua. 36(2):51-57. [Contribution in Japanese with English summary. Author observed males with serious or fatal injuries caused by conspecifics and postulated that fighting may be the primary cause of death, other than through human activity.]

Tochimoto, T. 1995. Ecological studies on the Japanese giant salamander, *Andrias japonicus*, in the Ichi River in Hyogo Prefecture IX. Reproductive ecology. Part 3. The swollen area surrounding the cloacal area. J. Japan Assoc. Zoos Aqua. 37(1):7-12. [Contribution in Japanese with English summary. Author observed some males with swollen cloacal areas during breeding season.]

Tochimoto, T. 1995. Ecological studies on the Japanese giant salamander, *Andrias japonicus*, in the Ichi River in Hyogo Prefecture X. An attempt to rebuild spawning places along the River. J. Japan Assoc. Zoos Aqua. 37(1):13-17 [contribution in Japanese with English summary].

Tochimoto, T. 1995. Ecology of giant salamander (part 3) - Eggs and larvae. Newsletter Himeji City Aquarium March. No. 26 [contribution in Japanese].

Tochimoto, T. 1996. Coexistence for both human and giant salamander. - River alteration and giant salamander. Newsletter Himeji City Aquarium Sept. No. 29 [contribution in Japanese].

Tochimoto, T. 1998. Creating a harmonious environment for man and giant salamander. Newsletter Himeji City Aquarium March. No. 32 [contribution in Japanese].

Tochimoto, T. September 2000. Siebold's salamanders. Newsletter Himeji City Aquarium No. 37:2-3 [in Japanese].

Tochimoto, T. 2001. Food habits of the Japanese salamander. Newsletter Himeji City Aquarium March. No. 38 [contribution in Japanese].

Tochimoto, T., and K. Shimizu. 1997. Regeneration on the Japanese giant salamander, *Andrias japonicus*, in captivity. J. Japan Assoc. Zoos Aqua. 38(3):65-68 [contribution in Japanese with English summary].

Wakabayashi, F. 1973. Conservation of the Giant Salamander in Hiroshima Prefecture, 1972. Asa Zoological Park Husbandry Records (2), 15-19. (Society) December 1973.

Wakabayashi, F. 1975. Conservation of the Giant Salamander in Hiroshima Prefecture: I. Breeding Behavior, II. Rescued Specimens in Captivity. Asa Zoological Park Husbandry Records (4), 1-9. (Society) December 1974.

Wakabayashi, F., K. Kuwabara, K. Ashikaga, T. Inoue, N. Suzuki, and J. Kobara. 1976. The study on the protection of Japanese giant salamander, *Megalobatrachus j. japonicus*, in Hiroshima Prefecture. 3. The relation between the breeding migration and the weir. J. Japan Assoc. Zoos Aqua. 18:31-36 [in Japanese. Asa Zoological Park].

Atagawa Tropical Garden and Alligator Farm (1958)

Waturu Kimura (1922-1981) was Proprietor of the private Atagawa Tropical and Alligator Garden, Shizuoka Prefecture, established in 1958. The collection of tropical plants is enormous with many types of orchids, bromeliads, water lilies, lotus, fruit trees and palms mostly held in large greenhouses. The crocodilians are no less impressive with 25 taxa represented. Notable breedings included broad-snouted caimans for many years, Cuban crocodiles in the 1970s, and Australian freshwater crocodiles. Gharial eggs were regularly collected in Nepal and hatched at Atagawa for reintroduction. Crocodilians are maintained in large, well-planted outdoor displays with a steady supply of warm spring-fed water. Several large pythons and boas, various chelonians and the Chinese giant salamander are on exhibit.

References

Kimura, W. 1968. Crocodiles of Palau Islands. Atagawa Tropical Garden and Alligator Farm. Research Report No. 1:1-44. [in Japanese with English summary. This account covered the former and present distributions of Philippine, New Guinean and salt water crocodiles.].

Kimura, W. 1969. Crocodiles in Cambodia. Atagawa Tropical Garden and Alligator Farm. Research Report No. 3:1-23. [In Japanese with English summary. This report covered a trip by Kimura to observe wild populations and captive breeding farms for crocodiles.]

Kimura, W. 1976. Artificial incubation of crocodile eggs. Atagawa Tropical Garden and Alligator Farm. Research Report No. 4:1-44. [English report covered egg incubation, morphology and development of neonates, and abnormalities.]

Kimura, W. 1978. Artificial incubation of gavials eggs. Atagawa Tropical Garden and Alligator Farm. Research Report No. 8:1-16. [in Japanese and English.].

Kimura, W., and H. Fukada. 1966. Crocodiles of the World. Atagawa Crocodile Vivarium, Atagawa [in Japanese with English summary].

China

There are over 100 zoos in China and the Chinese Association of Zoological Gardens (CAZG) in Beijing is the professional association coordinating information about zoos and aquariums in China. There are three main herpetological collections: Beijing, Chengdu and the private Black Dragon-Pool Park, Nanjing, First Tortoise and Turtle Museum.

Beijing (Peking) Zoological Gardens (1906)

History and Mission: In the 1960s, the Zoo began keeping amphibians and reptiles but there was no facility dedicated to them until 1979.

Facility and Collection: In 1979, a new reptile and amphibian building was constructed. It is an enormous structure with two and one-half levels. When the visitor faces the building and looks to the left, a large outdoor pool housing many Chinese alligators and turtles is apparent. Upon entering, one sees a large crocodilian exhibit on the left and since traffic flow moves in only one direction, a series of smaller displays (ca. 1.75 x 1.5m²) containing turtles, frogs and giant salamanders are encountered. The lower section on the right contains the nonvenomous snakes. These smaller glass-fronted exhibits contain live plants, gravel or sand substrates, tree braches and permanent pools. In the center of the building is a two-story glassed enclosure with large trees for Indian pythons. On the second floor, the venomous snakes are found. The staffing level is impressive, with 12 or 13 employees on call.

As can be anticipated with such a large facility to be filled, the collection is enormous as well. There are 34 chelonian taxa representing many families. In the family Emydidae, there are two species of pond turtles (*Chinemys*), five species of Asian box turtles (*Cuora*), four eyed turtle (*Sacalia quadriocellata*), Chinese stripe-necked turtle (*Ocadia sinensis*), giant Asian pond turtle (*Heosemys grandis*), Asian leaf turtle (*Cyclemys dentata*), yellow-bellied turtle (*Trachemys*), keeled box turtle (*Pyxidea mouhoti*), Chinese black-breasted box turtle (*Geoemyda spengleri*), and Coahuilan box turtle (*Terrapene coahuila*). In the family Testudinidae, many tortoise taxa are represented: radiated (*Geochelone radiata*), leopard (*G. pardalis*), Indian star (*G. elegans*), impressed (*Indotestudo impressa*), elongated (*I. elongata*), Central Asia (*Testudo horsfieldii*), gopher (*Gopherus polyphemus*) and pancake tortoises (*Malacochersus tornieri*). There are two soft shell turtles: Asian giant softshell turtle (*Pelochelys bibroni*), a species rarely seen in captivity and the wattleneck softshell (*Pelodiscus sinensis*). Common snapping turtles (*Chelydra serpentina*), big-headed turtles (*Platysternon megacephalum*), and Krefft's river turtle (*Emydura krefftii*) round out the collection.

Squamates are equally impressive. There are fewer lizards than snakes such as the tokay gecko, water monitor, Chinese skink, Chinese water dragon, and several agamids (*Calotes, Leiolepis*). The snake collection is varied, comprising approximately three dozen types. Reticulated and Indian pythons are featured. Ratsnakes are well represented with the Mandarin ratsnake (*Elaphe mandarina*), Taiwan ratsnake (*E. carinata*), frog-eating ratsnake (*E. rufodorsata*), steppe ratsnake (*E. dione*), black-banded trinket snake (*E. porphyracea*), twin-spotted ratsnake (*E. bimaculata*), radiated ratsnake (*E. radiata*), Russian ratsnake (*E.*

schrenckii), red-headed ratsnake (*E. moellendorffi*), Taiwan beauty snake (*E. taeniura*), Cantor's ratsnake (*Zaocys dhumnades*), and Indian and Oriental ratsnakes (*Ptyas korros* and *P. mucosus*). Other colubrids include the common wolf snake (*Lycodon aulicus*), Stejneger's bamboo snake (*Pseudoxenodon stejnegeri*), red banded snake (*Dinodon rufozonatus*), ringed water snake (*Sinonatrix annularis*) and eastern water snake (*S. pericarinata*), Asian tiger, checkered and striped keelbacks (*Rhabdophis tigrinus, Xenochrophis piscator* and *Amphiesma stolata*)), Bocourt's water snake (*Enhydris bocourti*), and three species of rainbow water snakes (*Enhydris*).

Elapids include the king and Asiatic cobra, banded and many-banded kraits (*Bungarus fasciatus* and *B. multicinctus*), and Amazon coral snake (*Micrurus spixii*). There are many pitvipers: Chinese habu (*Trimeresurus mucrosquamatus*), Chinese green tree viper (*T. stejnegeri*), Chinese mountain pitviper (*Ovophis monticola*), hundred-pace viper (*Deinagkistrodon acutus*), and Siberian pitviper (*Agkistrodon halys*). There are six rare and delicate Fea's viper (*Azemiops feae*) in the collection.

Six crocodilians are on display: Siamese (*Crocodylus siamensis*), mugger (*C. palustris*), estuarine (*C. porosus*), West African dwarf (*Osteolaemis tetraspis*), with two American and approximately 20 Chinese alligators.

Amphibians are also on exhibit: the fire-bellied newt (*Cynops orientalis*), Chinhai newt (*Echinotriton chianhaiensis*), Chinese giant salamander (*Andrias davidianus*), Chusan Island toad (*B. gargarizans*), black-spined toad (*B. melanostictus*), Gunther's Amoy frog (*Rana guentheri*), black-spotted frog (*R. nigromaculata*), and Schmacker's frog (*R. schmackeri*).

Staff and Scientific Achievements: Liu Liquan is the Curator of Reptiles and Amphibians.

Chengdu Zoological Gardens (1954) (Chengdu, China)

Facility and Collection: During the past 30 years, amphibians and reptiles were not well represented in the collection. The old facility was very small (280m²), simple, and not planned specifically for amphibians and reptiles. In 2001, a new building was designed and construction commenced. This new building was much larger (1600m²), with central air-conditioning. The new exhibits are naturalistic and represent the habitat of each species. The building opened in July 2001 and since that time, over 40 species have reproduced.

The collection is large and varied. The chelonian representatives include the Malayasia giant turtle (*Orlitia borneensis*), elongate, Asian brown, and impressed tortoises, Indonesian and Southeast Asian box turtles, yellow-headed temple turtle, spiny turtle, giant Asian pond turtle, Asian softshell turtle, black marsh turtle, and red-eared slider. Estuarine crocodiles and Chinese alligators are on exhibit. Water and Bengal monitors, Chinese crocodile lizard, Chinese water dragon, and several types of mountain lizards (*Japalura*) are displayed. The Burmese python is the sole python in the collection but ratsnakes include the Mandarin, red-headed, Taiwan beauty snake, Cantor's ratsnake and Indian and Oriental ratsnakes. Other colubrids include the red-banded snake, ringed water snake, and Bocourt's water snake (*Enhydris bocourti*). Venomous forms include the king cobra, many-banded krait, Russell's viper, hundred-pace viper, and Chinese green tree viper. Chinese giant salamanders round out the collection.

Staff and Scientific Achievements: Wang qiang oversees the collection.

Black Dragon-Pool Park, Nanjing, First Tortoise and Turtle Museum

In 1991, Zhou Jiufa and Zhou Ting, owners of the private Black Dragon-Pool Park, Nanjing, First Tortoise and Turtle Museum in China, published an interesting book called *Chinese Chelonians Illustrated*. The first part of this book lists known species in China with keys. The second part covers chelonians as food, medicine, commercial uses, habits, distribution, diseases and treatments. There are many photographs of the chelonians living in this important collection, as well as a number of artifacts with a turtle or tortoise theme. Zhou's box turtle (*Cuora zhoui*) was described by Zhao Er-mi in 1990, based on three specimens purchased in the markets in Nanning and Pingxiang, Guangxi Zhuang Autonomous Region; this rare turtle was named in honor of Zhou Jiufa.

India

Madras Snake Park (1969), Madras Crocodile Park Trust (1974) and Center for Herpetology (1991)

History and Mission: Romulus Whitaker, born in New York City, returned to India in 1967 where he pursued his interest in herpetology. His first venture was the Madras Snake Park on India's east coast where he encouraged Irula villagers to collect snakes which provided a source of income. Co-founder of the Madras Crocodile Bank in 1976, he was instrumental in developing captive breeding programs for India's three most endangered reptiles: mugger, gharial and saltwater crocodilians. Through his efforts, thousands of neonates were hatched, head-started, and released into the wild. Whitaker has developed formal associations with a number of conservation and scientific organizations, such as IUCN, AZA, Wildlife Conservation Society and a number of Indian groups. One important initiative was the creation in 1978 of the Irula Snake Catcher's Cooperative Society in India which encourages the local residents to capture venomous snakes for antivenin production. After venom extraction, the snakes are released. The Trust has evolved into the Centre for Herpetology, established in 1991.

Facility and Collection: Blessed with a constant water supply provided by an aquifer, a number of croco-

Fig. 155. Old Madras Snake Park in India in 1971. *Photograph provided by Romulus Whitaker.*

Fig. 156. New Madras Snake Park in India in 1973. *Photograph provided by Romulus Whitaker.*

Fig. 157. Romulus Whitaker (center) ropes "Jaws," a sixteen foot long estuarine crocodile (*Crocodylus porosus*) at Madras Crocodile Trust in India. *Photograph provided by Romulus Whitaker.*

dilians, turtles and water monitors have bred at Madras Crocodile Bank. In 1992, over 8000 crocodilians comprising 10 species were held and 10 percent of these were either released or used to establish captive breeding programs throughout India. Over 200 chelonians (26 species) were also maintained and many of these have bred at the facility.

Staff and Scientific Achievements: Jeffrey Lang, now at the University of North Dakota, studied a number of crocodilians at the Trust which resulted in many papers on behavior, sex determination and reproductive biology. Indraneil Das, author of many books on Asian herpetology, was the Scientific Officer in 1996. Harry Andrews, manager since 1983, has published a number of papers on reptiles at the Trust, some with Whitaker.

Publication: The journal *Hamadryad* began publication in 1976 and contains a number of herpetological papers.

Personal Reflections: When Rom last visited me in Washington, he was involved in a filmmaking project on king cobras with the National Geographic Society. As he explained his career and accomplishments over the years, I reallized that there is no question that the improved status of the crocodilians and other reptiles of India is almost due entirely to his vision and energy. I admire him greatly for his conservation ethic.

Historical Overview

Whitaker, R. 1992. Indian herpetology and the Madras Crocodile Bank, p. 1-25. 16th International Herpetological Symposium on Captive Propagation and Husbandry. International Herpetological Symposium.

Whitaker, Z. 1989. Snakeman. Story of a Naturalist. India Magazine Books, Bombay. [Zai Whitaker wrote this book about the life and times of Romulus Whitaker.]

Behavior

Lang, J. W., R. Whitaker, and H. Andrews. 1986. Male parental care in mugger crocodiles. National Geographic Research 2(4):519-525.

Captive Management

Whitaker, R. 1987. The management of crocodilians in India, p. 63-72. In G. J. W. Webb, S. C. Manolis, and P. J. Whitehead (eds.). Wildlife Management: Crocodiles and Alligators. Surrey Beatty and Sons, Sydney, Australia.

Reproduction

Andrews, H. V. 1990. Observations on the reproductive biology and growth of the water monitor (*Varanus salvator*) at the Madras Crocodilian Bank. Hamadryad 15:1-5.

Andrews, H. V. 1995. Sexual maturation in *Varanus salvator* (Laurenti, 1768), with notes on growth and reproductive effort. Herpetol. J. 5:189-194.

Lang, J. W., H. Andrews, and R. Whitaker. 1989. Sex determination and sex ratio in *Crocodylus palustris*. Amer. Zool. 29:935-952.

Whitaker, R. 1984. Captive breeding of crocodilians in India. Acta Zool. Path. Antverspiensia 78:309-318.

Whitaker, R., and Z. Whitaker. 1984. Reproductive biology of the mugger (*Crocodylus palustris*). J. Bombay Nat. Hist. Soc. 81:297-316.

Conservation

Andrews, H. V., and R. Whitaker. 1994. Status of the

saltwater crocodile (*Crocodylus porosus*) in North Andaman Island. Hamadryad 19:79-92.

Whitaker, R. 1982. Status of Asian crocodilians, p. 236-266. In D. Dietz, F. W. King, and R. J. Bryant. (eds.). Crocodiles. Proceedings of the Fifth Working Group Meeting of the IUCN/SSC Crocodile Specialist Group.

General

Das, I., and R. Whitaker. 1996. Bibliography of the king cobra (*Ophiophagus hannah*). Smithsonian Herpetol. Inf. Serv. No. 108.

Whitaker, R. 1978. Common Indian Snakes: A Field Guide. Macmillan of India, Delhi.

Whitaker, R. 1979. The Madras Snake Park: Its role in public education and reptile research. Inter. Zoo Yearb. 19:31-38.

Whitaker, R., and D. Basu. 1983. The gharial (*Gavialis gengeticus*): A review. J. Bombay Nat. Hist. Soc. 79:531-548.

Whitaker, R., and Z. Whitaker. 1978. A preliminary survey of the saltwater crocodile (*C. porosus*) in the Andamans. J. Bombay Nat. Hist. Soc. 75:43-49.

Whitaker, R., and Z. Whitaker. 1979. Preliminary crocodile survey - Sri Lanka.. J. Bombay Nat. Hist. Soc. 76:66-85.

Other Indian Zoo References

Baskar, N., N. Krishnakumar, and A. Manimozhi. 1999. Incubation, feeding and growth of Indian rock pythons. Inter. Zoo News 46(2):90-93 [study done at Arignar Anna Zoological Park, Madras, India].

Coote, J. G., 2001. A history of western herpetoculture before the 20th century, p. 19-47. In W. E. Becker (ed.). 25th International Herpetological Symposium on Captive Propagation and Husbandry. International Herpetological Symposium, Detroit MI. [The Calcutta Zoological Gardens was established in 1875. The next year, Ram Bramha Sanyal was hired and published a book in 1892 called *A Handbook of the Management of Animals in Captivity in Lower Bengal*. Beginning in 1895, Sanyal investigated the action of reputed antidotes to snake venom.]

Fayrer, J. 1874. The Thanatophidia of India; Being a Description of the Venomous Snakes of the Indian Peninsula, With an Account of the Influence of Their Poison on Life and a Series of Experiments. J. and A. Churchill, London UK. [In 1866, Sir Joseph Fayrer (1824-1907) tried to establish a zoo in Calcutta, India. He was President of the Asiatic Society of Bengal. Although he secured funding from Indian royals, the project never materialized.]

Loisel, G. 1912. Histoire des ménageries de l'antiquit, . . . nos jours (History of Menageries from Antiquity to Present Times). O. Doin et fils, Paris, France [zoological garden in Calcutta which had 307 reptiles in 1907].

Saxena, R. 1994. Reptiles in Gwalior Zoo (Madhya Pradesh). Cobra (Madras) 15:13. [The author published papers on reptiles of Madhav National Park and conservation of the Bengal monitor.]

Sekar, M., and M. Jagnnadha-Rao. 1995. Management of Indian rock python (*Python molurus*) in captivity at Arignar Anna Zoological Park, Vandalur. Cobra (Madras) 22:14-16.

Walker, S. 2001. Zoological gardens of India, p. 251-294. In V. N Kisling Jr. (ed.), Zoo and Aquarium History. Ancient Animal Collections to Zoological Gardens. CRC Press, Boca Raton, FL; London; New York; Washington DC. [Wajid Ali Shah kept a large menagerie in Calcutta which included tortoises and a large enclosure containing thousands of snakes, and Madras Museum, mentioned by Stanley Flower as having a collection of reptiles. What was interesting to me was the Trivandrum Zoo kept an Indian tigrine frog alive by force-feeding on fish from 1904 to 1912 when Flower saw the specimen. There was a reptile house at this zoo. Sally Walker described the Indian Crocodile Project, Madras Snake Park and Madras Crocodile Bank in detail.]

Tables

Table 1.1. Prices of live reptiles and amphibians in London in 1897 according to Reverend Gregory Bateman. It is clear that many species were available for sale from dealers such as A. Green, F. Young, and Robert Green to the private fancier. Aquaria, Vivaria and Fern Cases were manufactured by several companies and a sizeable number of terrarium plants were available. Equivalents in US currency.

Small Monitor Lizards — 5 shillings (38 cents)

Common Tegus — about £2.00 ($3.60)

Red Tegus — £3.00 to £8.00 ($4.50 to $12.00) depending upon size

Bearded Dragons —10 shillings to £1.00 (75 cents to $1.50)

Green Anoles — as low as half-a-crown (16 cents)

European Chameleons —3 shillings and 6 pence to 7 shillings and 6 pence (30 to 65 cents)

British Common Lizards — 4 to 6 pence each (6 to 9 cents)

Mud Tortoises — 3 shillings per dozen

Young Alligators and Crocodiles — available but price not given

Diamond Pythons — between 10 shillings and £3.00 (75 cents to $4.50)

African Pythons approximately 4 feet long — 25 shillings ($1.88) to £3.00 ($4.50)

Ball Pythons — same price as other pythons

Anacondas — £30.00 ($45.00)

Tuataras — £2.00 ($3.00)

Bull Frogs — 5 shillings, 6 pence to 15 shillings (40 cents to $1.15)

Horned Frogs — £1 to £1, 10 shillings ($ 1.50 to $2.26)

Peron's and Krefft's Tree Frogs — a few shillings

Golden Tree Frogs — 3 to 7 shillings (23 cents to 58 cents)

Painted Frogs — a few shillings

Midwife Toads — 3 to 4 shillings

Salamanders — 3 to 7 shillings (23 cents to 58 cents)

Newts — a few shillings

Giant Salamanders — up to £10 ($15.00)

Olms — 7 to 10 shillings (58 cents to 76 cents)

Aquaria, Vivaria and Fern Cases – available from E. Clifton, Eade and Son, and F. Barritt

Terrarium, Aquarium and Vivarium Plants Recommended and Available – approximately 35 species of Ferns, five of *Selaginella*, five of Spiderworts, five of Club-mosses, Houseleeks, Saxifrages, Ivies, Stonecrops, Orange and Lemon trees, Fuchsias, Geraniums, Myrtles, and dozens of aquatic plants

Table 2.1. Chronological List of Important Books Dealing with Medical Management of Amphibians and Reptiles Beginning in 1955.

Klingelhöffer, W. 1955-1959. Terrarienkunde (Terrarium Science). Alfred Kernen Verlag, Stuttgart.

Reichenbach-Klinke, H.-H. 1961. Krankheiten der Amphibien (Diseases of Amphibians). Gustav Fischer Verlag, Stuttgart.

Reichenbach-Klinke, H.-H. 1963. Krankheiten der Reptilien (Diseases of Reptiles). Gustav Fischer Verlag, Stuttgart/Jena.

Reichenbach-Klinke, H., and E. Elkan. 1965. The Principal Diseases of Lower Vertebrates. Academic Press, London and New York.

Nietzke, G. 1969, 1972. Die terrarientiere: Bau, technische Einrichtung und Bepflanzung der Terrarien: Haltung, Fütterung und Pflege der Terrientiere in zwei Bänden (Terrarium Animals: Construction, Technical Equipment, and Planning of Terraria: Care and Feeding of Terrarium Animals in Two Volumes). Eugen Ulmer, Stuttgart.

Frye, F. L. 1973. Husbandry, Medicine & Surgery in Captive Reptiles. VM Publishing, Bonner Springs KS.

Murphy, J. B., and B. L. Armstrong. 1978. Maintenance of rattlesnakes in captivity. Univ. Kansas Mus. Nat. Hist. Spec. Publ. no. 3:1-40. Lawrence KS.

Murphy, J. B., and J. T. Collins (eds.). 1980. Reproductive Biology and Diseases of Captive Reptiles. Society for the Study of Amphibians and Reptiles Contributions to Herpetology, volume 1. Lawrence KS.

Cooper, J. E., and O. F. Jackson. 1981. Diseases of the Reptilia. 2 Vols. Academic Press, London, New York.

Frye, F. L. 1981. Biomedical and Surgical Aspects of Captive Reptile Husbandry. Veterinary Medicine Publishing, Edwardsville KS.

Marcus, L. C. 1981. Veterinary Biology and Medicine of Captive Amphibians and Reptiles. Lea & Febiger, Philadelphia.

Klös, H.-G., and E. M. Lang. 1982. Handbook of Zoo Medicine: Diseases and Treatment of Wild Animals in Zoos, Game Parks, Circuses, and Private Collections. Van Nostrand Reinhold, New York. [Translation of *Zootierkrankheiten: Krankheiten von Wildtieren im Zoo, Wildpark, Zirkus und in Privathand sowie ihre Therapie*. Werner Frank discusses amphibians and reptiles.]

Murphy, J. B., and J. T. Collins. 1983. A Review of the Diseases and Treatments of Captive Turtles. AMS Publishing, Lawrence KS.

Hoff, G. L., F. L. Frye, and E. R. Jacobson (eds.). 1984. Diseases of Amphibians and Reptiles. Plenum Press, New York.

Ross, R. A., and G. Marzec. 1984. The Bacterial Diseases of Reptiles. Institute for Herpetological Research, Stanford CA.

Ippen, R., H. D. Schröder, and K. Elze. 1985. Handbuch der Zootierkrankheiten, Band 1, Reptilien (Handbook of the Diseases of Zoo Animals, Volume I, Reptiles). Akademie Verlag, Berlin.

Isenbügel, E., and W. Frank. 1985. Heimtierkrank-heiten (Diseases of Pets). Ulmer Verlag, Stuttgart.

Table 2.1. continued

Bels, V. L., and P. Van den Sande (eds.). 1986. Maintenance and reproduction of reptiles in captivity. Volume II. Diseases. Acta Zool. Path. Antverpiensia. No. 79. [Second volume included 10 papers by a number of prominent specialists.]

Obst, F. J., K. Richter, and U. Jacob. 1988. The Completely Illustrated Atlas of Reptiles and Amphibians for the Terrarium. T. F. H. Publications, Neptune City NJ.

Frye, F. L. 1991. Reptile Care. An Atlas of Diseases and Treatments. 2 Vols. Joint publishing venture by Krieger, Malabar, FL and T. F. H. Publications, Neptune City NJ.

Klingenberg, R. J. 1993. Understanding Reptile Parasites. Advanced Vivarium Systems, Lakeside CA.

Jarofke, D., and J. Lange. 1993. Tierärztliche Heimtierpraxis. Band 3. Reptilien: Krankheiten und Haltung (Veterinary Care of Animals Kept in the Home. Vol. 3: Reptiles: Diseases and Care). Verlag Paul Parey, Berlin, Hamburg.

Barnard, S. M., and S. J. Upton. 1994. A Veterinary Guide to the Parasites of Reptiles. Vol. 1 Protozoa, Vol. 2. Arthropods (Excluding Mites). Krieger, Malabar FL.

Warwick, C., F. L. Frye, and J. B. Murphy (eds.). 1994. Health and Welfare of Captive Reptiles. Chapman & Hall, New York.

Frye, F. L., and D. L. Williams. 1995. Self-Assessment Color Review of Reptiles and Amphibians. Iowa State University Press, Ames.

Barnard, S. M. 1996. Reptile Keeper's Handbook. Krieger, Malabar FL.

Köhler, G. 1996. Krankheiten der Reptilien und Amphibien (Diseases of Reptiles and Amphibians). Verlag Eugen Ulmer, Stuttgart.

Mader, D. R. 1996. Reptile Medicine and Surgery. W. B. Saunders, Philadelphia.

Ackerman, L. [1997]. The Biology, Husbandry, and Health Care of Captive Reptiles. T. F. H. Publications, Neptune City NJ [excellent overview].

Jarofke, D., and H.-J. Herrmann. 1997. Amphibien: Biologie – Haltung – Krankheiten – Bioindikation (Amphibians: Biology-Care-Diseases-Bio-indications). Ferdinand Enke Verlag, Stuttgart.

Vassiliev, D. B. 1999. Turtles: Husbandry, Diseases and Treatment in Captivity. Àkvarium, Moscow [English translation of title].

Vassiliev D. B., and A. Sokolov. 1999. Turtles, Lizards, Snakes: Husbandry, Care and Treatment in Captivity. Àkvarium, Moscow. [English translation of title.].

Wright, K. M., and B. R. Whitaker (eds.). 2001. Amphibian Medicine and Captive Husbandry. Krieger, Malabar FL [important reference].

Table 2.2. Chronological List of Seminal Papers and Books Dealing with Conservation Issues of Importance to Zoo Biologists.

Honegger, R. E. 1969. Bedrohte Amphibien und Reptilien (Endangered amphibians and reptiles). Zool. Gart., Leipzig (N.F.) 36 Hediger-Festschrift:173-185. [See Zürich Zoo account for René Honegger's papers on conservation problems with amphibians and reptiles.]

Wayre, P. 1969. The role of zoos in breeding threatened species of mammals and birds in captivity. Biol. Conserv. 2:47-49.

Foose, T. J. 1980. Demographic management of endangered species in captivity. Inter. Zoo Yearb. 20:154-166.

Senner, J. W. 1980. Inbreeding depression and the survival of zoo populations. In Conservation Biology: An Evolutionary Perspective. M. E. Soulé and B.A. Wilcox (eds.). Sinauer Assoc., Sunderland MA.

Foose, T. J. 1983. The relevance of captive populations to the conservation of biotic diversity, p. 374-401. In C. M. Schonewald-Cox, et al. (eds.), Genetics and Conservation. Benjamin Cummings, Menlo Park CA.

Hudson, R. 1983. The Species Survival Plan and its application to reptiles, p. 1-9. In P. J. Tolson (ed.), 7th International Herpetological Symposium on Captive Propagation and Husbandry. International Herpetological Symposium, Thurmont MD. [Rick Hudson, employed at the Ft. Worth Zoo, is currently Lizard Advisory Group Chair for the AZA, responsible for the development of Species Survival Plans.]

Kleiman, D. G., M. R. Stanley Price, and B. B. Beck. 1984. Criteria for reintroductions, p. 287-303. Creative Conservation: Interactive Management of Wild and Captive Animals. Chapman and Hall, London.

Rosenbaum, P. A. et al. 1984. The value and application of frozen tissue collections in the propagation, management and conservation of herpetofauna, p. 10-21. In R. A. Hahn (ser. ed.), 8th International Herpetological Symposium on Captive Propagation and Husbandry. International Herpetological Symposium, Thurmont MD [joint program between Audubon Park & Zoological Garden and Louisiana State University, New Orleans].

Conway, W. 1985. The Species Survival Plan and the Conference on Reproductive Strategies for Endangered Wildlife. Zoo Biol. 4:219-223.

Foose, T. J., U. S. Seal, and N. R. Flesness. 1985. Conserving animal genetic resources. IUCN Bull. 16:20-21.

Seal, U. S. 1985. The realities of preserving species in captivity, p. 71-96. In R. J. Hoage (ed.), Animal Extinctions. Smithsonian Institution Press, Washington DC.

Conway, W. G. 1986. The practical difficulties and financial implications of endangered species breeding programmes. Inter. Zoo Yearb. 24/25:210-219.

Seal. U.S. 1986. Goals of propagation programmes for the conservation of endangered species. Inter. Zoo Yearb. 24/25:174-179.

Soulé, M. et al. 1986. The Millennium Ark: How long a voyage, how many staterooms, how many passengers? Zoo Biol. 5:101-113.

Mace, G. M. 1986. Genetic management of small populations. Inter. Zoo Yearb. 24/25:167-174.

Glatson, A. R. 1986. Studbooks: The basis of breeding programmes. Int. Zoo Yearb. 24/25:162-167.

Table 2.2. continued

Lacy, R. C. 1987. Loss of genetic diversity from managed populations: Interesting effects of drift, mutation, immigration, selection, and population subdivision. Conserv. Biol. 1:143-158.

Flesness, N. R., and G. M. Mace. 1988. Population databases and zoological conservation. Inter. Zoo Yearb. 27:42-49.

Foose, T. J. 1988. Management of small populations. Inter. Zoo Yearb. 27:26-41.

Hutchins, M. 1988. On the design of zoo research programmes. Inter. Zoo Yearb. 27:9-19. [Michael Hutchins was employed at AZA. Account covered administration, coordination, animal welfare, preparation of manual, establishing priorities, protocols, health and safety regulations, integration of research and animal management.]

Ryder, O. A., J. H. Shaw, and C. M. Wemmer. 1988. Species, subspecies and *ex situ* conservation. Inter. Zoo Yearb. 27:134-140.

Schmitt, E. C. 1988. Effects of conservation legislation on the professional development of zoos. Inter. Zoo Yearb. 27:3-9.

Conway, W. 1989. The prospect for sustaining species and their evolution, p. 199-209. In D. Western, and M. Pearl. (eds.), Conservation for the Twenty-First Century. Oxford University Press, New York.

Wildt, D. E. 1989. Reproductive research in conservation biology: Priorities and aims for support. J. Zoo Wildl. Med. 20:391-395.

Haig, S., J. Ballou, and S. Derrickson. 1990. Management options for preserving genetic diversity: Reintroduction of Guam rails to the wild. Conserv. Biol. 4:290-300.

Hutchins, M., and R. Wiese. 1991. Beyond genetic and demographic management: The future of the Species Survival Plan and related AAZPA conservation efforts. Zoo Biol. 10:285-292.

Martin, D. L. 1991. Captive husbandry as a technique to conserve a species of special concern, the Yosemite toad, p. 17-32. In R. E. Staub (ed.), Captive Propagation and Husbandry of Reptiles and Amphibians. Northern California Herpetological Society. [Satellite colonies were started at Sacramento Zoo and Chaffee Zoological Gardens.]

Konig, C., and A. Schluter. 1991. Nachzucht der Balearen-Geburtshelferkröte *Alytes muletensis* (Sanchiz et Adrover 1979) im Rahmen eines Artenschutzprogrammes (Amphibia; Discoglossidae) ((Breeding the Balearic midwife toad *Alytes muletensis* (Sanchiz et Androver 1979) in the context of an endangered species program (Amphibia, Discoglossidae)). Jahresh. Ges. Naturkd. Württemb. (Annual Report of the Wuerttemberg Society for Natural History) 146:193-205.

Kleiman, D. G. 1992. Behavior research in zoos: Past, present, and future. Zoo Biol. 11:301-312.

Beck, B., M. Cooper, and B. Griffith. 1993. Infectious disease considerations in reintroduction programs for captive wildlife. J. Zoo Wildl. Med. 24:394-397.

International Union of Directors of Zoological Gardens (of the World Zoo Organization). 1993. The World Zoo Conservation Strategy: The Role of Zoos and Aquaria of the World in Global Conservation, p. 1-76. Chicago Zoological Society, Brookfield IL.

Table 2.2. continued

Thompson, S. D. 1993. Zoo research and conservation: Beyond sperm and eggs toward the science of animal management. Zoo Biol. 12:155-159 [virtues of applied vs. basic research in a zoo setting].

Willis, K., R. Wiese, and M. Hutchins. 1993. AAZPA Conservation Resource Guide. American Association of Zoological Parks and Aquariums, Bethesda MD.

Gibbons, J. W. 1994. Reproductive patterns of reptiles and amphibians: Considerations for captive breeding and conservation, p. 119-123. In J. B. Murphy, K. Adler, and J. T. Collins (eds.), Captive Management and Conservation of Amphibians and Reptiles. Society for the Study of Amphibians and Reptiles. Contributions to Herpetology, volume 11, Ithaca NY.

Marcellini, D. L. 1994. Collection-management, captive-breeding, and conservation programs in zoo herpetological collections, p. 397-400. In J. B. Murphy, K. Adler, and J. T. Collins (eds.), Captive Management and Conservation of Amphibians and Reptiles. Society for the Study of Amphibians and Reptiles. Contributions to Herpetology, volume 11, Ithaca NY.

Mittermeier, R. A., and J. L. Carr. 1994. Conservation of reptiles and amphibians: A global perspective, p. 27-35. In J. B. Murphy, K. Adler, and J. T. Collins (eds.), Captive Management and Conservation of Amphibians and Reptiles. Society for the Study of Amphibians and Reptiles. Contributions to Herpetology, volume 11, Ithaca NY.

Seal, U. S., T. J. Foose, and S. Ellis. 1994. Conservation Assessment and Management Plans (CAMPs) and Global Captive Action Plans (GCAPSs), p. 312-328. In P. J. S. Olney, G. M. Mace, and A. T. C. Feistner (eds.), Creative Conservation. Chapman and Hall, London.

Swingland, I. R. 1994. International conservation and captive management of tortoises, p. 99-107. In J. B. Murphy, K. Adler, and J. T. Collins (eds.), Captive Management and Conservation of Amphibians and Reptiles. Society for the Study of Amphibians and Reptiles. Contributions to Herpetology, volume 11, Ithaca NY.

Wayne, R. K. et al. 1994. Molecular genetics of endangered species, p. 92-117. In P. J. S. Olney, G. M. Mace, and A. T. C. Feistner (eds.), Creative Conservation. Chapman and Hall, London.

Wiese, R. J., and M. Hutchins. 1994. The role of zoos and aquariums in amphibian and reptile conservation, p. 37-45. In J. B. Murphy, K. Adler, and J. T. Collins (eds.), Captive Management and Conservation of Amphibians and Reptiles. Society for the Study of Amphibians and Reptiles. Contributions to Herpetology, volume 11, Ithaca NY.

Wiese, R. J., K. Willis, and M. Hutchins. 1994. Is genetic and demographic management conservation? Zoo Biol. 13:297-299. [Genetic and demographic management is not conservation.]

Anderson, D. E. 1995. Conservation potential of zoos and aquariums in developing worlds, p. 206-220. In C. M. Wemmer (ed.), The Ark Evolving. Zoos and Aquariums in Transition. Smithsonian Institution, Conservation and Research Center, Front Royal VA.

Burghardt, G. M., and M. A. Milostan. 1995. Ethological studies on reptiles and amphibians: Lessons for Species Survival Plans, p. 187-203. In E. F. Gibbons Jr., B. S. Durrant, and J. Demarest (eds.), Conservation of Endangered Species in Captivity: An Interdisciplinary Approach. State University of New York Press, Albany.

Table 2.2. continued

Conway, W. 1995. The conservation park: A new zoo synthesis for a changed world, p. 259-276. In C. M. Wemmer (ed.), The Ark Evolving. Zoos and Aquariums in Transition. Smithsonian Institution, Conservation and Research Center, Front Royal VA.

de Boer, L. E. M., and T. J. Foose. 1995. Towards global management of *ex situ* populations, p. 158-175. In C. M. Wemmer (ed.), The Ark Evolving. Zoos and Aquariums in Transition. Smithsonian Institution, Conservation and Research Center, Front Royal VA.

Hutchins, M., and W. G. Conway. 1995. Beyond Noah's ark: The evolving role of modern zoological parks and aquariums in field conservation. Inter. Zoo Yearb. 34:117-130.

Hutchins, M., K. Willis, and R. J. Wiese. 1995. Strategic collection planning: Theory and practice. Zoo Biol. 14:5-25. [Excellent overview of issues and recommendations for development of future initiatives.]

Karesh, W. B., and R. A. Cook. 1995. Environmental health in an international conservation program, p. 15-19. In R. E. Junge (ed.), Proceedings Joint Conference American Association of Zoo Veterinarians, Wildlife Disease Association, American Association of Wildlife Veterinarians, East Lansing, Michigan, August 12-17, 1995. American Association of Zoo Veterinarians, American Association of Wildlife Veterinarians & Wildlife Disease Association.

Konstant, W. 1995. Launching the ark and missing the boat: building local capacity for wildlife conservation, p. 192-205. In C. M. Wemmer (ed.), The Ark Evolving. Zoos and Aquariums in Transition. Smithsonian Institution, Conservation and Research Center, Front Royal VA.

Philippart, J. C. 1995. Is captive breeding an effective solution for the preservation of endemic species? Biol. Conserv. 72:281-295.

Robinson, M. H. 1995. Zoo and aquarium messages, meanings and contexts, p. 1-24. In C. M. Wemmer (ed.), The Ark Evolving. Zoos and Aquariums in Transition. Smithsonian Institution, Conservation and Research Center, Front Royal VA.

Rowan, A., and R. Hoage. 1995. Public attitudes toward wildlife: The awakening awareness, p. 32-60. In C. M. Wemmer (ed.), The Ark Evolving. Zoos and Aquariums in Transition. Smithsonian Institution, Conservation and Research Center, Front Royal VA.

Wemmer, C., and S. Thompson. 1995. A short history of scientific research in zoological gardens, p. 70-94. In C. M. Wemmer (ed.), The Ark Evolving. Zoos and Aquariums in Transition. Smithsonian Institution, Conservation and Research Center, Front Royal VA.

Ballou, J. D., and T. J. Foose. 1996. Demographic and genetic management of captive populations, p. 263-283. In D. G. Kleiman, M. E. Allen, K. V. Thompson, S. Lumpkin, and H. Harris (ed.), Wild Mammals in Captivity. University of Chicago Press, Chicago.

Balmford, A., G. M. Mace, and N. Leader-Williams. 1996. Designing the ark: Setting priorities for captive breeding. Conserv. Biol. 10(3):719-727.

Barrowclough, G. F., and N. R. Flesness. 1996. Species, subspecies, and races: The problem of units of management in conservation, p. 247-254. In D. G. Kleiman, M. E. Allen, K. V. Thompson, S. Lumpkin, and H. Harris (eds.), Wild Mammals in Captivity. University of Chicago Press, Chicago.

Table 2.2. continued

Ryder, O. A., and R. C. Fleischer. 1996. Genetic research and its application in zoos, p. 255-262. In D. G. Kleiman, M. E. Allen, K. V. Thompson, S. Lumpkin, and H. Harris (eds.), Wild Mammals in Captivity. University of Chicago Press, Chicago.

Schubert, A., and G. Santana. 1996. Conservation of the American crocodile (*Crocodylus acutus*) in the Dominican Republic, p. 425-433. In R. Powell, and R. W. Henderson (eds.), Contributions to West Indian Herpetology: A Tribute to Albert Schwartz. Society for the Study of Amphibians and Reptiles, Ithaca NY [description of head-starting program at ZOODOM (Santo Domingo Zoo)].

Snyder, N. F. R. et al. 1996. Limitations of captive breeding in endangered species recovery. Conserv. Biol. 10:338-348.

Gippoliti, S., and G. M. Carpaneto. 1997. Captive breeding, zoos, and good sense. Conserv. Biol. 11:806-807.

Hutchins, M., R. Wiese, and K. Willis. 1997. Priority-setting for ex situ conservation. Conserv. Biol. 11(3):593.

Kitchner, A. 1997. The role of museums and zoos in conservation biology. Ratel 24:97-104.

Montgomery, M. E. et al. 1997. Minimizing kinship in captive breeding programs. Zoo Biol. 16:377-389.

Meritt Jr., D. A. 1997. Species Survival Programs: Are they for everyone? Zoo Biol. 16:103-106.

Quinn, H. 1997. Expanding conservation partnerships in Venezuela: the Orinco [Orinoco] crocodile PHVA. AAZPA Reg. Conf. Proc. 1997:448-451.

Ralls, K. 1997. On becoming a conservation biologist: Autobiography and advice, p. 356-372. In J. R. Clemmons, and R. Buchholz (eds.), Behavioral Approaches to Conservation in the Wild. Cambridge University Press, Cambridge UK.

Snyder, N. F. R. et al. 1997. Limitations of captive breeding: Reply to Gippoliti and Carpaneto. Conserv. Biol. 11:808-810.

Cooper, J. E., C. J. Dutton, and A. F. Allchurch. 1998. Reference collections: Their importance and relevance to modern zoo management and conservation biology. Dodo, J. Jersey Wildl. Preserv. Trust 34:159-166.

Kalinowski, S. T., and P. W. Hedrick. 1998. An improved method for estimating inbreeding depression in pedigrees. Zoo Biol. 17:481-497.

Durrell, L. 1998. Strategic planning for species conservation by Jersey Wildlife Preservation Trust. Dodo, J. Jersey Wildl. Preserv. Trust 34:176-177.

Durrell, L., and J. J. C. Mallinson. 1998. The impact of an institutional review: A change of emphasis toward field conservation programmes. Inter. Zoo Yearb. 36:1-8. [Authors stress need to change from captive breeding, research and education in zoos to field projects with a conservation focus.]

Margan, S. H. et al. 1998. Single large or several small? Population fragmentation in the captive management of endangered species. Zoo Biol. 17:467-480.

Redford, K. H., and B. D. Richter. 1999. Conservation of biodiversity in a world of use. Conserv. Biol. 13:1246-1256.

Table 2.2. continued

Westley, F., U. Seal, and C. C. M. Clark. 1999. The Population and Habitat Viability facilitators' course: A retrospective. Dodo, J. Jersey Wildl. Preserv. Trust 35:124-133.

Kleiman, D. G. R. P. Reading, and F. Felleman. 2000. Improving the value of conservation projects. Conserv. Biol. 14:356-365.

Conway, W. G., et al. 2001. The AZA Field Conservation Resource Guide. Zoo Atlanta, Atlanta. [Papers are on many projects by zoo herpetologists.]

Hancocks, D. 2001. A Different Nature. The Paradoxical World of Zoos and Their Uncertain Future. University California Press, Berkeley, Los Angeles, London. [David Hancocks' basic message is that zoos, other than a few exceptions, have not been effective examples overall in terms of exhibitry, conservation, education and research.]

Frankham, R., J. D. Ballou, and D. A. Briscoe. 2002. Introduction to Conservation Genetics. Cambridge University Press, Cambridge UK. [Exceptional book summarizes the history of genetic management.]

Hutchins, M., B. Smith, and R. Ballard. 2003. In defense of zoos and aquariums: the ethical basis for keeping wild animals in captivity. J. Amer. Vet. Med. Assoc. 223:958-966. [Provides excellent overview of value and challenges of conservation programs.]

Table 2.3. Alphabetical List of Seminal Papers Addressing Allometric or Behavioral Alterations Caused by Captivity or Other Environmental Effects.

Andrews, R. M. 1976. Growth rate in island and mainland anoline lizards. Copeia 1976:477-482.

Arnold, S. J., and C. R. Peterson. 1989. A test for temperature effects on the ontogeny of shape in the gartersnake *Thamnophis sirtalis*. Physiol. Zool. 62:1316-1333.

Barnett, B., and T. D. Schwaner. 1985. Growth in captive born tiger snakes (*Notechis ater serventyi*) from Chappell Island: Implications for field and laboratory studies. Trans. R. Soc. S. Aust. 109 (Parts 1 & 2):31-36.

Blouin, M. S., and M. L. G. Loeb. 1991. Effects of environmentally induced development-rate variation on head and limb morphology in the green tree frog, *Hyla cinerea*. Amer. Natur. 138:717-728.

Bookstein, F. L. 1989. "Size and Shape": A comment on semantics. Syst. Zool. 38:173-180.

Case, T. J. 1978. A general explanation for insular body size trends in terrestrial vertebrates. Ecology 59:1-18.

Cock, A. G. 1966. Genetical aspects of metrical growth and form in animals. Q. Rev. Biol. 41:131-190.

Dunham, A. E., D. W. Tinkle, and J. W. Gibbons. 1978. Body size in island lizards: A cautionary tale. Ecology 59:1230-1238.

Emerson, S. B. 1986. Heterochrony and frogs: The relationship of a life history trait to morphological form. Amer. Natur. 127:176-183.

Emerson, S. B., and D. M. Bramble. 1993. Scaling, allometry, and skull design, p. 384-421. In J. Hanken, and B. K. Hall (eds.), The Skull, Vol. 3, Functional and Evolutionary Mechanisms. University of Chicago Press, Chicago.

Ford, N. B., and R. A. Seigel. 1989. Phenotypic plasticity in reproductive traits: Evidence from a viviparous snake. Ecology 70:1768-1774.

Forsman, A. 1991. Variation in sexual size dimorphism and maximum body size among adder populations: effects of prey size. J. Anim. Ecol. 60:253-267.

Forsman, A. 1996. An experimental test for food effects on head size allometry in juvenile snakes. Evolution 50:2536-2542.

Forsman, A., and R. Shine. 1997. Rejection of non-adaptive hypotheses for intraspecific variation in trophic morphology in gape-limited predators. Biol. J. Linn. Soc. 62:209-223.

Fox, W., C. Gordon, and M. H. Fox. 1961. Morphological effects of low temperatures during the embryonic development of the garter snake, *Thamnophis elegans*. Zoologica (New York) 46:57-71.

Graham, T. E., and V. H. Hutchison. 1979. Effect of temperature and photoperiod acclimatization on thermal preferences of selected freshwater turtles. Copeia 1979:165-169.

Madsen, T. 1987. Cost of reproduction and female life-history traits in a population of grass snakes, *Natrix natrix*, in southern Sweden. Oikos 49:129-132.

Table 2.3. continued

Madsen, T., and R. Shine. 1993. Phenotypic plasticity in body sizes and sexual size dimorphism in European grass snakes. Evolution 47:321-325.

Moore, W. J. 1965. Masticory function and skull growth. J. Zool. (London) 146:123-131.

Osgood, D. W. 1978. Effects of temperature on the development of meristic characters in *Natrix fasciata*. Copeia 1978:33-47.

Presch, W. 1989. Systematics and science: A comment. Syst. Zool. 38(2):181-190.

Reist, J. D. 1985. An empirical evaluation of several univariate methods that adjust for size variation in morphometric data. Can. J. Zool. 63:1429-1439.

Schwaner, T. D., and S. D. Sarre. 1988. Body size of tiger snakes in southern Australia, with particular reference to *Notechis ater serventyi* (Elapidae) on Chappell Island. J. Herpetol. 22:24-33.

Shine, R. 1990. Proximate determinants of sexual differences in adult body size. Amer. Nat. 135:278-283.

Shine, R. 1978. Sexual size dimorphism and male combat in snakes. Oecologia 33:269-277.

Thornhill, R., and A. P. Moller. 1998. The relative importance of size and asymmetry in sexual selection. Behav. Ecol. 9(6):546-551.

Vogt, T., and D. L. Jameson. 1970. Chronological correlation between change of weather and change of morphology of the Pacific tree frog in southern California. Copeia 1970:135-144.

Zimmer, C. 2003. Rapid evolution can foil even the best-laid plans. Science 300: 895.

Table 4.1. Sampling of Newspaper Headlines* Mentioning Grace Olive Wiley Between 1922-1937

"Museum Head Once Afraid of Bugs, Reptiles"

"Kansas Woman Finds Dynamite Effective as Snake Charmer"

"No Mad Rush for This Job"

"Herd of Snakes Comes to Party; Women Scamper"

"Mr. Lizard's Night Out"

"Monstrosities of Last Stage of Drunken Delirium Leer Lustfully in New Abode at University Farm"

"Museum Curator Gets Tube of Rare Snake-Bite Serum"

"Snake Family Sad; Stork Misses Them"

"Flirts with Death"

"Ailing Rattler Gets Taxicab Rides to Medical School to Look at Tuberculosis"

"Snakes Win New Friends at Traffic Club Meeting"

"A Siamese-Twin Turtle" (two-headed snapping turtle)

"Troup of Snakes to be Actors in Entertainment to Help Flood Sufferers"

"Pet Serpents are Guests of Business Women at Luncheon"

"Alligator Triplets Celebrate Birthday"

"They're Just Huck and Jimmy to Her" (picture of Wiley holding 5 ft. diamond back rattlesnake and Gila monster)

'Tumor' of Wiley's Python is Blanket"

"Rattlers' Nurse"

"Taming Poisonous Snakes Her Vocation"

"Rattler Bites Woman Keeper"

"Snake Hunting Has Charms for Woman Curator"

"Why Swat Flies? Take Them to Hungry Reptiles at Public Library"

"Children Rush Alligator Birthday Party"

"Rattler's Cold is Cured by Use of Oxygen Tent"

"Big Ben, Zoo Rattlesnake is Wooed by Woman's Love"

"Woman Delights in Giving Bath to Texas Rattlers"

"Moving Texas Rattlesnakes from Box to Glass Case All in Day's Work for Curator"

"Saving Lives of Rattlesnakes Woman's Job"

"Congratulations Besiege Albert and Ethel Rattler, Proud Parents of 10 Babies"

"Baby Rattlers at Museum Shed First Skin"

"Vipers Pass Summer as Girls' Pets"

"Mrs. Wiley Leaves"

"Woman's Kindness Wins Deadly Snakes at Brookfield Zoo"

Table 4.1. continued

"19 Snakes Got Away, Grace Wiley Fired as Snakes Escape"

"Bandy-Bandy Found Alive"

"Mrs. Grace Wiley Packs Her Reptile Friends"

"Snakes Depart. Long Beach Herpetologist Taking Collection to Hollywood"

* Published in *Minneapolis Daily Star, The Minneapolis Journal, The Minneapolis Evening Tribune, Saint Paul Dispatch, The Chicago Tribune, Chicago Daily News, Columbus, Ohio Citizen, Long Beach Press-Telegram, Kansas City Post, Evening Star-Washington D.C., New York World Telegram, The New York Times, Kansas City Journal, Springfield Sunday Bulletin, Mayville Exponent, Daily Journal Press, The Minnetonka Record, The Washburn Grist,* and other newspapers

Egyptian Cobra (*Naja haie*)

Appendix 1

Supplemental List of Living Herpetological Collections. Collection and Personnel Data Derived in Part from *AZA Membership Directory (2006)* and *International Zoo Yearbook* (1998: Volume 28)

Brazil: Parque Zoobotânico Museu Paraense Emilio Goeldi (1895)
Emilio Augusto Goeldi (1859–1917), herpetologist and Director of the Museu Paraense, published annual reports between 1894–1909, outlining the need for a representative live animal collection and overall expansion of the museum. He was instrumental in adding a zoological and botanical garden to the museum facility, which was later renamed in his honor (Museu Paraense, now known as Parque Zoobotânico Museu Paraense Emilio Goeldi) after he assumed the directorship in 1894. He published on a variety of Brazilian topics: mammals including deer, birds, mosquitoes, zoology and natural history, as well as South American Indians, archaeology and antiquities.
References: Adler, K. 1989:64; Ellis Jr., J. F., and G. A. Ellis. 2001; Goeldi, E. A. 1894-1895.

Canary Islands: Reptilandia (1985)
This private reptile exhibit, located on Gran Canaria Island, houses more than 150 species and over 1000 individual reptiles, mostly lizards such as Komodo dragons and West Indian ground iguanas. Bert Langerwerf and Jim Pether were originally involved with the operation but Langerwerf left to start Agama, Inc. in Alabama and Pether has continued alone. When the giant Gomeran Island lacertid lizard (*Gallotia gomerana*), thought to be extinct was rediscovered in 1999, Pether became involved in setting up a small captive colony on the island for potential breeding and eventual reintroduction.

Costa Rica: Zoo Nacional Simon Bolivar (1916)
The Zoo is located in the capital city of San Jose. There is an excellent exhibit of native amphibians, such as hylid and dendrobatid frogs, and many smaller pitvipers and colubrid snakes from Costa Rica. All of the terrariums, although small, are attractive with living plants. The National Aquarium in Baltimore assisted in developing this exhibit and collecting specimens for display.

Czech Republic: Suchdol Herpetological Station (1947)
Zdenek Vogel was not a biologist by training but devoted most of his life to popularizing herpetology and herpetoculture. The Suchdol Herpetological Station, begun in 1947, was situated in his private house in Suchdol near Prague (now part of Prague), and was disbanded after his death.
References: Vogel, Z. undated, 1964; Chapter 4.

Czech Republic: Zoologicka Zahrada Jihlava (1957)
Herpetologist Vladislav Jirousek is the Director and the Zoo specializes in reptiles. In 1991, he published a series of papers on the vivariums of western Europe in Akvarium Terrarium. The collection numbers approximately 28 taxa and approximately 100 individuals.
Reference: Jirousek, V. T. 1997.

Dominican Republic: Parque Zoologico Nacional (ZOODOM) (1974)
The Zoo specializes in Hispaniolan and Caribbean fauna. There is a successful head-starting program for conservation of the American crocodile in the Dominican Republic. The reptile collection numbers approximately 20 taxa and over 800 individuals.
References: Schubert, A., and G. Santana. 1996; Schürer, U. 1982.

Guatemala: Zoológico Nacional La Aurora (1892)
The Zoo is located in Guatemala City. There are small exhibits containing mostly indigenous amphibians and reptiles in the historical "Tea House," so-called because the President of Guatemala used to have daily tea parties at this site in the early days. A large outdoor enclosure contains aquatic turtles and small crocodilians.

Indonesia: Kebun Binatang Ragunan Zoo (1864)
The Zoo is located in Jakarta, Java. There is a small exhibit of mainly indigenous reptiles. In a large outdoor enclosure, Komodo dragons have reproduced.
Reference: Gaulstaun, B. 1973.

Indonesia: Kebun Binatang Surabaya (1916)
The Zoo is located in Surabaya, Java. In several large

enclosures, Komodo dragons have been bred and raised successfully. Director KMT A. Tirtodiningrat coauthored a paper on captive management with an analysis of growth of captive Komodo dragons produced at the Zoo. The young dragons at Gembira Loka are kept in large groups, resulting in differential growth rates from dragons which are maintained singly and fed less often in other zoos.
References: Walsh, T., et al. 1998, 2002.

Kenya: Nairobi Snake Park (1962)
The Snake Park is associated with the National Museum of Kenya. The collection, numbering about three dozen taxa, focuses on East African herpetofauna. There are a number of large outdoor exhibits housing a variety of snakes such as boomslangs. Smaller displays feature species such as Kenyan sand boas, many elapids and smaller vipers.
References: Cooper, J. E. 1973; Labuschagne, W., and S. Walker. 2001; Nares, P., and J. E. Cooper. 1971.

Malayasia: Zoo Negara Malayasia (National Zoological Park) (1957)
The Malaysian National Zoo is located in the region of Kuala Lumpur. There is an exhibit building for the Komodo dragon. The collection numbers approximately 60 taxa and 380 individuals.
References: Sims, K. J. 1971; Sims, K. J., and I. Singh. 1978.

Mexico: Zoologico Guadalajara (1988)
The Zoo, located in the state of Jalisco, has a large collection of amphibians and reptiles. Eduardo Fanti is the curator. The collection numbers approximately 78 taxa and 325 individuals.
Reference: Espinosa, L. D., et al. 1996.

Mexico: Zoologico Regional Miguel Alvarez Del Toro (ZOOMAT) (1942)
This institution, located in Tuxtla Gutiérrez in the state of Chiapas, is well known as for its display of local fauna in naturalistic exhibits. The facility is named after Miguel Alvarez del Toro (see Chapter 4 for his biography). Graciela Velasco S. is the curator. The collection numbers approximately 50 taxa and nearly 290 individuals.
Reference: Balek, J. 1993.

Poland: Miejski Ogrod Zoologiczny (1865/1948)
The Zoo was formerly located in Breslau, Germany, and now is in Wroclaw. The collection numbers approximately 150 taxa and nearly 700 individuals.

Reference: Felkle, L. 1964.

Poland: Miejski Ogrod Zoologiczny (1951)
The "Plock" Zoo has the second largest collection of reptiles in Poland. The curator of reptiles and fish is Jan Kopczyñski. The reptile collection numbers approximately 94 taxa and over 580 individuals.
References: Kopczynski, J. 1993; See Sz. Poradowski, 1958, Ogrod Zoologiczny w Plocku (The zoo in Plock), Przeglad Zoologiczny (Wroclaw) 2:53.

Singapore: Singapore Zoological Gardens/Night Safari (1971)
The reptile collection numbers approximately 60 taxa and over 400 individuals.

South Africa: Mandela National Zoological Gardens of South Africa (1899)
The Aquarium and Reptile Park is located in Pretoria. The curator is Chris de Beer. The collection numbers approximately 90 taxa and over 380 individuals.

South Africa: Port Elizabeth Snake Park (Museum established 1897)
Frederick William FitzSimons (born 1875) was the first Director of the Port Elizabeth Museum in 1906 and developed the Snake Park. He was assisted by his son, Vivian, and both of them published in herpetology. His younger brother, Desmond C. FitzSimons, started the Durban Snake Park. F. W. FitzSimons also wrote books on the natural history of South African mammals, including primates. Bill Branch, who has written a number of books and papers on South African herpetofauna, is the current curator of herpetology at the Museum. The facility has suffered hard times recently due to lack of funding.
References: Adler, K. 1989; Branch, W. R., and N. Schaefer. 1987; FitzSimons, F. W. 1921, 1930, 1932; Labuschagne, W., and S. Walker. 2001.

South Africa: Transvaal Snake Park (1961)
When the director Rod Patterson retired, this facility, located in Johannesburg, was sold and has been incorporated into a local casino and animal attraction. It still has some snakes, but has expanded into primates and birds and is no longer called TSP. Patterson published a book called *Reptiles of Southern Africa* in 1987 with an extensive section on reptiles in captivity, as well as several papers on reproduction in African pythons. Patterson and Meakin also published a book entitled *Snakes* the previous year.
References: Patterson, R. W. 1974, 1978, 1987;

Patterson, R. W., and P. R. Meakin. 1986.

Spain: Parc Zoologic de Barcelona (1894)
The Zoo is in the center of the city. Manuel Aresté is the curator. The collection numbers approximately 120 taxa and over 400 individuals.

Sri Lanka: National Zoological Gardens (1936)
The Zoo is located in Colombo. The collection numbers approximately 30 taxa and 360 individuals.

Sweden: Skansen-Akvariet (1891)
This aquarium specializes in endangered and threatened reptiles. The curator is Jan Wihman. The collection numbers approximately 36 taxa and over 150 individuals.
References: Larsson, H.-O., and S. Holm. 1983; Larsson, H.-O., and J. Wihman. 1989.

Switzerland: Tierpark Dählhölzli/Bern Zoo (1939)
A large outdoor exhibit with Greek vegetation and hibernacula has been constructed for marginated tortoises (*Testudo marginata*) The indoor terrarium has a state-of-the-art environmental system and visitors are able to look at recording devices showing environmental parameters within the displays. Madagascan species such as the tomato frog are featured.
Reference: Petzold, D. 2001.

Switzerland: Zoologischer Garten Basel (1874)
Head Keeper C. Stemmler observed that a captive water snake, probably covered with fish oil, started to swallow its own body; as the snake reacted to pain, it struggled more intensely which in turn led to more violent biting. There is a vivarium building, constructed in 1972, designed like a helix or snail shell, with a variety of herps on the top floor. The curator is Thomas Jermann. The collection numbers approximately 33 taxa and over 275 individuals.
Reference: Stemmler, C. 1937.

Thailand: Pasteur Institute Snake Farm (Queen Saovabbha Memorial Institute) (1923) and Bangkok Reptile Grove (1930)
The former produces antivenin for a number of Asian elapids and vipers commonly held in zoos.

United States of America: Abilene Zoological Gardens (1966)
The curator of small animals and herps was David Heckard. The lead supervisor is Becky Irby. The collection numbers approximately 55 taxa and 270 individuals.

United States of America: Albuquerque Biological Park (1925). See Rio Grande Zoo account.

United States of America: Buffalo Zoological Gardens (1875)
Frederick Paine was in charge of the department for many years and a major player in the AZA Puerto Rican crested toad project. The current curator is Kevin Murphy. The herpetological exhibit has been extensively remodeled in 2005. The collection numbers approximately 38 taxa and 163 individuals.
References: Connaughton, S. W., and F. L. Paine. 1989; Paine, F.L. 1984; Paine, F. L., and J. Weinheimer. 1984; Paine, F. L., et al. 1989; Radford, L., and F. L. Paine. 1989; Smith, C. F., and F. L. Paine. 1989.

United States of America: Caldwell Zoo (1953)
This private Zoo is located in Tyler, Texas. A small display focuses mainly on native herpetofauna, placed in elaborate naturalistic dioramas. A large exhibit featuring African Rift Lake cichlids below and an array of African reptiles above is beautiful. William Lamar was the first curator and Yvonne Stainback is currently the curator. The collection numbers approximately 50 taxa and 150 individuals.

United States of America: Cameron Park Zoo (1955)
This Zoo, known as the Cen-Tex Zoo in the past, is located in Waco, Texas. Initially, this small zoo was located in an inadequate setting with space at a premium. The reptile collection was squeezed into a small frame building. Now, the Zoo has been moved to a scenic spot along the Brazos River and a new building has been constructed for the herpetological collection. Terri Cox is the curator of programs and exhibits. Otis Powell is Animal Care Manager-herpetology. The collection numbers approximately 70 taxa and over 130 individuals.

United States of America: Cape May County Park Zoo (1977)
This Zoo is located in Cape May, New Jersey. In May 1998, a fire destroyed the original building and entire collection, except for some tortoises in outside displays. The curator Jonnie Johnson-Gove was hired to assemble a replacement collection focusing on endangered species and develop new exhibits in the re-

built facility. She has been studying the Pine Barrens treefrog (*Hyla andersoni*) in the wild and is involved in Proyecto Rana Dorada. The collection (total ca. 200 specimens) numbers 40 species of reptiles and 20 species of amphibians, including threatened F_3 Mexican axolotls (*Ambystoma mexicanum*) and endangered eastern tiger salamanders (*Ambystoma t. tigrinum*).

United States of America: Central Florida Zoological Park (1975)
The Zoo is located in Lake Monroe, Florida. General Curator Fred Antonio has published papers on pitviper reproduction. The collection numbers approximately 60 taxa and over 180 individuals.
Reference: Chiszar, D., et al. 1994.

United States of America: Chaffee Zoological Gardens (1929)
This Zoo in Fresno, California, incorporated computerized environmental chambers which were state-of-the art technology when installed in the reptile building. Ron Tremper was the curator during this period but left to develop a private "Center for Reptile and Amphibian Propagation" facility in Boerne, Texas. The late Sean McKeown followed Tremper as curator. Unfortunately, the Zoo administration has now abolished the curatorial position. The collection numbers approximately 50 taxa and over 109 individuals.
References: Baker, A. G. 1986; Chaffee, P. S. 1969; Knepper, D. 1993; McKeown, S. 1984, 1985, 1989, 1991, 1992, 1993, 1984; McKeown, S., and M. J. Miller. 1984; McKeown, S., et al. 1990; Tremper, R. L. 1981; Tremper, R. L., and G. R. Barnes. 1984.

United States of America: Cleveland Metroparks Zoo (1882)
Retired General Curator Hugh Quinn, formerly an SSAR President, has published papers on a variety of herpetological topics. The collection numbers approximately 60 taxa and over 180 individuals.

United States of America: Clyde Peeling's Reptiland (1964)
This private reptile exhibit is located in Allenwood, Pennsylvania. The curator is Chad Peeling. A number of highly successful traveling exhibits, such as FROGS! A CHORUS OF COLORS have been developed by the staff. The collection numbers approximately 60 taxa and nearly 190 individuals.
Reference: Behler, J. L. and D. A. Behler. 2005.

United States of America: El Paso Zoo (1940)
The collection numbers approximately 50 taxa and nearly 195 individuals.

United States of America: Ellen Trout Zoo (1967)
This Zoo is located in Lufkin, Texas. The director, Gordon Henley, has focused on crocodilians. Ben Roberts is collection manager-herpetology. The collection numbers approximately 80 taxa and over 200 individuals.

United States of America: Honolulu Zoo (1947)
Retired Director Paul Breese began his career at the San Diego Zoo, working as a reptile keeper with C. B. Perkins. Laurence Klauber hired both of them on a part-time basis to count rattlesnake scales; because their salaries were low, they were forced to do so for many hours in Klauber's home (where the only available space was the dark and dank basement). The Honolulu Zoo has been successful at reproducing several species of tortoises and the Komodo dragon. The late Sean McKeown was curator until moving back to the mainland; his replacement is Duane Meier. The collection numbers approximately 65 taxa and over 300 individuals.
Reference: McKeown, S., et al. 1982.

United States of America: Jacksonville Zoological Gardens (1914)
Greg Lepera is curator of herpetology. The collection numbers approximately 48 taxa and nearly 150 individuals.
References: Collins, D. E. 1984; Page, C. D., et al. 1988/1989, 1991.

United States of America: Los Angeles Zoo (1966)
When the building opened, the first curator was William Turner. His successor Harvey Fischer has retired and Russ Smith now is the curator. The building is large with many exhibits, viewable from outdoors. The rare Blomberg's toad was bred at the Zoo in 1975. The collection numbers approximately 90 taxa and over 280 individuals.
Reference: Smith, R. J., and H. M. Fischer. 1975.

United States of America: Memphis Zoo (1906)
The Zoo has an older reptile building with a large planted crocodilian display and a number of smaller exhibits. A new Komodo dragon facility has been constructed. Pathologist Joel Wallach was Director during the mid-1970s. Charles Beck has retired from the curatorial position but still works as a consultant at the zoo. The current curator Steve Reichling has pub-

lished many herpetological papers: dystocia in snakes, habits of the mangrove snake in captivity, reproduction in captive black pine snakes and Dumeril's boas, conservation status of the Lesser Antillean iguana, phenotypic consequences of incubation environment on cobra eggs, and many studies on the reproductive biology and current taxonomic status of the Louisiana pine snake. The collection numbers approximately 95 taxa and over 325 individuals.

References: Beck, C. 1978, 1986; Day, M., et al. 2000; Reichling, S. B. 1973, 1974, 1981, 1983, 1986, 1988, 1989, 1990, 1995; Reichling, S. B., and W. H. N. Gutzke. 1996; Reichling, S. B., and P. Louton. 1989; Wallach, J. D. 1975.

United States of America: Miami Metrozoo (1981)

Because the climate is moderate, there are large outdoor enclosures for crocodilians and chelonians. Two species have reproduced: Siamese and African slender-snouted crocodile. J. Steven Connors is the General Curator. His predecessor, William Ziegler, developed the impressive crocodilian collection. Komodo dragons have reproduced at the Zoo. The collection numbers approximately 35 taxa and over 135 individuals.

References: Condie, T., and P. Monseur. 1997; Magill, R. N. 1982, 1984.

United States of America: North Carolina Zoological Park (1974)

This Zoo is located in Asheboro and is owned and financed by the State of North Carolina. Indoor panoramic dioramas depicting the region include the southern cypress swamp and mixed coniferous and deciduous forest exhibits. The animals within the display are separated with clear barriers to give the illusion of a complete faunal and floral community. Non-native species are interspersed in other exhibits. Since much of the zoo was left in a natural state, many local amphibians and reptiles can be seen. The curator is John Groves. Composition of the collection was nearly 60 taxa of reptiles with 227 specimens and 17 taxa of amphibians numbering 86.

Reference: Tocidlowski et al. 1997.

United States of America: Rio Grande Zoo/Albuquerque Biological Park (1925)

This Zoo is located in Albuquerque, New Mexico. There is a diverse collection of amphibians and reptiles with some large varanid displays. The curator of herpetology/invertebrates is A. Dale Belcher. The collection numbers approximately 60 taxa and 200 individuals.

References: Belcher, A. D., et al. 1987; Jacobs, D. M., and A. D. Belcher. 1983.

United States of America: Riverbanks Zoological Park and Botanical Gardens (1974)

This Zoo is located in Columbia, South Carolina. The first director, herpetologist John Mehrtens, was responsible for much of the design and construction. In 1989, a combination aquarium and reptile building called Aquarium and Reptile Complex (ARC) was added and C. Scott Pfaff was named curator of herpetology. There are six large exhibit halls: South Carolina, desert, two tropical, coral reef, and rocky shore galleries. Three outdoor exhibits for tortoises and one for crocodilians have been completed. The staff has focused on the captive maintenance and breeding of leaf-tailed geckos of the genus *Uroplatus*, building captive colonies of endangered South Carolina amphibians, and breeding Australian pythons and rare tortoises. Several AZA studbooks and Taxon Management Accounts have been published: Malagasy leaf-tailed geckos, king cobra, false gharial and green tree monitor. Papers on the captive maintenance and reproduction of the broad-striped dwarf siren (*Pseudobranchus s. striatus*) and Oates' twig snake (*Thelotornis capensis oatesi*) are in print. The collection numbers approximately 88 taxa and over 400 individuals.

References: Foley, S. C. 1998; Pfaff, C. S., and K. Vause. 2002.

United States of America: St. Augustine Alligator Farm (1893)

This private operation has large outdoor exhibits for most of the crocodilian species. The extensive crocodilian collection owned by Arthur Jones was transferred here, including an enormous estuarine crocodile which has since died. Mark Wise was formerly curator but David Kledzik is now Assistant Curator of Reptiles. Wise published an important paper on techniques used for the capture and restraint of captive crocodilians, based on his extensive experience. The collection numbers approximately 60 taxa and over 1300 individuals, mostly crocodilians.

Reference: Wise, M. 1994.

United States of America: Sedgwick County Zoo (1971)

This Zoo is located in Wichita, Kansas. The comprehensive reptile and amphibian collection is exhibited in a large building which was constructed partially below ground level. Two large displays house giant tortoises and America alligators; a number of smaller

species are also shown. Karen Graham is the Curator of Amphibians and Reptiles. She has been involved in the Jamaican iguana head-starting program and published several articles and papers on amphibian decline. The collection numbers approximately 86 taxa and over 250 individuals.
References: Bryant, W. M. 1982; Grow, D. T. 1980.

United States of America: Steinhart Aquarium (1923)

This facility is located in San Francisco, California. There is a collection of amphibians and reptiles: Karl Switak and B. Ian Hiler were formerly in charge. In the past, an albino bull frog, yellow-legged frog, Pacific gopher snake, California kingsnake, and Northern Pacific rattlesnake were exhibited. The first captive hatching of bushmasters occurred. The collection numbers approximately 50 taxa and over 150 individuals.
References: Herald, E. S. 1949, 1955; Hiler, B. I. 1985; Morales, P., and F. Dunker. 2001; Switak, K. H. 1967, 1969.

United States of America: Tennessee Aquarium (1992)

The Tennessee Aquarium, located in Chattanooga, opened in May 1992 and features fauna and flora from the State and surrounding areas. It is the world's largest freshwater aquarium. A superb North American turtle collection was assembled. Some of the best naturalistic displays in a zoo setting were built by curator of herpetology Dave Collins and Greg George.
Reference: Collins, D., and G. George. 1992.

United States of America: Woodland Park Zoological Gardens (1899)

This Zoo, located in Seattle, Washington, has benefited from contributions by two herpetologists. Curator Ernie Wagner, now retired, published many papers on captive management and husbandry: snake mite control, parameters for breeding reptiles in captivity, thermal requirements of captive reptiles, husbandry of wild-caught emerald tree boas and removal of retained eggs from snakes. His studies on breeding the Burmese python, kingsnakes (*Lampropeltis* spp.), and poison dart frogs were important contributions. He discovered temperature dependent sex determination in leopard geckos. His successor, Frank Slavens, who has also retired, produced annual inventories of herps in zoos which were called *The Inventory of Live Reptiles and Amphibians in Captivity.*

These lists were critical for ensuring that proper pairings were chosen by allowing zoos to exchange stock. Dana Payne currently oversees the department. A paper on maintenance and captive breeding of the Solomon Island leaf frog (*Ceratobatrachus guentheri*) outlined the interesting reproductive behavior of this amphibian. A study on enclosure utilization and activity patterns of captive bog turtles was accomplished. There has been a long-term commitment to study and protect the Pacific pond turtle. The collection numbers approximately 63 taxa and over 300 individuals.
References: Slavens, F. L. 1989; Wagner, E. 1971, 1974, 1975, 1979, 1980, 1981, 1985, 1987; Wagner, E., and J. Foster. 1984; Wallace, M. C. 1978; Yoshimi, D. H. et al. 1996.

Literature Cited

Acharjyo, L. N., and C. G. Misra. 1980. Growth rate of Indian python *Python molurus molurus* in captivity with special reference to age at first egg-laying. J. Bombay Nat. Hist. Soc.7:344-350.

Acharjyo, L. N., and R. Misra. 1976. Aspects of reproduction and growth of the Indian python, *Python molurus molurus,* in captivity. Brit. J. Herpetol. 5:562-565.

Ackermann, H. 1976. Aus dem Vivarium (From the Vivarium). Vivarium Darmstadt Informationen 1976 (4):16-20. [This series includes articles by a number of authors.].

Ackerman, H. 1979. Aus dem Vivarium (From the Vivarium). Vivarium Darmstadt Informationen 1979 (3):2-7.

Ackermann, J., and E. J. Miller. 1992. Chromomycosis in an African bullfrog, *Pyxicephalus adspersus.* Bull. Assoc. Rept. Amphib. Vet. 2:8-9.

Adamy, F.-W. 1966. Ein Beitrag zur Therapie der Salmonellose bei Schlangen (An essay on the therapy of salmonellosis with serpents). Zool. Gart. (N.F.), Leipzig 32:67.

Adler, K. 1989. Herpetologists of the past, p. 5-141. In K. Adler (ed.), Contributions to the History of Herpetology. Society for the Study of Amphibians and Reptiles. Contributions to Herpetology, volume 5, Oxford OH.

Adler, K. 1994. The remarkable career of Roger Conant, p. 17- 23. In J. B. Murphy, K. Adler, and J. T. Collins (eds.), Captive Management and Conservation of Amphibians and Reptiles. Society for the Study of Amphibians and Reptiles. Contributions to Herpetology, volume 11, Ithaca NY.

Adler, K. 2004. In memoriam: Roger Conant (1909–2003) with reflections by some of Roger's many friends and colleagues. Herpetol. Rev. 35:101-107.

Alberts, A. C. 1995. Use of statistical models based on radiographic measurements to predict oviposition date and clutch size in rock iguanas (*Cyclura nubila*). Zoo Biol. 14:543-553.

Alberts, A. C. (ed.). 2000. West Indian Iguanas: Status Survey and Conservation Action Plan. IUCN-The World Conservation Union, Gland, Switzerland.

Alberts, A. C. 2002. Ten years of conservation research on Cuban rock iguanas. Herpetol. Rev. 33:119-120.

Alberts, A. C., D. C. Rostal, and V. A. Lance. 1994. Studies on the chemistry and social significance of chin gland secretions in the desert tortoise, *Gopherus agassizii.* Herpetol. Monogr. 8:116-124.

Alberts, A. C., A. M. Perry, J. M. Lemm, and J. A. Phillips. 1997. Effects of incubation temperature and water potential on growth and thermoregulatory behavior of hatchling rock iguanas (*Cyclura nubila*). Copeia 1997:766-776.

Allen, E. R. 1949. Observations on the feeding habits of the juvenile cantil. Copeia 1949:225-226.

Almandarz, E. 1969. Hatching and care of the bearded dragon, *Amphibolurus barbatus,* at Lincoln Park Zoo. Inter. Zoo Yearb. 9:50-51.

Almandarz, E. 1975. The use of chilled water to transfer adult crocodilians. Inter. Zoo Yearb. 15:171-172.

Altimari, W. 1998. Venomous snakes: a safety guide for reptile keepers. Herpetol. Circ. No. 26:1-24.

Alvarez del Toro, M. 1960. Reptiles de Chiapas. Instituto Zoológico del estado, Tuxtla Gutiérrez, Chiapas, Mexico.

Alvarez del Toro, M. 1969. Breeding the spectacled caiman, *Caiman crocodilus,* at Tuxtla Gutierrez Zoo. Inter. Zoo Yearb. 9:35-36.

Alvarez del Toro, M. 1973. Los reptiles de Chiapas. 2nd ed. Gobierno del estado, Tuxtla Gutiérrez, Chiapas, Mexico.

Alvarez del Toro, M. 1974. Los Crocodylia de Mexico (Estudio Comparativo). Instituto de Recursos Naturales Renovables, México, D. F.

Alvarez del Toro, M. 1983 [dated 1982]. Los reptiles de Chiapas. 3rd ed. Instituto de Historia Natural, Tuxtla Gutiérrez, Chiapas, Mexico.

Alving, W. R., and K. V. Kardong. 1994. Aging in snakes: Results of long-term captivity on rattlesnake (*Crotalus viridis oreganus*) predatory behavior. Zoo Biol. 13:537-544.

Amaral, A. d. 1929. Key to the rattlesnakes of the genus *Crotalus* Linné. Bull. Antivenin Inst. America 3(1):64.

Anderson, M. P. 1986. Anatomic hints on wild animals and the clinician. Proc. Amer. Assoc. Zoo Vet. 1986:125-126.

Anderson, N. L., R. F. Wack, L. Calloway, T. E. Hetherington, and J. B. Williams. 1999. Cardiopulmonary effects and efficacy of propofol as an anesthetic agent in brown tree snakes, *Boiga irregularis*. Bull. Assoc. Rept. Amphib. Vet. 9(2): 9-15.

Anstandig, L. M. 1983. The breeding and rearing of the Mexican beaded lizard, *Heloderma horridum*, at the Detroit Zoo, p. 64-73. In P. J. Tolson (ed.), 7th International Herpetological Symposium on Captive Propagation and Husbandry. International Herpetological Symposium, Thurmont MD.

Antonio, F. B. 1980. Mating behavior and reproduction of the eyelash viper (*Bothrops schlegeli*) in captivity. Herpetologica 36:231-233.

Antonio, F. B., and J. B. Barker. 1983. Inventory of phenotypic aberrancies in the eastern diamondback rattlesnake (*Crotalus adamanteus*). Herpetol. Rev. 14(4):108-110.

Araki, K., H. Nishi, and M. Matumoto. 1982. Breeding reticulate python, *Python reticulatus*, at Takarazuka Zool. and Bot. Gardens. J. Japan Assoc. Zoos Aqua. 24:1-5 [in Japanese].

Armstrong, B. L., and J. B. Murphy. 1979. The natural history of Mexican rattlesnakes. Univ. Kansas Mus. Nat. Hist. Spec. Publ.:1-88.

Arnett, J. R. 1979. Breeding the Fiji banded iguana, *Brachylophus fasciatus*, at Knoxville Zoo. Inter. Zoo Yearb.19:78-79.

Arnett, J. R., M. Goodwin, and H. M. Teagarden. 1992. *Trachyboa* Peters in captivity: An overview. Contributions in Herpetology, Special Publication of the Greater Cincinnati Herpetological Society :91-94.

Aronson, V. D. 1929. Spontaneous tuberculosis in snakes. J. Infect. Dis. 44:215-223.

Asa, C. S., G. D. London, R. R. Goellner, N. Haskell, G. Roberts, and C. Wilson. 1998. Thermoregulatory behavior of captive American alligators (*Alligator mississippiensis*). J. Herpetol. 32:191-197.

Ashley, B. D., and P. M. Burchfield. 1966. Maintaining a snake colony for venom collection. U.S. Army Medical Research Laboratory Report 696:1-27.

Astreiko, E.A., and S. P. Popovskaya. 1999. Investigations of the potential of breeding and raising of the African fat-tailed gecko *Hemitheconyx caudicinctus* under laboratory conditions. Herpetological Herald (L'vov, the Ukraine) 1:23-27.

Austin, W. A. 1962. Tuatara (*Sphenodon*) in captivity. Inter. Zoo Yearb. 4:124-125.

Avery, R. A. 1982. Field studies of body temperatures and thermoregulation, p. 93-166. In C. Gans and F. H. Pough (eds.). Biology of the Reptilia, Vol. 12. Physiology C. Physiological Ecology. Academic Press, London, New York.

Avery, R. A. 1994. The effects of temperature on cap-

tive amphibians and reptiles, p. 47-51. In J. B. Murphy, K. Adler, and J. T. Collins (eds.), Captive Management and Conservation of Amphibians and Reptiles. Society for the Study of Amphibians and Reptiles. Contributions to Herpetology, volume 11, Ithaca NY.

Backhaus, D. 1972. Unterschiede in der Giftwirkung von Sandrasselottern (*Echis carinatus*) und Kettenvipern (*Vipera russelii*) (Differences in the poison effects of saw-scaled vipers [*Echis carinatus*] and Russell's vipers [*Vipera russelii*]). Salamandra 8:177-178.

Backues, K. A., and E. C. Ramsey. 1994. Ovariectomy for treatment of follicular stasis in lizards. J. Zoo Wildl. Med. 25:111-116.

Baer, D. J. 1994. The nutrition of herbivorous reptiles, p. 83-90. In J. B. Murphy, K. Adler, and J. T. Collins (eds.), Captive Management and Conservation of Amphibians and Reptiles. Society for the Study of Amphibians and Reptiles. Contributions to Herpetology, volume 11, Ithaca NY.

Baker, A. G. 1986. High tech herpetology: an update of the Fresno Zoo's computerized environmental chambers, p. 73-76. In S. McKeown, F. Caporaso, and K. H. Peterson (eds.), 9th International Herpetological Symposium on Captive Propagation and Husbandry. Zoological Consortium Inc., Thurmont MD.

Balcar, M. 1996. Chov a odchov krokodyla nilskeho *Crocodylus niloticus* v Zoo Brno (Breeding and rearing of the Nile crocodile *Crocodylus niloticus* at Brno Zoo). Akvarium Terrarium 39(1):30-34.

Balek, J. 1993. The zoo in Tuxtla Guttierez. Akvarium Terrarium 36(3):38-40.

Ball, D. J. 1982. A brief history of the Reptile Department at the Zoological Society of London, p. 41-62. In D. L. Marcellini (ed.), 6th International Herpetological Symposium on Captive Propagation and Husbandry. International Herpetological Symposium, Thurmont MD.

Banks, C. B. 1983. Reproduction in two species of captive brown snakes, genus *Pseudonaja*. Herpetol. Rev. 14(3):77-79.

Banks, C. B. 1983. Breeding and growth of the plumed basilisk (*Basiliscus plumifrons*) at the Royal Melbourne Zoo. Brit. Herpetol. Soc. Bull. 8:26-30.

Banks, C. B. 1984. Breeding the taipan *Oxyuranus scutellatus* at the Royal Melbourne Zoo. Inter. Zoo Yearb. 23:159-162.

Banks, C. B. 1984 Reproductive history of a colony of captive common iguanas (*Iguana iguana*). In V.L.

Bels, and A.P.V. den Sande (eds.) Acta Zool. Path. Antverpiensia 78:101-14.

Banks, C. B. 1985. Breeding D'Albertis python at the Melbourne Zoo. Thylacynus 10 (4):17-21.

Banks, C. B. 1985. Observations on feeding and sloughing in a collection of captive snakes, p. 495-501. In G. Grigg, R. Shine, and H. Ehmann (eds.), Biology of Australian Frogs and Reptiles. Royal Society of New South Wales, Sydney, Australia.

Banks, C. B. 1989. Management of fully aquatic snakes. Inter. Zoo Yearb. 28:155-163.

Banks, C. B., T. Hawkes, J. R. Birkett, and M. Vincent. 1999. Captive management and breeding of the Striped Legless Lizard, *Delma impar.* Herpetofauna 29 (2):18-30.

Banning, G. H. 1933. Hancock Expedition of 1933 to the Galápagos Islands, General Report. Bull. Zool. Soc. San Diego 10:1-30.

Barker, D. G. 1984. Maintenance and reproduction of green tree monitors at the Dallas Zoo, p. 91-92. In R. A. Hahn (ser. ed.), 8th International Herpetological Symposium on Captive Propagation and Husbandry. International Herpetological Symposium, Thurmont MD.

Barker, D. G. 1984. Maintenance and reproduction of black-headed pythons at the Dallas Zoo, p. 106-108. In R. A. Hahn (ser. ed.), 8th International Herpetological Symposium on Captive Propagation and Husbandry. International Herpetological Symposium, Thurmont MD.

Barker, D. G., J. B. Murphy, and K. W. Smith. 1979. Social behavior in a captive group of Indian Pythons, *Python molurus* (Serpentes, Boidae) with formation of a linear social hierarchy. Copeia 1979:466-471.

Barker, I. K., and M. Cranfield. 1988/1989. *Schellackia* (*Lainsonia*) *sp.* in chuckwallas *Sauromalus obesus.* Proc. Amer. Assoc. Zoo Vet. 1988/1989:61.

Barnard, S. M., T. G. Hollinger, and T. A. Romaine. 1979. Growth and food consumption in the corn snake, *Elaphe guttata guttata* (Serpentes: Colubridae). Copeia 1979:739-741.

Barrie, M. T., E. Castle, and D. Grow. 1993. Diseases of chameleons at the Oklahoma City Zoological Park. Proc. Amer. Assoc. Zoo Vet. 1993:1-8.

Bartlett, A. D. 1896. Notes on the breeding of the Surinam water-toad (*Pipa americana*) in the Society's gardens. Proc. Zool. Soc. London 1896:595-597.

Bartlett, A. D. 1899. Wild Animals in Captivity Being an Account of the Habits, Food, Management and Treatment of the Beasts and Birds at the 'Zoo' with Reminiscences and Anecdotes by A. D. Bartlett/ Compiled and Edited by Edward Bartlett. Chapman and Hall, London.

Bartholomew, G. A. 1982. Physiological control of body temperature, p. 167-211. In C. Gans and F. H. Pough (eds.). Biology of the Reptilia, Vol. 12. Physiology C. Physiological Ecology. Academic Press, London, New York.

Barton, A. J., and W. B. Allen Jr. 1961. Observations on the feeding, shedding and growth rates of captive snakes (Boidae). Zoologica (New York) 46:83-87.

Baskar, N., N. Krishnakumar, and A. Manimozhi. 1999. Incubation, feeding and growth of Indian rock pythons. Inter. Zoo News 46(2):90-93.

Bateman, G. C. 1897. The Vivarium, Being a Practical Guide to the Construction, Arrangement, and Management of Vivaria, Containing Full Information as to all Reptiles Suitable as Pets, How and Where to Obtain Them, and How to Keep Them in Health. L. Upcott Gill, London.

Baumgartner, R., B. Hauser, A. Rübel, R. E. Honegger, and E. Isenbügel. 1987. Tubuläres Adenokarzinom im Dickdarm einer Netzpython (*Python reticulatus*) (Tubular adeno-cancer in the colon of a reticulate python (*Python reticulatus*)). Verhandlber. 29. Internat. Symp. Erkrankungen Zootiere (Illnesses of Zoo Animals), Cardiff, 1987:311-314.

Bechstein, J. M. 1797. Naturgeschichte; oder, Anleitung zur Kenntniss und Wartung der Säugethiere, Amphibien, Fische, Insecten und Würmer, welche man in der Stube halten kann (Natural History, or, Guide to the Knowledge and Care of Mammals, Amphibians, Fish, Insects and Worms Which Can Be Kept in the Home). C. W. Ettinger, Gotha.

Beck, C. 1978. Breeding the West African dwarf crocodile *Osteolaemus tetraspis* at Memphis Zoo. Inter. Zoo Yearb. 18:89-91.

Beck, C. 1986. Tagging the green turtle (*Chelonia mydas*) in Costa Rica, with side notes on flora and fauna observed, p. 31-33. In S. McKeown, F. Caporaso, and K. H. Peterson (eds.), 9th International Herpetological Symposium on Captive Propagation and Husbandry. Zoological Consortium, Inc., Thurmont MD.

Beck, K., M. R. Loomis, G. Lewbart, L. H. Spelman, and M. Papich. 1995. Preliminary comparison of plasma concentrations of gentamicin injected into the cranial and caudal limb musculature of the eastern box turtle (*Terrapene carolina carolina*). J. Zoo Wildl. Med. 26:165-168.

Beebe, W. 1946. Field notes on the snakes of Kartabo, British Guiana and Caripito. Zoologica (New York) 31:11-52.

Beebe, W. 1946. Field notes on the snakes of British Guiana and Venezuela. Amphisbaenidae and Scincidae. Zoologica (New York) 31(4).

Behler, J. 1977. A propagation program for Chinese alligators (*Alligator sinensis*) in captivity. Herpetol. Rev. 8(4):124-125.

Behler, J. 1978. Feasibility of the establishment of a captive-breeding population of the American crocodile. National Park Service Report T-509:1-94.

Behler, J. L. and D. A. Behler. 2005. Frogs! A Chorus of Colors. Sterling Publishing Co., Inc., New York.

Behler, J. L., and P. Brazaitis. 1974. Breeding the Egyptian cobra *Naja haje* at the New York Zoological Park. Inter. Zoo Yearb. 14:83-84.

Behler, J., P. Brazaitis, and T. Joanen. 1982. The Chinese alligator (*Alligator sinensis*), its status and propagation in captivity. Zool. Gart. (N.F.), Jena 52:73-77.

Belcher, A. D., G. Riordin, F. Groves, and H. Hunt. 1987. Captive propagation of the dwarf caiman *Paleosuchus palpebrosus* at the Rio Grande Zoo, p. 26-32. In R. Gowen (ed.), Captive Propagation and Husbandry of Reptiles and Amphibians. Northern California Herpetological Society.

Bell, B. D. 1985. Conservation status of the endemic New Zealand frogs, p. 449-458. In G. Grigg, R. Shine, and H. Ehmann (eds.), Biology of Australian Frogs and Reptiles. Royal Society of New South Wales, Sydney.

Bell, C. E. (ed.), 2001. Encyclopedia of the World's Zoos. Fitzroy Dearborn Publishers, Chicago, London.

Belluomini, H. E., and T. Veinert. 1967. Notes on breeding anacondas *Eunectes murinus* at Sao Paulo Zoo. Inter. Zoo Yearb. 7:181-182.

Bels, V. 1987. Observations of the courtship and mating behavior in the snake *Hydrodynastes gigas*. J. Herpetol. 21:350-352.

Bels, V. 1987. Analysis of the growth of *Dermochelys coriacea* (Reptilia: Testudines) in captivity, p. 157. In P. W. Scott, and A. G. Greenwood (eds.), Exotic Animals in the Eighties. Proceedings from the 25th Anniversary Symposium of the British Veterinary Zoological Society, 18th-20th April 1986. British Veterinary Zoological Society, place of publication not given.

Bels, V. L., and P. A. Van den Sande. 1986. Breeding the Australian carpet python *Morelia spilotes variegata* at Antwerp Zoo. Inter. Zoo Yearb. 24/25:231-238.

Benedict, F. C. 1932. The Physiology of Large Reptiles with Special Reference to the Heat Production of Snakes, Tortoises, Lizards, and Alligators. Carnegie Institution, Washington DC Publ. No. 425.

Benefield, G. E., R. D. Grimpe, and E. Olsen. 1981. Aspects of reproduction in western banded geckos *Coleonyx variegatus* at Tulsa Zoo. Inter. Zoo Yearb. 21:83-87.

Bennett, C., and D. Barnaby. 1989. The Reptiles of Belle Vue 1950 ~1977. A Curator's Viewpoint. ZSGM Publications, Timperly UK.

Bennett, E. T. 1829. The Tower Menagerie: Comprising the Natural History of the Animals Contained in that Establishment, with Anecdotes of Their Characters and History. Illustrated by Portraits of Each, Taken from Life, by William Harvey, and Engraved on Wood by Branston and Wright. Printed for R. Jennings, London.

Berg, J. 1901. Indische Dryophiden im Terrarium (Indian Dryophiden in the terrarium). Zool. Gart., Frankfurt a. M. 42:204-215.

Bernard, P. 1842-1843. Le Jardin des plantes : description complète, historique et pittoresque du Muséum d'histoire naturelle, de la ménagerie, des serres, des galeries de minéralogie et d'anatomie, et de la vallée suisse : moeurs et instincts des animaux, botanique, anatomie comparée : minéralogie, géologie, zoologie (The Jardin des plantes: A complete, historical and picturesque description of the Museum of Natural History, of the menagerie, the green-houses, the galleries of mineralogy and anatomy and of the Swiss Valley : mores and instincts of the animals, botany, comparative anatomy : mineralogy, geology, zoology). L. Curmer, Paris.

Biella, H.-J., P. Eckardt, and W. Ruosch. 1989. Zur Fortpflanzungsbiologie von *Bitis arietans* und *Bitis gabonica*, nebst Bemerkungen zum Nahrungsbedürfnis, Wachstum und zur Haltung beider Arten (Concerning the reproductive biology of *Bitis arietans* and *Bitis gabonica* along with observations regarding requirements for nourishment, growth, and care of both species). Zool. Gart. (N.F.), Jena 59:37-48.

Birchard, G. F., T. Walsh, R. Rosscoe, and C. L. Reiber. 1995. Oxygen uptake by Komodo Dragon (*Varanus komodoensis*) eggs: The energetics of prolonged development in a reptile. Physiol. Zool. 68:622-633.

Birkenmeier, E. 1972. Rearing a leathery turtle *Dermochelys coriacea* in captivity. Inter. Zoo Yearb. 12:204-207.

Birkett, J., and H. McCraken. 1992. Captive treatment of a Johnstone's crocodile *Crocodylus johnstoni*, under treatment for bilateral mandibular fractures, p. 87-93. In M. J. Uricheck (ed.), 15th International Herpetological Symposium on Captive Propagation and Husbandry. International Herpetological Symposium, Palo Alto CA.

Birkett, J. R., M. Vincent, and C. B. Banks. 1999. Captive management and rearing of the Roseate Frog, *Geocrinia rosea* at Melbourne Zoo. Herpetofauna 29 (2):49-56.

Blake, E. 1990. Cage design for the Trinidad stream frog (*Colostethus trinitatis*) at Edinburgh Zoo. R. Zool. Soc. Scotland Ann. Rep. No. 78. [1991]:55-57.

Blake, E., and G. Stewart. 1980. Maintenance and breeding of banded basilisk (*Basiliscus vittatus*). Annual Report Royal Zool. Soc. Scotland:21-24.

Blanchard, B. 1992. Management of the tuatara *Sphenodon* spp in the wild and in captivity. Inter. Zoo Yearb. 31:42-44.

Blanco, S., J. L. Behler, and F. Kostel. 1991. Propagation of the batagurine turtles *Batagur baska* and *Callagur borneoensis* at the Bronx Zoo. p. 63-65. In K. R. Beaman, F. Coporaso, S. McKeown, and M. D. Graff (eds.). Proceedings of the First International Symposium on Turtles and Tortoises: Conservation and Captive Husbandry. Chapman University, August 9-12, 1990. California Turtle and Tortoise Club, Van Nuys, CA.

Blody, D. A. 1983. Notes on the reproductive biology of the eyelash viper, *Bothrops schlegeli*, in captivity. Herpetol. Rev. 14(2):45-46.

Blody, D. A., and D. T. Mehaffey. 1989. The reproductive biology of the annulated boa *Corallus annulatus* in captivity. Inter. Zoo Yearb. 28:167-172.

Bloxam, Q. M. C. 1976. The maintenance and breeding of the Round Island skink *Leiolopisma telfairii* (Desjardins). Annual Rep. Jersey Wildl. Preserv. Trust 13:53-56.

Bloxam, Q M. C. 1982. Über Haltung und Zucht der Jamaika-Schlankboa (*Epicrates subflavus*) Stejneger (1901) im Jersey Zoo (Concerning the care of the Jamaican boa (*Epicrates subflavus*) Stejneger (1901) in the Jersey Zoo). Z. Köln. Zoo 25(4):133-137.

Bloxam, Q. M. C. 1983. A preliminary report on the captive management and reproduction of the Round Island boa, *Casarea dussumieri*, p. 115-117. In P. J. Tolson (ed.), 7th International Herpetological Symposium on Captive Propagation and Husbandry. International Herpetological Symposium, Thurmont MD.

Bloxam, Q. M. C. 1998. Distribution, density and abundance of the Madagascar flat-tailed tortoise *Pyxis planicauda*. Testudo 4(5):28-32.

Bloxam, Q. M. C., and S. J. Tonge. 1986. Breeding programmes for reptiles and snails at Jersey Zoo; an appraisal. Inter. Zoo Yearb. 24/25:49-56.

Bloxam, Q. M. C., and S. J. Tonge. 1995. Amphibians: suitable candidates for breeding-release programs. Biodiv. Conserv. 4:636-644.

Bloxam, Q. M. C., and S. Townson. 1980. Maintenance and breeding of *Phelsuma guentheri* (Boulenger 1885). The British Herpetological Society: The Care and Breeding of Captive Reptiles. p. 51-62.

Bloxam, Q. M. C., and M. Vokins. 1978. Breeding and maintenance of *Phelsuma guentheri* (Boulenger 1885) at the Jersey Zoological Park. Dodo, J. Jersey Wildl. Preserv. Trust 15:82-91.

Blunt, W. 1976. The Ark in the Park, the Zoo in the Nineteenth Century. Hamilton: Tryon Gallery, London.

Boardman, W. S. J., and M. D. Sibley. 1991. The captive management, diseases and veterinary care of tuatara. Proc. Amer. Assoc. Zoo Vet. 1991:159-167.

Bonath, K. 1979. Halothane inhalation anaesthesia in reptiles and its clinical control. Zoo Biol. 19:112-125.

Bonefield, J. 1979. Hatching the Argentine snake-necked turtle *Hydromedusa tectifera* at San Antonio Zoo. Inter. Zoo Yearb.19:55-58.

Boos, H. E. A. 1979. Some breeding records of Australian pythons, Pythonidae. Inter. Zoo Yearb. 19:87-89.

Bosch, H. 1991. Wirbelschwanzleguane im Löbbecke-Museum *Cyclura cychlura figginsi* (Ground iguanas in the Löbbecke Museum *Cyclura cychlura figginsi*). D. Aquar. Terrar.- Z. 44(1):29-31.

Bosch, H., and J. Boscheinen. 2001. Zur Geschichte der Herpetologie im „Aquazoo-Löbbecke-Museum" Düsseldorf. (About the history of herpetology in the „Aquazoo-Löbbecke-Museum" Düsseldorf.) In W. Rieck, G. T. Hallmann, and W. Bischoff (eds.), Die Geschichte der Herpetologie und Terrarienkunde im deutschsprachigen Raum - Mertensiella 12:281-283. Deutschen Gesellschaft für Herpetologie und Terrarienkunde e.V. (DGHT).

Bosch, H. and H. Werning. 1996. Green Iguanas and Other Iguanids. T.F.H. Publications, Inc., Neptune NJ.

Botting, D. 1999. Gerald Durrell. The Authorized Biography. Carroll & Graf Publishers, Inc., New York.

Boulenger, E. G. 1915. Notes on the feeding of snakes in captivity. Proc. Zool. Soc. London 1915:583-587.

Bourou, R., H. Tiandray, O.C. Razandrimamila-finiarivo, E. Bekarany, and J. Durbin. 2001. Comparative reproduction in wild and captive female ploughshare tortoises *Geochelone yniphora*. Dodo, J. Jersey Wildl. Preserv. Trust 37:70-79.

Boyer, D. M. 1995. Venomous reptiles: An overview of families, handling, restraint techniques, and emergency protocol. Proc. Assoc. Rept. Amphib. Vet. 1995:27-29, 83-90.

Boyer, D. M., and B. Baldwin. 1997. A simple method of preventing self-inflicted injury when feeding a dicephalic California kingsnake, *Lampropeltis getulus californiae*. Bull. Assoc. Rept. Amphib. Vet. 7(3):6-7.

Boyer, D. M., and T. H. Boyer. 1994. Tortoise care. Bull. Assoc. Rept. Amphib. Vet. 4(1):16-28.

Boyer, D. M., and W. E. Lamoreaux. 1983. Captive reproduction and husbandry of the pygmy mulga monitor, *Varanus gilleni* at the Dallas Zoo, p. 59-63. In P. J. Tolson (ed.), 7th International Herpetological Symposium on Captive Propagation and Husbandry. International Herpetological Symposium, Thurmont MD.

Boyer, D. M., and D. T. Roberts. 1998. The predatory strike of the temple viper *Tropidolaemus wagleri*. Snake 28:79-82.

Boyer, D., L. A. Mitchell, and J. B. Murphy. 1989. Reproduction and husbandry of the bushmaster *Lachesis muta muta* at the Dallas Zoo. Inter. Zoo Yearb. 28:190-194.

Boyer, D., C. M. Garrett, J. B. Murphy, H. M. Smith, and D. Chiszar, D. 1995. In the footsteps of Charles C. Carpenter: Facultative strike-induced chemosensory searching and trail following behavior of bushmasters (*Lachesis muta*) at Dallas Zoo. Herpetological Monographs 9:161-169.

Boylan, T. 1982. Reptile pits: For and against. Inter. Zoo News (No. 175) 29 (1) Sept/Oct:5-6.

Boylan, T. 1984. Breeding the rhinoceros iguana *Cyclura c. cornuta* at Sydney Zoo. Inter. Zoo Yearb. 23:144-148.

Boylan, T. 1985. Captive management of a population of rhinoceros iguanas *Cyclura cornuta cornuta* at Taronga Zoo, Sydney, p. 491-495. In: G. Grigg, R. Shine and H. Ehmann (eds.), Biology of Australian Frogs and Reptiles. Royal Society of New South Wales, Sydney.

Boylan, T. 1989. Problems associated with maintaining a healthy zoo reptile collection. Snake Keep. 3(10):5-10.

Boylan, T. 1989. Reproduction of the Fijian crested iguana *Brachylophus vitiensis* at Taronga Zoo, Sydney. Inter. Zoo Yearb. 28:126-130.

Bozhansky, A. T., and Kudrjavtsey, S. V. 1986. Ecological observations of the rare vipers of the Caucasus, p. 495-498. In Z. Rocek (ed.), Studies in Herpetology. Charles University, Prague, Czech Republic.

Bozhansky, A. T., and S. V. Kudrjavtsev. 1986. An experience of Caucasian viper breeding in captivity. Vestnik Zool.1986(3):78-81.

Bradley, T. A., and K. Wright. 2000. Captive care and breeding of White's tree frog, *Pelodryas caerulea*. Bull. Assoc. Rept. Amphib. Vet. 10(2):21-24.

Bradley, T. A., T. M. Norton, and K. S. Latimer. 1998. Hemogram values, morphological characteristics of blood cells and morphometric study of loggerhead sea turtles, *Caretta caretta*, in the first year of life. Bull. Assoc. Rept. Amphib. Vet. 8(3): 8-16.

Branch, W. R., and H. Erasmus. 1976. Reproduction in Madagascar ground and tree boas *Acrantophis madagascariensis* and *Sanzinia madagascariensis*. Inter. Zoo Yearb. 16:78-80.

Branch, W. R., and M. Griffin. 1996. Pythons in Namibia: Distribution, conservation, and captive breeding programs, p. 93-102. In P. D. Strimple (ed.), Advances in Herpetoculture. Special Publication of the International Herpetological Symposium, Inc., No. 1. International Herpetological Symposium, Inc., Des Moines IA.

Branch, W. R., and N. Schaefer. 1987. A new educational foyer to the Snake Park at the Port Elizabeth Museum complex. Inter. Zoo Yearb. 26: 309-314.

Brannian, R. E. 1984. A soft tissue laparotomy technique in turtles. Proc. Amer. Assoc. Zoo Vet. 1984:52.

Brannian, R. E., D. L. Graham, and D. M. Barnes. 1980. Mortality in captive green tree pythons. Proc. Amer. Assoc. Zoo Vet. 1980:63-65.

Brattstrom, B. 1963. A preliminary review of the thermal requirements in amphibians. Ecology 44:238-255.

Brattstrom, B. H. 1965. Body temperatures of reptiles. Amer. Midl. Nat. 73:376-422.

Brazaitis, P. 1969. Occurrence and ingestion of gastrolith in two crocodilians. Herpetologica 25:63-64.

Brazaitis, P. 1969. Determination of sex in living crocodilians. Brit. J. Herpetol. 4:54-58.

Brazaitis, P. 1973. The identification of living crocodilians. Zoologica (New York) 58:59-101.

Brazaitis, P. 1981. Maxillary regeneration in a marsh crocodile, *Crocodylus palustris*. J. Herpetol. 15:360-362.

Brazaitis, P. 1986. Management, reproduction, and growth of *Caiman crocodilus yacare* at the New

York Zoological Park, p. 389-397. Proc. 7th Working Meet., IUCN/SSC Crocodile Specialist Group, Caracas, Venezuela Oct. 21-28, 1984.

Brazaitis, P., and T. Joanen. 1984. Report on the status of the captive breeding program for the Chinese alligator *Alligator sinensis* in the United States, p. 117-121. Proc. 6th Working Meet., IUCN/SSC Crocodile Specialist Group, Victoria Falls, Zimbabwe and St. Lucia Estuary, Repub. South Africa, Sept. 19-30, 1982.

Brazaitis, P., and M. E. Watanabe. 1982. The Doppler, a new tool for reptile and amphibian hematological studies. J. Herpetol. 16:1-6.

Brazaitis, P., and M. E. Watanabe. 1983. Ultrasound scanning of Siamese crocodile eggs: Hello, are you in there? J. Herpetol. 17:286-287.

Brecke, B. J., J. B. Murphy, and W. Seifert. 1976. An inventory of reproduction and social behavior in captive Baird's ratsnake, *Elaphe obsoleta bairdi* (Yarrow). Herpetologica 32:339-341.

Breese, P. 2002. In Memoriam: Sean McKeown (1944-2002). Bull. Chicago Herpetol. Soc. 37:152.

Brodie Jr., E. D., I. Rehák, and T. Schöttler. 1992. Antipredator mechanisms of salamandrids: Observations on *Paramesotriton deloustali, Neurergus strauchii* and *Mertensiella luschani*, p. 83-87. In Z. Korsós and I. Kiss (eds.), Proc. Sixth Ord. Gen. Meet. S. E. H., Budapest 1991.

Broom, R. 1929. On the extinct Galapagos tortoise that inhabited Charles Island. Zoologica (New York) 9:313-320.

Brotzler, A. 1965. Mertens-Wasserwarane (*Varanus mertensi* Glauert 1951) züchteten in der Wilhelma (Mertens water monitor breeding in the Wilhelma). Freunde Köln Zoo 8(3):89.

Broughton, Lady, M. 1936. A modern dragon hunt on Komodo. National Geographic 70(3):321-331.

Bryant, B. R., L. Vogelnest, and F. Hulst. 1997. The use of cryosurgery in a diamond python, *Morelia spilota spilota*, with fibrosarcoma and radiotherapy in a common death adder, *Acanthophis antarcticus*, with melanoma. Bull. Assoc. Rept. Amphib. Vet. 7(3):9-12.

Bryant, W. M. 1982. Mycotic dermatitis in a collection of desert lizards. Proc. Amer. Assoc. Zoo Vet. 1982:4.

Buck, F., and E. Anthony. 1932. Wild Cargo. Simon and Schuster, New York.

Buley, K. R., and G. Garcia. 1997. The Recovery Programme for the Mallorcan midwife toad *Alytes muletensis* - An update. Dodo, J. Jersey Wildl. Preserv. Trust 33:80-90.

Buley, K. R., and C. Gonzalez. 2000. The Durrell Wildlife Conservation Trust and the Mallorcan Midwife Toad *Alytes muletensis* - Into the 21st century. Herpetol. Bull. 72:17-20.

Burchfield, P. M. 1975. Breeding the Colombian giant toad *Bufo blombergi* at Brownsville Zoo. Inter. Zoo Yearb. 15:89-90.

Burchfield, P. M. 1975. Hatching the radiated tortoise *Testudo radiata* at Brownsville Zoo. Inter. Zoo Yearb. 15:90-92.

Burchfield, P. M. 1975. Raising the fer-de-lance *Bothrops atrox* in captivity. Inter. Zoo Yearb. 15:173-174.

Burchfield, P. M. 1975. Adaptation of adult eastern diamondback rattlesnakes *Crotalus adamanteus* to captivity. Inter. Zoo Yearb. 15:174-175.

Burchfield, P. M. 1975. The bushmaster *Lachesis muta* in captivity. Inter. Zoo Yearb. 15:175-177.

Burchfield, P. M. 1977. Breeding the king cobra *Ophiophagus hannah* at Brownsville Zoo. Inter. Zoo Yearb. 17:136-140.

Burchfield, P. M. 1982. Additions to the natural history of the crotaline snake *Agkistrodon bilineatus taylori*. J. Herpetol. 16:376-382.

Burchfield, P. M., C. S. Doucette, and T. F. Biemler. 1980. Captive management of the radiated tortoise *Geochelone radiata* at Gladys Porter Zoo. Inter. Zoo Yearb. 20:1-6.

Burchfield, P. M., C. S. Hairston, and S. L. Huntress. 1987. Management of Galapagos tortoises (*Geochelone elephantopus*) at the Gladys Porter Zoo. AAZPA Annual Conf. Proc. 1987:151-157.

Burden, W. D. 1927. Dragon Lizards of Komodo. G. P. Putnam's Sons, New York.

Burghardt, G. M. 1977. Learning processes in reptiles, p. 555-681. In C. Gans and D. W. Tinkle (eds.). Biology of the Reptilia, Vol. 7. Ecology and Behaviour A. Academic Press, London, New York.

Burghardt, G. M., B. Ward, and R. Rosscoc. 1996. Problem of reptile play: environmental enrichment and play behavior in a captive Nile soft-shelled turtle, *Trionyx triunguis*. Zoo Biol. 15:223-238.

Burghardt, G. M., D. Chiszar, J. B. Murphy, John Romano Jr., T. Walsh and J. Manrod. 2002. Behavioral complexity, behavioral development, and play, p. 77-116. In J. B. Murphy, C. Ciofi, C. de La Panouse, and T. Walsh (eds.), Komodo Dragons. Biology and Conservation. Smithsonian Institution Press, Washington.

Burke, R. L. 1990. Conservation of the world's rarest tortoise. Conserv. Biol. 4(2):122-124.

Burns, R. 1995. Considerations in the euthanasia of reptiles, amphibians, and fish. Proc. Amer. Assoc.

Zoo Vet. 1995:243-249.

Burrage, B. R. 1965. Copulation in a pair of *Alligator mississipiensis*. Brit. J. Herpetol. 3:207-208.

Burton, F. 2000. Grand Cayman Iguana *Cyclura nubila lewisi*, p. 45-47. In A. C. Alberts (ed.), West Indian Iguanas: Status Survey and Conservation Action Plan. IUCN-The World Conservation Union, Gland, Switzerland.

Burton, M. S., 1991. Colorado frog watch. Proc. Amer. Assoc. Zoo Vet. 1991:81-84.

Burton, M. S., E. T. Thorne, A. Anderson, and D. R. Kwiatkowski. 1995. Captive management of the endangered Wyoming toad at the Cheyenne Mountain Zoo. Bull. Assoc. Rept. Amphib. Vet. 5(1):6-8.

Bush, M. 1978. Biological half-life of Gentamicin in gopher snakes. Amer. J. Vet. Res. 39(1):171-173.

Bush, M., and J. Smeller. 1978. Blood collection & injection techniques in snakes. Vet. Med./ Small Anim. Clin. February:211-214.

Bush, M., J. M. Smeller, P. Charache, and H. M. Solomon. 1976. Preliminary study of antibiotics in snakes. Proc. Amer. Assoc. Zoo Vet. 1976:50-54.

Buskirk, J. R. 1986. The endangered Egyptian tortoise *Testudo kleinmanni*: Status in Egypt and Israel, p. 35-51. In S. McKeown, F. Caporaso, and K. H. Peterson (eds.), 9th International Herpetological Symposium on Captive Propagation and Husbandry. Zoological Consortium, Inc., Thurmont MD.

Busono, M. S. 1974. Facts about the *Varanus komodoensis* at the Gembira Loka Zoo at Yogyakarta. Zool. Gart. (N.F.), Jena 44:62-63.

Butler, D. J. 1992. The role of zoos in the captive breeding of New Zealand's threatened fauna. Inter. Zoo Yearb. 31:4-9.

Cahill, L. W. 1971. Observations on a captive beaknosed snake *Scaphiophis albopunctatus* at the University of Ife Zoo. Inter. Zoo Yearb. 11:233-234.

Caillouet Jr., C. W., C. T. Fontane, and J. P. Flanagan. 1993. Captive rearing of sea turtles: Head starting Kemp's ridley *Lepidochelys kempii*. Proc. Amer. Assoc. Zoo Vet. 1993:8-12.

Calle, P. P., J. L. Behler, J. McDougal, S. M. Lee, I. Schumacher, and D. R. Brown. 1998. Mycoplasma survey of captive and free-ranging eastern box turtles (*Terrapene carolina carolina*) in New York. Proc. Amer. Assoc. Zoo Vet. 1998:285-287.

Camin, J. H. 1948. Mite transmission of a hemorrhagic septicemia in snakes. J. Parasitol. 34:345-354.

Camin, J. H. 1953. Observations on the life history and sensory behavior of the snake mite, *Ophionyssus natricis* (Gervais) (Acarina: Macronyssidae). Chicago Acad. Sci. Spec. Publ. No. 10:75 pp.

Camin, J. H., G. K. Clarke, L. H. Goodson, and H. R. Shuyer. 1964. Control of the snake mite, *Ophionyssus natricis* (Gervais) in captive reptile collections. Zoologica (New York) 49(2):65-79.

Campbell, H. W. 1973. Observations on the acoustic behavior of crocodilians. Zoologica (New York) 58(No. 1):1-11.

Campbell, J. A. 1972. Reproduction in captive Trans-Pecos ratsnakes, *Elaphe subocularis*. Herpetol. Rev. 4(4):129-130.

Campbell, J. A. 1972. Observations on Central American river turtles *Dermatemys mawi* at Fort Worth Zoo. Inter. Zoo Yearb. 12:202-204.

Campbell, J. A. 1973. A captive hatching *of Micrurus fulvius tenere* (Serpentes, Elapidae). J. Herpetol. 7:312-315.

Campbell, J. A., and D. R. Frost. 1993. Anguid lizards of the genus *Abronia*: Revisionary notes, descriptions of four new species, a phylogenetic analysis, and key. Bull. Amer. Mus. Nat. Hist. No. 216.

Campbell, J. A., and J. B. Murphy. 1984. Reproduction in five species of Paraguayan colubrids. Trans. Kansas Acad. Sci. 87((1-2):63-65.

Campbell, J. A., and H. R. Quinn. 1975. Reproduction in a pair of Asiatic cobras, *Naja naja* (Serpentes, Elapidae). J. Herpetol. 9:229-233.

Campbell, S. 1978. Lifeboats to Ararat. Times Books, New York.

Campden-Main, S. M., and L. K Campden-Main. 1982. A guide to the care of neonate Ophidia, p. 211-223. In D. L. Marcellini (ed.), 6th International Herpetological Symposium on Captive Propagation and Husbandry. International Herpetological Symposium, Thurmont MD.

Card, W. 1994. Notes on the natural history and husbandry of the temple viper (*Tropidolaemus wagleri*). Vivarium 5(5):22-24.

Card, W. 1994. Double clutching Gould's monitors (*Varanus gouldi*) and Gray's monitors (*Varanus olivaceus*) at the Dallas Zoo. Herpetol. Rev. 25(3):111-114.

Card, W., and D. Mehaffey. 1994. A radiographic sexing technique for *Heloderma suspectum*. Herpetol. Rev. 25(1):17-19.

Card, W., and J. B. Murphy. 2000. Lineages and histories of zoo herpetologists in the United States. Herpetol. Circ. No. 27:1-44.

Card, W., and D. T. Roberts. 1996. Incidence of bites from venomous reptiles in North American zoos. Herpetol. Rev. 27(1):15-16.

Card, W. C., D. T. Roberts, and R. A. Odum. 1998. Does zoo herpetology have a future? Zoo Biol. 17:453-462.

Carl, G., K. H. Peterson, and R. M. Hubbard. 1982. Reproduction in captive Aruba Island rattlesnakes. Herpetol. Rev. 13(3):89.

Carpenter, C. C., and G. W. Ferguson. 1977. Variation and evolution of stereotyped behavior in reptiles, p. 335-554. In C. Gans and D. W. Tinkle (eds.). Biology of the Reptilia, Vol. 7. Ecology and Behaviour A. Academic Press, London, New York.

Carpenter, C. C., and J. B. Murphy. 1978. Aggressive behavior in the Fiji Island lizard *Brachylophus fasciatus* (Reptilia, Lacertilia, Iguanidae). J. Herpetol. 12:251-252.

Carpenter, C. C., and J. B. Murphy. 1978. Tongue display by the common bluetongue (*Tiliqua scincoides*) Reptilia, Lacertilia, Scincidae). J. Herpetol. 12:428-429.

Carpenter, C. C., J. C. Gillingham, and J. B. Murphy. 1976. The combat ritual of the rock rattlesnake (*Crotalus lepidus*). Copeia 1976:764-780.

Carpenter, C. C., J. B. Murphy, and G. C. Carpenter. 1978. Tail luring in the death adder, *Acanthophis antarcticus*. J. Herpetol. 12:574-576.

Carpenter, C. C., J. B. Murphy, and L. A. Mitchell. 1978. Combat bouts with spur use in the Madagascan boa (*Sanzinia madagascariensis*). Herpetologica 34:207-212.

Carpenter, C. C., J. C. Gillingham, J. B. Murphy, and L. A. Mitchell. 1976. A further analysis of the combat ritual of the pygmy mulga monitor, *Varanus gilleni* (Reptilia: Varanidae). Herpetologica 32:35-40.

Casares, M. 1995. Untersuchungen zum Fortpflanzungsgeschehen bei Riesenschildkröten (*Geochelone elephantopus* und *G. gigantea*) und Landschildkröten (*Testudo graeca* und *T. hermanni*) anhand von Ultraschalldiagnostik und Steroidanalysen im Kot (Investigations concerning the reproductive process in giant tortoises [*Geochelone elephantopus* and *G. gigantea*] and land tortoises [*Testudo graeca* and *T. hermanni*] based on ultrasound diagnostic methods and steroid analyses in the Kot Zoological Garden). Zool. Gart. (N.F.), Jena 65:50-76.

Casares, M., R. E. Honegger, and A. Rübel. 1995. Management of giant tortoises *Geochelone elephantopus* and *Geochelone gigantea* at Zurich Zoological Gardens. Inter. Zoo Yearb. 34:135-143.

Casares, M., A. Rübel, M. Döbeli, R. E. Honegger, and E. Isenbügel. 1994. Non-invasive assessment of reproductive patterns in tortoises. Verh. ber. Erkrg. Zootiere 36:81-87.

Castle, E. 1990. Husbandry and breeding of chameleons *Chamaeleo* spp at Oklahoma City Zoo. Inter. Zoo Yearb. 29:74-84.

Castle, E. 1991. Captive reproduction and neonate husbandry of the Oustalet's chameleon, *C. oustaleti* at the Oklahoma City Zoological Park, p. 25-34. In A. W. Zulich (ed.), 14th International Herpetological Symposium on Captive Propagation and Husbandry. International Herpetological Symposium

Cayot, L. J., H. L. Snell, W. Llerena, and H. M. Snell. 1994. Conservation biology of Galápagos reptiles: Twenty-five years of successful research and management, p. 297-305. In J. B. Murphy, K. Adler and J. T. Collins (eds.), Captive Management and Conservation of Amphibians and Reptiles. Society for the Study of Amphibians and Reptiles. Contributions to Herpetology, volume 11, Ithaca NY.

Celler, M. 1977. Breeding of Nile crocodiles in zoo conditions. Przeglad zool. 21(1):62-73 [in Polish].

Cerda, A., and D. Waugh. 1992. Status and management of the Mexican box terrapin *Terrapene coahuila* at the Jersey Wildlife Preservation Trust. Dodo, J. Jersey Wildl. Preserv. Trust 28:126-142.

Chaffee, P. S. 1969. Artificial incubation of alligator eggs *Alligator mississippiensis* at Fresno Zoo. Inter. Zoo Yearb. 9:34.

Cherlin, V. A. 1985. The reproduction of *Echis multisquamatus* in the Tashkent Zoo, p. 232-235. Problems in Herpetology: 6th All-Union Herpetological Conference Leningrad: Nauka.

Chiszar, D., and C. W. Radcliffe. 1989. The predatory strike of the jumping viper (*Porthidium nummifera*). Copeia 1989:1037-1039.

Chiszar, D., and H. M. Smith. 2005. Some comments on our herpetological collaborations with zoos. Herpetol. Rev. 36:7-9.

Chiszar, D., J. B. Murphy, and H. M. Smith. 1993. In search of zoo-academic collaborations: a research agenda for the 1990's. Herpetologica 49:488-500.

Chiszar, D., H. M. Smith, and C. C. Carpenter. 1994. An ethological approach to reproductive success in reptiles, p. 147-173. In J. B. Murphy, K. Adler, and J. T. Collins (eds.), Captive Management and Conservation of Amphibians and Reptiles. Society for the Study of Amphibians and Reptiles. Contributions to Herpetology, volume 11, Ithaca NY.

Chiszar, D., H. M. Smith, and J. B. Murphy. 1992. Excitation and inhibition of reproductive events in captive reptiles, 95-102. In M. J. Uricheck (ed.),

15th International Herpetological Symposium on Captive Propagation and Husbandry. International Herpetological Symposium, Palo Alto CA.

Chiszar, D., D. Mehaffey, F. Antonio, and H. M. Smith. 1994. Strike-induced chemosensory searching in eastern green mambas (*Dendroaspis angusticeps*). Bull. Maryland Herpetol. Soc., 30, 149-156.

Chiszar, D., J. B. Murphy, C. W. Radcliffe, and H. M. Smith. 1989. Bushmaster (*Lachesis muta*) predatory behavior at Dallas Zoo and San Diego Zoo. Bull. Pyschon. Soc. 27:459-461.

Chiszar, D., D. Boyer, R. Lee, R., J. B. Murphy, and C. W. Radcliffe. 1990. Caudal luring in the southern death adder (*Acanthophis antarcticus*). J. Herpetol. 24:253-260.

Chiszar, D., W. T. Tomlinson, H. M. Smith, J. B. Murphy, and C. W. Radcliffe. 1995. Behavioural consequences of husbandry manipulations: Indicators of arousal, quiescence and environmental awareness, p. 186-204. In C. Warwick, F. L. Frye, and J. B. Murphy (eds.), Health and Welfare of Captive Reptiles. Chapman & Hall, New York.

Chiszar, D., B. O'Connell, R. Greenlee, B. Demeter, T. Walsh, J. Chiszar, K. Moran, and H. M. Smith. 1985. Duration of strike-induced chemosensory searching in long-term captive rattlesnakes at National Zoo, Audubon Zoo, and San Diego Zoo. Zoo Biol. 4:291-294.

Christie, W. 1982. Successful introduction technique of three incompatible male lace monitors at the Indianapolis Zoo, p. 206-210. In D. L. Marcellini (ed.), 6th International Herpetological Symposium on Captive Propagation and Husbandry. International Herpetological Symposium, Thurmont MD.

Chun, W. C. H., and K. R. Beaman. 2003. Sean McKeown (1944-2002). Herpetol. Rev. 34:9-11.

Claffey, O., and B. Johnson. 1983. Captive reproduction of the sheltopusik (*Ophisaurus apodus*). Inter. Zoo News 30(4):19-23.

Clarke, G. K. 1973. Longevity chronicle of a Great Plains rat snake, *Elaphe guttata emoryi*. HISS NJ 1(4):121-122.

Clum, N. J., M. P. Fitzpatrick, and E. S. Dierenfeld. 1996. Effects of diet on nutritional content of whole vertebrate prey. Zoo Biol. 15:525-537.

Coburn, J. 1975. Post-mortem removal and artificial hatching of rainbow lizard eggs, *Agama agama*. Inter. Zoo Yearb. 15:92-94.

Collins, D. E. 1984. Captive breeding and management of the Aldabra tortoise *Geochelone gigantea*, p. 76-84. In R. A. Hahn (ser. ed.), 8th International Herpetological Symposium on Captive Propagation and Husbandry. International Herpetological Symposium, Thurmont MD.

Collins, D. E. 1989. Western New York bog turtles perspectives on captive propagation, p. 17-23. In R. Gowen (ed.), Captive Propagation and Husbandry of Reptiles and Amphibians. Special Publication # 5 ed.: Northern California Herpetological Society.

Collins, D. E., and G. George. 1992. Establishing a North American turtle collection at the Tennessee Aquarium, p. 1-9. In 16th International Herpetological Symposium on Captive Propagation and Husbandry: International Herpetological Symposium.

Conant, R. 1933. Three generations of cottonmouths, *Agkistrodon piscivorus* (Lacépède). Copeia 1933:43.

Conant, R. 1934. Observations on the eggs and young of the black king snake, *Lampropeltis getulus nigra* (Yarrow). Copeia 1934:188-189.

Conant, R. 1938. A note on the eggs and young of *Leioheterodon madagascariensis* (Duméril & Bibron). Zoologica (New York) 23:389-392.

Conant, R. 1942. Notes on the young of three recently described snakes, with comments upon their relationships. Bull. Chicago Acad. Sci. 6:193-200.

Conant, R. 1957. Arthur Erwin Brown: "Custodian of the Garden" and naturalist of note. America's First Zoo 9(4):3 pp.

Conant, R. 1965. Notes on reproduction in two natricine snakes from Mexico. Herpetologica 21:140-144.

Conant, R. 1980. The reproductive biology of reptiles: An historical perspective, p. 3-18. In J. B. Murphy, and J. T. Collins (eds.), Reproductive Biology and Diseases of Captive Reptiles. Society for the Study of Amphibians and Reptiles. Contributions to Herpetology, volume 1, Lawrence KS.

Conant, R. 1993. The oldest snake. Bull. Chicago Herpetol. Soc. 28(4):77-78.

Conant, R. 1997. A Field Guide to the Life and Times of Roger Conant. Selva, Tyler TX.

Condie, T., and P. Monseur. 1997. Komodo dragons at Miami Metrozoo (*Varanus komodoensis*). Anim. Keepers Forum 24(8):353-356.

Connaughton, S. W., and F. L. Paine. 1989. Captive management and reproduction in the Venezuelan slider turtle *Pseudemys scripta chichiriviche* a new subspecies. Inter. Zoo Yearb. 28:62-65.

Connors, J. S. 1986. A captive breeding of the Great Basin gopher snake, *Pituophis malanoleucus deserticola*. Herpetol. Rev. 17(1):12-13.

Connors, J. S. 1993. A long-term breeding programme for the beaded lizard *Heloderma horridum* at Detroit Zoo. Inter. Zoo Yearb. 32:184-188.

Conway, W. 1968. How to exhibit a bullfrog: A bed-time story for zoo men. Curator 11:310-318.

Conway, W. 1973. How to exhibit a bullfrog: A bed-time story for zoo men. Inter. Zoo Yearb. 13:221-226.

Conway, W. 1985. The Species Survival Plan and the Conference on Reproductive Strategies for Endangered Wildlife. Zoo Biol. 4:219-223.

Cooper, J. E. 1973. Treatment of necrotic stomatitis at the Nairobi Snake Park. Inter. Zoo Yearb. 13:268-269.

Cooper, J. E., Q. M. C. Bloxam, and S. Tonge. 1998. Pathology of Round Island geckos *Phelsuma guentheri*: some unexpected findings. Dodo, J. Jersey Wildl. Preserv. Trust 34:153-158.

Cooper, W. E., Jr. 1989. Prey odor discrimination in the varanoid lizards *Heloderma suspectum* and *Varanus exanthematicus*. Ecology 81:250-258.

Cooper, W. E., Jr. 1989. Absence of prey odor discrimination by iguanid and agamid lizards in applicator tests. Copeia 1989:472-478.

Cooper, W. E., Jr. 2000. Correspondence between diet and food chemical discriminations by omnivorous geckos (*Rhacodactylus*). J. Chem. Ecol. 26:755-763.

Cooper, W. E., Jr. 2000. Food chemical discriminations by the omnivorous lizards *Tiliqua scincoides* and *Tiliqua rugosa*. Herpetologica 56:480-488.

Cooper, W. E., Jr., and J. J. Habegger. 2000. Lingually mediated discrimination of prey, but not plant chemicals, by the Central American anguid lizard, *Mesaspis moreletii*. Amphibia-Reptilia 22:81-90.

Cooper, W. E., Jr., and J. J. Habegger. 2000. Lingual and biting responses to food chemicals by some eublepharid and gekkonid geckos. J. Herpetol. 34:360-368.

Cooper, W. E., Jr., and R. Hartdegen. 1999. Discriminative response to animal, but not plant, chemicals by an insectivorous, actively foraging lizard *Scincella lateralis* and differential response to surface and internal prey cues. J. Chem. Ecol. 25:1531-1541.

Cooper, W. E., Jr., and R. Hartdegen. 2000. Lingual and biting responses to prey chemicals by ingestively naive lizards: Discrimination from control chemicals, time course, and effect of method of presentation. Chemoecology 10:51-58.

Cooper, W. E., Jr., C. S. DePerno, and J. Arnett. 1994. Prolonged post-bite elevation in tongue-flicking rate with rapid onset in the gila monster, *Heloderma suspectum*: Relation to diet and foraging and implications for evolution of chemosensory searching. J. Chem. Ecol. 20:2587-2601.

Coote, J. G., 2001. A history of western herpetoculture before the 20th century, p. 19-47. In W. E. Becker (ed.). 25th International Herpetological Symposium on Captive Propagation and Husbandry. International Herpetological Symposium, Detroit MI.

Cornish, C. J. 1894. Wild Animals in Captivity or, Orpheus at the Zoo and Other Papers. Macmillan and Co., New York [pp. 263-269.].

Corwin, W. 1986. The reproductive behavior of two Australian chelid turtles, *Emydura macquarii* and *Elseya latisternum*, at the Dallas Zoo, p. 121-124. In S. McKeown, F. Caporaso, and K. H. Peterson (eds.), 9th International Herpetological Symposium on Captive Propagation and Husbandry. Zoological Consortium, Inc., Thurmont MD.

Cover, J. F., Jr., 1982. Captive maintenance and propagation of the eyelash viper *Bothrops schlegeli*, p. 304-322. In D. L. Marcellini (ed.), 6th International HerpetologicalSymposium on Captive Propagation and Husbandry. International Herpetological Symposium, Thurmont MD.

Cover, J. F. Jr., 1992. Poison-dart frogs - wild & captive, p. 1-4. In M. J. Uricheck (ed.), 15th International Herpetological Symposium on Captive Propagation and Husbandry. International Herpetological Symposium, Palo Alto CA.

Cover, J. F., Jr., and D. M. Boyer. 1988. Captive reproduction of the San Francisco garter snake *Thamnophis sirtalis tetrataenia*. Herpetol. Rev. 19(2):29-30, 32-33.

Cover, J. F., Jr., S. L. Barnett, and R. L. Saunders. 1994. Captive management and breeding of dendrobatid and Neotropical hylid frogs at the National Aquarium in Baltimore, p. 267-273. In J. B. Murphy, K. Adler, and J. T. Collins (eds.), Captive Management and Conservation of Amphibians and Reptiles. Society for the Study of Amphibians and Reptiles. Contributions to Herpetology, volume 11, Ithaca NY.

Cowles, R. B., and C. M. Bogert. 1944. A preliminary study of the thermal requirements of desert reptiles. Bull. Amer. Mus. Nat. Hist. 83:265-296.

Cranfield, M. R., and T. K. Graczyk. 1995. An update on ophidian cryptosporidiosis. Proc. Amer. Assoc. Zoo Vet. 1995:225-230.

Cree, A., C. H. Daugherty, D. R. Towns, and B. Blanchard. 1994. The contribution of captive management to the conservation of tuatara (*Sphenodon*) in New Zealand, p. 377-385. In J. B. Murphy, K. Adler and J. T. Collins (eds.), Captive Management and Conservation of Amphibians and Reptiles. Society for the Study of Amphibians and Reptiles. Contributions to Herpetology, volume 11. Ithaca NY.

Cunningham, B. 1937. Axial Bifurcation in Serpents. Duke University Press, Durham NC.

Curl, D. A., J. C. Scoones, and M. K. Guy. 1985. The Madagascar tortoise *Geochelone yniphora*: Current status and distribution. Biol. Conserv. 34:35-54.

Czernay, S., and G. Praedicow. 1988. Haltung und Zucht der Spornschildkröte (*Testudo* [*Geochelone*] *sulcata*) im Thüringer Zoopark Erfurt (Care and breeding of the African spurred tortoise (*Testudo* [*Geochelone*] *sulcata*) in the Thuringia Zoo at Erfurt). Zool. Gart. (N.F.), Jena 58:281-305.

Daltry, J. C., Q. Bloxam, G. Cooper, M. L. Day, J. Hartley, M. Henry, K. Lindsay, and B. E. Smith. 2001. Five years of conserving the 'world's rarest snake', the Antiguan racer *Alsophis antiguae*. Oryx 35:119-127.

Daly, J. W., Jr., H. M. Garraffo, T. F. Spande, C. Jaramillo, and A. S. Rand. 1994. Dietary source for skin alkaloids of poison frogs (Dendrobatidae). J. Chem. Ecol. 20:943-955.

Daly, J. W., Jr., S. I. Secunda, H. M. Garraffo, T. F. Spande, A. Wisnieski, C. Nishihira, and J. F. Cover, Jr. 1992. Variability in alkaloid profiles in neotropical poison frogs (Dendrobatidae): genetic versus environmental determinants. Toxicon 30:887-898.

Darlington, A. F., and R. B. Davis. 1990. Reproduction in the pancake tortoise, *Malacochersus tornieri*, in captive collections. Herpetol. Rev. 21(1):16-18.

Darwin, C. 1871. The Descent of Man and Selection in Relation to Sex [reprinted by The Modern Library, New York, undated].

Darwin, C. 1872. The Expression of the Emotions in Man and Animals [reprinted in 1965 by University of Chicago Press, Chicago].

Daszkiewicz, P. 2001. 'Moult of the Serpens [*SIC*], Their Laying, Their Dissection'. An interesting document for the history of European herpetology by Georg Segerus, physician to the Polish kings. Herpetol. Bull. 78:3-6.

Dathe, F. 1990. Das Haus für Krokodile, Kolibris und Schildkröten im Tierpark Berlin (The building for crocodiles, hummingbirds and turtles in the Berlin Tierpark). Aquar. Terrar. 37(10):362-364.

Dathe, F. 1997. Pflege und Vermehrung der Roten Speikobra (*Naja pallida* Boulenger, 1896) im Tierpark Berlin-Friedrichsfelde (Care and reproduction of the red spitting cobra [*Naja pallida* Boulenger, 1896] in the Berlin Tierpark). Milu 9:144-150.

Dathe, F. 2000. Die Haltung von Stumpfkrokodilen, *Osteolaemus tetraspis* Cope, 1861, im Tierpark Berlin-Friedrichsfelde von 1956 bis 2000 (The care of dwarf crocodiles, *Osteolaemus tetraspis* Cope, 1861, in the Berlin-Friedrichsfelde Tierpark from 1956 to 2000). Milu 10:63-70.

Dathe, F. 2001. Pflege und Vermehrung von Amboina-Scharnierschildkröten, *Cuora amboinensis* (Daudin, 1802), im Tierpark (Care and reproduction of Amboina-hinge turtles, *Cuora amboinensis* [Daudin, 1802] in the Tierpark). Milu 10:443-452.

Dathe, F. 2003. Pflege und Vermehrung der Ägyptischen Landschildkröte, *Testudo kleinmanni* (Lortet, 1883), im Tierpark Berlin-Friedrichsfelde (Care and reproduction of the Egyptian land tortoise, *Testudo kleinmanni* [Lorlet, 1883] in the Tierpark Berlin-Friedrichsfelde). Milu 11:156-169.

Dathe, F., and K. Dedekind. 1985. Pflege und Zucht von Bairds Erdnattern (*Elaphe obsoleta bairdi* Yarrow, 1880) im Tierpark Berlin (Care and breeding of Baird's ratsnakes [*Elaphe obsoleta bairdi* Yarrow, 1880] in the Tierpark Berlin). Milu 6:11-17.

Dathe, F., and K. Dedekind. 1988. Haltung und Zucht von Hufeisennattern (*Coluber hippocrepis* Linné, 1758) im Tierpark Berlin (Care and breeding of horse shoe snakes [*Coluber hippocrepis* Linné, 1758] in the Tierpark Berlin). Aquar. Terrar. 35:349-351.

Dathe, F., and K. Dedekind. 1991. Erfahrungen bei der Haltung und Zucht von Venezuela-Baumsteigern, *Colostethus trinitatis* (Boulenger, 1889) im Tierpark Berlin (Experiences gained in the care and breeding of Venezuelan tree climbers *Colostethus trinitatis* [Boulenger, 1889] in the Tierpark Berlin). Amphibienforschung und Vivarium (Amphibian Research and Design). Schleusingen :22-24.

Dathe, F., and K. Dedekind. 1996. Pflege und Vermehrung von Madagaskar-Hakennattern (*Leioheterodon madagascariensis* Duméril & Bibron, 1854) im Tierpark Berlin-Friedrichsfelde (Care and propagation of Madagascar hog-nosed snakes [*Leioheterodon madagascariensis* Duméril & Bibron, 1854] in the Berlin Tierpark). Zool. Gart. (N.F.), Jena 66:69-76.

Dathe, F., and K. Dedekind. 2002. Pflege und Vermehrung von Prachtskinken, *Riopa fernandi* (Burton, 1836), im Tierpark Berlin-Friedrichsfelde (Care and reproduction of fire skinks, *Riopa fernandi* [Burton, 1936], in the Berlin-Friedrichsfelde Zoo). Zool. Gart. (N.F.) 72:360-371.

Dathe, H. 1942. Zur Geburt eines Panthergeckos (*Gekko gecko* L.) (About the birth of a panther gecko [*Gekko gecko* L.]). Zool. Gart. (N.F.), Leipzig 14:211-212.

Dathe, H. 1960. Schwanz-Regeneration beim Brillenkaiman (Tail regeneration in caimans). Natur. u. Volk 90:289-292.

Davenport, M. 1979. Review of husbandry and propagation procedures for captive crocodilians, p. 2-5. In R. A. Hahn (ser. ed.), 1st & 2nd International Herpetological Symposium on Captive Propagation and Husbandry. International Herpetological Symposium, Thurmont MD.

Davenport, M. L. 1982. Notes of *Alligator sinensis* and *Crocodylus cataphractus* (reproduction and aging). Herpetol. Rev. 13(3):94-95.

Davenport, M. 1995. Evidence of possible sperm storage in the caiman, *Paleosuchus palpebrosus*. Herpetol. Rev. 26(1):14-15.

David, R. 1970. Breeding the mugger crocodile and water monitor *Crocodylus palustris* and *Varanus salvator* at Ahmedabad Zoo. Inter. Zoo Yearb. 10:116-117.

Davis, R., R. Darling, and A. Darlington. 1986. Ritualized combat in captive Dumeril's monitors, *Varanus dumerili*. Herpetol. Rev. 17(4):85-88.

Davis, S. 1979. Husbandry and breeding of the red-footed tortoise *Geochelone carbonaria* at the National Zoological Park, Washington. Inter. Zoo Yearb. 19:50-53.

Day, M., M. Breuil, and S. M., Reichling. 2000. Species account: Lesser Antillean iguana, *Iguana delicatissima* . In A.C. Alberts (ed.), West Indian Iguanas: Status Survey and Conservation Action Plan. Gland, Switzerland: IUCN-The World Conservation Union.

De Courcy, C. 2001. Zoological gardens of Australia, p. 181-214. In V. N Kisling Jr. (ed.), Zoo and Aquarium History. Ancient Animal Collections to Zoological Gardens. CRC Press, Boca Raton, FL; London; New York; Washington DC.

Dedekind, K. 1977. Beobachtungen beim Schlupf und bei der Aufzucht eines Grünen Leguans, *Iguana iguana*, im Tierpark Berlin (Observations made during hatching and raising of a common iguana, *Iguana iguana*, in the Berlin Tierpark). Zool. Gart. (N.F.), Jena 47:413-418.

Dedekind, K., and H.-G. Petzold. 1982. Zur Haltung und Zucht der Hinterindischen Wasseragame (*Physignathus cocincinus* Cuvier, 1829) im Tierpark Berlin (Keeping and breeding of the water dragon [*Physignathus cocincinus* Cuvier, 1829] in Berlin Tierpark). Zool. Gart. (N.F.), Jena 52:29-45.

Demeter, B. J. 1976. Observations on the care, breeding, and behaviour of the giant day gecko *Phelsuma madagascariensis* at the National Zoological Park, Washington. Inter. Zoo Yearb. 16:130-133.

Demeter, B. J. 1981. Captive maintenance and breeding of the Chinese water dragon (*Physignathus cocincinus*) at the National Zoological Park. p. 122-132. In R. A. Hahn (ser. ed.), 5th International Herpetological Symposium on Captive Propagation and Husbandry, Thurmont MD.

Demeter, B. J. 1986. Combat behavior in the Gila monster (*Heloderma suspectum cinctum*). Herpetol. Rev. 17(1):9-11.

Demeter, B. J., and G. F. Birchard. 1994. Long-term breeding program with the giant day gecko (*Phelsuma madagascariensis*), p. 311-315. In J. B. Murphy, K. Adler, and J. T. Collins (eds.), Captive Management and Conservation of Amphibians and Reptiles. Society for the Study of Amphibians and Reptiles. Contributions to Herpetology, volume 11, Ithaca NY.

Demeter, B. J., and D. M. Marcellini. 1981. Courtship and aggressive behavior of the streak lizard *Gonatodes vittatus* in captivity. Herpetologica 37:250-256.

Densmore, L. D., E. D. Pliler, and H. E. Lawler. 1994. A molecular approach for determining genetic variation in captive and natural populations of the Piebald Chuckwalla (*Sauromalus varius*), p. 343-351. In J. B. Murphy, K. Adler, and J. T. Collins (eds.), Captive Management and Conservation of Amphibians and Reptiles. Society for the Study of Amphibians and Reptiles. Contributions to Herpetology, volume 11, Ithaca NY.

Deschanel, J. P. 1978. Reproduction of anacondas *Eunectes murinus* at Lyons Zoo. Inter. Zoo Yearb. 18:98-99.

Dickinson, J. L., and W. D. Koenig. 2003. Desperately seeking similarity. Science 300: 1887-1889.

Dickson, J. 1992. Maintenance and breeding the pancake tortoise at Bristol Zoo. Inter. Zoo News 39(3):29-34.

Dierenfeld, E. S., O. A. Leontyeva, W. B. Karesh, and P. P. Calle. 1999. Circulating à-tocopherol and retinol concentrations in free-ranging and zoo turtles and tortoises. Proc. Amer. Assoc. Zoo Vet. 1999:329-331.

Ditmars, R. L. 1910. Reptiles of the World; The Crocodilians, Lizards, Snakes, Turtles and Tortoises of the Eastern and Western Hemispheres, by

Raymond L. Ditmars. With a Frontispiece in Color, and Nearly 200 Illustrations, From Photographs Taken by the Author. Sturgis and Walton, New York.

Ditmars, R. L. 1933. Reptiles of the World; The Crocodilians, Lizards, Snakes, Turtles and Tortoises of the Eastern and Western Hemispheres. Macmillan, New York [revised edition].

Ditmars, R. L. 1951. The Reptiles of North America. Doubleday, Garden City NY.

Dobbs, J. S. 1967. The feeding of dead food to a king cobra *Ophiophagus hannah*. Inter. Zoo Yearb. 7:229.

Dobroruka. L. J. 1973. Pellet formation in the Japanese giant salamander *Megalobatrachus japonicus*. Inter. Zoo Yearb. 13:158.

Dodd, C. K. Jr., and R. A. Seigel. 1991. Relocation, repatriation and translocation of amphibians and reptiles: Are they conservation strategies that work? Herpetologica 47:336-350.

Doering, Z. 1994. From reptile houses to reptile discovery centers. Institutional Studies Smithsonian Institution Report 94-4.

Doi, T., H. Tsukamoto, and S. Aoyama. 1999. An attempt at breeding the Japanese giant salamander, *Andrias japonicus,* under indoor artificial conditions. J. Japan. Assoc. Zoos Aquar. 40(2):56-64 [description in Japanese with English summary].

Donlan, J., et al. 2005. Re-wilding North America. Nature 436:913-914.

Dowling, H. G. 1960. Thermistors, air-conditioners, and insecticides. Anim. Kingdom 63(6):206-207.

Dowling, H. G., and P. Brazaitis. 1966. Size and growth of captive crocodilians. Inter. Zoo Yearb. 6:265-270.

Duméril, A. H. A. 1842. Sur le développement de la chaleur dans les oeufs des serpents, et sur l'influence attribuée à l'incubation de la mere; par M. Duméril (Concerning the production of heat in the eggs of snakes, and the influence attributed to the incubation of the mother; by M. Duméril). Comptes Rendus 14:193-210.

Duméril, A. H. A. 1852. Influence exercée sur la temperature des Ophidiens par l'échauffement du milieu qu'ils habitant (Effects on the temperature of ophidians by heating their ambient environment). Annales d. Sci. Nat. Zool. 3rd ser.:17-22.

Duméril, A. H. A. 1854-55. Notice historique sur la ménagerie des reptiles du Muséum d'histoire naturelle et observations qui y ont été recueillies (Historical notice on the menagerie of reptiles of the Museum of Natural History and observations collected there). Archives du Muséum d'histoire naturelle 7:193-320.

Duméril, A. H. A. 1858-61. Deuxième notice (Second notice). Archives du Muséum d'histoire naturelle (Archives of the Museum of Natural History) 10:429-460 [longevities of specimens in Menagerie].

Duméril, A. H. A. 1858. (Minutes). Comptes Rendus 47:525-526.

Dunn, R. W. 1976. Breeding the elongate tortoise *Testudo elongata* at Melbourne Zoo. Inter. Zoo Yearb. 16:73-74.

Dunn, R. W. 1977. Notes on the breeding of Johnstone's crocodile, *Crocodylus johnsoni*. Inter. Zoo Yearb. 17:130-31.

Dunn, R. W. 1979. Breeding Children's pythons *Liasis childreni* at Melbourne Zoo. Inter. Zoo Yearb. 19:89-90.

Dunn, R. W. 1981. Breeding the estuarine crocodile *Crocodylus porosus* at Melbourne Zoo. Inter. Zoo Yearb. 21:79-81.

Dunn, R. W. 1981. Further observations on the captive reproduction of Johnstone's crocodile *Crocodylus johnsoni* at Melbourne Zoo. Inter. Zoo Yearb. 21:82-83.

Dunn, R. W., Banks, C. B., and J. R. Birkett. 1987. Exhibiting and breeding the Arafura File Snake, *Acrochordus arafurae*. Inter. Zoo Yearb. 26:98-103.

Durbin, J., V. Rzjafetra, D. Reid, and D. Razandrizanakanirina. 1996. Local people and Project Angonoka - conservation of the ploughshare tortoise in north-western Madagascar. Oryx 30:113-120.

Durrell, L., B. Groombridge, S. Tonge, and Q. Bloxam. 1989. *Geochelone yniphora,* plow share tortoise, ploughshare tortoise, angulated tortoise, Angonoka, p. 99-102. In I. R. Swingland and M. W. Klemens (eds.). The Conservation Biology of Tortoises. Occasional Papers of IUCN Species Survival Commission (SSC) No. 5. IUCN-The World Conservation Union, Gland, Switzerland.

Duval, J. J. 1982. Recommendations for the captive management of West Indian rock iguanas, (*Cyclura*), p. 181-196. In D. L. Marcellini (ed.), 6th International Herpetological Symposium on Captive Propagation and Husbandry. International Herpetological Symposium, Thurmont MD.

Duval, J. J., and W. D. Christie. 1990. Husbandry of the Cuban ground iguana *Cyclura n. nubila* at the Indianapolis Zoo. Inter. Zoo Yearb. 29:65-69.

Edwards, J. 1996. London Zoo from Old Photographs 1852-1914. John Edwards, London.

Edwards, M. S. 1969. Notes on some tropidophid snakes in captivity. Inter. Zoo Yearb. 9:53-54.

Eidenmüller, B., and R. Wicker. 1991. Einige Beobachtungen bei der Pflege und Nachzucht von *Varanus (Odatria) timorensis similis* Mertens, 1958 (Some observations regarding the care and breeding of *Varanus [Odatria] timorensis similis* Mertens, 1958). Salamandra 27:187-193.

Ellis, J. F., Jr., and G. A. Ellis. 2001. Zoological gardens of South America, p. 351-368. In V. N. Kisling (ed.), Zoo and Aquarium History. Ancient Animal Collections to Zoological Gardens. CRC Press, Boca Raton, FL; London; New York; Washington DC.

Elsey, R. M., T. Joanen, L. McNease, and V. Lance. 1990a. Growth rate and plasma corticosterone levels in juvenile alligators maintained at different stocking densities. J. Exper. Zool. 255:30-36.

Elsey, R. M., T. Joanen, L. McNease, and V. Lance. 1990b. Stress and plasma corticosterone levels in the American alligator: Relationships with stocking density and reproductive success. Comp. Biochem. Physiol. 95A:55-63.

Engelmann, W.-E. 1984. Zum Eiablageverhalten der Kükennatter (*Elaphe obsoleta quadrivittata*) (Concerning egg-laying behavior of the chicken snake [*Elaphe obsoleta quadrivittata*]). Zool. Gart. (N.F.), Jena 54:209-210.

Engelmann, W.-E. 1986. Zum Brutpflegeverhalten des Dunklen Tigerpythons (*Python molurus bivittatus)* (Concerning the breeding behavior of the dark tiger python [*Python molurus bivittatus*]). Elaphe, Aquaristisch-terraristische Beiträge (No. 4):64-65.

Engelmann, W.-E. 2001. Keeping and breeding of dwarf crocodiles (*Osteolaemus tetraspis*) in the Leipzig Zoo Aquarium. Bull. Inst. Oceanogr. Monaco 20:331-336.

Engelmann, W.-E., and F. J. Obst. 1982. Snakes: Biology, Behavior and Relationship to Man. Exeter Books, New York.

Espinosa, L. D., M. del Socorro Morales, and M. I. Linares. 1996. Cytogenetic technique for karyotyping in *Heloderma horridum horridum* in the Guadalajara Zoo. Proc. Amer. Assoc. Zoo Vet. 1996:235-241.

Ettling, J. 1991. Captive husbandry and reproduction of the blood python at St. Louis Zoological Park, p. 107-110. In R. E. Staub (ed.), Captive Propagation and Husbandry of Reptiles and Amphibians. Northern California Herpetological Society.

Ettling, J. 1996. Natural history, husbandry, and captive reproduction of mountain vipers (*Vipera bornmuelleri* and *Vipera wagneri*), p. 139-144. In P. D. Strimple (ed.), Advances in Herpetoculture.

Special Publication of the International Herpetological Symposium, Inc., No. 1. International Herpetological Symposium, Inc., Des Moines IA.

Ettling, J., and A. Marfisi. 2002. Male combat in two species of mountain vipers, *Montivipera raddei* and *M. wagneri,* p. 163-166. In G. W. Schuett, M. Höggren, M. E. Douglas, and H. W. Greene (eds.), Biology of the Vipers. Eagle Mountain Publishing, Eagle Mountain CO.

Evans, L. T., and J. V. Quaranta. 1951. A study of the social behavior of a captive herd of giant tortoises. Zoologica (New York) 36:171-181.

Ewert, M. A., R. L. Hatcher, and J. M. Goode. 2004. Sex determination and ontogeny in *Malacochersus tornieri*, the pancake tortoise. J. Herpetol. 38:291-295.

Felkle, L. 1964. On some cases of reproduction of snakes in the Zoological Garden, Wroclaw. Przegl. Zool. 4:365-369 [Polish with English summary].

Ferguson, G. W. 1994. Old world chameleons in captivity: Growth, maturity, and reproduction of Malagasy panther chameleons (*Chamaeleo pardalis*), p. 323-331. In J. B. Murphy, K. Adler, and J. T. Collins (eds.), Captive Management and Conservation of Amphibians and Reptiles. Society for Study of Amphibians and Reptiles. Contributions to Herpetology, volume 11, Ithaca NY.

Ferguson, G. W., J. B. Murphy, A. Raselemananana, and J.-B. Ramananmanjato. 2004. The Panther Chameleon: Color Variation, Natural History, Conservation and Captive Management. Krieger, Malabar FL.

Ferguson, G. W., J. R. Jones, W. H. Gehrmann, S. H. Hammack, L. G. Talent, R. D. Hudson, E. S. Dierenfeld, M. P. Fitzpatrick, F. L. Frye, M. F. Holick, T. C. Chen, Z. Lu, T. S. Gross, and J. J. Vogel. 1996. Indoor husbandry of the panther chameleon *Chamaeleo (Furcifer) pardalis:*Effects of dietary vitamins A and D and ultraviolet irradiation on pathology and life-history traits. Zoo Biol. 15:279-299.

Fischer, J. v. 1872. Über Schildkröten in der Gefangenschaft (Concerning turtles in captivity). Zool. Gart., Frankfurt a. M. 13:65-71, 13:116-120, 13:137-141.

Fischer, J. v. 1872. Über Krankheiten bei Schildkröten, deren Pflege und Verhütung (Treatment and prevention of illnesses in turtles). Zool. Gart., Frankfurt a. M. 13:193-202, 13:243-248.

Fischer, J. v. 1872. Über einige Schildkrötenarten in der Gefangenschaft und über Chelonier im

Allgemeinen (Concerning several species of turtles in captivity and about turtles in general). Zool. Gart., Frankfurt a. M. 13:321-327, 13:364-369.

Fischer, J. v. 1873. Über Lebensweise, Haltung und Pflege einiger Schildkröten (Habits, maintenance and care of some turtles). Zool. Gart., Frankfurt a. M. 14:241-252.

Fischer, J. v. 1873. Die Reptilien und Amphibien des St. Petersburger Gouvernements (The reptiles and amphibians of the Government of St. Petersburg). Zool. Gart., Frankfurt a. M. 14:324-328.

Fischer, J. v. 1874. Aufzucht junger Schildkröten (Raising young turtles). Zool. Gart., Frankfurt a. M. 15:262-264.

Fischer, J. v. 1879. Mein neues heizbares Terrarium für Reptilien (My new heatable terrarium for reptiles). Zool. Gart., Frankfurt a. M. 20:353-358.

Fischer, J. v. 1880. Die Ringelagame (*Oplurus torquatus*) in der Gefangenschaft (The ringed agama [*Oplurus torquatus*] in captivity). Zool. Gart., Frankfurt a. M. 21:16-20.

Fischer, J. v. 1882. Das Chamäleon (*Chamaeleo vulgaris*), sein Fang und Versandt, seine Haltung und seine Fortpflanzung in der Gefangenschaft (The chameleon [*Chamaeleo vulgaris*], its capture and transport, its care, and its reproduction in captivity). Zool. Gart., Frankfurt a. M. 23:4-13, 39-48, 70-82. [3 parts].

Fischer, J. v. 1882. Der Cap'sche Dornschweif (*Uromastix capensis auct.*) in der Gefangenschaft (The Cape spiny lizard [*Uromastix capensis auct.*] in captivity). Zool. Gart., Frankfurt a. M. 23:181-184.

Fischer, J. v. 1882. Die Stummelschwanz-Eidechse (*Trachydosaurus asper*) in der Gefangenschaft (The Stummelschwanz lizard [*Trachydosaurus asper*] in captivity). Zool. Gart., Frankfurt a. M. 23:206-210.

Fischer, J. v. 1882. Der Leguan (*Iguana tuberculata* Laur.) in der Gefangenschaft (The iguana [*Iguana tuberculata* Laur.] in captivity). Zool. Gart., Frankfurt a. M. 23:236-241.

Fischer, J. v. 1882. Fortpflanzung der Walzeneidechse (*Gongylus ocellatus* Wagl.) in der Gefangenschaft (Reproduction of the cylindrical lizard [*Gongylus ocellatus* Wagl.] in captivity). Zool. Gart., Frankfurt a. M. 23:241-243

Fischer, J. v. 1882. Die braune Peitschen- oder Baumschlange (*Oxybelis aeneus* Wagler) in der Gefangenschaft (The brown eastern coachwhip or tree snake [*Oxybelis aeneus* Wagler] in captivity). Zool. Gart., Frankfurt a. M. 23:331-336.

Fischer, J. v. 1883. Der australische Laubfrosch, *Pelotryas coeruleus* White (= *Hyla cyanea**) Daudin, in der Gefangenschaft (The Australian tree frog, *Pelotryas coeruleus* White [= *Hyla cyanea**] Daudin, in captivity). Zool. Gart., Frankfurt a. M. 24:21-25.

Fischer, J. v. 1883. Die Panther-Kröte in der Gefangenschaft (*Bufo pantherinus* Cichenot = *B. mauritanicus* Schlegel) (The panther toad [*Bufo pantherinus* Cichenot = *B. mauritanicus* Schlegel] in captivity). Zool. Gart., Frankfurt a. M. 24:43-45.

Fischer, J. v. 1884. Das Terrarium, seine Bepflanzung und Bevölkerung (The Terrarium, Its Plantings and Population). Mahlau & Waldschmidt, Frankfurt am Main. [Reprint of this work was published in 1989 by BINA Verlag für Biologie und Natur, Berlin (BINA Publisher for Biology and Nature, Berlin.)]

Fischer, J. v. 1884. Die Girondennatter in der Gefangenschaft (*Coronella girundica* Daud.) (The Gironde snake [*Coronella girundica* Daud.] in captivity). Zool. Gart., Frankfurt a. M. 25:145-148.

Fischer, J. v. 1884. Die Zwergschleiche (*Ablepharus pannonicus* Fitzinger) in der Gefangenschaft (The pygmy lateral fold lizard [*Ablepharus pannonicus* Fitzinger] in captivity). Zool. Gart., Frankfurt a. M. 25:314-316.

Fischer, J. v. 1885. Der veränderliche Schleuderschwanz (*Uromastix acanthinurus* Bell) in der Gefangenschaft (The changeable Schleuderschwanz, [*Uromastix acanthinurus* Bell] in captivity). Zool. Gart., Frankfurt a. M. 26:269-278. [The German word Schleuderschwanz means something like "throw- tail."]

Fischer, J. v. 1886. Die Brillensalamandrine (*Salamandrina perspicillata* Savi) in der Gefangenschaft (The Brillensalamandrine [*Salamandrina perspicillata* Savi] in captivity). Zool. Gart., Frankfurt a. M. 27:14-23.

Fischer, J. v. 1887. Der Blasius'sche Triton (*Triton Blasii* De l'Isle) und über die Haltung der europäischen Tritonen im allgemeinen (The Blasius Triton [*Triton Blasii* De l'Isle] and the care of European tritons in general). Zool. Gart., Frankfurt a. M. 28:11-20.

Fischer, J. v. 1887. Der Apotheker-Skink (*Scincus officinalis* Laur.) (The common skink [*Scincus officinalis* Laur.]). Zool. Gart., Frankfurt a. M. 28:309-314.

Fischer, J. v. 1888. Die Tüpfelechse, *Eremias pardalis* Dum. u. Bibron (The spotted lizard, *Eremias pardalis* Dum. and Bibron). Zool. Gart., Frankfurt a. M. 29:5-8.

Fischer, J. v. 1888. Nachtrag zur Naturgeschichte des

veränderlichen Schleuderschwanzes, *Uromastix acanthinurus* Bell (Appendix to the changeable Schleuderschwanz *Uromastix acanthinurus* Bell). Zool. Gart., Frankfurt a. M. 29:97-108.

Fischer, J. v. 1888. Der Bou-Rioun (*Lacerta pater* Lataste) und seine Verwandtschaft mit der Perleidechse (*L. ocellata* Daudin) und der Smaragdeidechse (*L. viridis* Daudin) (The Bou-Rioun [*Lacerta pater* Lataste] and its relationship to the eyed lizard [*L. ocellata* Daudin) and the emerald lizard (*L. viridis* Daudin]). Zool. Gart., Frankfurt a. M. 29:265-273.

Fischer, J. v. 1889. Die Wurmschleiche (*Trogonophis Wiegmanni* Kaup.) (The Worm Lizard – [*Trogonophis Wiegmanni* Kaup.]). Zool. Gart., Frankfurt a. M. 30:49-53.

Fitch, H. S. 1980. Reproductive strategies in reptiles, p. 25-31. In J. B. Murphy and J. T. Collins (eds.). Reproductive Biology and Diseases of Captive Reptiles. Society for the Study of Amphibians and Reptiles Contributions to Herpetology, volume 1. Lawrence KS.

Fitzinger, L. J. 1830. Über die Girafe (About the giraffe). Wiener Zeitschrift (Vienna Periodical) no. 118, 9/30/1828:957-961.

Fitzinger, L. J. 1830. Über den Krankheitszustand und Tod der Girafe in der k. k. Menagerie zu Schönbrunn (About the illness and death of the giraffe in the Imperial-Royal Menagerie at Schönbrunn). Isis, Jena 23(4):368-372.

Fitzinger, L. J. 1853. "Versuch einer Geschichte der Menagerien des Österreichisch-Kaiserlichen Hofes." Wien, 1853. "Aus dem März - und Aprilhefte des Jahrganges 1853 des Sitzungsberichtes der mathem.-naturw. Classe der kais. Akademie der Wissenschaften, besonders abgedruckt." ("Attempted History of the Zoological Gardens of the Austria Imperial Court." Vienna, 1853. "From the March and April Issues, 1853, of the Conference Reports of the Mathematics Natural History Class of the Imperial Academy of Sciences Printed Specially") Reptiles, pp. 109, 110, 135-147, 152, 188, 189.

Fitzinger, L. J. 1855. Bericht an die kaiserl. Akademie der Wissenschaften über die von dem Herrn Consulatverweser Dr. Theodor Heuglin für die kaiserl. Menagerie zu Schönbrunn mitgebrachten lebenden Thiere (Report to the Imperial Academy of Sciences about the living animals brought for the Imperial Academy at Schönbrunn by Consul Dr. Theodor Heuglin). Sitzungsberichte der mathematisch-naturwissen-schaftlichen Classe der kaiserlichen Akademie der Wissenschaften, Wien (Conference Report of the Mathematical-Natural Science Class of the Imperial Academy of Sciences, Vienna) 17(1-3):242-253.

Fitzinger, L. J. 1860. Die Ausbeute der österreichischen Naturforscher an Säugethieren und Reptilien während der Weltumsegelung Sr. Majestät Fregatte Novara (The exploration by Austrian natural scientists of mammals and reptiles during the World Tour of His Majesty's Frigate Novara). Sitzungsberichte der mathematisch-naturwissenschaftlichen Classe der kaiserlichen Akademie der Wissenschaften, Wien (Conference Report of the Mathematical-Natural Science Class of the Imperial Academy of Sciences) Section 1, Vienna 42:383-416.

Fitzinger, L. 1868. Geschichte des k. k. Hof-Naturalien-Cabinetes zu Wien. Periode unter Franz II. (Franz I. Kaiser von Österreich) bis zu Ende des Jahres 1815 (History of the Royal-Imperial Court Natural Science Cabinet in Vienna. Periods under Francis II [Francis I, Emperor of Austria] up to the year 1815). Sitzungsberichte der mathematisch-naturwissen-schaftlichen Classe der kaiserlichen Akademie der Wissenschaften, Wien (Conference Reports of the Mathematical-Natural Science Class of the Imperial Academy of Sciences) Section 1, Vienna 57:1013-1092.

Fitzinger, L. J. 1875. Die Kaiserliche Menagerie zu Schönbrunn. Eine populäre Schilderung sämmtlicher Thiere derselben (The Imperial Menagerie at Schönbrunn. A popular description of all animals kept there). Wilhelm Braumüller, Vienna.

Fitzinger, L. 1879. Der langhaarige gemeine Ferkelhase (Cavia Cobaya, longipilis.) (The long-haired guinea pig [Cavia Cobaya, longipilis]). Sitzungsberichte der mathematisch-naturwissenschaftlichen Classe der kaiserlichen Akademie der Wissenschaften, Wien (Conference Reports of the Mathematical-Natural Science Class of the Imperial Academy of Sciences) Section 1, Vienna 80:431-438.

FitzSimons, F. W. 1921. The Snakes of South Africa, Their Venom and Treatment of Snakebite. T. M. Miller, Cape Town, S.A. [Book was originally published in 1910.]

FitzSimons, F. W. 1921. Snakes and the Treatment of Snake Bite. T. M. Miller, Cape Town, S.A.

FitzSimons, F. W. 1930. Pythons and Their Ways. G. G. Harrap, London.

FitzSimons, F. W. 1932. Snakes. Hutchinson, London.

Flanagan, J. P., and G. M. Harwell. 1983. Pathobiology

and management of chronic regurgitation in snakes. Proc. Amer. Assoc. Zoo Vet. 1983:208-209.

Flower, S. S. 1896. Notes on a collection of reptiles and batrachians made in the Malay Peninsula in 1895-96; with a list of the species recorded from that region. Proc. Zool. Soc. London 1896:856-914.

Flower, S. S. 1899. Notes on a second collection of reptiles made in the Malay Peninsula and Siam, from November 1896 to September 1898, with a list of the species recorded from those countries. Proc. Zool. Soc. London 1899:600-696.

Flower, S. S. 1899. Notes on a second collection of batrachians made in the Malay Peninsula and Siam, from November 1896 to September 1898, with a list of the species recorded from those countries. Proc. Zool. Soc. London 1899:885-916.

Flower, S. S. 1900. Notes on the fauna of the White Nile and its tributaries. Proc. Zool. Soc. London 1900:950-973.

Flower, S. S. 1903. Zoological Gardens. Giza, near Cairo, Egypt. Plan and Guide-book. 2nd Edition. Al-Mokattam Printing Office, Cairo.

Flower. S. S. 1914. Ministry of Public Works, Egypt. Zoological Service. Report on a Zoological Mission to India in 1913. Government Press, Cairo.

Flower, S. S. 1921. Exhibition of, and remarks upon, living specimens of *Testudo leithii* and *T. ibera*. Proc. Zool. Soc. London 1921:645.

Flower, S. S. 1923. On additions to the snake fauna of Egypt. Proc. Zool. Soc. London 1923:1079-1083.

Flower, S. S. 1924. Exhibition of, and remarks upon, a remarkable tortoise of the genus *Testudo*. Proc. Zool. Soc. London 1924:920-921.

Flower, S. S. 1925. Contributions to our knowledge of the duration of life in vertebrate animals.– II. Batrachians. Proc. Zool. Soc. London 1925:269-289.

Flower, S. S. 1925. Contributions to our knowledge of the duration of life in vertebrate animals. III. Reptiles. Proc. Zool. Soc. London 1925:911-981.

Flower, S. S. 1927. Loss of memory accompanying metamorphosis in amphibians. Proc. Zool. Soc. London 1927:155-156.

Flower, S. S. 1928. Exhibition of living specimens of the great African tortoise *Testudo sulcata*. Proc. Zool. Soc. London 1928: 654.

Flower, S. S. 1928. Hints on the transport of animals. Proc. Zool. Soc. London 1928:631-692.

Flower, S. S. 1936. Further notes on the duration of life in animals. II. Amphibians. Proc. Zool. Soc. London 1936:369-394.

Flower, S. S. 1937. Further notes on the duration of life in animals. III. Reptiles. Proc. Zool. Soc. London 1937, 107 (ser. A):1-39.

Foley, S. C. 1998. Notes on the captive maintenance and reproduction of Oates' twig snake (*Thelotornis capensis oatesi*). Herpetol. Rev. 29(3).

Forbes, W. A. 1881. Observations on the incubation of the Indian python (*Python molurus*), with special regard to the alleged increase of temperature during the process. Proc. Zool. Soc. London 1881:960-967.

Fowler, M. E. 1979. Cobra snakebite in a keeper. Proc. Amer. Assoc. Zoo Vet.:8-11.

Fox, W., C. Gordon, and M. H. Fox. 1961. Morphological effects of low temperatures during the embryonic development of the garter snake, *Thamnophis elegans*. Zoologica (New York) 46(No. 5):57-71.

Frank, N., and E. Ramus. 1995. A Complete Guide to Scientific and Common Names of Reptiles and Amphibians of the World. N G Publishing, Pottsville PA.

Frazier, J., and G. Peters. 1981. The call of the Aldabran tortoise (*Geochelone gigantea*) (Reptilia, Testudinidae). Amphibia-Reptilia 2:165-179.

Freed, P. 1989. Delayed fertilization in the African egg eating snake, *Dasypeltis scabra*, including a brief bibliography of delayed fertilization in other species of snakes. Bull. Chicago Herpetol. Soc. 24:239-240.

Frelier, P. F., L. Sigler, and P. E. Nelson. 1985. Mycotic pneumonia caused by *Fusarium moniliforme* in an alligator. Sabouraudia 23(6):399-402.

Freytag, G. E., and H.-G. Petzold. 1978. Ein weiterer Beitrag zur Kenntnis der Gattung *Paramesotriton*, insbesondere des nordvietnamesischen Wassermolches *Paramesotriton deloustali* (Bourret 1934) (*Amphibia: Caudata: Salamandridae*) (An additional contribution to knowledge of the species *Paramesotriton*, especially of the North Vietnamese water moloch *Paramesotriton deloustali* [Bourret 1934] [*Amphibia: Caudata: Salamandridae*]). Salamandra 14(3):117-125.

Fritz, U. 1990. Haltung und Nachzucht der Jamaika-Schmuckschildkröte *Trachemys terrapen* (Lacépède, 1788) und Bemerkungen zur Fortpflanzungsstrategie von neotropischen Schmuckschildkröten der Gattung *Trachemys* (Care and breeding of the Jamaica pseudemid turtle *Trachemys terrapen* (Lacépède, 1788) and notes on the reproductive strategy of Neotropical pseudemid turtles of the genus *Trachemys*). Salamandra 26(1):1-18.

Fritz, U. 1993. Bemerkungen über das Werbeverhalten der Riesen-Schlangenhalsschildkröte (*Chelodina expansa*) (Notes on the courtship behavior of the giant snake necked turtle [*Chelodina expansa*]). Herpeto-fauna (Weist.) 15 No. 83:6-9.

Fritz, U., and D. Jauch. 1989. Haltung, Balzverhalten und Nachzucht von Parkers Schlangen-schildkröte *Chelodina parkeri* Rhodin & Mittermeier, 1976 (Mating behavior and reproduction of Parker's snake neck turtle *Chelodina parkeri* Rhodin & Mittermeier, 1976). Salamandra 25:1-13.

Fritz, U., D. Jauch, and H. Jes. 1991. Langzeit-Beobachtungen bei der Haltung und Nachzucht der Rotbauch-Spitzkopfschildkröte (*Emydura albertisii*) (Long-term observations concerning the care and breeding of the red-bellied sharp-snouted turtle [*Emydura albertisii*]). Z. Köln. Zoo 34(4):131-139.

Frolov, V. E. 1981. Fortpflanzung des Wundergeckos, *Teratoscincus s. scincus* (Schleg., 1858) im Moskauer Zoopark (Reproduction of the skink gecko, *Teratoscincus s. scincus*[(Schleg., 1858] in the Moscow Zoo). Zool. Gart. (N.F.), Jena 51:263-266.

Frolov, V. E. 1987. On *Teratoscincus scincus* reproduction in captivity. Vestnik Zool. 1987:86-87.

Frolov, V. Å., and S. V. Kudryavtsev. 1985. Keeping and breeding of Russian rat snakes at the Moscow Zoo. Sokhranim dikhikh zhivotnikh, Almaty, Kazakhstan, ð. 83-88.

Frolov, V. E., and L. A. Sanoyan. 1986. Maintenance and reproduction of the water dragon in the Moscow Zoo, p. 165-168. In A. F. Kovshar' (ed.), Sokhranimdikikh zhivotnykh (Let Us Preserve Wild Animals). Almar-Ata: Kaynar.

Frolov, V. Å., A. V. Korolev, and S.V. Kudryavtsev. 1982. Methods for captive maintenance of soft-shell turtles at Moscow Zoo. Razvedenie I sozdanie novikh populyatsii redkikh I tsennykh vidiv zhivotnykh, Àshkhabad, pð.177-180.

Frolov, V. Å., S. V. Kudryavtsev, and A. Gvozdarev. 1989. Keeping and breeding of Siamese crocodile in captivity. Ekologia I okhrana dikikh zhivotnykh, Ìîscow, p. 49–54.

Frolov, V. E., S. V. Kudryavtsev, S. I. Sapelkin, and V. A. Igolkina. 1986. Reproduction of the Paraguayan anaconda in the Moscow Zoo, p. 121-123. In A. F. Kovshar'(ed.), Sokhranim dikikh zhivotnykh (Let Us Preserve Wild Animals). Almar-Ata: Kaynar.

Fujitani, T., K. Nishio, and H. Hashikawa. 1998. Breeding the Amazonian poison frog, *Dendrobates ventrimaculatus*. Animals and Zoos 50(3. No.576):74-77 [in Japanese with English summary].

Furrer, S. C., J. M. Hatt, H. Snell, C. Marquez, R. E. Honegger, and A. Rübel. 2004. Comparative study on the growth of juvenile Galapagos giant tortoises (*Geochelone nigra*) at the Charles Darwin Research Station (Galapagos Islands, Ecuador) and Zoo Zurich (Zurich, Switzerland). Zoo Biol. 23:177-183.

Gadow, H. 1901. Amphibia and Reptiles. Macmillan, England [reprinted in 1968 by Wheldon & Wesley, Codicote UK].

Gamble, K. C., T. P. Alvarado, and C. L. Bennett. 1996. Plasma Itraconazole pharmacokinetics in spiny lizards (*Sceloporus* spp.). Proc. Amer. Assoc. Zoo Vet. 1996:245-246.

Gans, C., and A. M. Taub. 1964. Precautions for keeping venomous snakes in captivity. Curator 7(3):196-205.

Garner, M. M., D. Collins, and J. O. Joslin. 1995. Diseases of Solomon Island leaf frogs (*Ceratobatrachus guentheri*) at the Woodland Park Zoo: A retrospective study of seventy five necropsy cases. Proc. Amer. Assoc. Zoo Vet. 1995:236-237.

Garner, M. M., D. Collins, and J. Joslin. 1995. Vertebral chondrosarcoma in a corn snake. Proc. Amer. Assoc. Zoo Vet. 1995:332-333.

Garrett, C. M. 2005. Herpetological research in zoos: A contemporary assessment. Herpetol. Rev. 36:103-106.

Garrett, C. M., and B. E. Smith. 1994. Perch color preference in juvenile green tree pythons. Zoo Biol. 13:45-50.

Garrick, L. D. 1975. Structure and pattern of the roars of Chinese alligators (*Alligator sinensis* Fauvel). Herpetologica 31:26-31.

Gaulstaun, B. 1973. Eiablagen des Komodowarans (*Varanus komodoensis*) im Zoologischen und Botanischen Garten Jakarta (The depositing of eggs of the Komodo dragon [*Varanus komodoensis*] in the Zoological and Botanical Garden of Jakarta). Zool. Gart. (N.F.), Leipzig 43:136-139.

Gehrmann, W. H. 1971. Influence of constant illumination on thermal preference in the immature water snake, *Natrix erthythrogaster transversa*. Physiol. Zool. 44:84-89.

Gehrmann, W. H. 1987. Ultraviolet irradiances of various lamps used in animal husbandry. Zoo Biol. 6:117-127.

Gehrmann, W. H. 1994. Light requirements of captive amphibians and reptiles, p. 53-59. In J. B. Murphy, K. Adler, and J. T. Collins (eds.), Captive Management and Conservation of Amphibians and Reptiles. Society for the Study of Amphibians and Reptiles. Contributions to Herpetology, volume 11,

Ithaca NY.

Gens, E. 1861. Promenade au jardin zoologique d'Anvers par Eugène Gens. J.-E. Buschmann, Anvers (Antwerp, Belgium).

Gensch, W. 1969. Breeding boa hybrids *Constrictor c. constrictor* x *C. c. imperator* at Dresden Zoo. Inter. Zoo Yearb. 9:52.

George, G. 1987. An overview of the genus *Graptemys* with techniques in captive maintenance, p. 17-25. In R. Gowen (ed.), Captive Propagation and Husbandry of Reptiles and Amphibians. Northern California Herpetological Society.

Gerlach, J., and L. Canning. 1998. Taxonomy of Indian Ocean giant tortoises (*Dipsochelys*). Chelonian Conserv. Biol. 3:3-19.

Gewalt, W. 1977. Einige Bemerkungen über Fang, Transport und Haltung des Goliathfrosches (*Conraua goliath* Boulenger) (Some remarks on the catching, transport and keeping of the giant frog [*Conraua goliath* Boulenger]). Zool. Gart. (N.F.), Jena 47:161-192.

Gibson, R. C. 1993. A short study of captive Jamaican iguanas *Cyclura collei* at Hope Zoo. Dodo, J. Jersey Wildl. Proc. 29:156-167.

Gibson, R. C. 1997. Conservation of the Antiguan racer *Alsophis antiguae*: the captive component. International Herpetological Symposium on Captive Propagation and Husbandry. International Herpetological Symposium.

Gibson, R. C., and K. R. Buley. 1996. Captive management and breeding of Madagascar spiny iguanas *Oplurus cuvieri cuvieri* Gray, 1831. Dodo, J. Jersey Wildl. Preserv. Trust 32:137-143.

Gibson, R. C., and K. R. Buley. 2004. Maternal care and obligatory oophagy in *Leptodactylus fallax*: A new reproductive mode in frogs. Copeia 2004:128-135.

Gibson, R. C., and K. R. Buley. 2004. Biology, captive husbandry, and conservation of the Malagasy flat-tailed tortoise, *Pyxis planicauda* Grandidier, 1867. Herpetol. Rev. 35:111-116.

Giles, J. R., and J. D. Kelly. 1992. Conservation and research programme: proposals by the Zoological Parks Board of New South Wales. Inter. Zoo Yearb. 31:1-4.

Gillespie, D., B. L. Dresser, and E. J. Maruska. 1988/1989. Sexing goliath frogs *Gigantorana goliath* by laparoscopy. Proc. Amer. Assoc. Zoo Vet. 1989:62.

Gillingham, J. C., and T. R. Miller. 1991. Reproductive ethology of the tuatara *Sphenodon punctatus*: Applications to captive breeding. Inter. Zoo Yearb. 30:157-164.

Gillingham, J. C., C. C. Carpenter, and J. B. Murphy. 1983. Courtship, male combat and dominance in the western diamondback rattlesnake, *Crotalus atrox*. J. Herpetol. 17:265-270.

Gillingham, J. C., C. C. Carpenter, B. J. Brecke, and J. B. Murphy. 1977. Courtship and copulatory behavior of the Mexican milk snake, *Lampropeltis triangulum sinaloae* (Colubridae). Southwest. Natur. 22:187-194.

Glazebrook, R. S., and R. S. F. Campbell. 1990. A survey of the diseases of marine turtles in northern Australia. 2. Oceanarium-reared and wild turtles. Dis. Aquat. Org. 9(2):97-104.

Glenn, J. L., R. C. Straight, and D. Tuttle. 1973. The influence of Ketamine HCL on snakes during shipment. HISS NJ 1(1):29-30.

Goeldi, E.A. Relatorio 1894-1895, Boletim do Museu Paraense Tomo I. (fasc. 1-4), Belem, Brazil, 1894-1896, 221 (1894-1895 report).

Goetz, B. G. R., and B. W. Thomas. 1994. A study on the captive maintenance of tuatara (*Sphenodon punctatus*). Auckland, Manaaki Whenua-Landcare Research, Native Plants and Animals Division.

Goetz, B. G. R., and B. W. Thomas. 1994. Use of annual growth and activity patterns to assess management procedures for captive tuataras (*Sphenodon punctatus*). New Zealand J. Zool. 21(4):473-485.

Golding, R. R. 1965. Growth rate of a captive Indian python *Python molurus bivittatus*. Inter. Zoo Yearb. 5:166-167.

Goode, J. M. 1979. Notes on captive reproduction in *Echis colorata* (Serpentes: Viperidae). Herpetol. Rev. 10(3):94.

Goode, J. M. 1988. Reproduction and growth of the chelid turtle *Phrynops* (*Mesoclemmys*) *gibbus* at the Columbus Zoo. Herpetol. Rev. 19(1):11-13.

Goode, J. M. 1991. Breeding semi-aquatic and aquatic turtles at the Columbus Zoo, p. 66-76. In K. R. Beaman, F. Coporaso, S. McKeown, and M. D. Graff (eds.), Proceedings of the First International Symposium on Turtles and Tortoises: Conservation and Captive Husbandry. Chapman University, August 9-12, 1990. California Turtle and Tortoise Club, Van Nuys CA.

Goode, J. M. 1994. Reproduction in captive Neotropical musk and mud turtles (*Staurotypus triporcatus, S. salvini*, and *Kinosternon scorpiodes*), p. 275-295. In J. B. Murphy, K. Adler, and J. T. Collins (eds.), Captive Management and Conservation of Amphibians and Reptiles. Society for the Study of Amphibians and Reptiles. Contributions to Herpetology, volume 11, Ithaca NY.

Grant, C. 1947. Dr. Charles Haskins Townsend. Herpetologica 4:38-40.

Gray, C. W., L. C. Marcus, W. G. McCarten, and T. Sappington. 1966. Amoebiasis in the Komodo dragon *Varanus komodoensis*. Inter. Zoo Yearb. 6:279-283.

Green, J. 1981. Second hatching of the American alligator *Alligator mississippiensis* at the Australian Reptile Park, Gosford. Inter. Zoo Yearb. 21:76-77.

Greene, H. W. 1983. Dietary correlates of the origin and radiation of snakes. Amer. Zool. 23:431-441.

Greene, H. W. 1997. Snakes. The Evolution of Mystery in Nature. University of California Press, Berkeley, Los Angeles, London.

Greenhall, A. M. 1936. The care of the bushmaster and of certain lizards in the New York Zoological Park. Herpetologica 1:66-67.

Gregory, P. T. 1982. Reptilian hibernation, p. 53-154. In C. Gans and F. H. Pough (eds.). Biology of the Reptilia, Vol. 13. Physiology D. Physiological Ecology. Academic Press, London, New York.

Groves, F. 1957. Eggs and young of the corn snake in Maryland. Herpetologica 13:79-80.

Groves, F. 1960. The eggs and young of *Drymarchon corais couperi*. Copeia 1960:51-53.

Groves, F. 1973. Reproduction and venom in Blanding's tree snake *Boiga blandingi*. Inter. Zoo Yearb. 13:106-108.

Groves, J. D. 1994. Husbandry and reproduction of the prehensile-tailed skink, *Corucia zebrata*, p. 317-322. In J. B. Murphy, K. Adler, and J. T. Collins (eds.), Captive Management and Conservation of Amphibians and Reptiles. Society for the Study of Amphibians and Reptiles. Contributions to Herpetology, volume 11, Ithaca NY.

Groves, J. D., and J. R. Mellendick. 1973. Breeding the Madagascan boa *Sanzinia madagascariensis* at Baltimore Zoo. Inter. Zoo Yearb. 13:106.

Grow, D. T. 1980. Reptile reproduction and husbandry of the orange striped poison dart frog, (*Phyllobates vittutus*), at the Sedgwick County Zoo, Wichita, Kansas, p. 47-51. In R. A. Hahn (ser. ed.), 3rd International Herpetological Symposium on Captive Propagation and Husbandry. International Herpetological Symposium, Thurmont MD.

Grow, D. T., and J. Branham. 1996. Reproductive husbandry of the Gila monster (*Heloderma suspectum*), In P. D. Strimple (ed.), Advances in Herpetoculture. Special Publication of the International Herpetological Symposium, Inc., No. 1. International Herpetological Symposium, Inc., Des Moines IA.

Grow, D. T., S. Wheeler, and B. Clark. 1988. Reproduction of the amethystine python *Python amethystinus kinghorni* at the Oklahoma City Zoo. Inter. Zoo Yearb. 27:241-244.

Grummt, W. 2001. *Der Zoologische Garten*, p. 1440-1442. In C. E. Bell (ed.), Encyclopedia of the World's Zoos. Fitzroy Dearborn Publishers, Chicago, London.

Guggisberg, C. A. W. 1972. Crocodiles: Their Natural History, Folklore and Conservation. Stackpole Books, Harrisburg PA.

Guillery, P. 1993. The Buildings of the London Zoo. Royal Commission on the Historical Monuments of England, London.

Gumprecht, A., D. M. Boyer, and K. Tepedelen. 2002. Die Grubenottern der Gattung *Trimeresurus* (sensu lato) Lacépède. Teil I: Anmerkungen zur Biologie, Haltung und Nachzucht von *Ermia mangshanensis* (Zhao, 1990). (The pitvipers of genus *Trimeresurus* [sensu lato] Lacépède. Part I: Notes on the biology, husbandry and breeding of *Ermia mangshanensis* [Zhao, 1990]). Sauria 24 (4):3-11.

Gupta, B. K. 1998. On reproduction and captive breeding of ranids. Animal Keepers' Forum 25(12):464-468.

Haagner, G. V., and D. R. Morgan. 1989. The captive propagation of the eastern green mamba *Dendroaspis angusticeps* . Inter. Zoo Yearb. 28:195-199.

Haagner, G. V., and D. R. Morgan. 1993. The maintenance and propagation of the black mamba *Dendroaspis polylepis* at the Manyeleti Reptile Centre, Eastern Transvaal. Inter. Zoo Yearb. 32:191-196.

Haast, W. E. 1969. Hatching rhinoceros iguanas *Cyclura cornuta* at the Miami Serpentarium. Inter. Zoo Yearb. 9:49.

Hairston, C. 1991. Some observations concerning shell growth in captive Madagascan radiated tortoises, p. 44-52. In A. W. Zulich (ed.), 14th International Herpetological Symposium on Captive Propagation and Husbandry. International Herpetological Symposium.

Hairston, C., and P. M. Burchfield. 1989. Management and reproduction of the Galapagos tortoise *Geochelone elephantopus* at the Gladys Porter Zoo. Inter. Zoo Yearb. 28:70-77.

Hairston, C. S., and P. M. Burchfield. 1992. The reproduction and husbandry of the water monitor *Varanus salvator* at the Gladys Porter Zoo, Brownsville. Inter. Zoo Yearb. 31:124-130.

Hall, B. J. 1978. Notes on the husbandry, behaviour, and breeding of captive tegu lizards *Tupinambis tequixin*. Inter. Zoo Yearb. 18:91-95.

Hall, G., D. Groth, and J. Wetherall. 1992. Application of DNA profiling to the management of endangered species. Inter. Zoo Yearb. 31:103-108.

Hallman, G. 2001. Das Darmstädter Vivarium (The Darmstadt Vivarium). In W. Rieck, G. T. Hallmann, and W. Bischoff (eds.), Die Geschichte der Herpetologie und Terrarienkunde im deutschsprachigen Raum (History of Herpetology and Terrarium Science in German speaking areas) Mertensiella 12:279-280. Deutschen Gesellschaft für Herpetologie und Terrarienkunde e.V. (DGHT).

Hamilton, G., and G. R. Phelps. 1992. Zoos and conservation education. Inter. Zoo Yearb. 31:97-98.

Hammack, S. H. 1989. Reproduction of the Colombian milk snake *Lampropeltis triangulum andesiana* at the Dallas Zoo. Inter. Zoo Yearb. 28:172-177.

Hara, K., and F. Kikuchi. 1978. Breeding the West African dwarf crocodile *Osteolaemus tetraspis tetraspis* at Ueno Zoo, Tokyo. Inter. Zoo Yearb. 18:84-87.

Hardy, D. L. 1985. Venomous snakebite in North America: Some current views on management. Proc. Amer. Assoc. Zoo Vet.:102-104.

Hartdegen, R. W., and D. Chiszar. 2001. Discrimination of prey-derived chemical cues by the Gila monster (*Heloderma suspectum*) and lack of effect of a putative masking odor. Amphibia-Reptilia 22:249-253.

Hartdegen, R. W., D. Chiszar, and J. B. Murphy. 1999. Observations on the feeding behavior of captive black tree monitors, *Varanus beccari*. Amphibia-Reptilia 20:330-332.

Hartley, J. 1963. Notes on tuataras (*Sphenodon punctatus*) in captivity. Inter. Zoo Yearb. 5:170-171.

Harwell, G. M. 1982. Esophageal foreign body in a Kemp's ridley sea turtle. Proc. Amer. Assoc. Zoo Vet. 1982:3.

Harwell, G. M. 1986. Problems associated with captive management of crocodilians. Proc. Amer. Assoc. Zoo Vet. 1986:122.

Harwell, G. M., and H. Quinn. 1982. The Houston toad - problems associated with the captive propagation of amphibians. Proc. Amer. Assoc. Zoo Vet. 1982:1-2.

Hatt, J.-M., and R. E. Honegger. 1997. Erfahrungen und Beobachtungen bei der Aufzucht von Galapagos-Riesenschildkröten *Geochelone* (*elephantopus*) *nigra* im Zoo Zürich (Experiences and observations in raising Galapagos giant tortoises *Geochelone* [*elephantopus*] *nigra* in the Zürich Zoo). Verh. ber. Erkrg. Zootiere 38:131-135.

Hauser, B., F. Mettler, and R. E. Honegger. 1977. Knochenstoffwechselstörungen bei Seychellen-Riesenschildkröten (*Testudo* (*Geochelone*) *gigantea*) (Bone metabolism disturbances in Seychelles giant tortoises [*Testudo* (*Geochelone*) *gigantea*]). Verh. ber. Erkrg. Zootiere 19:121-125.

Heatwole, H. 1977. Habitat selection in reptiles, p. 137-155. In C. Gans and D. W. Tinkle (eds.). Biology of the Reptilia, Vol. 7. Ecology and Behaviour A. Academic Press, London, New York.

Hediger, H. 1958. Kleine Tropenzoologie (Small Tropical Zoology). Aufl. Acta trop., Suppl. I, Basel, Switzerland. 1:1-224.

Hediger, H. 1964. Wild Animals in Captivity, Dover, New York.

Hediger, H. 1968. The Psychology and Behaviour of Animals in Zoos and Circuses. Dover, New York.

Hediger, H. 1969 (released one year later due to publication delays). Man and Animal in the Zoo; Zoo Biology. Routledge & Kegan Paul, London.

Heichler, L., and J. B. Murphy. 2004. Johann Matthäus Bechstein: The father of herpetoculture. Herpetol. Rev. 35:8-13.

Heinroth, O. 1941. Wie Brüllt der Alligator? (How does the alligator roar?). Zool. Gart. (N.F.), Leipzig 13:284-288.

Held, S. P. 1985. Maintenance, exhibition, and breeding of the tailed frog, *Ascaphus truei* in a zoological park. Herpetol. Rev. 16(2):48-51.

Helman, R. G., M. T. Barrie, and C. H. Gardiner. 1998. Parasitic conjunctivitis and lacrimal adenitis in two tiger salamanders, *Ambystoma tigrinum mavortium*. Bull. Assoc. Rept. Amphib. Vet. 8(1):9-12.

Henderson, R. W., and R. A. Sajdak. 1996. Diets of West Indian racers (Colubridae: Alsophis): Composition and biogeographic implications, p. 327-338. In R. Powell, and R. W. Henderson (eds.), Contributions to West Indian Herpetology: A Tribute to Albert Schwartz. Society for the Study of Amphibians and Reptiles, Contributions to Herpetology, volume 12, Ithaca NY.

Herald, E. S. 1949. Effects of DDT-oil solutions upon amphibians and reptiles. Herpetologica 5(6):117-120.

Herald, E. S. 1955. A longevity record for the Texas blind salamander. Herpetologica 11(3):192.

Herman, D. W. 1979. Breeding the Jaliscan milk snake *Lampropeltis triangulum arcifera* at Atlanta Zoo.

Inter. Zoo Yearb. 19:96-97.

Herman, D. W. 1979. Captive reproduction in the scarlet kingsnake, *Lampropeltis triangulum elapsoides* (Holbrook). Herpetol. Rev. 10(4):115.

Herman, D. W. 1991. Captive husbandry of the eastern *Clemmys* group at Zoo Atlanta, p. 54-62. In K. R. Beaman, F. Coporaso, S. McKeown, and M. D. Graff (eds.), Proceedings of the First International Symposium on Turtles and Tortoises: Conservation and Captive Husbandry. Chapman University, August 9-12, 1990. California Turtle and Tortoise Club, Van Nuys CA.

Herman, D. W. 1993. Reproduction and management of the southeast Asian spiny turtle (*Heosemys spinosa*) in captivity. Herpetol. Nat. Hist. 1:97-100.

Herman, D. W., and G. A. George. 1986. Research, husbandry, and propagation of the bog turtle *Clemmys muhlenbergi* (Schoepff) at the Atlanta Zoo, p. 125-135. In S. McKeown, F. Caporaso, and K. H. Peterson (eds.), 9th International Herpetological Symposium on Captive Propagation and Husbandry. Zoological Consortium, Inc., Thurmont MD.

Herrmann, H.-W. 1998. *Holaspis guentheri laevis* (East African fringe-tailed forest lizard) Reproduction. Herpetol. Rev. 29(4):238.

Herrmann, H.-W. 1998. Der Hühnerfresser *Spilotes pullatus* im Kölner Aquarium am Zoo - Haltung und Reproduktion (The black and yellow rat snake *Spilotes pullatus* in the Aquarium of the Cologne Zoo — Upkeep [husbandry] and reproduction). Z. Köln. Zoo 41:139-144.

Herrmann, H.-W. 1999. Husbandry and captive breeding of the water monitor, *Varanus salvator* (Reptilia: Sauria: Varanidae) at the Cologne Aquarium (Cologne Zoo), In H.-G. Horn, and W. Böhme (eds.), Advances in Monitor Research II. - Mertensiella 11:95-103. Deutschen Gesellschaft für Herpetologie und Terrarienkunde e.V. (DGHT).

Herron, J. C., L. H. Emmons, and J. E. Cadle. 1990. Observations on reproduction in the black caiman *Melanosuchus niger*. J. Herpetol. 24:314-316.

Hick, U. 1965. Die Tuatara, ein Relikt aus längst vergangener Zeit (The tuatara, a relic of the ancient past). Freunde Köln. Zoo 8(3):75-82.

Hiler, B. I. 1985. An overview of the amphibian collection at the Steinhart Aquarium as an introduction to amphibian care, p. 145-152. In R. L. Gray (ed.), Captive Propagation and Husbandry of Reptiles and Amphibians. Northern California Herpetological Society & Bay Area Amphibian and Reptile Society.

Hilf, M., R. A. Wagner, and V. L. Yu. 1990. A prospective study of upper airway flora in healthy boid snakes and snakes with pneumonia. J. Zoo Wildl. Med. 21(3):318-325.

Hilf, M., D. Swanson, R. A. Wagner, and V. Yu. 1989. Pharmacokinetics of Gentamicin and Piperacillin in blood pythons: New dosing regimen, p. 87-90. In M. J. Uricheck (ed.), 13th International Herpetological Symposium on Captive Propagation and Husbandry. International Herpetological Symposium.

Hill, C. 1946. Playtime at the zoo. Zoo-Life 1:24-26.

Hiller, A. 1984. Tierärztliche Erfahrungen bei der Zucht der Europäischen Sumpfschildkröte (*Emys orbicularis*) (Veterinary experience gathered in the breeding of the swamp turtle [*Emys orbicularis*]). Zool. Gart. (N.F.), Jena 54:128-130.

Hinsche, G. 1928. Kampfreaktionen bei einheimischen Anuren (Fight reactions of domestic frogs). Biol. Zentralbl. 48:577-617.

Hinsche, G. 1939. Über die Entwicklund von Haltungs- und Bewegungsreaktionen (Concerning development of reactions to care and movement). W. Rous, Arch. Entwicklungsmech. 139:724-731.

Hinsche, G. 1941. Domestikationsmerkemale bei Anuren (Characteristics of domestication in frogs). Zool. Anz. 12:26.

Hirsch, U. 1980. Riesenschildkröten (*Testudo* [*Geochelone*] *elephantopus*) auf Galapagos (Giant tortoises [*Testudo* (*Geochelone*) *elephantopus*] on Galapagos). Z. Köln. Zoo 23(4):111-117.

Hitchiner, J. A. 1987. Reproduction in captive eyelash vipers, *Bothrops schlegeli*. Herpetol. Rev. 18(3):55.

Hoessle, C. 1963. A breeding pair of western diamondback rattlesnakes, *Crotalus atrox*. Bull. Philadelphia Herpetol. Soc. 11(3-3):65-66.

Hoessle, C. 1969. Display of tuatara (*Sphenodon punctatus*) at St. Louis Zoological Park. Inter. Zoo Yearb. 9:32-33.

Hollamby, S., D. Murphy, and C. A. Schiller. 2000. An epizootic of amoebiasis in a mixed species collection of juvenile tortoises. Bull. Assoc. Rept. Amphib. Vet. 10(1):9-15.

Holmback, E. 1981. Reproduction of the brown caiman *Caiman crocodilus fuscus* at the San Antonio Zoo. Inter. Zoo Yearb. 21:77-79.

Holmback, E. 1984. Parthenogenesis in the Central American night lizard *Lepidophyma flavimaculatum* at San Antonio Zoo. Inter. Zoo Yearb. 23:157-158.

Holmback, E. 1987. Captive reproduction of the New Guinea side-necked turtle *Emydura australis albertisii* at the San Antonio Zoo. Inter. Zoo Yearb. 26:94-98.

Holmstrom, W. F., Jr. 1981. Observations on the reproduction of the yellow anaconda *Eunectes notaeus* at New York Zoological Park. Inter. Zoo Yearb. 21:92-94.

Holmstrom, W. F., Jr., and J. L. Behler. 1981. Post-parturient behavior of the common anaconda, *Eunectes murinus*. Zool. Gart. (N.F.), Jena 51:353-356.

Holyoake, A., B. Waldman, and N. J. Gemmell. 2001. Determining the species status of one of the world's rarest frogs: A conservation dilemma. Animal Conserv. 4:29-35.

Honegger, R. E. 1970. Beitrag zur Fortpflanzungsbiologie von *Boa constrictor* und *Python reticulatus* (Reptilia, Boidae) (Paper on the reproductive biology of *Boa constrictor* and *Python reticulatus* [Reptilia, Boidae]). Salamandra 6:73-79.

Honegger, R. E. 1972. Zoo Breeding and Crocodile Bank. Proc. 1st Working Meeting Croc. Specialists, IUCN Publ. Ser. Suppl. Paper 32:86-97.

Honegger, R. E. 1975. The crocodilian situation in European zoos. Inter. Zoo Yearb. 15:277-283.

Honegger, R. E. 1975. Beitrag zur Kenntnis des Wickelskinkes *Corucia zebrata* (Paper concerning knowledge of the prehensile-tailed skink *Corucia zebrata*). Salamandra 11:27-32.

Honegger, R. E. 1978. Geschlechtsbestimmung bei Reptilien (Determination of gender in reptiles). Salamandra 14:69-79.

Honegger, R. E. 1979. Marking amphibians and reptiles for future identification. Inter. Zoo Yearb. 19:14-22.

Honegger, R. E. 1982. Breeding crocodiles in captivity, a retrospect. Proc. 5th Working Meeting Croc. Spec. Group SSC/IUCN. Florida State Mus., Gainesville, 12-16 Aug 1980.

Honegger, R. E. 1984. Beiträge zur Biologie von *Hydrodynastes gigas* im Terrarium (Serpentes, Colubridae) (Contributions to the biology of *Hydrodynastes gigas* in the terrarium [Serpentes, Colubridae]), p. 237-244. In V. L. Bels, and A. P. Van den Sande (eds.), Maintenance and Reproduction of Reptiles in Captivity. Acta Zool. Path. Antverpiensia 78.

Honegger, R. E. 1985. Additional notes on the breeding and captive management of the prehensile-tailed skink (*Corucia zebrata*). Herpetol. Rev. 16(1):21-23.

Honegger, R. E. 1986. Zur Pflege und langjährigen Nachzucht von *Siebenrockiella crassicollis* (Gray, 1831) (Concerning the care and long-term reproduction of *Siebenrockiella crassicollis* [Gray, 1831]). Salamandra 22:1-10.

Honegger, R. E. 1998. Beitrag zu Haltung und Zucht der Skorpionskrustenechse, *Heloderma horridum*, im Zoo Zürich (Contribution concerning upkeep and breeding of the Mexican beaded lizard, *Heloderma horridum*, in the Zürich Zoo). Zool. Gart. (N.F.), Jena 68:287-299.

Honegger, R. E., and H. Heusser. 1969. Beiträge zum Verhaltensinventar des Bindenwarans (*Varanus salvator*) (Contributions regarding the behavior inventory of the two-banded monitor [*Varanus salvator*]). Zool. Gart. (N.F.), Leipzig 36 Hediger-Festschrift: 251-260.

Honegger, R. E., and H. Hunt. 1990. Breeding crocodiles in zoological gardens outside the species range, with some data on the general situations in European zoos, p. 200-228. Crocodiles: Proc. 10th Working Meeting of the Crocodile Specialist Group, Gainesville, Florida. IUCN. The World Conservation Union Publ. N.S., Gland, Switzerland.

Honegger, R. E., and A. Rübel. 1991. Aufzucht und Erkrankungen der ersten in Europa nachgezogenen Galapagos-Schildkröten (*Geochelone elephantopus*) (Rearing and illnesses of the first Galapagos tortoises bred in Europe [*Geochelone elephantopus*]). In: 4. Int. Coll. Pathol. Med. Rept. Amph., Dtsche Vet. med. Ges. (German Veterinary Medical Society), Bad Nauheim: 225-243.

Honegger, R. E., C. Schneider, and E. Zimmermann. 1985. Notizen zur Aufzucht von Schmuckhornfröschen *Ceratophrys ornata* (Bell, 1843) (Notes on the breeding of the ornate horned frog *Ceratophrys ornata* [Bell, 1843]). Salamandra 21:70-80.

Honeyman, V. L., K. G. Merhen, I. K. Barker, and G. J. Crawshaw. 1992. *Bordetella* septicemia and chlamydiosis in eyelash leaf frogs (*Ceratobatrachus guentheri*). Proc. Amer. Assoc. Zoo Vet. 1992:168.

Hopkins, C. 1992. Zoo education: Into the 1990s. Inter. Zoo Yearb. 31:99-103.

Hopley, C. G. 1882. Snakes: Curiosities & Wonders of Serpent Life. Griffith & Farran, London.

Horn, H.-G., and G. J. Visser. 1989. Review of reproduction of monitor lizards *Varanus* spp. in captivity. Inter. Zoo Yearb. 28:140-150.

Horn, H.-G., and G. J. Visser. 1991. Basic data on the biology of monitors. Mertensiella 2:176-187.

Horn, H.-G., and G. J. Visser. 1997. Review of reproduction of monitor lizards *Varanus* spp in captivity

II. Inter. Zoo Yearb. 35:227-246.

Hosono, H. 1982. Successful breeding of Florida king snake, *Lampropeltis getulus floridana* at Kyoto Zoo. J. Japan. Assoc. Zoos Aquar. 24(3):57-60 [In Japanese].

Howard, C. J. 1980. Breeding the flat-tailed day gecko, *Phelsuma laticauda*, at Twycross Zoo. Inter. Zoo Yearb. 20:193-196.

Howard, C. J. 1980. Notes on the maintenance and breeding of the common iguana (*Iguana iguana iguana*) at Twycross Zoo, p. 47-50. In S. Towson, N. J. Millichamp, D. G. D. Lucas, and A. J. Millwood, (eds.). The Care and Breeding of Captive Reptiles. The British Herpetological Society; London.

Hudson, R. 1983. The Species Survival Plan and its application to reptiles, p. 1-9. In P. J. Tolson (ed.), 7th International Herpetological Symposium on Captive Propagation and Husbandry. International Herpetological Symposium, Thurmont MD.

Hudson, R. 2002. The Turtle Survival Alliance: A unified response to the global turtle crisis. Communiqué December 2002:18-20.

Huey, R. B. 1982. Temperature, physiology, and the ecology of reptiles, p. 25-91. In C. Gans and F. H. Pough (eds.). Biology of the Reptilia, Vol. 12. Physiology C. Physiological Ecology. Academic Press, London, New York.

Huff, T. A. 1976. Breeding the Cuban boa *Epicrates angulifer* at the Reptile Breeding Foundation. Inter. Zoo Yearb. 16:81-82.

Huff, T. A. 1978. Breeding the Puerto Rican boa *Epicrates inornatus* at the Reptile Breeding Foundation. Inter. Zoo Yearb. 18:96-97.

Huff, T. A. 1979. Breeding the Jamaican boa, *Epicrates subflavus*, in captivity: a five year review. AAZPA Regional Conference Proc.: 339-345.

Huff, T. A. 1984. The husbandry and propagation of the Madagascar ground boa, *Acrantophis dumerili* in captivity with notes on other Malagasy boids. Acta Zool. Path. Antverpiensia 78:255-270.

Huish, R. 1830. The Wonders of the Animal Kingdom Exhibiting Delineations of the Most Distinguished Wild Animals in the Various Menageries of the Country. Thomas Kelly, London.

Hunt, R. H. 1969. Breeding of spectacled caiman *Caiman c. crocodilus* at Atlanta Zoo. Inter. Zoo Yearb. 9:36-37.

Hunt, R. H. 1973. Breeding Morelet's crocodile *Crocodylus moreletii* at Atlanta Zoo. Inter. Zoo Yearb. 13:103-105.

Hunt, R. H. 1975. Maternal behavior in the Morelet's crocodile, *Crocodylus moreletii*. Copeia 1975:763-764.

Hunt, H. R. 1987. Aggressive behavior by adult Morelet's crocodiles, *Crocodylus moreletii* toward young. Herpetologica 33:195-201.

Hutchison, V. H., H. G. Dowling, and A. Vinegar. 1966. Thermoregulation in a brooding female Indian python, *Python molurus bivittatus*. Science 151:694-696.

Igolkina, V. A. 1989. Breeding the Lebetine viper ssp *Vipera lebetina turanica* and *Vipera l. obtusa* at Leningrad Zoo. Inter. Zoo Yearb. 28:183-189.

Indiviglio, F. 1997. Newts and Salamanders. Everything About Selection, Care, Nutrition, Diseases, Breeding, and Behavior. Barron's Educational Series, Inc., Hauppauge NY.

Ippen, R. 1980. Ein Beitrag zu den Mykosen der Schlangen (A contribution concerning mycosis in snakes). Milu 5(3):386-396.

Ippen, R., and A. Konstantinov. 1981. Durch Vitamin-A-Mangel bedingte Nierenveränderungen bei einem Ganges-Gavial (*Gavialis gangeticus*) (Kidney changes in an Indian gavial caused by lack of vitamin A [*Gavialis gangeticus*]), p. 127-131. In R. Ippen, and H.-D. Schröder (eds.), Erkrankungen der Zootiere; Verhandlungsbericht des 23. Internationalen Symposiums über die Erkrankungen der Zootiere vom 24. Juni bis 28. Juni in Halle/Saale. Akademie Verlag. Berlin -1981 (Illnesses of Zoo Animals. Conference Report of the 23rd International Symposium Concerning Illnesses of Zoo Animals, June 24-28 in Halle/Saale).

Irwin, S. 1994. Notes on the behaviour and diet of *Varanus teriae* Sprackland, 1991. Mem. Queensland Museum. 35(1):128.

Irwin, S. 1996. Courtship, mating and egg-deposition by the captive perentie *Varanus giganteus* at the Queensland Reptile and Fauna Park. Thylacynus 21(1):8-11.

Irwin, S., and S. Thomson. 1995. The first successful captive breeding of alligator snapping turtles *Macroclemys temminckii*. Thylacinus 20(1):6-9.

Isenbügel, E., and W. Frank. 1985. Heimtierkrankheiten (Diseases of Domestic Animals). E. Ulmer Verlag, Stuttgart.

IUCN. 1996. Red List of Threatened Animals. Edited by J. Baillie and B. Groombridge. International Union for Conservation of Nature and Natural Resources.

Ivanyi, C. S. 1989. Captive reproduction of the black-eared frog *Leptodactylus melanonotus* at the Arizona-Sonora Desert Museum, p. 195-199. In M. J. Uricheck (ed.), 13th International Herpetological Symposium on Captive Propagation and Husbandry. International Herpetological Symposium.

Jacobs, D. M., and A. D. Belcher. 1983. Notes on captive reproduction in the West African gaboon viper, *Bitis gabonica rhinoceros*, at the Rio Grande Zoological Park, p. 103-107. In P. J. Tolson (ed.), 7th International Herpetological Symposium on Captive Propagation and Husbandry. International Herpetological Symposium, Thurmont MD.

Jacobson, E., S. Clubb, and R. L. Napolitano. 1981. Amoebiasis in red-footed tortoises, *Geochelone carbonaria*. Proc. Amer. Assoc. Zoo Vet. 1981:16.

Jakob-Hoff, R. M. 1992. Recent developments in the Australasian Species Management Program. Inter. Zoo Yearb. 31:12-19.

Janacek, J. 1976. An exceptionally large puff adder brood *Bitis arietans*. Inter. Zoo Yearb. 16:85-86.

Jardine, D. R. 1981. First successful captive propagation of Schneider's smooth-fronted caiman, *Paleosuchus trigonatus*. Herpetol. Rev. 12(2):58-60.

Jennison, G. 1909. Description of the reproduction of a pair of Seba pythons in the Belle Vue Zoological Gardens, Manchester UK and exhibition of fertilized eggs by the Secretary of the London Zoological Society. Proc. Zool. Soc. London 1909: 392-393.

Jennison, G. 1910. Letter by George Jennison to the Secretary of the London Zoological Society describing reproduction and behavior of a pair of pine snakes in the Belle Vue Zoological Gardens, Manchester UK. Proc. Zool. Soc. London 1910:539.

Jes, H. 1971/1972. Jahresbericht Aquarium 1971 (Aquarium Annual Report for 1971). Z. Köln. Zoo 14:137-145.

Jes, H. 1973. Jahresbericht Aquarium 1972 (Aquarium Annual Report for 1972). Z. Köln. Zoo 16:19-28.

Jes, H. 1974. Jahresbericht Aquarium 1973 (Aquarium Annual Report for 1973). Z. Köln. Zoo 17(2):55-64.

Jes, H. 1975. Jahresbericht Aquarium 1974 (Aquarium Annual Report for 1974). Z. Köln. Zoo 18:15-23.

Jes, H. 1989. Treatment of amoebiasis in turtles. Inter. Zoo Yearb. 28:60-61.

Jes, H. 1997. 25 Jahre Aquarium (25 years Aquarium). Z. Köln. Zoo 40(1):35-38.

Jirousek, V. T. 1997. Chov a rozmnozovani cinskych uzovek rodu *Elaphe* v Jihlavske Zoo (Care and breeding of Chinese rat snakes of the genus *Elaphe* in Jihlava Zoo). Akvarium Terrarium 40(5):47-49.

Johnson, R. R. 1984. Breeding the Bell's horned frog (*Ceratophrys ornata*): An alternative to hormonally induced reproduction, p. 22-32. In R. A. Hahn (ser. ed.), 8th International Herpetological Symposium on Captive Propagation and Husbandry. International Herpetological Symposium, Thurmont MD.

Johnson, R. R. 1988. Combat and courtship of the Eastern massasauga rattlesnake: comparison of field and captive behaviour, p. 71-78. In M. J. Uricheck (ed.), 13th International Herpetological Symposium on Captive Propagation and Husbandry. International Herpetological Symposium.

Jones, J. R., G. W. Ferguson, W. H. Gehrmann, M. F. Holick, T. C. Chen, and Z. Lu. 1996. Vitamin D nutritional status influences voluntary behavioral photoregulation in a lizard, p. 49-55. In M. F. Holick and E. G. Jung (eds.), Biological Effects of Light 1995. Walter de Gruyter, Berlin.

Jones, M. L. 1965. The Komodo Dragon. Chronological list of the Komodo Dragon lizard (*Varanus komodoensis*] exhibited outside Indonesia 1926-1964. Inter. Zoo News (July) 12(3):92-93.

Jordan, T., D. M. Schleser, and D. T. Roberts. 1992. Captive reproduction of *Eurycea neotenes*, the Comal Springs salamander at the Dallas Aquarium. AAZPA Reg. Conf. Proc.1992:771-773.

Judd, H. L., G. A. Laughlin, J. P. Bacon, and K. Benirschke. 1976. Circulating androgen and estrogen concentrations in lizards (*Iguana iguana*). Gen. Comp. Endocrinol. 30:391-395.

Judd, H. L., J. P. Bacon, D. Rüedi, J. Girard, and K. Benirschke. 1977. Determination of sex in the Komodo dragon *Varanus komodoensis*. Inter. Zoo Yearb. 17:208-209.

Kagawa, K. 1994. News from Japan. Animals and Zoos 46/4:23.

Kamelin, E., and Y. Lukin. 2000. Keeping and breeding of Central Asia cobra (*Naja oxiana* Eichwald) in terrarium. Research in Zoological Parks 12.

Kamelin, E., Y. Lukin, and K. Mil'to. 1997. Hybridization of *Vipera schweizeri* (Werner, 1935) and *Vipera lebetina obtusa* (Dvigubsky, 1832). Russian J. Herpetol. 4:75-78.

Kardon, A. 1979. A note on captive reproduction in three Mexican milk snakes *Lampropeltis triangulum polyzona*, *L. t. nelsoni* and *L. t. sinaloae*. Inter. Zoo Yearb. 19:94-96.

Kardon, A. 1981. Captive reproduction in Geoffrey's side-necked turtle *Phrynops geoffroanus geoffroanus*. Inter. Zoo Yearb. 21:71-72.

Kauffeld, C. F. 1942. Care of frog tadpoles. Animaland 9(3)

Kauffeld, C. F. 1957. Snakes and Snake Hunting. Hanover House, Garden City NY.

Kauffeld, C. F. 1969. Snakes: The Keeper and the Kept. Doubleday, Garden City NY.

Kaverkin, J., S. V. Mamet, and M. Toriba. 1994. Husbandry and reproduction of the Eublepharid gecko *Goniurosaurus kuroiwae splendens* in captivity. Akamata 10:23-26

Kawata, K. 1991. Japan's Species Survival Programme gets off the ground. Inter. Zoo News 38/1.

Kawata, K. 2003. Reptiles in Japanese collections. Part 1: Chelonians, 1998. Inter. Zoo News 50:265-275.

Kawata, K. 2004. Reptiles in Japanese collections. Part 2: Squamates and crocodilians, 1999. Inter. Zoo News 51:73-86.

Kawata, K., and J. B. Murphy. 2004. John E. Werler (1922-2004). Herpetol. Rev. 35: 35:213-215.

Keeling, C. H. 1983. The Life and Death of Belle Vue. Clam Publications, Guilford UK.

Keeling, C. H. 1984. Where the Lion Trod: A Study of Forgotten Zoological Gardens. Clam Publications, Guilford UK.

Keeling, C. H. 1992. A Short History of British Reptile Keeping. Clam Publications, Guilford UK.

Keen, R. 1995. The study of the seven Aldabra giant tortoises (*Geochelone gigantea*) of Paignton Zoological and Botanical Gardens 1986-1994. Ratel 22(4):128-139.

Kerbert, C. 1904. Zur Fortpflanzung von *Megalobatrachus maximus* Schlegel (*Cryptobranchus japonicus* v.d. Hoeven) (Concerning the reproduction of *Megalobatrachus maximus* Schlegel [*Cryptobranchus japonicus* v.d. Hoeven]). Zool. Anz. (Zoological Information) 27:305-320.

Kimura, W. 1976. Artificial incubation of crocodile eggs. Atagawa Tropical Garden and Alligator Farm. Research Report No. 4:1-44.

Kimura, W. 1978. Artificial incubation of gavials eggs. Atagawa Tropical Garden and Alligator Farm. Research Report No. 8 :1-44 [in Japanese and English].

King, F. W., and J. S. Dobbs. 1975. Crocodilian propagation in American zoos and aquaria. Inter. Zoo Yearb. 15:272-277.

Kirsche, W. 1976. Beitrag zur Biologie der Sternschildkröte (*Testudo elegans* Schoepff) (A contribution to the biology of the star tortoise [*Testudo elegans* Schoepff]). Zool. Gart. (N.F.), Jena 46:66-81.

Kisling Jr, V. N. (ed.). 2001. Zoo and Aquarium History. Ancient Animal Collections to Zoological Gardens. CRC Press, Boca Raton, FL; London; New York; Washington DC.

Klauber, L. M. 1956. Rattlesnakes: Their Habits, Life Histories, and Influence on Mankind. Univ. California Press, Berkeley.

Klemmer, K. 1967. Observation on the sea snake *Laticauda laticaudata* in captivity. Inter. Zoo Yearb. 7:229-231.

Klemmer, K. 2001. Hans-Günter Petzold (1931-1982) In W. Rieck, G. T. Hallmann, and W. Bischoff (eds.), Die Geschichte der Herpetologie und Terrarienkunde im deutschsprachigen Raum (History of Herpetology and Terrarium Science in German-Speaking Areas) - Mertensiella 12:549-551. Deutschen Gesellschaft für Herpetologie und Terrarienkunde e.V. (DGHT).

Klingelhöffer, W. 1955-1959. Terrarienkunde (Terrarium Science). Alfred Kernen Verlag, Stuttgart.

Knapp, C. R. 2000. Home range and intraspecific interactions of a translocated iguana population (*Cyclura cychlura inornata* Barbour and Noble). Caribbean J. Sci. 36:250-257.

Knapp, C. R. 2001. Status of a translocated *Cyclura* iguana colony in the Bahamas. J. Herpetol. 35:239-248.

Knapp, C. R., and C. L. Malone. 2003. Patterns of reproductive success and genetic variability in a translocated iguana population. Herpetologica 59:195-202.

Knapp, C. R., S. Buckner, A. Feldman, and L. Roth. 1999. Status update and empirical field observations of the Andros rock iguana, *Cyclura cychlura cychlura*. Bahamas J. Sci. 7:2-5.

Knepper, D. 1993. Production of the azure dart-poison frog *Dendrobates azureus* at the Chaffee Zoological Gardens at Fresno, p. 1-5. In M. Bumgardner (ed.), Captive Propagation and Husbandry of Reptiles and Amphibians. Special Publication # 7 ed. Northern California Herpetological Society.

Kobara, J. 1985. The Giant Salamander. Dobutsu-sha, Tokyo.

Koch-Isenburg, L. 1971. Vivarium Darmstadt, moderne Heimstatt für Tiere (Vivarium Darmstadt, a modern homestead for animals). Reba Verlag, Darmstadt, Germany.

Konig, C., and A. Schluter. 1991. Nachzucht der Balearen-Geburtshelferkröte *Alytes muletensis* (Sanchiz et Adrover 1979) im Rahmen eines Artenschutzprogrammes (Amphibia; Discoglossidae) (Breeding the Balearic midwife toad *Alytes muletensis* [Sanchiz et Androver 1979] in the context of an endangered species program [Amphibia, Discoglossidae]). Jahresh. Ges. Naturkd. Württemb. (Annual Report of the Wuerttemberg Society for Natural History) 146:193-205.

Kopczynski, J. 1993. Breeding and exhibiting the monocellate cobra *Naja kaouthia* at Plock Zoo. Inter. Zoo Yearb. 32:197-204.

Korinek, M. 1997. O chovu koralovek *Lampropeltis triangulum campbelli* (Breeding the milk snake *Lampropeltis triangulum campbelli*). Akvarium Terrarium 40(7):41-43.

Korinek, M. 1997. Uzovka *Hydrodynastes gigas* v terariu Zoo Olomouc (The false water cobra *Hydrodynastes gigas* in the terrarium of Olomouc Zoo). Akvarium Terrarium 40(12):40-43.

Korolev, A. V., S. V. Kudryavtsev, and V. E. Frolov. 1984. Some special aspects of the husbandry of soft shell turtles (Reptilia, Testudines, Trionychidae) at the Moscow Zoo, p. 54-58. In R. Hahn (ser. ed.), 7th International Herpetological Symposium on Captive Propagation and Husbandry. Zoological Consortium, Inc., Thurmont MD.

Korzhov, A. V., S. V. Kudryavtsev, V. A. Latyshev, and D. B. Vassiliev. 1995. Keeping and breeding in captivity the Indonesian scrub python *Liasis a. amethistinus* (Schneider 1901) at the Moscow Zoo. Russian J. Zool. 2:68-70.

Koto, N. 1969. Zoo News, Animals and Zoos 21/2.

Kramer, L., B. L. Dresser, and E. J. Maruska. 1983. Sexing aquatic salamanders by laparoscopy. Proc. Amer. Assoc. Zoo Vet. 1983:192-194.

Kramer, M., and U. Fritz. 1989. Courtship of the turtle, *Pseudemys nelsoni*. J. Herpetol. 23:84-86.

Krefft, P. 1908. Das Terrarium (The Terrarium). Fritz Pfenningstorff, Berlin.

Krogh, A. 1916. The Respiratory Exchange of Animals and Man. Longmans, Green, London.

Kuchling, G., and J. P. Dejose. 1989. A captive breeding operation to rescue the critically endangered Western swamp turtle *Pseudemydura umbrina* from extinction. Inter. Zoo Yearb. 28:103-109.

Kuchling, G., and O.C. Razandrimamilafiniarivo. 1999. The use of ultrasound scanning to study the relationship of vitellogenesis, mating, egg production and follicular atresia in captive ploughshare tortoises *Geochelone yniphora*. Dodo, J. Jersey Wildl. Preserv. Trust 35:109-115.

Kuchling, G., J. P. Dejose, A. A. Burbidge, and S. D. Bradshaw. 1992. Beyond captive breeding: the western swamp tortoise *Pseudemydura umbrina* recovery programme. Inter. Zoo Yearb. 31: 37-41.

Kudryavtsev, S. V. 1985. On reproductive biology of Malayan pit viper (*Callaselasma rhodostoma*). Voprosy herpetologii, Òàshkent, p. 114-115.

Kudryavtsev, S. V. 1986. Reproduction of *Bothrops moojeni* in the Moscow Zoo, p. 155-158. In A. F. Kovshar' (ed.), Soderzhaniye i razvedeniye dikikh zhivotnykh (Maintaining and Breeding Wild Animals). Almar-Ata: Kaynar.

Kudryavtsev, S. V., and V. E. Frolov. 1984. On the reproductive biology of the Russian rat snake *Elaphe s. schrencki* (Squamata: Colubridae), p. 134-138. In R. Hahn (ser. ed.), 7th International Herpetological Symposium on Captive Propagation and Husbandry. Zoological Consortium, Inc., Thurmont MD.

Kudryavtsev, S. V., and V. E. Frolov. 1984. The third generation of the bamboo pit viper *Trimeresurus gramineus* (Squamata: Crotalidae) at the Moscow Zoo and an attempt at sex determination, p. 139-144. In R. Hahn (ser. ed.), 7th International Herpetological Symposium on Captive Propagation and Husbandry. Zoological Consortium, Inc., Thurmont MD.

Kudryavtsev, S. V., and S. V. Mamet. 1991. Husbandry and propagation of the Radde's viper *Vipera raddei raddei* Boett. Herpetol. Rev. 22(3):96.

Kudryavtsev, S. V., and V. I. Odinchenko. 1986. Second generation of the Paraguayan anaconda in the Moscow Zoo, p. 158-161. In A. F. Kovshar' (ed.), Sokhranim dikikh zhivotnykh (Let Us Preserve Wild Animals). Almar-Ata: Kaynar.

Kudryavtsev, S. V., and D. B. Vassiliev. 1998. Reproduction of *Shinisaurus crocodilurus* Ahl, 1930 at the Moscow Zoo. Russian J. Zool. 5:15-16.

Kudryavtsev, S. V., V. M. Makeev, and V. E. Frolov. 1986. The second generation of *Agkistrodon piscivorus* in captivity. Vestnik Zool. 1986:79-80.

Kursh, H. 1965. Cobras in his Garden. Harvey House, Irvington-on-Hudson NY.

Labuschagne, W., and S. Walker. 2001. Zoological gardens of Africa, p. 331-349. In V. N. Kisling (ed.), Zoo and Aquarium History. Ancient Animal Collections to Zoological Gardens. CRC Press; Boca Raton, FL; London; New York; Washington DC.

Lamarre-Picquot, M. 1835. Ophiologie. L'Institut 3:70.

Lamarre-Picquot, M. 1842. Troisième Mémoire sur l'incubation et autres phénomènes observés chez les ophidians; par M. Lamarre-Picquot (Third memorandum on incubation and other phenomena observed in ophidians; by M. Lamarre-Picquot). Comptes Rendus 14:164.

Lance, V. A. 1990. Stress in reptiles, p. 461-466. In A.

Epple, C. G. Scanes, and M. H. Stetson (eds.), Progress in Comparative Endocrinology. Wiley-Liss, New York.

Lance, V. A., and R. M. Elsey. 1986. Stress-induced suppression of testosterone secretion in male alligators. J. Exper. Zool. 239:241-246.

Lance, V. A., and R. M. Elsey. 1999. Plasma catecholamines and plasma corticosterone following restraint stress in juvenile alligators. J. Exper. Zool. 283(6):559-565.

Lance, V. A., and D. Lauren. 1984. Circadian variation in plasma corticosterone in the American alligator, *Alligator mississippiensis*, and the effects of ACTH injections. Gen. Comp. Endocrinol. 54:1-7.

Lance, V. A., D. C. Rostal, J. S. Grumbles, and L. Morici. 1995. Endocrine profiles of the reproductive cycle of male and female desert tortoises, p. 45-49. In G. Aquirre, E. D. McCoy, and H. Mushinsky (eds.), Proc. N. American Tortoise Conference, Durango, Mexico: Soc. Herpetol. Mex.

Lange, J. 1981. Beitrag zur Zucht von Smaragd-Baumsteigerfröschen (*Dendrobates auratus*) (Contribution concerning the breeding of emerald tree-climbing frogs [*Dendrobates auratus*]). Z. Köln. Zoo 40(1):6-8.

Lange, J. 1997. Die Haltung von Brückenechsen (*Sphenodon punctatus*) im Zoo-Aquarium Berlin (The care of tuatara [*Sphenodon punctatus*] in the Zoo-Aquarium in Berlin). Z. Köln. Zoo 40 (2):47-52.

Langebaek, R. 1979. Observations on the behaviour of captive *Phelsuma guentheri* during the breeding season at Jersey Wildlife Preservation Trust. Dodo, J. Jersey Wildl. Preserv. Trust 16:75-83.

Larsson, H.-O., and S. Holm. 1983. Tiere in ihrem Milieu zeigen-Wunsch oder Möglichkeit? (Exhibiting animals in their environment--wish or possibility?). Z. Köln. Zoo 26(3):105-108.

Larsson, H.-O., and J. Wihman. 1989. Breeding the Cuban crocodile *Crocodylus rhombifer* at Skansen Aquarium. Inter. Zoo Yearb. 28:110-113.

Laszlo, J. 1975. Probing as a practical method of sex recognition in snakes. Inter. Zoo Yearb. 15:178-179.

Lawler, H. E., and J. L. Jarchow. 1986. A captive management plan for large iguanine lizards using the Isla San Esteban chuckwalla *Sauromalus varius* Dickerson as a model, p. 137-164. In R. Hahn (ser. ed.), 9th International Herpetological Symposium on Captive Propagation and Husbandry. Zoological Consortium, Thurmont MD.

Lawler, H. E., and C. Norris. 1980. Breeding the Haitian giant galliwasp, (*Diploglossus warreni*) (SAURIA: ANGUINIDAE) at the Knoxville Zoological Park, p. 73-79. In R. A. Hahn (ser. ed.), 3rd International Herpetological Symposium on Captive Propagation and Husbandry. International Herpetological Symposium, Thurmont MD.

Lawler, H. E., and W. K. Wintin. 1987. Captive management and propagation of the reticulated Gila monster *Heloderma suspectum suspectum* Cope, p. 48-56. In R. Gowen (ed.), Captive Propagation and Husbandry of Reptiles and Amphibians. Northern California Herpetological Society.

Lawler, H. E., T. R. Van Devender, and J. L. Jarchow. 1994. Ecological and nutritional management of the endangered Piebald Chuckwalla (*Sauromalus varius*) in captivity, p. 333-341. In J. B. Murphy, K. Adler, and J. T. Collins (eds.), Captive Management and Conservation of Amphibians and Reptiles. Society for the Study of Amphibians and Reptiles. Contributions to Herpetology, volume 11, Ithaca NY.

Lederer, G. 1931. Erkennen wechselwarme Tiere ihren Pfleger? (Do cold-blooded animals recognize their keeper?). Wschr. Aquar.-Terrark. 28:636-638.

Lederer, G. 1931. Ein weiterer Beitrag zur Ethologie der Segelechse (*Hydrosaurus amboinensis* Schloss.) (A further contribution to the ethology of the sailfin lizard [*Hydrosaurus amboinensis* Schloss.]). Zool. Gart. (N.F.), Leipzig 4:277-279.

Lederer, G. 1941. Zur Haltung des China-Alligators (*Alligator sinensis* Fauvel) (Concerning care of the China alligator [*Alligator sinensis* Fauvel]). Zool. Gart. (N.F.), Leipzig 13:255-263.

Lederer, G. 1942. Fortpflanzung und Entwicklung von *Eunectes notaeus* Cope (*Boidae*) (Reproduction and development of *Eunectes notaeus* Cope [*Boidae*]). Zool. Anz. 139:162-176.

Lederer, G. 1942. Der Drachenwaran (*Varanus komodoensis* Ouwens) (The dragon monitor [*Varanus komodoensis* Ouwens]). Zool. Gart. (N.F.), Leipzig 14:227-244.

Lederer, G. 1944. Nahrungserwerb, Entwicklung, Paarung und Brutfürsorge von *Python reticulatus* (Schneider) (Food supply, development, mating and brood care of *Python reticulatus* [Schneider]). Zool. Jb. Anat. 68:363-398.

Lederer, G. 1949. Die Vipernatter, *Natrix maura* (L.) (The viperine snake, *Natrix maura* [L.]). Zool. Gart. (N. F.), Leipzig 16:74-93.

Lederer, G. 1950. Ein Bastard von *Elaphe guttata* (Linné)-Männchen x *Elaphe quadrivittata quadrivittata* (Holbrook)-Weibchen und dessen Rückkreuzung mit der mütterlichen Ausgangsart (A bastard of *Elaphe guttata* [Linné] male x *Elaphe quadrivittata quadrivittata* [Holbrook] female and their reverse mating with the maternal species of origin). Zool. Gart. (N. F.), Leipzig 17:235-242.

Lederer, G. 1956. Fortpflanzungsbiologie und Entwicklung von *Python molurus molurus* (Linné) und *Python molurus bivittatus* (Kühl) (Reproductive biology and development of *Python molurus molurus* [Linné] and *Python molurus bivittatus* [Kuhl]). D. Aquar. Terrar.-Z. 9:243-248.

Lee, S., K. C. Zippel, L. Ramos, and J. Searle. 2006. Captive breeding programme for the Kihansi spray toad (*Nectophrynoides asperginis*) at the Wildlife Conservation Society. Bronx, New York. Inter. Zoo Yearb. 40:241-253.

Legge, R. E. 1969. Further notes on the mating behaviour of American alligators *Alligator mississippiensis* at Belle Vue Zoo, Manchester. Inter. Zoo Yearb. 9:35.

Leloup, P. 1964. Observations sur la reproduction du *Dendroaspis jamesoni kaimosae* (Loveridge) (Observations on the reproduction of *Dendroaspis jamesoni kaimosae* [Loveridge]). Bull. Soc. r. Zool. Anvers No. 33:13-27.

Leloup, P. 1975. Observations sur la reproduction de *Bothrops moojeni* Hoge en captivit, (Observations on the reproduction of *Bothrops moojeni* Hoge in captivity). Acta Zool. Path. Antverpiensia No. 62:173-201.

Leloup, P. 1980. Liber amicorum Walter van den Bergh. Lannoo, Tielt.

Leutscher, A. 1952. Vivarium Life. A Manual on Amphibians, Reptiles and Cold-water Fish. Cleaver-Hume Press Ltd, London.

Liesegang, A., J.-M. Hatt, J. Nijboer, R. Forrer, M. Wanner, and E. Isenbügel. 2001. Influence of different dietary calcium levels on the digestibility of CA, Mg, and P in captive-born juvenile Galapagos giant tortoises (*Geochelone nigra*). Zoo Biol. 20:367-374.

Lindsey, P. 1979. Combat behavior in the dusky pygmy rattlesnake, *Sistrurus miliarius barbouri*, in captivity. Herpetol. Rev. 10(3):93.

Linnetz, E. H., D. K. Nichols, B. J. Demeter, and J. M. Scarlett. 1996. High prevalence of gout at necropsy in giant day geckoes (*Phelsuma madagascariensis*) at the National Zoological Park. Proc. Amer. Assoc. Zoo Vet. 1996:223-227.

Liu, S.-K., and F. W. King. 1971. Microsporidiosis in the tuatara. J. Amer. Vet. Med. Assoc. 159:1578-1582.

Lloyd, M. L. 1992. Ophidian paramyxovirus. Historical overview and current recommendations, p. 1-8. 16th International Herpetological Symposium on Captive Propagation and Husbandry. International Herpetological Symposium.

Lloyd, M. L. 1992. Reptilian dsytocias: The common causes and relative treatment success, p. 1-9. 16th International Herpetological Symposium on Captive Propagation and Husbandry. International Herpetological Symposium.

Lloyd, M. L. 1999. Crocodilian anesthesia, p. 205-216. In M. E. Fowler, and R. E. Miller (eds.), Zoo and Wild Animal Medicine: Current Therapy. W. B. Saunders, Philadelphia, London.

Lloyd, M. L., and J. Flanagan. 1991. Recent developments in ophidian paramyxovirus research and recommendations on control. Proc. Amer. Assoc. Zoo Vet. 1991:151-155.

Lloyd, M. L., and P. J. Morris. 1999. Phlebotomy techniques in crocodilians. Bull. Assoc. Rept. Amphib. Vet. 9(3):12-13.

Lloyd, M. L., T. Reichard, and R. A. Odum. 1994. Gallamine reversal in Cuban crocodiles (*Crocodylus rhombifer*) using neostigmine alone versus neostigmine with hyaluronidase. Proc. Amer. Assoc. Zoo Vet. 1994:117-120.

Loehr, V. J. T. 1999. Husbandry, behavior, and captive breeding of the Namaqualand speckled padloper (*Homopus signatus signatus*). Chelonian Conserv. Biol. 3(3):468-473.

Loisel, G. 1912. Histoire des ménageries de l'antiquité à nos jours (History of Menageries from Antiquity to Present Times). O. Doin et fils, Paris.

Loomis, M., and R. Smith. 1987. Fertility following caesarean section in an Aruba Island rattlesnake *Crotalus unicolor*. Inter. Zoo Yearb. 26:187-188.

Loop, M. S. 1976. Auto shaping: A simple technique for teaching a lizard to perform a visual discrimination task. Copeia 1976:574-576.

Louis, E., S. K. Davis, and J. N. Derr. 1990. The Galapagos tortoise (*Geochelone elephantopus*): taxonomic relationships among subspecies. Proc. Amer. Assoc. Zoo Vet. 1990:16-18.

Louwman, J. W. W. 1982. Breeding the six-footed tortoise *Geochelone emys* at Wassenaar Zoo. Inter. Zoo Yearb. 22:153-156.

Loveridge, J. P. 1979. The immobilisation and anaesthesia of crocodilians. Inter. Zoo Yearb. 19:103-112.

Lowe, C. H., M. D. Robinson, and V. D. Roth. 1971. A population of (*Phrynosoma ditmarsi*) from Sonora, Mexico. J. Arizona Acad. Sci. 6:275-277.

Lucia Da Silveira, C. 1986. Nota sobre nascimento e crescimento de *Rhinoclemmys punctularis* (aperema) na Fundacão Rio Zoo (Notes concerning the birth and growth of *Rhinoclemmys punctularis* (aoerema) at the Fundaqão Rio Zoo). Arq. Soc. Zool. Bras. 10th Congr. (Rio de Janeiro Zoological Foundation, Brazilian Arqueolog. Zoological Society, 10th Congress):2-3.

Lucia Da Silveira, C., and C. A. F. Andre. 1986. Notas preliminares sobre lesoes no plastrao por fungos e bacterias em *Phrynops gibbus* (Preliminary notes concerning lesions to the plastron of *Phrynops gibbus* caused by fungi and bacteria). Arq. Soc. Zool. Bras. No. 6:33-36.

Lukin, Y. A. 1977. The reproduction of the blunt-nosed viper (*Vipera lebetina*) in captivity, p. 98-99. In I. S. Darevskij (ed.), Problems in Herpetology: 4th All-Union Herpetological Conference. Leningrad: Nauka.

MacBride, E. W. 1931-1932. Obituary of Miss Joan Procter (1897-1931). Proc. Linn. Soc. London, p. 183-185.

Mágdefrau, H. 1997. Biologie, Haltung und Zucht der Krokodilschwanz-Höckerechse (*Shinisaurus crocodilurus*) (Biology, care and breeding of the Chinese crocodile lizard (*Shinisaurus crocodilurus*)). Z. Köln. Zoo 40 (2):55-60.

Magill, R. N. 1982. Breeding the Siamese crocodile *Crocodylus siamensis* at Miami Metrozoo. Inter. Zoo Yearb. 22:156-158.

Magill, R. N. 1984. Breeding the African slender-snouted crocodile *Crocodylus cataphractus* at Miami Metrozoo. Inter. Zoo Yearb. 23:139-143.

Mallinson, J. J. C. 1998. Collaboration for conservation between the Jersey Wildlife Preservation Trust and countries where species are endangered. Inter. Zoo Yearb. 27:176-191.

Mamet, S. V. 1989. On reproductive biology of ratsnakes of *longissima*-complex from Southern Azerbaidzhan. Voprosy gerpetologii, Kiev, p. 150.

Mamet, S. V., and S. V. Kudryavtsev. 1996. Breeding Latifi's viper (*Vipera latifiii*) at Moscow Zoo. Snake 27:147-148.

Mamet, S. V., and S. V. Kudryavtsev. 1997. Captive propagation of the Mandarin rat snake (*Elaphe mandarina*) at Moscow Zoo. Asiatic Herpetol. Res. 7:85-86.

Mamet, S. V., and S. V. Kudryavtsev. 1997. Caucasian viper: Husbandry and reproduction in captivity. Russian J. Zool. 4:198-199.

Mamet, S. V., and S. V. Kudryavtsev. 1997. Notes on the reproductive biology of rare Asiatic ratsnake *Elaphe persica*. Russian J. Zool. 4:200-202.

Mannion, W. 1996. Maintenance and captive reproduction of the green python *Morelia viridis* at the Queensland Reptile and Fauna Park. Thylacynus 21(3b):14-17.

Mannix, J. 1954. Married to Adventure. Hamish Hamilton, London.

Marcellini, D. L. 1977. The function of the vocal display of the lizard *Hemidactylus frenatus* (Sauria: Gekkonidae). Anim. Behav. 25:414-417.

Marcellini, D. L. 1977. Acoustic and visual display behavior of gekkonid lizards. Amer. Zool. 17:251-260.

Marcellini, D. L. 1978. The acoustic behavior of lizards, p. 287-300. In N. Greenberg, and P. D. MacLean (eds.), Behavior and Neurology of Lizards. National Institute of Mental Health, Rockville, MD.

Marcellini, D. L., and T. A. Jenssen. 1988. Visitor behavior in the National Zoo's Reptile House. Zoo Biol. 7:329-338.

Marcellini, D. L., and T. A. Jenssen. 1991. Avoidance learning by the curly-tailed lizard, *Leiocephalus schreibersi*: implications for anti-predator behavior. J. Herpetol. 25:238-241.

Marcellini, D. L., and J. B. Murphy. 1998. Education in a zoological park or aquarium: An ontogeny of learning opportunities. Herpetologica 54 (Suppl.):S12-S16.

Marcellini, D. L., and A. Peters. 1982. Preliminary observations on endogenous heat production after feeding in *Python molurus*. J. Herpetol. 16:92-95.

Marks, S. K., and S. B. Citino. 1990. Hematology and serum chemistry of the radiated tortoise (*Testudo radiata*). J. Zoo Wildl. Med. 21:342-344.

Martin, D. L. 1991. Captive husbandry as a technique to conserve a species of special concern, the Yosemite toad, p. 17-32. In R. E. Staub (ed.), Captive Propagation and Husbandry of Reptiles and Amphibians. Northern California Herpetological Society.

Martin de Camilo, J. E., C. S. Asa, J. A. Ettling, A. M. Hampton, and N. Haskell. 1999. Comparative follicular dynamics between the Brazilian rainbow boa (*Epicrates cenchria cenchria*) and the ball python (*Python regius*). Proc. Amer. Assoc. Zoo Vet. 1999:95-96.

Maruska, E. J. 1986. Amphibians: Rreview of zoo breeding programmes. Inter. Zoo Yearb. 24/25:56-65.

Maruska, E. J. 1989. Husbandry and propagation of salamander species at the Cincinnati Zoo, p. 1-9. In R. Gowen (ed.), Captive Propagation and Husbandry of Reptiles and Amphibians. Northern California Herpetological Society.

Maruska, E. J. 1994. Procedures for setting up and

maintaining a salamander colony, p. 229-242. In J. B. Murphy, K. Adler, and J. T. Collins (eds.), Captive Management and Conservation of Amphibians and Reptiles. Society for the Study of Amphibians and Reptiles. Contributions to Herpetology, volume 11, Ithaca NY.

Matsuoka, Y. 1994. Captive breeding and growth of Egyptian tortoises. Animals and Zoos 46(2. No. 528):44-47 [in Japanese].

Mautino, M., and C. D. Page. 1993. Biology and medicine of turtles and tortoises. Vet. Clinics North America Small Anim. Pract. 23:1251-1270.

Mayeaux, M. H. 1994. Symptoms of gas-bubble trauma in two species of turtles *Chelydra serpentina* and *Apalone spinifera*. Herpetol. Rev. 25(1):19.

Mays, S. R., and K. H. Peterson. 1996. Aggregative behavior and sibling recognition in larvae of the Houston toad (*Bufo houstonensis*), p. 33-38. In P. D. Strimple (ed.), Advances in Herpetoculture. Special Publication of the International Herpetological Symposium, Inc., No. 1. International Herpetological Symposium, Inc., Des Moines IA.

McCrady, W. B., J. B. Murphy, C. M. Garrett, and D. T. Roberts. 1994. Scale variation in a laboratory colony of amelanistic diamondback rattlesnakes (*Crotalus atrox*). Zoo Biol. 13:95-106.

McCrystal, H. K., and J. L. Behler. 1982. Husbandry and reproduction of captive giant ameiva lizards *Ameiva ameiva* at the New York Zoological Park. Inter. Zoo Yearb. 22:159-163.

McDougal, J., and C. Castellano. 1996. Husbandry and captive breeding of the Travancore tortoise (*Indotestudo travancorica*) at the Wildlife Conservation Society, In P. D. Strimple (ed.), Advances in Herpetoculture. Special Publication of the International Herpetological Symposium, Inc., No. 1. International Herpetological Symposium, Inc., Des Moines IA.

McGeorge, I. 1997. The successful breeding of the sunbeam snake *Xenopeltis unicolor* Reinwardt, 1826 at Chester Zoo. Brit. Herpetol. Soc. Bull. 61:1-5.

McKeown, S. 1978. Hawaiian Reptiles and Amphibians. Oriental Publ. Co., Honolulu HI.

McKeown, S. 1982. Wild status and captive management of Indian Ocean *Phelsuma* with special reference to the Mauritius lowland forest day gecko (*Phelsuma guimbeaui*), p. 157-170. In D. L. Marcellini (ed.) 6th International Herpetological Symposium on Captive Propagation and Husbandry. International Herpetological Symposium, Thurmont MD.

McKeown, S. 1984. Captive maintenance and propa-
gation of Indian Ocean day geckos (genus *Phelsuma*). Bull. Chicago Herpetol. Soc. 19:55-63.

McKeown, S. 1984. Captive maintenance and propagation of Indian Ocean day geckos (Genus *Phelsuma*). Bull. Chicago Herpetol. Soc. 19:55-63.

McKeown, S. 1985. The ecosystem approach--new survival strategies for managing and displaying reptiles and amphibians in zoos, p. 1-5. In R. L. Gray (ed.), Captive Propagation and Husbandry of Reptiles and Amphibians: Northern California Herpetological Society & Bay Area Amphibian and Reptile Society.

McKeown, S. 1989. The first captive breeding of the Madagascar ground boa (*Acrantophis madagascariensis*) in North America from long term captive adults, p. 69-76. In R. Gowen (ed.), Captive Propagation and Husbandry of Reptiles and Amphibians. Special Publication # 5 ed. Northern California Herpetological Society.

McKeown, S. 1989. Breeding and maintenance of the Mauritius lowland forest day gecko *Phelsuma g. guimbeaui* at Fresno Zoo. Inter. Zoo Yearb. 28:116-122.

McKeown, S. 1991. Managing and breeding tortoises in captivity, p. 111-116. In R. E. Staub (ed.), Captive Propagation and Husbandry of Reptiles and Amphibians. Special Publication # 6 ed. Northern California Herpetological Society.

McKeown, S. 1992. Day geckos (genus *Phelsuma*): Comments on the wild status and captive management of selected species. J. Herpetol. Assoc. Africa 40:80-81.

McKeown, S. 1992. The brown tree snake, *Boiga irregularis*, in Guam. Controlling the spread of the "reptilian mongoose" of the South Pacific, p. 1-12. In 16th International Herpetological Symposium on Captive Propagation and Husbandry: International Herpetological Symposium.

McKeown, S. 1993. The General Care and Maintenance of Day Geckos. The Herpetocultural Library- Special edition. Advanced Vivarium Systems, Lakeside CA.

McKeown, S. 1996. Field Guide to the Reptiles and Amphibians in the Hawaiian Islands. Diamond Head Publ., Los Osos CA.

McKeown, S., and M. J. Miller. 1984. A brief note on the natural history, captive maintenance and propagation of the Seychelles giant skin-sloughing gecko, *Ailuronyx seychellensis*, p. 96-102. In R. A. Hahn (ser. ed.), 8th International Herpetological Symposium on Captive Propagation and Husbandry. International Herpetological Symposium

Thurmont MD.

McKeown, S., J. O. Juvik, and D. E. Meier. 1982. Observations on the reproductive biology of the land tortoises *Geochelone emys* and *Geochelone yniphora* in the Honolulu Zoo. Zoo Biol. 1:223-235.

McKeown, S., D. E. Meier, and J. O. Juvik. 1991. The management and breeding of the Asian forest tortoise (*Manouria emys*) in captivity. In K. R. Beaman, F. Coporaso, S. McKeown, and M. D. Graff (eds.), Proceedings of the First International Symposium on Turtles and Tortoises: Conservation and Captive Husbandry. Chapman University, August 9-12, 1990. Van Nuys, CA: California Turtle and Tortoise Club.

McLain, J. M. 1983. Reproduction in captive Malagasy tree boas, *Sanzinia madagascariensis* (Serpentes: Boidae), p. 124-131. In P. J. Tolson (ed.), 7th International Herpetological Symposium on Captive Propagation and Husbandry. International Herpetological Symposium, Thurmont MD.

McNamara, T. S., P. C. Charles, Y. Kress, K. Weidenheim, and W. Holmstrom. 1999. Crystalline inclusions associated with vomeronasal organ pathology in red-eyed tree frogs (*Agalychnis callidryas*). Proc. Amer. Assoc. Zoo Vet. 1999:27-29.

Mebs, D. 1965. Zur Pathologie und Therapie einer bei Wasserschildkröten häufig auftretenden Augenerkrankung (Concerning the pathology and therapy of an illness of the eyes frequently found in water turtles). Zool. Gart. (N.F.), Leipzig 31:304-309.

Mehrtens, J. M. 1950. The saw-scaled viper in captivity. Herpetologica 6:202.

Miller, E. A., R. J. Montali, E. C. Ramsey, and B. A. Rideout. 1992. Disseminated chromoblastomycosis in a colony of ornate-horned frogs (*Ceratophrys ornata*). J. Zoo Wildl. Med. 23:433-438.

Miller, S. M., B. Koike, and C. Lobue. 1997. Treatment of an esophageal foreign body in a Kemp's ridley sea turtle, *Lepidochelys kempii*. Bull. Assoc. Rept. Amphib. Vet. Vet. 7(3):13-15.

Miller, T. J. 1983. Notes on a breeding of the red-eyed tree frog, *Agalychnis callidryas*, p. 34-41. In P. J. Tolson (ed.), 7th International Herpetological Symposium on Captive Propagation and Husbandry. International Herpetological Symposium, Thurmont MD.

Minton, S. 1975. Snakebite in zoos. Inter. Zoo Yearb. 15:179-185.

Misra, P. R., D. Kumar, G. M. Patnaik, R. P. Raman, and A. Sinha. 1993. Bacterial isolates from apparently healthy and diseased crocodiles (*Gavialis gangeticus*). Indian Vet. J. 70:375-376.

Mitchell, L. A. 1986. Comments on the maintenance and reproduction of *Hydrosaurus pustulatus* at the Dallas Zoo, p. 185-186. In S. McKeown, F. Caporaso, and K. H. Peterson (eds.), 9th International Herpetological Symposium on Captive Propagation and Husbandry. Zoological Consortium, Inc., Thurmont MD.

Mitchell, L. A. 1990. Reproduction of Gould's monitors (*Varanus gouldii*) at the Dallas Zoo. Bull. Chicago Herpetol. Soc. 25(1):8-9.

Mitchell, P. C. 1929. Centenary History of the Zoological Society of London. London Zoological Society, London.

Mitchell, P. C., and R. I. Pocock. 1907. On the feeding of reptiles in captivity, with observations on the fear of snakes by other vertebrates. Proc. Zool. Soc. London 1907:785-794.

Montali, R. J., E. E. Smith, M. Davenport, and M. Bush. 1975. Dermatophilosis in Australian bearded lizards. J. Amer. Vet. Med. Assoc. 167:553-555.

Montanucci, R. R. 1984. Breeding, captive care and longevity of the shorthorned lizard *Phrynosoma douglassi*. Inter. Zoo Yearb. 23:148-156.

Montanucci, R. R. 1989. The reproduction and growth of *Phrynosoma ditmarsi* in captivity. Zoo Biol. 8:139-149.

Moore, D., and J. McLaughlin. 1981. Survey of boid care in American zoos, p. 144-147. In C. B. Banks, and A. A. Martin (eds.), Proceedings of the Melbourne Herpetological Symposium May 19-21, 1980. Zoological Board of Victoria, The Royal Melbourne Zoological Gardens, Parkville, Victoria, Australia.

Moore, F. L. 1994. Complexity in the control of amphibian reproduction: Reproductive behavior and physiology of a salamander, p. 125-131. In J. B. Murphy, K. Adler, and J. T. Collins (eds.), Captive Management and Conservation of Amphibians and Reptiles. Society for the Study of Amphibians and Reptiles. Contributions to Herpetology, volume 11, Ithaca NY.

Morales, P., and F. Dunker. 2001. Fish tuberculosis, *Mycobacterium marinum*, in a group of Egyptian spiny-tailed lizards, *Uromastyx aegyptius*. J. Herpetol. Med. Surg. 11(3):27-30.

Morgan, D. R. 1993. *Homopus signatus*, speckled padloper, reproduction. J. Herpetol. Assoc. Afr. 42:34.

Morley, T. P. 1992. Eggs and incubation in the Australian lizards *Amphibolurus nobbi* and *Eremiascincus richardsoni*. Trans. Royal Soc. South Australia

116:147-148.

Morris, D. 1968. Must we have zoos? A famous zoologist answers: Yes but . . . Life magazine 65 (No. 18), November 1968:78-86.

Morris, P. J., and A. C. Alberts. 1996. Determination of sex in white-throated monitors (*Varanus albigularis*), Gila monsters (*Heloderma suspectum*), and beaded lizards (*H. horridum*) using two-dimensional ultrasound imaging. J. Zoo Wildl. Med. 27:371-377.

Morris, P. J., and C. Henderson. 1998. Gender determination in mature Gila monsters, *Heloderma suspectum*, and Mexican beaded lizards, *Heloderma horridum*, by ultrasound imaging of the ventral tail. Bull. Assoc. Rept. Amphib. Vet. 8(4):4-5.

Morris, P. J., L. A. Jackintell, and A. C. Alberts. 1996. Predicting gender of subadult Komodo dragons (*Varanus komodoensis*) using two dimensional ultrasound imaging and plasma testosterone concentration. Zoo Biol. 15:341-348.

Mowbray, L. S. 1965. Hawaiian monk seals *Monachus schauinslandi* and green turtles *Chelonia mydas*, at Waikiki Aquarium. Inter. Zoo Yearb. 5:146-147.

Mulder, J. B., J. J. Hauser, and J. J. Perry. 1979. Surgical removal of retained eggs from a kingsnake (*Lampropeltis getulus*). J. Zoo Anim. Med. 10:22-23.

Müller, P. 1970. Notes on reptile breeding at Leipzig Zoo. Inter. Zoo Yearb. 10:104-105.

Mumaw, L. M. 1992. ARAZPA: developing the Australasian zoo industry as a conservation resource. Int. Zoo Yearb. 31:9-12.

Murphy, J. B. 1971. Notes on the care of the ridge-tailed monitor *Varanus acanthurus brachyurus* at Dallas Zoo. Inter. Zoo Yearb. 11:230-231.

Murphy, J. B. 1972. Notes on Indo-Australian varanids in captivity. Inter. Zoo Yearb. 12:199-202.

Murphy, J. B. 1976. Pedal luring in the leptodactylid frog, *Ceratophrys calcarata* Boulenger. Herpetologica 32:339-341.

Murphy, J. B. 1977. An unusual method of immobilizing avian prey by the dog-tooth cat snake, *Boiga cynodon*. Copeia 1977:182-184.

Murphy, J. B. 2005. The sad tale of two zoos. Herpetol. Rev. 36:6-7.

Murphy, J. B. 2005. Chameleons: Johann von Fischer and other perspectives. Soc. Study Amphib. Rept. Herp. Circ. 33.

Murphy, J. B. 2006. Wild and ferocious reptiles in the Tower of London. Herpetol. Rev. 37:10-13.

Murphy, J. B., and B. L. Armstrong. 1978. Maintenance of rattlesnakes in captivity. Univ. Kansas Mus. Nat. Hist. Spec. Publ. no. 3:1-40.

Murphy, J. B., and D. G. Barker. 1980. Courtship and copulation of the Ottoman viper (*Vipera xanthina*) with special reference to use of hemipenes. Herpetologica 36:165-170.

Murphy, J. B., and J. A. Campbell. 1987. Captive maintenance, p. 165-183. In R. A. Seigel, J. T. Collins, and S. S. Novak (eds.), Snakes: Ecology and Evolutionary Biology. Macmillan, New York.

Murphy, J. B., and W. Card. 1998. A glimpse into the life of a zoo herpetologist. Herpetol. Rev. 29(2):85-90.

Murphy, J. B., and D. Chiszar. 1989. Herpetological masterplanning for the 1990s. Inter. Zoo Yearb. 28:1-7.

Murphy, J. B., and R. K. Guese. 1977. Reproduction in the Hispaniolan boa, *Epicrates fordii fordii*, at Dallas Zoo. Inter. Zoo Yearb. 17:132-133.

Murphy, J. B., and D. E. Jacques. 2005. Grace Olive Wiley: Zoo curator with safety issues. Herpetol. Rev. 36:365-367.

Murphy, J. B., and D. E. Jacques. 2006. Death by snakebite: The entwined histories of Grace Olive Wiley and Wesley H. Dickinson. Bull. Chicago Herpetol. Soc. Spec. Suppl.

Murphy, J. B., and W. E. Lamoreaux. 1978. Mating behavior in three Australian chelid turtles (Testudines: Pleurodira: Chelidae). Herpetologica 34:398-405.

Murphy, J. B., and L. A. Mitchell. 1974. Ritualized combat behavior of the pygmy mulga monitor lizard, *Varanus gilleni* (Sauria: Varanidae). Herpetologica 30:90-97.

Murphy, J. B., and L. A. Mitchell. 1984. Miscellaneous notes on the reproductive biology of reptiles. 6. Thirteen varieties of the genus *Bothrops* (Serpentes, Crotalidae). Acta Zool. Path. Antverpiensia 78:199-214.

Murphy, J. B., and L. A. Mitchell. 1984. Breeding the aquatic box turtle at Dallas Zoo. Inter. Zoo Yearb. 23:135-137.

Murphy, J. B., and J. A. Shadduck. 1976. Reproduction in the eastern diamondback rattlesnake, *Crotalus adamanteus* in captivity, with comments regarding a teratoid birth anomaly. Brit. J. Herpetol. 5:727-733.

Murphy, J. B., D. G. Barker, and B. W. Tryon. 1978. Miscellaneous notes on the reproductive biology of reptiles. 2. Eleven species of the family Boidae, genera *Candoia, Corallus, Epicrates* and *Python*. J. Herpetol. 12:385-390.

Murphy, J. B., C. C. Carpenter, and J. C. Gillingham. 1978. Caudal luring in the green tree python,

Chondropython viridis (Reptilia, Serpentes, Boidae). J. Herpetol. 12:117-119.

Murphy, J. B., J. T. Collins, and K. Adler. 1997. Herpetology versus herpetoculture. Cooperation or competition? Reptiles 5(3):72-79.

Murphy, J. B., W. E. Lamoreaux, and D. G. Barker. 1981. Miscellaneous notes on the reproductive biology of reptiles. 4. Eight species of the family Boidae, genera *Acrantophis, Aspidites, Candoia, Liasis* and *Python*. Trans. Kansas Acad. Sci. 85:96-119.

Murphy, J. B., W. E. Lamoreaux, and C. C. Carpenter. 1978. Threatening behavior in the angle-headed dragon, *Goniocephalus dilophus* (Reptilia, Lacertilia, Agamidae). J. Herpetol. 12:455-460.

Murphy, J. B., H. Quinn, and J. A. Campbell. 1977. Observations on the breeding habits of the aquatic caecilian *Typhlonectes compressicauda*. Copeia 1977:66-69.

Murphy, J. B., B. W. Tryon, and B. J. Brecke. 1978. An inventory of reproduction and social behavior in captive gray-banded kingsnakes, *Lampropeltis mexicana alterna* (Brown). Herpetologica 34:84-93.

Murphy, J. B., C. Ciofi, C. de La Panouse, and T. Walsh (eds.). 2002. Komodo Dragons. Biology and Conservation. Smithsonian Institution Press. Washington DC.

Murphy, J. B., J. E. Rehg, P. F. A. Maderson, and W. B. McCrady. 1987. Scutellation and pigmentation defects in a laboratory colony of western diamondback rattlesnakes *Crotalus atrox*: mode of inheritance. Herpetologica 43:292-300.

Mutschmann, F. 1998. Nachweis von *Chlamydia psittaci* - Infektionen bei Amphibien mittels eines spezifischen Immunofluoreszenztests (Evidence of *Chlamydia psittaci* infections in amphibians, obtained by means of a specific immunofluorescence test [IFT]). Berliner und Muenchener Tierärzliche Wochenschrift (Berlin and Munich Veterinary Weekly) 111(5):187-189.

Nace, G. W. 1977. Breeding amphibians in captivity. Inter. Zoo Yearb. 17:44-50.

Nares, P., and J. E. Cooper. 1971. The Nairobi Snake Park. Inter. Zoo Yearb. 11:254.

Naulleau, G. 1973. Rearing the asp viper *Vipera aspis* in captivity. Inter. Zoo Yearb. 13:108-111.

Naulleau, G., and B. van den Brule. 1981. Feeding, growth, molt and venom production in the Russell's viper *Vipera russelli* in captivity. Inter. Zoo Yearb. 21:163-172.

Neiffer, D. L., S. K. Marks, E. C. Klein, and N. J. Brady. 1998. Shell lesion management in two loggerhead sea turtles, *Caretta caretta*, with employment of PC-7 epoxy paste. Bull. Assoc. Rept. Amphib. Vet. 8(4):12-17.

Neitman, K. 1983. Captive husbandry of the tuberculate geckos of the genus *Coleonyx*, p. 74-77. In P. J. Tolson (ed.), 7th International Herpetological Symposium on Captive Propagation and Husbandry. International Herpetological Symposium, Thurmont MD.

Netten, H., and F. Zuurmond. 1985. Offspring of the common snapping turtle *Chelydra serpentina* in the reptile zoo Iguana. Lacerta 44(3):42-43.

Newman, D. G. 1982. Breeding tuataras, *Sphenodon punctatus*, in captivity, p. 277-283. In D. Newman (ed.), New Zealand Herpetology. New Zealand Wildlife Service, Wellington.

Newman, D. G., I. G. Crook, and L. R. Moran 1979. Some recommendations on the captive maintenance of tuataras *Sphenodon punctatus* based on observations in the field. Inter. Zoo Yearb. 19:68-74.

Nichols, D. K. 2000. Amphibian respiratory diseases. Vet. Clin. North America: Exotic Animal Practice 3:551-554.

Nichols, D. K. 2003. Tracking down the killer chytrid of amphibians. Herpetol. Rev. 34:101-104.

Nichols, D. K., and E. W. Lamirande. 1994. Use of methohexital sodium as an anesthetic in two species of colubrid snakes. Proc. Amer. Assoc. Zoo Vet. 1994:161-162.

Nichols, D. K., A. P. Pessier, and J. E. Longcore. 1998. Cutaneous chytridiomycosis in amphibians: an emerging disease? Proc. Amer. Assoc. Zoo Vet. 1998:269-271.

Nichols, D. K., A. J. Smith, and C. H. Gardiner. 1996. Dermatitis of anurans caused by fungal-like protiists. Proc. Amer. Assoc. Zoo Vet. 1996:220-222.

Nichols, D. K., E. W. Lamirande, A. P. Pessier, and J. E. Longcore. 2000. Experimental transmission and treatment of cutaneous chytridiomycosis in poison dart frogs (*Dendrobates auratus* and *Dendrobates tinctorius*). Proc. Amer. Assoc. Zoo Vet. 2000:42-44.

Nichols, D. K., E. W. Lamirande, A. P. Pessier, and J. E. Longcore. 2001. Experimental transmission of cutaneous chytridiomycosis in dendrobatid frogs. J. Wildl. Dis. 37:1-11.

Nietzke, G. 1969, 1972. Die terrarientiere: Bau, technische Einrichtung und Bepflanzung der Terrarien: Haltung, Fütterung und Pflege der Terrientiere in zwei Bänden (Terrarium Animals: Construction, Technical Equipment, and

Planning of Terraria: Care and Feeding of Terrarium Animals in Two Volumes). Eugen Ulmer, Stuttgart.

Noegel, R. P. 1989. Husbandry of the West Indian rock iguanas *Cyclura* at Life Fellowship Bird Sanctuary, Seffner. Inter. Zoo Yearb. 28:131-135.

Noegel, R. P., and G. A. Moss. 1989. Breeding the Galapagos tortoise *Geochelone elephantopus* at Life Fellowship Bird Sanctuary, Seffner. Inter. Zoo Yearb. 28:78-83.

Nuitja, I. N. S., and I. Uchida (1982). Preliminary studies on the growth and food consumption of the juvenile loggerhead turtle (*Caretta caretta* L.) in captivity. Aquaculture 27:157-160.

Obst, F. J., K. Richter, and U. Jacob. 1988. The Completely Illustrated Atlas of Reptiles and Amphibians for the Terrarium. T.F.H. Publications, Neptune City NJ.

Ocholi, R. A., and L. U. Enurah. 1989. Salmonellosis in a captive crocodile (*Crocodylus niloticus*) due to *Salmonella choleraesuis*. J. Zoo Wildl. Med. 20:377-378.

Odum, R. A. 1984. Water quality, an often overlooked parameter for the amphibian enclosure, p. 33-58. In R. A. Hahn (ser. ed.), 8th International Herpetological Symposium on Captive Propagation and Husbandry. International Herpetological Symposium, Thurmont MD.

Odum, R. A., J. M. McLain, and T. C. Sheley. 1983. Hormonally induced breeding and rearing of White's tree frog, *Litoria caerulea* (Anura: Pelodryadidae), p. 42-53. In P. J. Tolson (ed.), 7th International Herpetological Symposium on Captive Propagation and Husbandry. International Herpetological Symposium, Thurmont MD.

Oka, K., H. Takayama, H. Nakamura, and Y. Michimori. 1983. Keeping the leather-back turtle, *Dermochelys coriacea*. J. Japan Assoc. Zoos Aquar. 25(3):67-70 [in Japanese].

Olexa, A. 1975. Breeding of the Indian cobra, *Naja naja* (Linnaeus, 1758) at Prague Zoo. Gazella (Praha) 1(No. 2):45-46.

Oliver, J. A. 1955. The Natural History of North American Amphibians and Reptiles, D. Van Nostrand, Princeton NJ.

Oliver, J. A. 1956. Reproduction in the king cobra, *Ophiophagus hannah* Cantor. Zoologica (New York) 41:145-157.

Oliver, J. A. 1958. Snakes in Fact and Fiction. Macmillan, New York.

Orlov, N., and S. Ryabov. 2002. Breeding of black man-

grove snake *Boiga dendrophila gemmicincta* (Duméril, Bibron et Duméril, 1854) [Serpentes: Colubridae: Colubrinae] from Sulawesi Island (Indonesia). Russian J. Herpetol.9:77-79.

Osman, H. 1967. A note on the breeding behaviour of the Komodo dragons *Varanus komodoensis* at Jogjakarta Zoo. Inter. Zoo Yearb. 7:181.

Otis, V. S., and J. L. Behler. 1973. The occurrence of Salmonellae and *Edwardsiella* in the turtles of the New York Zoological Park. J. Wildl. Dis. 9:4-6.

Ovezmukhammedov, A., V. S. Glebezdin, M. I. Dobrynin, and D. Annaev. 1975. On the fauna of parasites of *Python reticulatus* from the Ashkhabad Zoo. Izvestiya Akad. Nauk. turkmen. SSR (Biol.) 1975(3):59-64.

Owen, C. 1742. An essay towards a natural history of serpents: in two parts. I. The first exhibits a general view of serpents, in their various aspects . . . II. The second gives a view of most serpents that are known in the several parts of the world . . . III. To which is added a third part; containing six dissertations upon the following articles, as collateral to the subject. 1. Upon the primeval serpent in paradise. 2. The fiery serpents that infested the camp of Israel. 3. The brazen serpent erected by Moses. 4. The divine worship given to serpents by the nations. 5. The origin and reason of that monstrous worship. 6. Upon the adoration of different kinds of beasts by the Egyptians . . . The whole intermix'd with variety of entertaining digressions, philosophical and historical. Printed for the author, London.

Owens, D. W., J. R. Hendrickson, V. Lance, and I. P. Callard. 1978. A technique for determining sex of immature *Chelonia mydas* using radioimmunoassay. Herpetologica 34:270-273.

Packard, G. C., and J. A. Phillips. 1994. The importance of the physical environment for the incubation of reptile eggs, p. 195-208. In J. B. Murphy, K. Adler, and J. T. Collins (eds.), Captive Management and Conservation of Amphibians and Reptiles. Society for the Study of Amphibians and Reptiles. Contributions to Herpetology, volume 11, Ithaca NY.

Page, C. D., M. Mautino, H. Derendorf, and W. Mechlinski. 1991. Multiple-dose pharmacokinetics of ketoconazole administered orally to gopher tortoises (*Gopherus polyphemus*). J. Zoo Wildl. Med. 22:191-198.

Page, C. D., M. Mautino, J. R. Meyer, and W. Mechlinski. 1988/1989. Preliminary pharmacokinetics of ketoconazole in gopher tortoises *Gopherus polyphemus*. Proc. Amer. Assoc. Zoo Vet. 1988/1989: 63.

Paine, F.L. 1984. The husbandry, management and reproduction of the Puerto Rican crested toad (*Bufo lemur*), p. 59-75. In R. A. Hahn (ser. ed.), 8th International Herpetological Symposium on Captive Propagation and Husbandry. International Herpetological Symposium, Thurmont MD.

Paine, F. L., and J. Weinheimer. 1984. A method for tube-feeding anurans developed for *Rhacophorus viridis*. Inter. Zoo Yearb. 23:204-205.

Paine, F. L., J. D. Miller, G. Crawshaw, B. Johnson, R. Lacy, C. F. Smith III, and P. J. Tolson. 1989. Status of the Puerto Rican crested toad *Peltophyrne lemur*. Inter. Zoo Yearb. 28:53-58.

Panouse, de La, R., and C. Pellier. 1973. Ponte d'un python réticulé (*Python reticulatus*) élevé en terrarium, et incubation des oeufs (Oviposition by a (*Python reticulatus*) raised in the terrarium, and incubation of the eggs). Bull. Mus. Nat. Hist. (Paris) No. 105 (Zool. 79):37-48.

Patterson, R. W. 1974. Hatching the African python, *Python sebae*, in captivity. Inter. Zoo Yearb. 14:81-82.

Patterson, R. W. 1978. Hatching the Anchieta's dwarf python *Python anchietae* in captivity. Inter. Zoo Yearb. 18:99-101.

Patterson, R. W. 1987. Reptiles of Southern Africa. C. Struik, Cape Town, S.A.

Patterson, R. W., and P. R. Meakin. 1986. Snakes. Struik Pocket Guide Series. C. Struik, Cape Town, S.A.

Paulraj, S., S. Subbarayalu Naidu, and J. Pakkiaraj. 1987. Rearing the olive ridley *Lepidochelys olivacea* in artificial sea water. Inter. Zoo Yearb. 26:90-94.

Pawley, R. 1965. In the Chicago Zoological Park -- Housekeeping for marine iguanas. Anim. Kingdom 68:146-150.

Pawley, R. 1966. Observation on the care and nutrition of a captive group of marine iguanas *Amblyrhynchos cristatus*. Inter. Zoo Yearb. 6:107-115.

Pawley, R. 1969. Further notes on a captive colony of marine iguanas *Amblyrhynchos cristatus* at Brookfield Zoo, Chicago. Inter. Zoo Yearb. 9:41-44.

Pawley, R. 1972. Notes on reproduction and behaviour of the green crested basilisk *Basiliscus plumifrons* at Brookfield Zoo, Chicago. Inter. Zoo Yearb. 12:141-144.

Pawley, R. 1982. Husbandry of the eastern coral snake (*Micrurus fulvius*) at Brookfield Zoo, p. 294-303.

In D. L. Marcellini (ed.), 6th International Herpetological Symposium on Captive Propagation and Husbandry. International Herpetological Symposium, Thurmont MD.

Pawley, R. 1988. Blomberg toad, *Bufo blombergi*, reproduction at Brookfield Zoo: Paucity to profusion. Bull. Chicago Herpetol. Soc. 23(4):53-54.

Payne, D. 1989. Breeding the West African dwarf crocodile at Woodland Park Zoological Gardens, p. 37-42. In R. Gowen (ed.), Captive Propagation and Husbandry of Reptiles and Amphibians. Northern California Herpetological Society.

Peaker, M. 1969. Some aspects of the thermal requirements of reptiles in captivity. Inter. Zoo Yearb. 9:3-8.

Pederzani, H.-A. 2000. Erinnerung an Werner Krause (1905-1970) (Commemoration for Werner Krause (1905-1970)). Milu 10:230-235.

Pedrono, M., and A. Sarovy. 1998. First release of ploughshare tortoises *Geochelone yniphora*. Dodo, J. Jersey Wildl. Preserv. Trust 34:173-174.

Pedrono, M., and A. Sarovy. 2000. Trial release of the world's rarest tortoise *Geochelone yniphora* in Madagascar. Biol. Conserv. 95:333-342.

Pedrono, M., L. L. Smith, A. Sarovy, R. Bourou, and H. Tiandray. 2001. Reproductive ecology of the ploughshare tortoise (*Geochelone yniphora*). J. Herpetol. 35:151-156.

Peel, C. V. A. 1903. The Zoological Gardens of Europe. Their History and Chief Features. F. E. Robinson & Co., London.

Pérez-Higareda, G., H. M. Smith, and R. B. Smith. 1985. A new species of *Tantilla* from Veracruz, Mexico. J. Herpetol. 19:290-292.

Pérez-Higareda, G., A. Rangel-Rangel, D. Chiszar, and H. M. Smith. 1995. Growth of Morelet's crocodile (*Crocodylus moreletii*) during the first three years of life. Zoo Biol. 14:173-177.

Perry, G., R. Habani, and II. Mendelssohn. 1993. The first captive reproduction of the desert monitor *Varanus griseus griseus* at the Research Zoo of Tel Aviv University. Inter. Zoo Yearb. 32:188-190.

Perry, J. J., and D. A. Blody. 1986. Courtship and reproduction in captive Cretan vipers, *Vipera lebetina schweizeri*. Herpetol. Rev. 17(2):41-42.

Pestinsky, B. V. 1939. Materials for venomous snakes in biology: their capture and keeping, p. 4-62. The Works of Uzbek Zoo. Tashkent: Gostechizdat [in Russian].

Peters, U. W. 1969. Some observations on the captive breeding of the Madagascan tortoise *Testudo radiata* at Sydney Zoo. Inter. Zoo Yearb. 9:29.

Peters, U. W. 1979. Second generation breeding of the cantil *Agkistrodon bilineatus* at Taronga Zoo. Inter. Zoo Yearb.19:100-101.

Peters, U. W. 1982. The breeding of endangered reptiles, a success story. Zool. Gart. (N.F.), Jena 52:21-28.

Peters, U. W., and E. P. Finnie. 1979. First breeding of the Aldabra tortoise *Geochelone gigantea* at Sydney Zoo. Inter. Zoo Yearb.19:53-55.

Peterson, K. H. 1982. Reproduction of captive *Crotalus m. mitchelli* and *Crotalus durissus* at the Houston Zoological Garden, p. 323-328. In D. L. Marcellini (ed.), 6th International Herpetological Symposium on Captive Propagation and Husbandry. International Herpetological Symposium, Thurmont MD.

Peterson, K. H. 1982. Reproduction in captive *Heloderma suspectum*. Herpetol. Rev. 13(4):122-124.

Peterson, K. H. 1992. Reproduction of the yellow-lined palm viper *Bothriechis lateralis* Peters. Contributions in Herpetology, Special Publication of the Greater Cincinnati Herpetological Society :65-69.

Peterson, K. H., and R. A. Odum. 1986. Reproduction and notes on the maintenance of arboreal and terrestrial *Bothrops* at Houston Zoological Gardens, p. 187-197. In S. McKeown, F. Caporaso, and K. H. Peterson (eds.), 9th International Herpetological Symposium on Captive Propagation and Husbandry. Zoological Consortium, Thurmont MD.

Petzold, D. 2001. Tierpark Dählhölzli/Bern Zoo, p. 1224-1228. In C. E. Bell (ed.), Encyclopedia of the World's Zoos. Fitzroy Dearborn Publishers, Chicago, London.

Petzold, H.-G. 1959. Gewöllbildung bei Krokodilen (Formation of pellets in crocodiles). Zool. Anz. 163:76-82.

Petzold, H.-G. 1962. Successful breeding of *Leiocephalus carinatus*. Inter. Zoo Yearb. 4:97-98.

Petzold, H.-G. 1963. Beobachtungen und Erfahrungen bei der Zucht des Tokehs (*Gekko gecko*) (L.) (Observations and experiences in breeding the Tokay gecko [*Gekko gecko*] [L.]). Zool. Gart. (N.F.), Leipzig 27:194-202.

Petzold, H.-G. 1963. Notizen zur Fortpflanzungsbiologie und Jugendentwicklung zweier Grubenottern (*Serpentes: Crotalidae: Crotalus atrox* und *Agkistrodon p. piscivorus*) (Notes on the reproductive biology and youthful development of two pit vipers [*Serpentes: Crotalidae: Crotalus atrox* and *Agkistrodon p. piscivorus*]). Bijdr. Dierk. (Amsterdam) 33:61-69.

Petzold, H.-G. 1967. Notizen zur Gewöllbildung bei einem Bindenwaran (*Varanus salvator*) und einige allgemeine Bemerkungen über Reptiliengewölle (Notes on pellet formation in a two-banded monitor [*Varanus salvator*] and several general comments about reptile pellets). Zool. Gart. (N.F.), Leipzig 34:134-138.

Petzold, H.-G. 1967. Some remarks on the breeding biology and keeping of *Tretanorhinus variabilis*, a water snake of Cuba. Herpetologica 23:242-246.

Petzold, H.-G. 1968. Zur Fortpflanzungsbiologie asiatischer Kobras (*Naja naja*) (Concerning the reproductive biology of Asiatic cobras [*Naja naja*). Zool. Gart. (N.F.), Leipzig 36:133-146.

Petzold, H. G. 1969. Observations on the reproductive biology of the American ringed snake *Leptodeira annulata* at East Berlin Zoo. Inter. Zoo Yearb. 9:54-56.

Petzold, H.-G. 1978. Zur Ökologie und Fortpflanzungsbiologie der Kuba-Schlanknatter, *Alsophis cantherigerus* (Bibron 1840) im Terrarium (Concerning the ecology and reproductive biology of the Cuban racer, *Alsophis cantherigerus* [Bibron 1840] in the terrarium). Zool. Gart. (N.F.), Jena 48:155-163.

Petzold, H.-G. 1980. Statistisches über die Geburtsgewichte von Kreuzottern (*Vipera b. berus*) (Statistical data concerning birth weights of the common viper [*Vipera b. berus*]). Milu 5:443-448.

Petzold, H.-G. 1984. Aufgaben und Probleme bei der Erforschung der Lebensäusserungen der Niederen Amnioten (Reptilien) (Tasks and Problems Connected with Research into the Life Expressions of the Lower Amniotic Animals [Reptiles]). This book (Nr. 38) is in the series "Berliner Tierpark-Buch," published by Bina in Berlin [Note: Nachdruck aus Milu, Bd. 5, Heft 4/5:485-786 (1982). Translated version by Lucian Heichler and J. B. Murphy. 2007. The Lives of Captive Reptiles. SSAR Contrib. Herpetol.].

Petzold, H.-G., and P. H. Stettler. 1972. Zur Haltung und Fortpflanzungsbiologie der Indischen Streifennatter, *Natrix (Amphiesma) stolata* (Boie 1827) (Concerning the care and reproductive biology of the Indian striped snake *Natrix [Amphiesma] stolata* [Boie 1827]). Zool. Gart. (N.F.), Leipzig 41:192-195.

Pfaff, C. S., and K. Vause. 2002. Captive reproduction and growth of the broad-striped dwarf siren (*Pseudobranchus s. striatus*). Herpetol. Rev. 33(1).

Pickering, D. C. 1990. Captive management of the pancake tortoise *Malacochersus tornieri*, at the Tulsa Zoological Park. AAZPA Reg. Conf. Proc.

1990:373-376.

Poglayen-Neuwall, I. 1983. Geglückte Zucht der Panther-Schildkröte (*Geochelone pardalis babcocki*) (Successful breeding of the leopard tortoise [*Geochelone pardalis babcocki*]). Zool. Gart. (N.F.), Jena 53:217-225.

Post, M. J. 2000. The captive husbandry and reproduction of the Hosmer's skink *Egernia hosmeri*. Herpetofauna 30(2):2-6.

Preece, D. 1998. The captive management and breeding of poison-dart frogs, family Dendrobatidae at Jersey Wildlife Preservation Trust. Dodo, J. Jersey Wildl. Preserv. Trust 34:103-114.

Pritchard, P. C. H. 1979. "Head starting" and other conservation techniques for marine turtles Cheloniidae and Dermochelyidae. Inter. Zoo Yearb. 19:38-42.

Pritchard, P. C. H. 1996. The Galápagos Tortoises. Nomenclatural and Survival Status. Chelonian Research Monographs No. 1. Chelonian Research Foundation, Lunenburg MA.

Pritchard, P. C. H., and P. Trebbau. 1984. Turtles of Venezuela. Society for the Study of Amphibians and Reptiles. Contributions to Herpetology, volume 2, Ann Arbor MI.

Procter, J. B. 1922. A study of the remarkable tortoise, *Testudo loveridgeii*, Blgr., and the morphology of the chelonian carapace. Proc. Zool. Soc. London 1923:483-526.

Procter, J. B. 1928. On a living Komodo dragon *Varanus komodoensis* Ouwens, exhibited at the Scientific Meeting, October 23rd, 1928. Proc. Zool. Soc. London 1928:1017-1019.

Quinn, H. R. 1979. Reproduction and growth of the Texas coral snake *(Micrurus fulvius tenere)*. Copeia 1979:453-463.

Quinn, H. R. 1980. Captive propagation of endangered Houston toads. Herpetol. Rev. 11(4):109.

Quinn, H. R., and T. G. Hulsey. 1978. Growth of the cobra *Naja naja* (Serpentes, Elapidae). Herpetol. Rev. 9(4):138-139.

Quinn, H. R., and G. A. Mengden. 1984. Reproduction and growth of the Houston toad, *Bufo houstonensis* (Bufonidae). Southwest. Nat. 29:189-195.

Quinn, H. R., and K. Neitman. 1978. Reproduction in the snake *Boiga cynodon* (Reptilia, Serpentes, Colubridae). J. Herpetol. 12:255-256.

Quinn, H. R., and H. Quinn. 1993. Estimated number of snake species that can be managed by Species Survival Plans in North America. Zoo Biol. 12:243-255.

Rabb, G. B. 1960. On the unique sound production of the Surinam toad, *Pipa pipa*. Copeia 1960:368-369.

Rabb, G. B. 1967. An issue about pipid frogs. Brookfield Bandarlog No. 37.

Rabb, G. B. 1973. Evolutionary aspects of the reproductive behavior of frogs, p. 213-227. In J. L. Vial (ed.), Evolutionary Biology of the Anurans. University of Missouri Press, Columbia MO.

Rabb, G. B. 1993. Heini Hediger - A pioneer in the science of animal behavior. Zool. Gart. (N.F.), Jena 63(3):163-167.

Rabb, G. B., and M. S. Rabb. 1960. On the mating and egg-laying behavior of the Surinam toad, *Pipa pipa*. Copeia 1960:271-276.

Rabb, G. B., and M. S. Rabb. 1963. On the behavior and breeding biology of the African pipid frog *Hymenochirus boettgeri*. Z. Tierpsych. 20:215-240.

Rabb, G. B., and M. S. Rabb. 1963. Additional observations on breeding behavior of the Surinam toad, *Pipa pipa*. Copeia 1963:636-642.

Rabb, G. B., and R. Snedigar. 1960. Observations on the breeding and development of the Surinam toad, *Pipa pipa*. Copeia 1960:40-44.

Radcliffe, C. W., D. Chiszar, K. Estep, J. B. Murphy, and H. M. Smith. 1986. Observations on pedal luring in leptodactylid frogs. J. Herpetol. 20:300-306.

Radford, L., and F. L. Paine. 1989. The reproduction and management of the Dumeril's monitor *Varanus dumerili* at the Buffalo Zoo. Inter. Zoo Yearb. 28:153-155.

Ralls, K., K. Brugger, and J. Ballou. 1979. Inbreeding and juvenile mortality in small populations of ungulates. Science 206:1101-1103.

Ramsay, E. C., G. B. Daniels, D. A. Bemis, and B. W. Tryon. 1996. *Salmonella arizona* osteomyelitis in a colony of Arizona ridge-nosed rattlesnakes (*Crotalus w. willardi*). Proc. Amer. Assoc. Zoo Vet. 1996:218-219.

Raphael, B. L. 1980. Sand impaction in a Galapagos tortoise. Proc. Amer. Assoc. Zoo Vet. 1980:67.

Raphael, B. L. 1993. Amphibians. Vet. Clinics North Amer. Small Anim. Pract. 23:1271-1286.

Raphael, B. L., S. B. James, and R. A. Cook. 1999. Evaluation of vitamin D concentrations in *Uromastyx* spp. with and without radiographic evidence of dystrophic mineralization. Proc. Amer. Assoc. Zoo Vet. 1999:20-22.

Raphael, B. L., M. Papich, and R. A. Cook. 1994. Pharmacokinetics of enrofloxacin after a single intramuscular injection in Indian star tortoises (*Geochelone elegans*). J. Zoo Wildl. Med. 25:88-94.

Raphael, B. L., M. W. Klemens, P. Moehlman, E.

Dierenfeld, and W. B. Karesh. 1994. Blood values in free-ranging pancake tortoises (*Malacochersus tornieri*). J. Zoo Wildl. Med. 25:63-67.

Raphael, B. L., P. P. Calle, N. Gottdenker, S. James, W. R. Karesh, M. J. Linn, T. McNamara, and R. A. Cook. 1997. Clinical significance of *Cryptosporidia* in captive and free-ranging chelonians. Proc. Amer. Assoc. Zoo Vet. 1997:19-20.

Read, B. 1996. Training zoo professionals for Studbook and Species Survival Plan programs. Zoo Biol. 14:149-157.

Rehák, I. 1990. Rozmnožení kubánského hada *Tropidophis feicki* v zajetí (Captive reproduction in the Cuban snake *Tropidophis feicki*). Gazella (Praha)17:111-114 [in Czech, English summary].

Rehák, I. 1991. Reprodukèní biologie hroznýška *Eryx j. jaculus* (Reproductive biology in javelin sand boa, *Eryx j. jaculus*). Gazella (Praha)18:55-60. [in Czech, English summary.].

Rehák, I. 1991. Natural history and captive breeding in *Paramesotriton deloustali*, p. 53-70. In A. W. Zulich (ed.), 14th International Herpetological Symposium on Captive Propagation and Husbandry. International Herpetological Symposium.

Rehák, I. 1993. Breeding the Ceylon python (*Python molurus "pimbura"*) in Prague Zoo. Akvarium Terrarium 36(12):33-38.

Rehák, I. 1994. Rozmnožování a etologie leguánu kubánských, *Cyclura nubila*, v Zoo Praha (Breeding and ethology of the Cuban ground iguanas, *Cyclura nubila*, at Prague Zoo). Gazella (Praha) 21:61-78. [in Czech, English summary.].

Rehák, I. 1994. Chov a rozmnožování krajt tygrovitých, *Python molurus molurus*, a krajt pestrých, *Python curtus breitensteini*, v Zoo Praha (Captive care and breeding of the Indian pythons, *Python molurus molurus*, and blood pythons, *Python curtus breitensteini*, at Zoo Prague). Gazella (Praha) 21:79-86. [in Czech, English summary.].

Rehák, I.. 1995 (1997). Chovný program pro leguána kubánského (*Cyclura nubila*) v Zoo Praha s poznámkami ke strategii ochrany rodu *Cyclura* (Breeding program for the Cuban ground iguana, *Cyclura nubila*, at the Prague Zoo, with notes on the conservation strategy of the genus *Cyclura*). Pøíroda (Praha) 2:45-47. [in Czech, English summary.].

Rehák, I. 1996. První rozmnožení leguána nosorohého, *Cyclura cornuta*, v Èeské republice (The first captive breeding of the rhinoceros iguana, *Cyclura cornuta*, in Czech Republic). Gazella (Praha) 23:101-108. [in Czech, English summary.].

Rehák, I. 1996. K reprodukèní biologii varana mangrovového, *Varanus indicus*, v pražské Zoo (On the reproductive biology of the mangrove monitor in Prague Zoo). Gazella (Praha) 23:109-116. [in Czech, English summary.].

Rehák, I. 1999. Captive breeding of the caiman lizard, *Dracaena guianensis*. Herpetofauna (Australia) 29(2):57-60.

Rehák, I., and B. Král. 1993. Chovný program pro leguána kubánského, *Cyclura nubila* v Zoo Praha, s poznámkami k ekologii a etologii druhu na Kubì (Breeding programme for the Cuban ground iguana, *Cyclura nubila*, in Zoo Prague, with notes to ecology and ethology of this species in Cuba). Gazella (Praha) 20:83-94 [in Czech, English summary].

Rehák, I., and P. Velenský. 1997. Druhá generace leguánù kubánských, *Cyclura nubila*, narozená v Zoo Praha (Second generation of Cuban iguanas, *Cyclura nubila*, born in Prague Zoo). Gazella (Praha) 24:93-107 [in Czech, English summary].

Rehák, I., and P. Velenský. 2001. Biologie a chov leguána kubánského (*Cyclura nubila*) v lidské péèi (The biology and breeding of the Cuban ground iguana [*Cyclura nubila*] in captivity). Gazella (Praha) 28:129-208 [in Czech, English summary].

Reichenbach, H. 2002. Lost menageries - why and how zoos disappear (Part 1). Inter. Zoo News 49:153-163.

Reichenbach-Klinke, H.-H. 1961. Krankheiten der Amphibien (Diseases of Amphibians). Gustav Fischer Verlag, Stuttgart.

Reichenbach-Klinke, H.-H. 1963. Krankheiten der Reptilien (Diseases of Reptiles). Gustav Fischer Verlag, Stuttgart/Jena.

Reichling, S. B. 1973. Habits of *Boiga dendrophila* in captivity. Bull. New York Herpetol. Soc. 10:38-39.

Reichling, S. B. 1974. A new record-size *Natrix erythrogaster flavigaster* x *neglecta*. Bull. Chicago Herpetol. Soc. 93:16.

Reichling, S. B. 1981. Reproduction in captive black pine snakes, *Pituophis melanoleucus lodingi*. Herpetol. Rev.13:41.

Reichling, S. B. 1983. Reproduction of Dumeril's boas at the Memphis Zoo. AAZPA Reg. Conf. Proc. 1983:236-242.

Reichling, S. B. 1986. The endangered pine snakes of the gulf coast: Their ecology and husbandry. AAZPA Reg. Conf. Proc. 1986:119-125.

Reichling, S. B. 1988. Reproduction in captive Louisiana pine snakes. Herpetol. Rev. 19(4):77-78.

Reichling, S. B. 1988. A simple technique for the treatment of dystocia in snakes. Companion Anim. Pract. 2:42-44.

Reichling, S. B. 1989. Reproductive biology and current status of the Louisiana pine snake, *Pituophis melanoleucus ruthveni*, p. 95-98. In M. J. Uricheck (ed.), 13th International Herpetological Symposium on Captive Propagation and Husbandry.

Reichling, S. B. 1990. Reproductive traits of the Louisiana pine snake *Pituophis melanoleucus ruthveni* (Serpentes: Colubridae). Southwest. Natur. 35:221-222.

Reichling, S. B. 1990. The taxonomic status of the Louisiana pine snake (*Pituophis melanoleucus ruthveni*) and its relevance to the evolutionary species concept. J. Herpetol. 29:186-198.

Reichling, S. B. 1995. Conservation status of the Lesser Antillean iguana (*Iguana iguana*). Reptiles Magazine 2:40-50.

Reichling, S. B., and W. H. N. Gutzke. 1996. Phenotypic consequences of incubation environment in the African elapid genus *Aspidelaps*. Zoo Biol. 15:301-308.

Reichling, S. B., and P. Louton. 1989. Geographic distribution: *Rhadinaea flavilata*. Herpetol. Rev. 20:76.

Reid, D. 1995. Observations on hatchling and juvenile captive-bred angonoka (*Geochelone yniphora*) in Madagascar. Dodo, J. Jersey Wildl. Preserv. Trust 31:112-119.

Reid, D., L. Durrell, and G. Rakotobearison. 1989. The captive breeding project for the angonoka *Geochelone yniphora* in Madagascar. Dodo, J. Jersey Wildl. Preserv. Trust 26:34-48.

Richter, U. 1989. Troparium Hamburg: Nachzucht der Fransenschildkröte (*Chelus fimbriatus*) (Hamburg Troparium: breeding the matamata turtle [*Chelus fimbriatus*]). Aquar.-u. Terra.-Z. 42(2):99-100.

Rideout, B. A., R. J. Montali, L. G. Phillips, and C. N. Gardiner. 1987. Mortality of captive tortoises due to viviparous nematodes of the genus *Proatractis* (Family Atractidae). J. Wildl. Dis. 23:103-108.

Ridley, H. N. 1906. The Menagerie at the Botanic Gardens. J. Straits Branch Royal Asiatic Soc. No. 46:133-194.

Rieck, W. 2001. Johann Matthaeus Bechstein (1757-1822). In W. Rieck, G. T. Hallmann, and W. Bischoff (eds.), Die Geschichte der Herpetologie und Terrarienkunde im deutschsprachigen Raum (History of Herpetology and Terrarium Science in German speaking areas) Mertensiella 12:415-418. Deutschen Gesellschaft für Herpetologie und Terrarienkunde e.V. (DGHT).

Rieck, W. 2001. Johann von Fischer (1850-1901). In W.

Rieck, G. T. Hallmann, and W. Bischoff (eds.), Die Geschichte der Herpetologie und Terrarienkunde im deutschsprachigen Raum (History of Herpetology and Terrarium Science in German speaking areas) Mertensiella 12:444-445. Deutschen Gesellschaft für Herpetologie und Terrarienkunde e.V. (DGHT).

Rieck, W. 2001. Werner Schröder (1907-1985). In W. Rieck, G. T. Hallmann, and W. Bischoff (eds.), Die Geschichte der Herpetologie und Terrarienkunde im deutschsprachigen Raum (History of Herpetology and Terrarium Science in German-Speaking Areas) - Mertensiella 12:592-594. Deutschen Gesellschaft für Herpetologie und Terrarienkunde e.V. (DGHT).

Risley, D. 1989. Breeding the Namib gecko *Chondrodactylus angulifer* at London Zoo. Inter. Zoo Yearb. 28:113-115.

Robeck, T. R., D. C. Rostal, P. M. Burchfield, D. W. Owens, and D. C. Kraemer. 1990. Ultrasound imaging of reproductive organs and eggs in Galapagos tortoises, (*Geochelone elephantopus* spp. Zoo Biol. 9:349-359.

Roberts, D. T., and S. H. Hammack. 1995. Captive reproduction and husbandry of the speckled forest-pitviper *Bothriopsis taeniata* (Wagler) at the Dallas Zoo. Snake 27:53-55.

Roberts, D. T., D. M. Schleser, and T. L. Jordan. 1995. Notes on the captive husbandry and reproduction of the Texas salamander *Eurycea neotenes* at the Dallas Aquarium. Herpetol. Rev. 26(1):23-25.

Robinson, P. T., C. F. von Essen, K. Benirschke, J. E. Meier, and J. P. Bacon. 1978. Radiation therapy for treatment of an intraoral malignant lymphoma in an Indian rock python. J. Amer. Vet. Radiol. 19(3):92-95.

Roca, V., G. Garcia, E. Carbonell, C. Sanchez-Acedo, and E. DelCacho. 1998. Parasites and conservation of *Alytes muletensis* (Sanchiz et Androver, 1977) (Anura: Discoglossidae). Rev. Esp. Herp. 12:91-95.

Rokosky, E. J. 1941. Notes on new-born jumping vipers, *Bothrops nummifera*. Copeia 1941:267.

Rolker, A. W. 1956. The story of the snake, p. 274-283. In B. Aymar (ed.), Treasury of Snake Lore. Greenberg Publisher, New York.

Ross, R. 1973. Successful mating and hatching of Children's python, *Liasis childreni*. HISS NJ 1(6):181-182.

Ross, R., and R. Larman. 1977. Captive breeding in two species of python *Liasis albertisii* and *L. mackloti*. Inter. Zoo Yearb. 17:133-136.

Rosscoe, R., and W. Holmstrom. 1996. Successful incubation of eggs from matamatas (*Chelus fimbriatus*), p. 47-50. In P. D. Strimple (ed.), Advances in Herpetoculture. Special Publication of the International Herpetological Symposium, Inc., No. 1. International Herpetological Symposium, Inc., Des Moines IA.

Rostal, D. C., J. S. Grumbles, V. A. Lance, and J. R. Spotila. 1994. Non-lethal sexing techniques for hatchling and immature desert tortoises (*Gopherus agassizii*). Herpetol. Monogr. 8:83-87.

Rostal, D. C., V. A. Lance, J. S. Grumbles, and A. C. Alberts. 1994. Seasonal reproductive cycle of the desert tortoise (*Gopherus agassizii*) in the eastern Mojave desert. Herpetol. Monogr. 8:72-82.

Rostal, D. C., T. R. Robeck, J. S. Grumbles, P. M. Burchfield, and D. W. Owens. 1998. Seasonal reproductive cycle of the Galápagos tortoise (*Geochelone nigra*) in captivity. Zoo Biol. 17:519-524 [study done at Zoo].

Roth, V. D. 1971. Food habits of Ditmars' horned lizard with speculations on its type locality. J. Arizona Acad. Sci. 6:278-281.

Roth, V. D. 1997. Ditmars' horned lizard (*Phrynosoma ditmarsi*) or the case of the lost lizard. Sonoran Herpetologist 10(1): 2-6.

Rübel, A., and R. E. Honegger. 2001. Hediger, Heini 1908-1992. Swiss zoologist, animal psychologist, and zoo director, p. 547-550. In C. E. Bell (ed.), Encyclopedia of the World's Zoos. Fitzroy Dearborn Publishers, Chicago, London.

Rundquist, E. 1980. Reproduction and captive maintenance of Koch's day gecko, (*Phelsuma madagascariensis kochi*), p. 80-83. In R. A. Hahn (ser. ed.), 3rd International Herpetological Symposium on Captive Propagation and Husbandry. International Herpetological Symposium, Thurmont MD.

Rundquist, E. M. 1993. Captive amphibian culture: Some observations and suggestions for future veterinary investigations. Proc. Amer. Assoc. Zoo Vet. 1993:19-22.

Russ, I.G., L. Vogelnest, F. Hulst, and D. Blyde. 1995. Assisted breeding at Taronga and Western Plains Zoos, p. 438-442. In R. E. Junge (ed.), Proceedings Joint Conference American Association of Zoo Veterinarians, Wildlife Disease Association, American Association of Wildlife Veterinarians, East Lansing, Michigan, August 12-17, 1995. American Association of Zoo Veterinarians, American Association of Wildlife Veterinarians & Wildlife Disease Association.

Russell, M. J. 1996. Notes on the natural history, captive husbandry, and reproduction of crocodile skinks (*Tribolonotus gracilis*) at the Dallas Zoo, In P. D. Strimple (ed.), Advances in Herpetoculture. Special Publication of the International Herpetological Symposium, Inc., No. 1. International Herpetological Symposium, Inc., Des Moines IA.

Ryabov, S. A. 1997. On the extraordinary high productivity in *Elaphe radiata*. Russian J. Herpetol.4:203-204.

Ryabov, S. A. 1999. Breeding of Stimson's python *Antaresia stimsoni orientalis* in Russia. Herpetological Herald (L'vov, the Ukraine)1:28-30.

Ryabov, S. A. 2001. Persian Ratsnake *Elaphe persica* (Werner, 1913): Natural History, Keeping and Breeding in Captivity. In: "Litteratura Serpentium" 21:137-142.

Ryabov, S. A., and S. P. Popovskaya. 2000. Some comparative data on the breeding of four subspecies of Stripe-tailed rat snake *Elaphe taeniura*. Russian J. Herpetol. 7:81-84.

Ryboltovsky, E. 1999. The Annamese flying frog. Inter. Zoo News 46(7):392-397.

Ryboltovsky, E. 1999. A wonderful frog from Vietnam: biology, management and breeding. Inter. Zoo News 46(6):347-352.

Ryder, O. A., L. G. Chemnick, S. F. Schafer, and A. L. Shima. 1989. Individual DNA fingerprints from Galapagos tortoises *Geochelone elephantopus*. Inter. Zoo Yearb. 28:84-87.

Sajdak, R. A. 1983. Herpetological research in zoos: A literature survey, 1977-1981. Zoo Biol. 2:149-152.

Sasaki, C. 1887. Some notes on the giant salamander of Japan (Cryptobranchus Japonicus, Van der Hoeven.). J. College Sci. Imperial Univ. Tokyo 1:269-274.

Schafer, S. F. 1981. Rearing the Asiatic tree frog *Rhacophorus leucomystax* at the San Diego Zoo, p. 147-150. In R. A. Hahn (ser. ed.), 5th International Herpetological Symposium on Captive Propagation and Husbandry. International Herpetological Symposium, Thurmont MD.

Schafer, S. F., and C. O. Krekorian. 1983. Agnostic behavior of the Galapagos tortoise, *Geochelone elephantopus*, with emphasis on the relationship to saddle-backed shell shape. Herpetologica 39:448-456.

Scheffel, S. 2001. Werner Krause (1905-1970). In W. Rieck, G. T. Hallmann, and W. Bischoff (eds.), Die Geschichte der Herpetologie und Terrarienkunde im deutschsprachigen Raum - Mertensiella 12:500-

501. Deutschen Gesellschaft für Herpetologie und Terrarienkunde e.V. (DGHT).

Scherpner, C. 1975. The crocodiles and alligators, p. 124-146. In B. Grzimek, H. Hediger, K. Klemmer, O. Kuhn, and H. Wermuth (eds.), Grzimek's Animal Life Encyclopedia, Reptiles. Van Nostrand Rheinhold, New York.

Scherpner, C. 2001. Wilhelm Klingelhöffer (1871-1953). In W. Rieck, G. T. Hallmann, and W. Bischoff (eds.), Die Geschichte der Herpetologie und Terrarienkunde im deutschsprachigen Raum (History of Herpetology and Terrarium Science in German speaking areas) Mertensiella 12:490-493. Deutschen Gesellschaft für Herpetologie und Terrarienkunde e.V. (DGHT).

Scherpner, C. 2001. Gustav Lederer (1892-1962). In W. Rieck, G. T. Hallmann, and W. Bischoff (eds.), Die Geschichte der Herpetologie und Terrarienkunde im deutschsprachigen Raum (History of Herpetology and Terrarium Science in German speaking areas) Mertensiella 12:517-518. Deutschen Gesellschaft für Herpetologie und Terrarienkunde e.V. (DGHT).

Scherren, H. 1905. The Zoological Society of London. Cassell, London.

Schildger, B.-J., and R. Wicker. 1987. Endoskopische Geschlechtsbestimmung bei *Trachydosaurus rugosus* (G. 1827) (Sauria: Scincidae) (Endoscopic gender determination in *Trachydosaurus rugosus* [G. 1827] [Sauria: Scincidae]). Salamandra 14:63-79.

Schmidt, A. A., and R. Wicker. 1977. Weitere Beobachtungen bei der Nachzucht des Zipfelfrosches *Megophrys nasuta* (Amphibia, Salientia, Pelobatidae) (Additional observations in breeding the horned frog *Megophrys nasuta* [Amphibia, Salientia, Pelobatidae]). Salamandra 13:43-48.

Schmidt, H. 2001. Günther Nietzke (1911-1998). In W. Rieck, G. T. Hallmann, and W. Bischoff (eds.), Die Geschichte der Herpetologie und Terrarienkunde im deutschsprachigen Raum (History of Herpetology and Terrarium Science in German speaking areas) Mertensiella 12:500-501. Deutschen Gesellschaft für Herpetologie und Terrarienkunde e.V. (DGHT).

Schneider, K. M. 1941. Einige Gefangenschaftsbeobachtungen über die Fortpflanzung des Hechtalligators (*Alligator mississippiensis* DAUDIN) (Some observations in captivity about the reproduction of the pike-snouted alligator [*Alligator mississippiensis* DAUDIN]). Zool. Gart. (N.F.), Leipzig 13:275-284.

Schoener, T. W. 1977. Competition and the niche, p. 35-136. In C. Gans and D. W. Tinkle (eds.). Biology of the Reptilia, Vol. 7. Ecology and Behaviour A. Academic Press, London, New York.

Schomberg, G. 1957. British Zoos. A Study of Animals in Captivity. Allan Wingate London.

Schramm, B. G., M. Casares, and V. A. Lance. 1999. Steroid levels and reproductive cycle of the Galápagos tortoise, *Geochelone nigra*, living under seminatural conditions on Santa Cruz Island (Galápagos). Gen. Comp. Endocrinol. 114:108-112.

Schramm, B. G., M. Casares, and V. A. Lance. 2000. Ultrasound scanning of ovaries and eggs in Galápagos tortoises, *Geochelone nigra*, on Santa Cruz Island, Gàlápagos. Chelonian Conserv. Biol. 3:706-713.

Schramm, B. G., V. A. Lance, and M. Casares. 1999. Reproductive cycles of male and female giant tortoises (*Geochelone nigra*) on the Galápagos Islands by plasma steroid analysis and ultrasound scanning. Linnaeus Fund Research Report. Chelonian Conserv. Biol. 3:523-528.

Schubert, A., and G. Santana. 1996. Conservation of the American crocodile (*Crocodylus acutus*) in the Dominican Republic, p. 425-433. In R. Powell, and R. W. Henderson (eds.), Contributions to West Indian Herpetology: A Tribute to Albert Schwartz. Society for the Study of Amphibians and Reptiles, Contributions to Herpetology, volume 12, Ithaca NY.

Schultschik, G. 2001. Leopold Josef Fitzinger (1802-1884). In W. Rieck, G. T. Hallmann, and W. Bischoff (eds.), Die Geschichte der Herpetologie und Terrarienkunde im deutschsprachigen Raum (History of Herpetology and Terrarium Science in German speaking areas) Mertensiella 12:445-448. Deutschen Gesellschaft für Herpetologie und Terrarienkunde e.V. (DGHT).

Schürer, U. 1982. ZOODOM, der Nationalzoo der Dominikanischen Republik und einige Bemerkungen über die Fauna des Landes (ZOODOM, the National Zoo of the Dominican Republic and some comments about the fauna of that country). Z. Köln. Zoo 25:59-63.

Schürer, U., and M. Bürger. 1998. Nachzucht bei den Neuguinea-Krokodilen (*Crocodylus novaeguineae* Schmidt, 1928) im Zoologischen Garten Wuppertal (Reproduction of New Guinea crocodiles [*Crocodylus novaeguineae* Schmidt, 1928] in the Wuppertal Zoological Garden). Zool. Gart. (N.F.) 68:332-336.

Schwammer, G. V. 2001. Tiergarten Schönbrunn.

Schönbrunn Zoological Garden, p. 1216-1220. In C. E. Bell (ed.), Encyclopedia of the World's Zoos. Fitzroy Dearborn Publishers, Chicago, London.

Sclater, P. L. 1862. Notes on the incubation of *Python sebae*, as observed in the Society's Gardens. Proc. Zool. Soc. London 1862:365-368.

Sclater, P. L. 1872. List of the vertebrate animals now or lately living in the Gardens of the Zoological Society of London. Printed for the Society, London.

Sclater, P. L. 1895. Note on the breeding of the Surinam water-toad (*Pipa surinamensis*) in the Society's Reptile-House. Proc. Zool. Soc. London 1895:86-88.

Sekar, M., and M. Jagnnadha-Rao. 1995. Management of Indian rock python (*Python molurus*) in captivity at Arignar Anna Zoological Park, Vandalur. Cobra (Madras) 22:14-16.

Sharbel, T. F., and D. M. Green. 1989. Easy to build riffle tanks for aquatic amphibians. Can. Assoc. Herpetol. Bull. 3(1):3-5.

Sharbel, T. F., and D. M. Green. 1992. Captive maintenance of the primitive New Zealand frog, *Leiopelma hochstetteri*. Herpetol. Rev. 23(3):77-79.

Sharma, R. C. 1998. Fauna of India-Reptilia (Testudines and Crocodilia) Vol. I. Publication Division by the Director, Zoological Society of India, Calcutta.

Shaw, C. E. 1954. Captive-bred Cuban iguanas, *Cyclura macleayi macleayi*. Herpetologica 10:73-78.

Shaw, C. E. 1959. Notes on the eggs, incubation and young of *Chamaeleo basiliscus*. Brit. J. Herpetol. 2:182-185.

Shaw, C. E. 1961. Breeding the Galapagos tortoise, *Testudo elephantopus*. Inter. Zoo Yearb. 3:102-104.

Shaw, C. E. 1962. Eine neue Generation Galapagos-Riesen (A new generation of Galapagos giants). Freunde Köln. Zoo 5(1):18-20.

Shaw, C. E. 1962. Sea snakes at San Diego Zoo. Inter. Zoo Yearb. 4:49-52.

Shaw, C. E., and S. Campbell. 1974. Snakes of the American West. Alfred A. Knopf, New York.

Shepstone, H. J. 1932. Wild Beasts To-Day. Being an Account of the World's Leading Zoological Gardens, the Catching, Transportation and Doctoring of Wild Animals, the Rearing of Them on Farms, and the Work of Conserving the Rarer Species in Parks and Reservations. Macmillan, New York.

Sherbrooke, W. C., B. E. Martin, and C. H. Lowe. 1998. *Phrynosoma ditmarsi* (rock horned lizard). Herpetol. Rev. 29(2):110-111.

Sherriff, D. 1989. The inland bearded dragon, *Pogona vitticeps* and its maintenance and breeding in captivity. Reptiles: Proceedings of the 1988 U.K. Herpetological Societies. p. 27-38.

Shibuya, Y. 1978. Breeding of the Surinam Toad. Animals and Zoos (Official Bulletin of Ueno Zoological Garden, Tokyo) 30:4-5.

Shiihara, S., and M. Samejima. 1995. Breeding Ishikawa's frog in captivity. J. Japan. Assoc. Zoos Aquar. 36(4):101-105.

Sievert, L., and V. H. Hutchison. 1988. Light versus heat: Thermoregulatory behavior in a nocturnal lizard (*Gekko gecko*). Herpetologica 44:266-273.

Sievert, L., and V. H. Hutchison. 1989. The parietal eye and thermoregulatory behavior of *Crotaphytus collaris* (Squamata: Iguanidae). Comp. Biochem. Physiol. 94A(2):339-343.

Sievert, L. M., and V. H. Hutchison. 1991. The influence of photoperiod and position of a light source on behavioral thermoregulation in *Crotaphytus collaris* (Squamata: Iguanidae). Copeia 1991:105-110.

Simmons, J. E. 1977. Reproduction of the Chinese red snake, *Dinodon rufozonatum* (Cantor) in captivity. Herpetol. Rev. 8(2):32.

Sims, K. J. 1971. Exhibit building for the Komodo dragon *Varanus komodoensis*. Inter. Zoo Yearb. 11:76-77.

Sims, K. J., and I. Singh. 1978. Breeding the West African dwarf crocodile *Osteolaemus tetraspis tetraspis* at Kuala Lumpur Zoo, with observations on nest construction. Inter. Zoo Yearb. 18:83-84.

Sinervo, B. 1994. Manipulations of clutch and offspring size in lizards: mechanistic, evolutionary, and conservation considerations, p. 183-193. In J. B. Murphy, K. Adler, and J. T. Collins (eds.), Captive Management and Conservation of Amphibians and Reptiles. Society for the Study of Amphibians and Reptiles. Contributions to Herpetology, volume 11, Ithaca NY.

Sinervo, B., and J. Clobert. 2003. Morphs, dispersal behavior, genetic similarity, and the evolution of cooperation. Science 300: 1949-1951.

Slavens, F. L. 1989. The Inventory of Live Reptiles and Amphibians in Captivity: A brief history. Inter. Zoo Yearb. 28:7-9.

Sloan, N. 1978. Dr. James A. Oliver — Herpetologist and Educator — Retired. Bull. New York Herpetol. Soc. 14(2):2.

Smith, C. F., and F. L. Paine. 1989. A system for rearing tadpoles at the Buffalo Zoological Gardens. Inter. Zoo Yearb. 28:58-59.

Smith, G. M., and C. W. Coates. 1938. Fibro-epithelial growths of the skin in large marine turtles, *Chelonia mydas* (Linnaeus). Zoologica (New York)

23:93-98.

Smith, R. J., and H. M. Fischer. 1975. Breeding and rearing the Colombian giant toad *Bufo blombergi* at Los Angeles Zoo. Inter. Zoo Yearb.15:87-89.

Smith, L. L., D. Reid, B. Robert, M. Joby, and S. Clément. 1999. Status and distribution of the angonoka tortoise (*Geochelone yniphora*) of western Madagascar. Biol. Conserv. 91:23-33.

Snedigar, R., and E. J. Rokosky. 1949. Notes on newborn Gaboon vipers. Copeia 1949:145-146.

Snedigar, R., and E. J. Rokosky. 1950. Courtship and egg-laying of captive *Testudo denticulata*. Copeia; 1950:46-48.

Snider, A. T. and J. K. Bowler. 1992. Longevity of reptiles and amphibians in North American collections. Second Edition. Soc. Study Amphib. Rept. Herp. Circ. No. 21:1-40.

Soetbeer, F. 1898. Ueber die Körperwärme der poikilothermen Wirbelthiere (Concerning the body warmth of poikilothermic vertebrates). Arch. f. expt. Pathol. u Pharm. 40: 53-80.

Spelman, L. H., R. C. Cambre, T. Walsh, and R. Rosscoe. 1996. Anesthetic techniques in Komodo dragons (*Varanus komodoensis*). Proc. Amer. Assoc. Zoo Vet. 1996:247-250.

Spratt, D. M. J. 1989. Operation Curieuse: A conservation programme for the Aldabra giant tortoise *Geochelone gigantea* in the Republic of Seychelles. Inter. Zoo Yearb. 28:66-69.

Stamper, M. A., and B. R. Whitaker. 1994. Medical observations and implications on "healthy" sea turtles prior to release into the wild. Proc. Amer. Assoc. Zoo Vet. 1994:182-185.

Stamper, M. A., M. Papich, G. Lewbart, D. Plummer, S. May, and M. Stoskopf. 1997. Single dose pharmacokinetics of ceftazidime in loggerhead sea turtles (*Caretta caretta*). Proc. Amer. Assoc. Zoo Vet. 1997:16.

Stamps, J. A. 1977. Social behavior and spacing patterns in lizards, p. 265-334. In C. Gans and D. W. Tinkle (eds.). Biology of the Reptilia, Vol. 7. Ecology and Behaviour A. Academic Press, London, New York.

Stancel, C. F., E. S. Dierenfeld, and P. A. Schoknecht. 1998. Calcium and phosphorus supplementation decreases growth, but does not induce pyramiding, in young red-eared sliders, *Trachemys scripta elegans*. Zoo Biol. 17:17-24.

Stanger, M., M. Clayton, R. Schodde, J. Wombey, and I. Mason. 1998. CSIRO List of Australian Vertebrates: A Reference with Conservation Status. CSIRO, Collingwood.

Stearns, B. C. 1985. Captive husbandry and propagation of tortoises, p. 113-129. In R. L. Gray (ed.), Captive Propagation and Husbandry of Reptiles and Amphibians. Northern California Herpetological Society & Bay Area Amphibian and Reptile Society.

Stearns, B. C. 1987. Captive husbandry and propagation of the Aldabra giant tortoise *Geochelone gigantea* at the Institute for Herpetological Research. Inter. Zoo Yearb. 27:98-103.

Stearns, B. C. 1989. The captive status of the African spurred tortoise *Geochelone sulcata*: Recent developments. Inter. Zoo Yearb. 28:97-98.

Stejneger, L. 1906. A new lizard of the genus *Phrynosoma,* from Mexico. Proc. U. S. National Mus. 29(1437):565-567.

Stemmler, C. 1937. Zur Gefangenschaftsbiologie von *Natrix natrix* und *Natrix tesselatus* (Concerning the captive biology of *Natrix natrix* and *Natrix tesselatus*). Zool. Gart. (N.F.), Leipzig 9:229-230.

Stemmler-Morath, C. 1956. Beitrag zur Gefangenschafts- und Fortpflanzsbiologie von *Python molurus* L. (Contributions concerning the biology of captivity and reproduction of *Python molurus* L.). Zool. Gart. (N.F.), Leipzig 21:347-364.

Stetter, M. D., and R. A. Cook. 1994. Normal and pathological ultrasonographic anatomy of amphibians. Proc. Amer. Assoc. Zoo Vet. 1994:76-78.

Stetter, M. D., B. Raphael, F. Indiviglio, and R. A. Cook. 1996. Isoflurane anesthesia in amphibians: Comparison of five application methods. Proc. Amer. Assoc. Zoo Vet. 1996:255-257.

Stirnberg, E. 1997. Die Haltung und Zucht des Australischen Buntwarans (*Varanus varius*) im Bochumer Tierpark (Care and breeding of the Australian lace monitor [*Varanus varius*] in the Bochum Zoo. Z. Köln. Zoo 40 (2):63-67.

Stoskopf, M. K., A. Wisnieski, and L. Pieper. 1985. Iodine toxicity in poison arrow frogs. Proc. Amer. Assoc. Zoo Vet. 1985:86-88.

Stunkard, H. W., and C. P. Gandal. 1961. Nematodes and cestodes from the Australian monitor, *Varanus gouldii*. Zoologica (New York) 46:161-166.

Stunkard, H. W., and C. P. Gandal. 1966. A digenetic trematode, *Parahaplometroides basiliscae* Thatcher, 1963, from the mouth of the crested lizard, *Basiliscus basiliscus*. Zoologica (New York) 51:91-95.

Suedmeyer, W. K., D. S. Gillespie, and L. Pace. 1997. Chromomycosis in a marine toad, *Bufo marinus*. Bull. Assoc. Rept. Amphib. Vet. 7(3):13-15.

Sugiura, H. 1977. The thriving Indian python. Animals

& Zoos (Tokyo) 29: 6-8.

Swingland, I. R. 1994. International conservation and captive management of tortoises, p. 99-107. In J. B. Murphy, K. Adler, and J. T. Collins (editors), Captive Management and Conservation of Amphibians and Reptiles. Society for the Study of Amphibians and Reptiles. Contributions to Herpetology, volume 11, Ithaca, New York.

Swingland, I. R., and M. W. Klemens (eds.). 1989. The Conservation Biology of Tortoises. Occasional Papers of IUCN Species Survival Commission (SSC) No. 5. IUCN-The World Conservation Union, Gland, Switzerland.

Switak, K. H. 1967. Notes on albino reptiles and amphibians at Steinhart Aquarium. Inter. Zoo Yearb. 7:228.

Switak, K. H. 1969. First captive hatching of bushmasters *Lachesis muta*. Inter. Zoo Yearb. 9:56-57.

Takahashi, H. 1967. Breeding of the Indian python. Animals and Zoos 19. No. 10. (No. 213):325, 330-331 [in Japanese with English summary].

Tarbet, S. A. 1983. Reproduction of the D'Albert's python, *Python albertisii* at the Oklahoma City Zoo, p. 132-133. In P. J. Tolson (ed.), 7th International Herpetological Symposium on Captive Propagation and Husbandry. International Herpetological Symposium, Thurmont MD.

Taylor, S. K., E. S. Williams, K. Mills, A. Boerger-Fields, E. T. Thorne, D. R. Kwiatkowski, S. L. Anderson, and M. S. Burton. 1994. The Wyoming toad (*Bufo hemiophrys baxteri*): A review of causes of mortality in the captive population. Proc. Amer. Assoc. Zoo Vet. 1994:75.

Teare, J. A., R. S. Wallace, and M. Bush. 1991. Pharmacology of gentamicin in the leopard frog (*Rana pipiens*). Proc. Amer. Assoc. Zoo Vet. 1991:128-131.

Teichner, O. 1978. Breeding the West African dwarf crocodile *Osteolaemus tetraspis tetraspis* at Metro Toronto Zoo. Inter. Zoo Yearb. 18:88-89.

Thomson, S., S. Irwin, and T. Irwin. 1998. Harriet the Galapagos tortoise: Disclosing one and a half centuries of history. Reptilia. March/April (2):48-51.

Thorogood, J., and I. W. Whimster. 1979. The maintenance and breeding of the leopard gecko *Eublepharis macularius* as a laboratory animal. Inter. Zoo Yearb. 19:74-78.

Throp, J. L. 1969. Notes on breeding the Galapagos tortoise (*Testudo elephantopus*) at Honolulu Zoo. Inter. Zoo Yearb. 9:30-31.

Throp, J. L. C. 1975. Notes on the management and reproduction of the Galápagos tortoise at the Ho-nolulu Zoo, p. 39-42. In R. D. Martin (ed.), Breeding Endangered Species in Captivity. Academic Press, London, New York, San Francisco.

Timmis, W. H. 1969. Observations on Pacific boas *Candoia* spp. at Sydney Zoo. Inter. Zoo Yearb. 9: 53.

Tintinger, V. 1987. Breeding the tuatara, *Sphenodon punctatus*, at Auckland Zoo. Inter. Zoo Yearb. 26:183-186.

Tochimoto, T. September 2000. Siebold's salamanders. Newsletter Himeji City Aquarium No. 37:2-3.

Tocidlowski, M. E., M. R. Loomis, C. L. Merrill, and J. F. Wright. 1997. Teratoma in desert grassland whiptail (*Cnemidophorus uniparens*). Proc. Amer. Assoc. Zoo Vet. 1997:23-24.

Tolson, P. J. 1980. Comparative reproductive behavior of four species of snakes of the boid genus *Epicrates*, p. 87-97. In R. Hahn (ed.), 4th International Herpetological Symposium on Captive Propagation and Husbandry. International Herpetological Symposium, Thurmont MD.

Tolson, P. J. 1982. Captive propagation and husbandry of the Cuban boa, *Epicrates angulifer*, p. 285-293. In D. L. Marcellini (ed.), 6th International Herpetological Symposium on Captive Propagation and Husbandry. International Herpetological Symposium, Thurmont MD.

Tolson, P. J. 1989. Breeding the Virgin Islands boa *Epicrates monensis granti* at the Toledo Zoological Gardens. Inter. Zoo Yearb. 28:163-167.

Tolson, P. J. 1992. The reproductive biology of the Neotropical boid genus *Epicrates* (Serpentes: Boidae), p. 165-178. In W. C. Hamlett (ed), Reproductive Biology of South American Vertebrates. Springer-Verlag, New York.

Tolson, P. J. 1994. The reproductive management of the insular species of *Epicrates* (Serpentes: Boidae) in captivity, p. 353-357. In J. B. Murphy, K. Adler, and J. T. Collins (eds.), Captive Management and Conservation of Amphibians and Reptiles. Society for the Study of Amphibians and Reptiles. Contributions to Herpetology, volume 11, Ithaca NY.

Tolson, P. J., and V. A. Teubner. 1987. The role of social manipulation and environmental cycling in propagation of the boid snake genus *Epicrates*: Lessons from the field and laboratory. AAZPA Reg. Conf. Proc. 1987 :606-613.

Tolson, P. J., C. P. Black, T. A. Riechard, and T. Slovis. 1983. Determination of litter size and embryo viability in the Cuban boa, *Epicrates angulifer*, by the use of imaging ultrasonography, p. 118-123. In P. J. Tolson (ed.), 7th International Herpetological

Symposium on Captive Propagation and Husbandry. International Herpetological Symposium, Thurmont MD

Tonge, S., and Q. Bloxam. 1989. Breeding the Mallorcan midwife toad *Alytes muletensis* in captivity. Inter. Zoo Yearb. 28:45-53.

Townsend, C. H. 1925. The Galapagos tortoises in their relation to the whaling industry. A study of old logbooks. Zoologica (New York) 4:55-135.

Townsend, C. H. 1929. On the extinct Galapagos tortoise that inhabited Charles Island. Zoologica (New York) 9:313-320.

Townsend, C. H. 1931. Growth and age in the giant tortoise of the Galapagos. Zoologica (New York) 9:459-474.

Townsend, C. H. 1937. Growth of Galápagos tortoises, *Testudo vicina*, from 1928 to 1937. Zoologica (New York) 22:289-292.

Townsend, C. R. 1979. Establishment and maintenance of colonies of parthogenetic whiptail lizards *Cnemidophorus* spp. Inter. Zoo Yearb. 19:80-86.

Townsend, C. R., and C. J. Cole. 1985. Additional notes on requirements of captive whiptail lizards (*Cnemidophorus*) with emphasis on ultraviolet radiation. Zoo Biol. 4:49-55.

Townson, S. 1980. Observations on the reproduction of the Indian python in captivity, with special reference to the interbreeding of the two subspecies, *Python molurus molurus* and *Python molurus bivittatus*. The British Herpetological Society: The Care and Breeding of Captive Reptiles p. 69-80.

Tremper, R. L. 1981. A note on the breeding of the neotropical rat snake *Elaphe flavirufa flavirufa* at the Fresno Zoo. Inter. Zoo Yearb. 21:95-96.

Tremper, R. L., and G. R. Barnes. 1984. An overview of the Center for Reptile and Amphibian Propagation. Bull. Chicago Herpetol. Soc.19:69-70.

Troncone, L. R. P., and P. F. Silveira. 2001. Predatory behavior of the snake *Bothrops jararaca* and its adaptation to captivity. Zoo Biol. 20:399-406.

Tryon, B. W. 1976. Second generation reproduction and courtship behavior in the Trans-Pecos ratsnake, *Elaphe subocularis*. Herpetol. Rev. 7(4): 156-157.

Tryon, B. W. 1978. Reproduction in a pair of captive Arizona ridge-nosed rattlesnakes, *Crotalus willardi willardi* (Reptilia, Serpentes, Crotalidae). Bull. Maryland Herpetol. Soc. 14(2):83-88.

Tryon, B. W. 1979. Reproduction in captive forest cobras, *Naja melanoleuca*. J. Herpetol. 13:499-503.

Tryon, B. W. 1980. Observations on reproduction in the West African dwarf crocodile with a description of parental behavior, p. 167-185. In J. B. Murphy, and J. T. Collins (eds.), Reproductive Biology and Diseases of Captive Reptiles. Society for the Study of Amphibians and Reptiles. Contributions to Herpetology, volume 1, Lawrence KS.

Tryon, B. W. 1984. Snake hibernation: In and out of the zoo. Brit. Herpetol. Soc. Bull. 10:22-29.

Tryon, B. W., and T. G. Hulsey. 1976. Notes on reproduction in captive *Lampropeltis triangulum nelsoni* (Serpentes: Colubridae). Herpetol. Rev. 7(4):161-162.

Tryon, B. W., and T. G. Hulsey. 1977. Breeding and rearing the bog turtle *Clemmys muhlenbergii* at the Fort Worth Zoo. Inter. Zoo Yearb. 17:125-130.

Tryon, B. W., and J. B. Murphy. 1982. Miscellaneous notes on the reproductive biology of reptiles. 5. Thirteen varieties of the genus *Lampropeltis*, species *mexicana, triangulum* and *zonata*. Trans. Kansas Acad. Sci. 85:96-119.

Tryon, B. W., and C. W. Radcliffe. 1977. Reproduction in captive Lower California rattlesnakes, *Crotalus enyo enyo* (Cope). Herpetol. Rev. 8(2):34-36.

Tschambers, B. 1949. Birth of the Australian blue-tongued lizard, *Tiliqua scincoides* (Shaw). Herpetologica 5:141-142.

Turner, F. B. 1977. The dynamics of populations of squamates, crocodilians and rhynchocephalians, p. 157-264. In C. Gans and D. W. Tinkle (eds.). Biology of the Reptilia, Vol. 7. Ecology and Behaviour A. Academic Press, London, New York.

Tyler, M. J. 1996. The Action Plan for Australian Frogs. Wildlife Australia, Canberra.

Tytle, T. L. 1994. Husbandry of captive day geckos (*Phelsuma*): Past, present, and future, p. 307-310. In J. B. Murphy, K. Adler, and J. T. Collins (eds.), Captive Management and Conservation of Amphibians and Reptiles. Society for the Study of Amphibians and Reptiles. Contributions to Herpetology, volume 11, Ithaca NY.

Uchida, I. 1996. Indoor breeding of loggerhead turtles. Animals and Zoos 48/4:12-16 [English summary].

Ullrey, D. E. 1996. Skepticism and Science: Responsibilities of the comparative nutritionist. Zoo Biol. 15:449-453.

Ungureanu, C. et al. 1972. Hemorrhagic septicemia in tortoises and snakes in captivity at the Zoological Garden of Bucharest. Arch. Vet., 8: Fasc. 2 :85-96.

Upton, S. J., P. S. Freed, D. A. Freed, C. T. McAlllister, and S. R. Goldberg. 1992. Testicular myxosporidiasis in the flat-backed toad, *Bufo maculatus* (Amphibia; Bufonidae), from Cameroon, Africa. J. Wildl. Dis. 28:326-329.

Valenciennes, A. 1841. Observations faites pendant l'incubation d'une femelle du Python à deux raies (*Python bivittatus, Kuhl.*) pendant les mois de mai et de juin 1841; par M. Valenciennes (Observations made during incubation of a female python on two occasions in [*Python bivittatus, Kuhl.*] during the months of May and June 1841; by M. Valenciennes). Comptes Rendus 13:126-133.

Van Aperen, W. 1969. Notes on the breeding of bearded dragons *Amphibolurus barbatus* at Melbourne Zoo. Inter. Zoo Yearb. 9:51-52.

Van Devender, T. R. 2002. The Sonoran Desert Tortoise: Natural History, Biology, and Conservation. University of Arizona Press, Tucson.

Vandeventer, T. L., and M. Schmidt. 1977. Caesarean section on a western gaboon viper *Bitis gabonica rhinoceros*. Inter. Zoo Yearb. 17:140-143.

Van Mierop, L. H. S., and S. M. Barnard. 1976. Observations on the reproduction of *Python molurus bivittatus*. J. Herpetol. 10:333-340.

Van Mierop, L. H. S., and S. M. Barnard. 1976. Thermoregulation in a brooding female *Python molurus bivittatus* (Serpentes: Boidae). Copeia 1976:398-400.

Van Mierop, L. H. S., and S. M. Barnard. 1978. Further observations on thermoregulation in the brooding female *Python molurus bivittatus* (Serpentes: Boidae). Copeia 1978:615-621.

Van Mierop, L. H. S., T. Walsh, and D. L. Marcellini. 1982. Reproduction of *Chondropython viridis* (Reptilia, Serpentes, Boidae), p. 265-274. In D. L. Marcellini (ed.), 6th International Herpetological Symposium on Captive Propagation and Husbandry. International Herpetological Symposium, Thurmont MD.

Vassiliev, D. B. 1996. Keeping and breeding in captivity Rinkhals (*Hemachatus haemachatus* Lacépède) in Moscow Zoo. Snake 27:145-146.

Velayan, S., and S. Ambu. 1991. A study of bacterial infections in reptiles at the National Zoo, Malaysia. Snake 23:84-89.

Veltre, T. 1996. Menageries, metaphors, and meanings, p. 19-29. In R. J. Hoage and W. A. Diess (eds.), New Worlds, New Animals. From Menagerie to Zoological Park in the Nineteenth Century. Johns Hopkins University Press, Baltimore and London.

Vetesi, F., M. Dobos-Kovacs, J. Ujhelyi, M. Janisch, and L. Horvath, 1981. Pockenartiger Ausschlag bei Kaimanen (*Caiman sclerops*) (Pockmark-resembling rash in caimans [*Caiman sclerops*]), p. 359-364. In R. Ippen, and H.-D. Schröder (eds.), Erkrankungen der Zootiere; Verhandlungsbericht des 23. Internationalen Symposiums über die Erkrankungen die Zootiere vom 24. Juni bis 28. Juni in Halle/Saale. (Diseases of Zoo Animals/Conference Report of the 23rd International Symposium about Diseases of Zoo Animals, June 24-28 in Halle/Saale) Akademie Verlag, Berlin-1981.

Vinegar, A., V. H. Hutchison, and H. G. Dowling. 1970. Metabolism, energetics, and thermoregulation during brooding of snakes of the genus *Python* (Reptilia, Boidae). Zoologica (New York) 55:19-48.

Visser, G. J. 1981. Breeding the white-throated monitor *Varanus albigularis* at Rotterdam Zoo. Inter. Zoo Yearb. 21:87-91.

Visser, G. J. 1984. Husbandry and reproduction of the sail-tailed lizard *Hydrosaurus amboinensis* (Schlosser, 1768) (Reptilia: Sauria: Agamidae) at Rotterdam Zoo. Acta Zool. Path. Antverpiensia 78: 129-148.

Visser, G. J. 1985. Notizen zur Brutbiologie des Gelbwarans *Varanus* (*Empagusia*) *flavescens* (Hardwicke & Gray, 1827) im Zoo Rotterdam (Notes concerning the breeding biology of the yellow monitor *Varanus* (*Empagusia*) *flavescens* in the Rotterdam Zoo). Salamandra 21:161-168.

Visser, G. J. 1989. Chinese krokodilstaarthagedissen (*Shinisaurus crocodilurus*) in Diergaarde Blijdorp (Chinese crocodile lizard (*Shinisaurus crocodilurus* in the Blijdorp Zoo). Lacerta 47(4):98-105.

Visser, G. J. 1989. Chinese crocodile lizards *Shinisaurus crocodilurus* in Rotterdam Zoo. Inter. Zoo Yearb. 28:136-139.

Visser, G. J., and H. A. Zwartepoorte. 1989. Breeding the leopard tortoise *Geochelone pardalis babcocki* at Rotterdam Zoo. Inter. Zoo Yearb. 28:98-102.

Vogel, D. 1992. Haltung und Zucht von *Laemanctus serratus* über mehrere Generationen im Exotarium des Zoologischen Garten Frankfurt am Main (Care and breeding of *Laemanctus serratus* over several generations in the Exotarium of the Zoo in Frankfurt on the Main). Iguana Rundschreiben [circular] 1992(2):15-17.

Vogel, P., R. Nelson, and R. Kerr. 1996. Conservation strategy for the Jamaican iguana, *Cyclura collei*, p. 395-406. Contributions to West Indian Herpetology: A Tribute to Albert Schwartz. Society for the Study of Amphibians and Reptiles, Contributions to Herpetology, volume 12. Ithaca NY.

Vogel, Z. Undated. Reptile Life. Spring Books, London.

Vogel, Z. 1964. Reptiles and Amphibians. Their Care and Behaviour. The Viking Press, New York.

Vogt, P. 1974. Breeding and rearing the Colombian giant toad *Bufo blombergi* at Krefeld Zoo. Inter. Zoo Yearb. 14:87-90.

Vokins, A. M. A. 1977. Breeding the red footed tortoise *Geochelone carbonaria* (Spix 1824). Dodo, J. Jersey Wildl. Preserv. Trust 14:73-80.

Vyas, R. 1996. Captive breeding of the Indian rock python (*Python molurus molurus*). Snake 27:127-134.

Wagner, E. 1971. A method of snake mite control. Inter. Zoo Yearb. 11:234-235.

Wagner, E. 1974. Breeding the leopard gecko *Eublepharis macularius* at Seattle Zoo. Inter. Zoo Yearb. 14:84-86.

Wagner, E. 1975. Breeding the Burmese python (*Python molurus bivittatus*) at Seattle Zoo. Inter. Zoo Yearb. 16:83-85.

Wagner, E. 1979. Breeding king snakes *Lampropeltis* spp. Inter. Zoo Yearb. 19:98-100.

Wagner, E. 1979. Some parameters for breeding reptiles in captivity, p. 47-53. In R. A. Hahn (ser. ed.), 1st & 2nd International Herpetological Symposium on Captive Propagation and Husbandry. International Herpetological Symposium, Thurmont MD.

Wagner, E. 1980. Gecko husbandry and reproduction, p. 115-117. In J. B. Murphy, and J. T. Collins (eds.), Reproductive Biology and Diseases of Captive Reptiles. Society for the Study of Amphibians and Reptiles. Contributions to Herpetology, volume 1, Lawrence KS.

Wagner, E. 1981. Thermal requirements in captive reptiles. Proc. Amer. Assoc. Zoo Vet. 1981:10-11.

Wagner, E. 1985. Captive husbandry of wild-caught emerald tree boas, p. 109-111. In R. L. Gray (ed.), Captive Propagation and Husbandry of Reptiles and Amphibians. Northern California Herpetological Society & Bay Area Amphibian and Reptile Society.

Wagner, E. 1987. Breeding arrow poison frogs and looking for ways to reduce the labor, p. 11-16. In R. Gowen (ed.), Captive Propagation and Husbandry of Reptiles and Amphibians. Special Publication # 4 ed.: Northern California Herpetological Society.

Wagner, E., and J. Foster. 1984. Removing retained eggs from snakes, p. 112-113. In R. A. Hahn (ser. ed.) 8th International Herpetological Symposium on Captive Propagation and Husbandry. International Herpetological Symposium, Thurmont MD.

Wake, M. H. 1994. Caecilians (Amphibia: Gymnophiona) in captivity, p. 223-228. In J. B. Murphy, K. Adler, and J. T. Collins (eds.), Captive Management and Conservation of Amphibians and Reptiles. Society for the Study of Amphibians and Reptiles. Contributions to Herpetology, volume 11, Ithaca NY.

Walder, R. 1990. Chelonian management at the Lowry Park Zoological Garden. AAZPA Reg. Conf. Proc. 1990:394-398.

Walker, S. 2001. Zoological gardens of India. In V. N. Kisling Jr. (ed.), Zoo and Aquarium History. Ancient Animal Collections to Zoological Gardens. Boca Raton, FL; London; New York; Washington DC.

Wallace, M. C. 1978. Enclosure utilization and activity patterns of captive bog turtles, p. 391-407. In C. Crockett and M. Hutchins (eds.), Applied Behavioral Research at the Woodland Park Zoological Gardens, Seattle, Washington 1977. Pika Press, Seattle WA.

Wallach, J. D. 1966. Hypervitaminosis D in green iguanas. J. Amer. Vet. Med. Assoc. 149:912-914.

Wallach, J. D. 1975. Herpatecology - Observations on the captive reptile and its environment. Proc. Amer. Assoc. Zoo Vet. 1975:48-65.

Wallach, J. D. 1975. Partial gastrectomy in a Timor python *Python timorensis*. J. Zoo Anim. Med. 6(2):13-15.

Wallach, J. D., and C. Hoessle. 1968. Steatitis in captive crocodilians. J. Amer. Vet. Med. Assoc. 153:845-847.

Wallach, J. D., and C. Hoessle. 1968. Fibrous osteodystrophy in green iguanas. J. Amer. Vet. Med. Assoc. 153:863-865.

Wallach, J. D., C. Hoessle, and J. Bennett. 1967. Hypoglycemic shock in captive alligators. J. Amer. Vet. Med. Assoc. 151:893-896.

Walsh, L. T. 1977. Husbandry and breeding of *Chondropython viridis*. National Association for Sound Wildlife Programs 1(2):10-22.

Walsh, L. T. 1980. Further notes on the husbandry, breeding, and behavior of (*Chondropython viridis*), National Zoological Park, p. 102-111. In R. A. Hahn (ser. ed.), 3rd International Herpetological Symposium on Captive Propagation and Husbandry. International Herpetological Symposium, Thurmont MD.

Walsh, T. L. 1994. Husbandry of long-term captive populations of boid snakes (*Epicrates, Corallus,* and *Chondropython*), p. 359-362. In J. B. Murphy, K. Adler, and J. T. Collins (eds.), Captive Management and Conservation of Amphibians and Reptiles. Society for the Study of Amphibians and Reptiles.

Contributions to Herpetology, volume 11, Ithaca NY.

Walsh, T. L., and B. Davis. 1983. Husbandry and breeding of the Brazilian rainbow boa, *Epicrates cenchria*, at the National Zoological Park, p. 109-114. In P. J. Tolson (ed.), 7th International Herpetological Symposium on Captive Propagation and Husbandry. International Herpetological Symposium, Thurmont MD.

Walsh, T. L., and S. W. Davis. 1978. Observations on the husbandry and breeding of the rufous-beaked snake, *Rhamphiophis oxyrhynchus rostratus* at the National Zoological Park. Bull. Maryland Herpetol. Soc. 14(2):75-78.

Walsh, T. L., and J. B. Murphy. 2003. Observations on the husbandry, breeding and behaviour of the Indian python, *Python molurus molurus* at the National Zoological Park, Washington, DC. Inter. Zoo Yearb. 38:145-152.

Walsh, L. T., and R. Rosscoe. 1993. Komodo monitors hatch at the National Zoo. Vivarium 4(5):13.

Walsh, L. T., R. Rosscoe, and G. F. Birchard. 1993. Dragon Tales: The history, husbandry, and breeding of Komodo monitors at the National Zoological Park. Vivarium 4(6):23-26.

Walsh, T., R. Rosscoe and J. B. Murphy. 1998. 21st Century conservation of the Komodo dragon: A lost world relic in modern times. Reptile & Amphibian Magazine. July/August:48-55.

Walsh, T., D. Chiszar, G. F. Birchard, and KMT A. Tirtodiningrat. 2002. Captive management and growth, p. 178-195. In J. B. Murphy, C. Ciofi, C. de La Panouse, and T. Walsh (eds.), Komodo Dragons. Biology and Conservation. Smithsonian Institution Press, Washington DC.

Walsh, T., D. Chiszar, E. Wikramanayake, H. M. Smith, and J. B. Murphy. 1999. The thermal biology of captive and free ranging wild Komodo dragons, *Varanus komodoensis* (Reptilia: Sauria: Varanidae). Advances in Monitor Research II-Mertensiella 11:239-245.

Watanabe, M. E. 1986. A change of fortune for the Chinese alligator. Anim. Kingdom 89(5):34-39.

Watson, G. 1969. Notes on the care of Mastigure lizards *Uromastix acanthinurus* at Jersey Zoo. Inter. Zoo Yearb. 9:49-50.

Weigel, J. 1988. The Australian Reptile Park Story. Weigel Photoscript, Gosford, New South Wales.

Weigel, J. 1988. Care of Australian Reptiles in Captivity. Reptile Keeper Association, Gosford, New South Wales.

Weigel, J. 1989. Maintenance and breeding of the superb dragon *Diporiphora superba* at the Australian Reptile Park, Gosford. Inter. Zoo Yearb. 28:122-126.

Weigel, J. 1990. Australian Reptile Park's Guide to Snakes of South-East Australia. Australian Reptile Park Gosford, New South Wales.

Weigel, J. 1992. The conservation role of private zoos in Australia: The Australian Reptile Park experience. Inter. Zoo Yearb. 31:24-29.

Weigel, J. 1992. Maintenance and breeding of Australian elapid snakes, p. 41-48. In M. J. Uricheck (ed.), 15th International Herpetological Symposium on Captive Propagation and Husbandry. International Herpetological Symposium, Palo Alto CA.

Weigel, J. 1992. Reptiles and the law in Australia: How protective wildlife legislation can impede herpetology, p. 75-86. In M. J. Uricheck (ed.), 15th International Herpetological Symposium on Captive Propagation and Husbandry. International Herpetological Symposium, Palo Alto CA.

Wemmer, C. 1991. Rambling thoughts on zoo biology and research. AAZPA 1991 Regional Conference Proceedings, pp. 138-146.

Wenker, C. J., M. Bart, F. Guscetti, J.-M. Hatt, and E. Isenbügel. 1999. Periarticular hydroxyapatite deposition disease in two red-bellied short-necked turtles (*Emydura albertisii*). Proc. Amer. Assoc. Zoo Vet. 1999:23-26.

Werler, J. E., and J. A. Campbell. 2004. New lizard of the genus *Diploglossus* (Anguidae: Diploglossinae) from the Tuxtlan Faunal Region, Veracruz, Mexico. Southwest. Natur. 49:327-333.

West, G. 2001. Endoscopic hepatic biopsy in Coahuilan box turtles, *Terrapene coahuila*. Bull. Assoc. Rept. Amphib. Vet. 11(2):28-29.

Weygoldt, P. 1976. Beobachtungen zur Fortpflanzungsbiologie der Wabenkröte *Pipa carvalhoi* Miranda Ribeiro (Observations concerning the reproductive biology of the Surinam toad *Pipa carvalhoi* Miranda Ribeiro). Z. Köln. Zoo 19(3):77-84.

Wheeler, S. 1990. Husbandry of the spiny-tailed agamas *Uromastyx acanthinurus* and *U. aegyptius* at Oklahoma City Zoo. Inter. Zoo Yearb. 29:70-74.

Wheeler, S. 1991. Additional notes on the captive reproduction of the spiny-tail agama *Uromastyx acanthinurus* at the Oklahoma City Zoo, p. 35-38. In A. W. Zulich (ed.), 14th International Herpetological Symposium on Captive Propagation and Husbandry. International Herpetological Symposium.

Wheler, C. L., and J. E. Fa. 1995. Enclosure utilization and activity of Round Island geckos (*Phelsuma guentheri*). Zoo Biol. 14:361-369.

Whitaker, B. R., and H. Krum. 1999. Medical management of sea turtles in aquaria, p. 217-231. In Fowler, M. E. and R. E. Miller (eds.), Zoo and Wild Animal Medicine: Current Veterinary Therapy. W. B. Saunders, Philadelphia, London.

Whitaker, B. R., and S. L. Poyton. 1993. Protozoans in dendrobatid frogs: Identification, clinical assessment, and indications for treatment. Proc. Amer. Assoc. Zoo Vet. 1993:13-15.

Whitaker, R. 1992. Indian herpetology and the Madras Crocodile Bank, p. 1-25. 16th International Herpetological Symposium on Captive Propagation and Husbandry. International Herpetological Symposium.

White, G. 1906. The Natural History of Selborne. J. M. Dent & Co., New York.

Whiteside, D. P., and M. M. Garner. 2001. Thyroid adenocarcinoma in a crocodile lizard, *Shinisaurus crocodilurus*. Bull. Assoc. Rept. Amphib. Vet. 11(1):13-16.

Whitworth, J. 1971. Notes on the growth and mating of American alligators *Alligator mississippiensis* at the Cannon Aquarium, Manchester Museum. Inter. Zoo Yearb. 11:144.

Whitworth, J. 1974. Notes on the growth of an African python *Python sebae* and an Indian python *Python molurus* at the Cannon Aquarium and Vivarium, Manchester Museum. Inter. Zoo Yearb. 14:140-141.

Wicker, R. 1984. Beobachtungen bei mehrjähriger Zucht von *Phrynops geoffroanus geoffroanus* (Schweigger, 1812) (Testudines: Chelidae) (Observations during several years of breeding *Phrynops geoffroanus geoffroanus* [Schweigger, 1812] [Testudines: Chelidae]). Salamandra 20:185-191.

Widholzer, F. L., B. Borne, and T. Tesche. 1986. Breeding the broad-nosed caiman *Caiman latirostris* in captivity. Inter. Zoo Yearb. 24/25:226-230.

Wiese, R. J., and M. Hutchins. 1994. The role of zoos and aquariums in amphibian and reptilian conservation, p. 37-45. In J. B. Murphy, K. Adler, and J. T. Collins (eds.), Captive Management and Conservation of Amphibians and Reptiles. Society for the Study of Amphibians and Reptiles. Contributions to Herpetology, volume 11, Ithaca NY.

Wiley, G. O. 1929. Notes on the Texas rattlesnake in captivity with special reference to the birth of a litter of young. Bull. Antivenin Inst. Amer. 3(1):8-14.

Wilke, H. 1984. Breeding the pancake tortoise *Malacochersus tornieri* at Frankfurt Zoo. Inter. Zoo Yearb. 23:137-139.

Wise, M. 1994. Techniques for the capture and restraint of captive crocodilians, p. 401-405. In J. B. Murphy, K. Adler, and J. T. Collins (eds.), Captive Management and Conservation of Amphibians and Reptiles. Society for the Study of Amphibians and Reptiles. Contributions to Herpetology, volume 11, Ithaca NY.

Wisniewski, P. J., and L. M. Paull. 1986. Breeding and rearing the European fire salamander *Salamandra s. salamandra*. Inter. Zoo Yearb. 24/25:223-226.

Wood, D. 1967. Breeding tuataras (*Sphenodon punctatus*) at Auckland Zoo. Inter. Zoo Yearb. 7:178-179.

Wood, L. N. 1944. Raymond L. Ditmars. His Exciting Career with Reptiles, Animals and Insects. Julian Messner, New York.

Worrell, E. 1958. Song of the Snake. Angus and Robertson, Sydney, Australia.

Worrell, E. 1961. Dangerous snakes of Australia and New Guinea . . . written and illustrated by Eric Worrell. Angus and Robertson, Sydney, Australia.

Worrell, E. 1963. Reptiles of Australia: Crocodiles, turtles, tortoises, lizards, snakes; describing all Australian species, their appearance, their haunts, their habits. Angus and Robertson, Sydney, Australia.

Worrell, E. c1963. Dangerous snakes of Australia and New Guinea; a handbook for bushmen, bushwalkers, mission workers, servicemen ... on the identification and venoms of Australian snakes, with directions for first-aid treatment of snake-bite and the use of antivenenes. Angus and Robertson, Sydney, Australia.

Worrell, E. 1966. Australian wildlife; best-known birds, mammals, reptiles, plants of Australia and New Guinea. Angus and Robertson, Sydney, Australia.

Worrell, E. 1966. Australian snakes, crocodiles, tortoises, turtles, lizards. Angus and Robertson, Sydney, Australia.

Wright, C. 1981. Breeding the American alligator *Alligator mississippiensis* at the Tulsa Zoological Park. Inter. Zoo Yearb. 21:73-75.

Wright, K. M. 1992. Hematology and plasma chemistries of prehensile-tailed skinks (*Corucia zebrata*). J. Zoo Wildl. Med. 23:429-432.

Wright, K. M. 1992. Dysecdysis in boid snakes with neurologic diseases. Bull. Assoc. Rept. Amphib. Vet. 2(2):7.

Wright, K. M. 1993. Medical management of the Solomon Island prehensile-tailed skink, *Corucia zebrata*. Bull. Assoc. Rept. Amphib. Vet. 3(1):9-17.

Wright, K. M. 1993. Captive husbandry of the Solomon Island prehensile-tailed skink, *Corucia zebrata*. Bull. Assoc. Rept. Amphib. Vet. 3(1):18-21.

Wright, K. M. 1994. Amputation of the tail of a two-toed amphiuma, *Amphiuma means*. Bull. Assoc. Rept. Amphib. Vet. 4(1):5.

Wright, K. M. 1996. Cover your asp: training keepers to safely handle venomous snakes. AAZPA Reg. Conf. Proc. 1996:384-385.

Wright, K. M. 1996. Amphibian husbandry and medicine, p. 436-459. In D. R. Mader (ed.), Reptile Medicine and Surgery. W. B. Saunders Co., Philadelphia, London.

Wright, K. M., and T. Minott. 1999. Individual identification of captive Mexican caecilians. Herpetol. Rev. 30(1):32-33.

Wright, K. M., and S. Skeba. 1992. Hematology and plasma chemistries of captive prehensile-tailed skinks (*Corucia zebrata*). J. Zoo Wildl. Med. 23:429-432.

Wright, K. M., and B. R. Whitaker. 2001. Amphibian Medicine and Captive Husbandry. Krieger, Malabar FL.

Wright, K. M., C. Pugh, and L. Walker. 1995. Ultrasonographic sexing of helodermatid lizards, p. 239-242. In R. E. Junge (ed.), Proceedings Joint Conference American Association of Zoo Veterinarians, Wildlife Disease Association, American Association of Wildlife Veterinarians, East Lansing, Michigan, August 12-17, 1995. American Association of Zoo Veterinarians, American Association of Wildlife Veterinarians & Wildlife Disease Association, place of publication not listed.

Wright, K. M., B. Toddes, and S. Donoghue. 1997. To form the more perfect stool: feeding trials on the Galapagos tortoise (*Geochelone nigra*) and Aldabra tortoise (*Geochelone gigantea*) population at the Philadelphia Zoological Garden. Proc. Amer. Assoc. Zoo Vet. 1997:11-15.

Yadav, R. N. 1969. Breeding the mugger crocodile *Crocodylus palustris* at Jaipur Zoo. Inter. Zoo Yearb. 9:33.

Yadav, R. N. 1979. A further report on breeding the mugger crocodile *Crocodylus palustris* at Jaipur Zoo. Inter. Zoo Yearb. 19:66-68.

Yanaga, T., A. Nakagawa, T. Furuya, and Y. Yasui. 1984. Artificial incubation and rearing of Leopard tortoise, *Testudo pardalis* at Miyazaki Safari Park. J. Japan. Assoc. Zoos Aquar. 26(1):8-12 [in Japanese].

Yoshimi, D. H., D. A. Payne, and F. L. Slavens. 1996. Maintenance and captive breeding of the Solomon Island leaf frog (*Ceratobatrachus guentheri*), p. 23-32. In P. D. Strimple (ed.), Advances in Herpetoculture. Special Publication of the International Herpetological Symposium, No. 1. Des Moines IA.

Yoshioka, S., and M. Samejima. 1989. Breeding of the loggerhead turtle, *Caretta caretta* L. on the Nagasakibana coast of Kagoshima Prefecture. J. Japan. Assoc. Zoos Aquar. 31(2):51-55 [in Japanese with English summary].

Zachiesche, W., A. Konstantinov, R. Ippen, and Z. Mladenov. 1988. Lymphoid leukemia with presence of type C virus particles in a four-lined chicken snake (*Elaphe obsoleta quadrivittata*). Erkr. Zootiere No. 30:275-278.

Zhou Jiufa and Zhou Ting. 1991. Chinese Chelonians Illustrated. Jiangsu Science and Technology Publishing House, Nanjing, China.

Zimmermann, E. 1983. Breeding Terrarium Animals. T. F. H. Publications, Neptune City NJ.

Zimmermann, E., and H. Zimmermann. 1994. Reproductive strategies, breeding, and conservation of tropical frogs: Dart-poison frogs and Malagasy poison frogs., p. 255-266. In J. B. Murphy, K. Adler, and J. T. Collins (eds.), Captive Management and Conservation of Amphibians and Reptiles. Society for the Study of Amphibians and Reptiles. Contributions to Herpetology, volume 11, Ithaca NY.

Zimmermann, H. 1992. Nachzucht und Schutz von *Mantella crocea*, *Mantella viridis* und vom madagassischen Goldfröschchen *Mantella aurantiaca* (Breeding and protection of *Mantella crocea*, *Mantella viridis* and of Madagascar gold frogs *Mantella aurantiaca*). Z. Köln. Zoo 35(4):165-171.

Zimmermann, H. 1994. Erhaltungszuchtprogramme für Anuren am Beispiel von *Dendrobates variabilis*, einem seltenen Vertreter der Pfeilgiftfrösche Südamerikas (Preservation breeding programs for frogs, using the example of *Dendrobates variabilis*, a rare representative of the poison dart frogs of South America). Z. Köln. Zoo 37(4):143-151.

Zimmermann, H., and E. Zimmermann. 1981. Sozialverhalten, Fortpflanzungsverhalten und Zucht der Färberfrösche *Dendrobates histrionicus* und *D. lehmanni* sowie einiger anderer Dendrobatiden (Social behavior, reproductive behavior and breeding of the poison dart frogs *Dendrobates histrionicus* and *Dendrobates lehmanni* as well as of several other dendrobatids). Z. Köln. Zoo 24(3):83-99.

Zimmermann, H., and E. Zimmermann. 1987. Mindestanforderungen für eine artgerechte Haltung einiger tropischer Anurenarten (Minimum requirements for the species-appropriate care of some tropical frog species). Z. Köln. Zoo 30(2):61-71.

Zippel, K. 2002. Conserving the Panamanian golden frog: Proyecto Rana Dorada. Herpetol. Rev. 33:11-12.

Zippel, K., and A. Snider. 2001. The Detroit Zoo makes a bold statement for amphibians. Communiqué March 2001:5-6, 51.

Zuckerman, L. 1976. Golden Days. Historical Photographs of the London Zoo. Gerald Duckworth, London.

Zug, G. R., L. J. Vitt, and J. P. Caldwell. 2001. Herpetology. An Introductory Biology of Amphibians and Reptiles. Academic Press, San Diego, New York, London.

Zwart, P., and C. C. Van de Watering. 1969. Disturbance of bone formation in the common iguana (*Iguana iguana* L): Pathology and etiology. Acta Zool. Path. Antverpiensia No. 48:333-356.

Zwart, P., S. A. Goedegebuure, and L. Lambrechts. 1994. Relationship between intestinal tract and mineralization of skeleton in reptiles, with reference to normal development of the carapace in young tortoises. Erkrg. Zootiere 36:315-319.